Minimally Invasive MAXILLOFACIAL SURGERY

Maria J. Troulis, DDS, MSc
Associate Professor of Oral and Maxillofacial Surgery
Director, Residency Training Program in OMFS
Harvard School of Dental Medicine
Massachusetts General Hospital
Boston, Massachusetts

Leonard B. Kaban, DMD, MD
Department Head and Walter C. Guralnick Professor of Oral and Maxillofacial Surgery
Harvard School of Dental Medicine
Massachusetts General Hospital
Boston, Massachusetts

PMPH-USA

2013
PEOPLE'S MEDICAL PUBLISHING HOUSE—USA
SHELTON, CONNECTICUT

People's Medical Publishing House-USA
2 Enterprise Drive, Suite 509
Shelton, CT 06484
Tel: 203-402-0646
Fax: 203-402-0854

E-mail: info@pmph-usa.com

13 14 15 16/QG/9 8 7 6 5 4 3 2 1

ISBN-13 978-1-60795-012-7
ISBN-10 1-60795-012-X
eISBN-13 978-1-60795-220-6

Printed in the United States of America by IBT Global.
Editor: Jason Malley; Development Editor: Linda H. Mehta; Book Interior and Cover designer: Mary McKeon

Library of Congress Cataloging-in-Publication Data

Troulis, Maria J.
Minimally invasive maxillofacial surgery / Maria J. Troulis, DDS, MSc, Associate Professor of Oral and Maxillofacial Surgery, Director, Residency Training Program in OMFS, Harvard School of Dental Medicine, Massachusetts General Hospital, Boston, Massachusetts, Leonard B. Kaban, DMD, MD, Department Head, and Walter C. Guralnick, Professor of Oral and Maxillofacial Surgery, Harvard School of Dental Medicine, Massachusetts General Hospital, Boston, Massachusetts.

pages cm
ISBN 978-1-60795-012-7 (alk. paper)
1. Mouth--Endoscopic surgery. 2. Face--Endoscopic surgery. 3. Maxilla--Endoscopic surgery. I. Kaban, Leonard B. II. Title.

RK529.T76 2013
617.5'22059--dc23

2012041189

Sales and Distribution

Canada
McGraw-Hill Ryerson Education
Customer Care
300 Water St
Whitby, Ontario L1N 9B6
Canada
Tel: 1-800-565-5758
Fax: 1-800-463-5885
www.mcgrawhill.ca

Foreign Rights
John Scott & Company
International Publisher's Agency
P.O. Box 878
Kimberton, PA 19442
USA
Tel: 610-827-1640
Fax: 610-827-1671

Japan
United Publishers Services Limited
1-32-5 Higashi-Shinagawa
Shinagawa-ku, Tokyo 140-0002
Japan
Tel: 03-5479-7251
Fax: 03-5479-7307
Email: kakimoto@ups.co.jp

United Kingdom, Europe, Middle East, Africa
McGraw Hill Education
Shoppenhangers Road
Maidenhead
Berkshire, SL6 2QL
England
Tel: 44-0-1628-502500
Fax: 44-0-1628-635895
www.mcgraw-hill.co.uk

Singapore, Thailand, Philippines, Indonesia, Vietnam, Pacific Rim, Korea
McGraw-Hill Education
60 Tuas Basin Link
Singapore 638775
Tel: 65-6863-1580
Fax: 65-6862-3354
www.mcgraw-hill.com.sg

Australia, New Zealand
Elsevier Australia
Locked Bag 7500
Chatswood DC, NSW 2067
Australia
Tel: 161 (2) 9422-8500
Fax: 161 (2) 9422-8562
www.elsevier.com.au

Brazil
SuperPedido Tecmedd
Beatriz Alves, Foreign Trade Department
R. Sansao Alves dos Santos, 102 | 7th floor
Brooklin Novo
São Paolo 04571-090
Brazil
Tel: 55-16-3512-5539
www.superpedidotecmedd.com.br

India, Bangladesh, Pakistan, Sri Lanka, Malaysia
CBS Publishers
4819/X1 Prahlad Street 24
Ansari Road, Darya Ganj, New Delhi-110002
India
Tel: 91-11-23266861/67
Fax: 91-11-23266818
Email:cbspubs@vsnl.com

People's Republic of China
People's Medical Publishing House
International Trade Department
No. 19, Pan Jia Yuan Nan Li
Chaoyang District
Beijing 100021
P.R. China
Tel: 8610-67653342
Fax: 8610-67691034
www.pmph.com/en/

Dedication

Leonard Kaban (left), John Yeonopolus (center), and Maria Troulis (right) at a MGH luncheon, in 2001, honoring Mr. Yeonopolus for his support of the MGH Oral and Maxillofacial Surgery Minimally Invasive Surgery Program.

This book is dedicated to the memory of Mr. John James Yeonopolus and to the Miles and Gertrude W. Hanson Foundation, Bank of America, N.A., Trustee. Without the financial support received from John and the Foundation, the work done in our research program, much of which is described in this book, would not have been possible.

John "Yanni" Yeonopolus (1918–2010) was born in Pueblo, Colorado, to Greek immigrant parents, working in steel mills and the railroads, just after the turn of the 20th century. He was a graduate of Colorado College, majoring in finance, and worked many jobs to support his family and his own education. He attended and received a graduate degree in business administration from Stanford University. He ultimately rose to become the chief financial officer of Colorado Interstate Gas and Oil Company, a post he served in for 15 years. He then became the chief financial officer of Algonquin Gas and remained in that position until he retired in 1988 at age 70.

Mr. Yeonopolus enjoyed a 55-year marriage to Bernice Besse Yeonopolus (1921–1999) of Broken Arrow, Colorado. They resided in Colorado Springs for many years and had three children. They moved to Weston, Massachusetts, in 1972 and became dear friends of Gertrude and Miles Hanson. Upon Mrs. Hanson's death, Mr. Yeonopolus founded and was the first Trustee of the Miles and Gertrude W. Hanson Foundation. As the Hansons had no children, Mr. Yeonopolus ensured that their memory would be honored by making and overseeing philanthropic gifts from the Foundation to help people through medical research.

In March 2009, John met and was the beneficiary of minimally invasive surgery by Maria Troulis and Leonard Kaban at the Massachusetts General Hospital (MGH). Maria became a close friend and ultimately Mr. Yeonopolus's "favorite doctor." Thereafter, John and the Foundation have continued to generously support our research efforts at MGH.

The editors would like to acknowledge and thank Mr. Yeonopolus, the Miles and Gertrude W. Hanson Foundation, Bank of America, N.A., Trustee, and Ms. Augusta Haylock and Lisa Von Lichtenberg for their long-term friendship and support.

In the giving spirit of John, some final famous words of advice from John to Maria over the years: "Everything in moderation!" "Remember to ask for what you want, because, in life, you don't get what you deserve, but rather what you ask for!" Finally, "Save your money." This, of course, would be accompanied by a big hug.

Maria J. Troulis, DDS, MSc
Boston, Massachusetts
2012

Contents

Contributors

Zachary Abramson, DMD [14]
Former Fellow, Skeletal Biology Research Center
Department of Oral and Maxillofacial Surgery
Massachusetts General Hospital
Boston, Massachusetts

Meredith August, MD, DMD [21]
Associate Professor of Oral and Maxillofacial Surgery
Harvard School of Dental Medicine
Massachusetts General Hospital
Boston, Massachusetts

Arnulf Baumann, MD, DMD, PhD [4]
Associate Professor
University Clinic of Craniofacial, Maxillofacial, and Oral Surgery
Medical University of Vienna
Vienna, Austria

Gabriel Bendahan, DDS, MSc [14]
Orthodontist, Private Practice
Paris, France
Former Orthodontic Fellow
Harvard School of Dental Medicine
Boston, Massachusetts

Michael S. Block, DMD [16]
Clinical Professor and Director of Research
Department of Oral and Maxillofacial
Louisiana State University
The Center for Dental Reconstruction
Metairie, Lousiana

Carl Bouchard, DDS [12, 14, 19]
Associate Professor
University of Laval
Quebec, Canada
Former Fellow, Skeletal Biology Research Center
Department of Oral and Maxillofacial Surgery
Massachusetts General Hospital
Boston, Massachusetts

Alvaro Figueroa, DDS, MS [18]
Instructor, Orthodontics
Rush University School of Medicine
Rush University Hospital
Chicago, Illinois

Batya Goldwasser, DMD, MD [3]
Former Fellow, Skeletal Biology Research Center
Department of Oral and Maxillofacial Surgery
Massachusetts General Hospital
Boston, Massachusetts

Christopher J. Haggerty, DDS, MD [16]
Lakewood Oral and Maxillofacial Surgery Specialists
Lee's Summit, Missouri

Seth J. Isaacs, MD [6]
Private Practice in Otolaryngology
3219 Clifton Avenue
Cincinnati, Ohio

Leonard B. Kaban, DMD, MD [1, 2, 9, 10, 12, 13, 14, 19, 20]
Department Head and Walter C. Guralnick Professor of Oral and Maxillofacial Surgery
Harvard School of Dental Medicine
Massachusetts General Hospital
Boston, Massachusetts

Robert Kellman, MD [3, 6]
Professor and Chair of Otolaryngology and Communication Sciences
Upstate University Hospital
SUNY Upstate Medical University
Syracuse, New York

King Kim, DMD [7]
Private Practice in Maxillofacial Surgery
New Image Maxillofacial
Melbourne, Florida

Adjunct Clinical Professor
Department of Oral and Maxillofacial Surgery
Nova Southeastern Medical Center
Broward General Medical Center
Fort Lauderdale, Florida

Joseph P. McCain, DMD [7, 9]
Private Practice
Miami Oral and Maxillofacial Surgery
Miami, Florida
Chief of Oral and Maxillofacial Surgery
Baptist Health Systems
Clinical Associate Professor, OMS
FIU Medical School
Miami, Florida
Adjunct Professor of Oral and Maxillofacial Surgery
Nova Southeastern School of Dental Medicine
Fort Lauderdale, Florida

Michael Miloro, DMD, MD [5]
Professor and Head
Program Director
Department of Oral and Maxillofacial Surgery
University of Illinois at Chicago
Chicago, Illinois

Oded Nahlieli, DMD [8]
Chairman, Oral and Maxillofacial Surgery Department
Barzilai Medical Center
Ashkelon, Israel
Faculty of Medicine
Ben-Gurion University of the Negev
Be'er Sheva, Israel

Bonnie Padwa, DMD, MD [13, 17, 19]
Associate Professor
Department of Oral and Maxillofacial Surgery
Harvard School of Dental Medicine
Massachusetts General Hospital
Boston, Massachusetts

Maria Papadaki, DDS, MD [9, 10]
Former Fellow and Instructor in Oral and Maxillofacial Surgery
Harvard School of Dental Medicine
Massachusetts General Hospital
Boston, Massachusetts

Robert J. Paresi, Jr., MD, MPH [18]
Mercy Regional Plastic Surgery, Skin & Laser Center
Mercy Walworth Hospital and Medical Center
Lake Geneva, Wisconsin

John W. Polley, MD [18]
Professor of Plastic Surgery
Rush University Medical Center
Chicago, Illinois

Edward B. Seldin, DMD, MD [11, 13]
Associate Professor of Oral and Maxillofacial Surgery
Harvard School of Dental Medicine
Massachusetts General Hospital
Boston, Massachusetts

Basel Sharaf, MD, DDS [19]
Chief, Plastic and Reconstructive Surgery
Buffalo General Hospital
Clinical Assistant Professor
University at Buffalo
School of Medicine and Biomedical Sciences
Buffalo, New York
Former Resident, Oral and Maxillofacial Surgery
Massachusetts General Hospital
Boston, Massachusetts

James Shoukas, MD [18]
Private Practice
Lake Mary Plastic Surgery
Lake Mary, Florida

Ryan J. Smart, DMD, MD [19]
Resident in Oral and Maxillofacial Surgery
Massachusetts General Hospital
Boston, Massachusetts

Suzanne U. Stucki-McCormick, MS, DDS [15]
Department of Oral and Maxillofacial Surgery
Tri-City Medical Center
Oceanside, California

Maria J. Troulis, DDS, MSc [1, 2, 9, 10, 12, 13, 14, 19]
Associate Professor of Oral and Maxillofacial Surgery
Director, Residency Training Program in OMFS
Harvard School of Dental Medicine
Massachusetts General Hospital
Boston, Massachusetts

W. Bradford Williams, DMD, MD [2]
Former Fellow and Resident
Department of Oral and Maxillofacial Surgery
Massachusetts General Hospital
Boston, Massachusetts

Krishna C. Yeshwant, MD [11]
Clinical Instructor in Medicine
Brigham and Women's Hospital
Boston, Massachusetts
Former Fellow in Oral and Maxillofacial Surgery
Skeletal Biology Research Center
Massachusetts General Hospital
Boston, Massachusetts

Preface

Since its beginnings, minimally invasive surgery, with its decreased morbidity, shortened hospital stay, and quicker recovery, has been especially well accepted by the patients whom it benefits. Surgeons of many disciplines, often driven by the public, have also embraced minimally invasive surgery, and it has become the standard of care for many conditions.

This book was conceived in 1996, when David Perrott and I had the idea to develop a totally buried, miniature, automated distraction device that could be placed using minimally invasive endoscopic techniques. The book was born when we hired Dr. Maria Troulis as our second AO/Synthes/Massachusetts General Hospital (MGH) Fellow in Pediatric Craniomaxillofacial Surgery. During her two-year fellowship, Maria worked tirelessly with our group to develop a minipig model for distraction osteogenesis and to develop the technique for endoscopic access to the ramus condyle unit for placement of the miniature devices to come.

We used a 1.5-cm submandibular incision and subperiosteal dissection to develop an optical cavity that permitted "*en face*" exposure and operation for the mandible. The rationale was that *en face* exposure produced a working field and orientation familiar to surgeons. This would be easier to teach and master than working parallel to the field of interest as occurs with intraoral techniques.

In truth, the projects would have never progressed without Dr. Troulis's skill, hard work, and devotion. With the help of Oded Nahielili, we also learned the technique of sialendoscopy to round out our minimally invasive surgery program.

From 1996 to the present, we have continued to use the 1.5-cm submandibular incision and subperiosteal dissection as the basis for development of endoscopic procedures for open reduction and internal fixation of condylar and ramus fractures, condylectomy and placement of costochondral grafts, vertical ramus osteotomy, coronoidectomy, and biopsy or excision of certain soft tissue and bone lesions of the ramus. Sialendoscopy has become the standard of care for obstructive salivary gland disease. A series of papers describing the distraction model, biology of healing of the mandibular distraction wound, biomechanical strength of the distraction wound, outcomes studies using these techniques on the ramus condyle units, outcomes studies of distraction, and outcomes of sialendoscopy have been published and are described in the chapters of this book.

During this time, the field of minimally invasive maxillofacial surgery and reconstruction has continued to grow and expand. Dr. Troulis has developed and led the programs in minimally invasive reconstructive surgery and sialendoscopy at the MGH. Hence, this volume has been conceived, born, and completed to bring the knowledge and excitement for minimally invasive surgery to our colleagues and their patients. The editors have treated over 500 patients by minimally invasive surgical procedures with excellent patient satisfaction, excellent outcomes, and minimal morbidity, all of which have encouraged us to continue in this endeavor. It has been said that the use of minimally invasive techniques results in less training for our residents in the standard techniques and exposures. Learning from our general surgery colleagues, our answer to this criticism is that we should be training our residents for the future rather than just for the past.

This book has been completed through the efforts of the two editors and a multidisciplinary group of 24 expert clinicians (17 oral and maxillofacial surgeons, 2 otolaryngologists, 3 plastic surgeons, and 2 orthodontists) and includes chapters on minimally invasive reconstruction and orthognathic surgery of the ramus/condyle unit (Chapters 1 and 14), management of maxillofacial trauma (Chapters 2–5), minimally invasive management of the maxillary sinus (Chapter 6), TMJ (Chapter 7), sialoendoscopy (Chapters 8 and 9), distraction osteogenesis (Chapters 10–19), and minimally invasive management of tumors and jaw cysts (Chapters 20 and 21).

We are forever grateful to our wonderful group of contributors who worked very hard to conform to our format and suffered through our multiple revisions. We are convinced that we now have a complete, well illustrated, and readable volume. We also thank the publisher and the editors Jason Malley and Linda Mehta for their help and cooperation.

We would like to acknowledge our patients for whom we all work so hard to accomplish the best possible result with the least morbidity. A successful outcome for a patient in need is the ultimate reward for a surgeon.

Finally, we would like to acknowledge the late Robert V. Walker, Professor of OMFS at the University of Texas Southwestern and Parkland Hospital. Dr. Walker has been a pioneer and leader in OMFS for more than 50 years. It was an honor to have him accept the invitation to write a foreword to this book. Unfortunately, he passed away before he could complete the manuscript. We thank him for his support of our pursuit of minimally invasive technologies. We are including his letter to Dr. Troulis for our readers to see along with this preface.

Leonard B. Kaban, DMD, MD, FACS
Boston, Massachusetts
November 2012

Correspondence

SOUTHWESTERN
THE UNIVERSITY OF TEXAS
SOUTHWESTERN MEDICAL CENTER
AT DALLAS

5323 Harry Hines Blvd.
Dallas, Texas 75390-9109

Robert V. Walker, D.D.S.
Professor Emeritus
Division of Oral and Maxillofacial Surgery

Date: Sept. 22, 200.

To: Dr. Maria J. Troulis – Dept. of OMFS
Massachusetts Gen. Hosp.

ear Dr. Troulis,

I am a bit delayed in getting off
ote telling you how impressed I am with
our work on the endoscopic assistance wit
andibular condyle fractures. There is nothing
asy about condylar fracture management
gardless of the approach, but everyone wan
o do the very best that can be done with
uch fractures. Hopefully, you will be able
o refine the endoscopic approach to fit most
urgeons' skills — & validate with good
linical & radiographic follow-up to show
hy the method has merit. I wish you the
ery best at your work.

It was a compliment to be included on
he symposium with you — & I was delighted
ith the few minutes to chat with you afterwards

Sincerely,
Robert V. Walker

(214)648-3526 FAX (214)648-2918
E-mail: yvette.martinez@utsouthwestern.edu

SOUTHWESTERN
THE UNIVERSITY OF TEXAS
SOUTHWESTERN MEDICAL CENTER
AT DALLAS

5323 Harry Hines Blvd.
Dallas, Texas 75390-9109

Robert V. Walker, D.D.S.
Professor Emeritus
Division of Oral and Maxillofacial Surgery

Date: 31 March 2011

To: Jason Malley, Exec. Ed., PMPH-USA

Dear Mr. Malley,

I have terminal pancreatic cancer, and I am very ill at the moment. I will be unable to write the forward for what has to be a truly marvelous book by Dr. Troulis and Dr. Kaban, a commitment made to these fine surgeons over two years ago. I am simply unable to write creatively any longer. I appreciate the great compliment of doing the forward, but I must decline.

Sincerely,
Robert V. Walker

(214)648-3526 FAX (214)648-2918
E-mail: yvette.martinez@utsouthwestern.edu

Endoscopic Reconstruction of Ramus Condyle Unit

1

Maria J. Troulis, DDS, MSc / Leonard B. Kaban, DMD, MD

OUTLINE

INTRODUCTION

Until the mid-19th century, surgeons were limited to performing only rapid and life-saving procedures because of high complication rates and excessive mortality.[1,2] With the development of anesthesia, the adoption of aseptic technique, the discovery of antibiotics and advances in understanding the principles of fluid and electrolyte balance, elective procedures (such as orthognathic surgery) to improve the quality of life became feasible.[1–3]

Simon Hullihen, DDS, MD, reported the first "orthognathic" surgical procedure in the *American Journal of Dental Science* in 1849. He described a mandibular ostectomy for the correction of a deformation consisting of an anterior open bite[4] resulting from a burn scar contracture of the neck. In 1921, Gunther Cohn-Stock[4] performed the first maxillary osteotomy for correction of a malocclusion. Obwegeser, in 1955 (in German literature) and in 1957 (in the English literature), described the sagittal split osteotomy of the mandible.[5,6,7] This procedure represented a significant advance because the operation could be used for the correction of a variety of deformities such as mandibular prognathism, mandibular retrognathism, open bite, and asymmetry. The current bilateral sagittal split osteotomy (BSSO) has evolved with modifications and has become a commonly performed orthognathic surgical operation.

In the 1970s and 1980s, orthognathic surgery progressed because of William Bell's work delineating the blood supply of the maxillofacial skeleton.[8] Recently, orthognathic surgeons have attempted to decrease the perioperative morbidity associated with orthognathic surgery. Rigid internal fixation (RIF), controlled hypotensive anesthesia,[9,10] administration of perioperative steroids,[11] and patient positioning have decreased patient morbidity by eliminating the need for maxillomandibular fixation, decreasing blood loss/risk of transfusion, decreasing edema and decreasing discomfort. The combination of these techniques has decreased the length of hospital stay required for recovery.[12,13]

Advances in technology and instrumentation have allowed surgeons to develop and refine minimally invasive procedures to replace standard, maximally invasive operations in many surgical specialties.[3] Initially, minimally invasive surgery was synonymous with endoscopy. It has now evolved to be the discipline of surgical innovation combined with technology.[1,2,4] The impetus for minimally invasive surgery comes from both patients and surgeons who appreciate the benefits of endoscopic techniques: less swelling, less pain, shorter hospital stay and earlier return to work. This is cost effective and "state-of-the-art medical care."[14]

In this chapter we describe the endoscopic technique for access to the mandibular ramus condyle unit (RCU). The principle of this technique is to use very small submandibular incisions with minimal dissection to facilitate insertion of an endoscope. The surgeon is then able to visualize the anatomy of the ramus condyle region and to perform a variety of mandibular osteotomies, condylectomy, coronoidectomy, repair of fractures, and to biopsy and excise pathology. Mandibular distraction devices can also be inserted and fixed using endoscopic access.

In part II of this book, distraction osteogenesis is discussed in detail. The indications for mandibular distraction, as a minimally invasive alternative to the standard sagittal split osteotomy, include patients with a third molar bud present, patients older than 40 years of age (risk of inferior alveolar nerve sensory deficit), and patients who require mandibular advancement greater than 10 mm (see Chapter 14, "Mandibular Advancement—DO versus BSSO"). Patients who would benefit from maxillary distraction, as an alternative to standard LeFort I osteotomy and acute movement, are those who require advancement greater than 10 mm and those who have significant soft-tissue scarring from multiple procedures as in cleft lip/palate repair and reconstruction (see Chapter 17, "Lefort 1 DO").

ENDOSCOPIC TECHNIQUE FOR ACCESS TO THE RAMUS CONDYLE UNIT: "HOW TO"

Surface Landmarks

The first step and a critical part of the operation is to accurately identify and mark the surface landmarks. They are drawn on the skin with malachite green or any other suitable marking ink that will not run or easily fade.

The midline of the face is identified and drawn. The skeletal landmarks are palpated and marked. These include: 1) entire inferior border; 2) antigonial notch; 3) angle of the mandible; 4) posterior border of the mandible; 5) temporomandibular joint region; 6) zygomatic arch; 7) anterior border of the ramus; and 8) superior border of the mandible. The mandible is opened and closed to better identify the condyle and the temporomandibular joint (Figure 1). Once the surface landmarks are identified and drawn, the planned osteotomy and/or the pathological region is also marked. Then, a line from the area of interest, parallel to the posterior border and extending to the area of the incision in the submandibular triangle, is drawn. A 1.5 cm incision, parallel to the neck skin creases, is marked at the angle of the mandible. (Figure 1b) It is important for the incision to hug the angle so that the endoscope is parallel to the anterior/posterior borders of the vertical ramus of the mandible when inserted.[15]

Patient Positioning and Equipment Set-up

After induction of anesthesia, endotracheal intubation and identification and drawing of surface landmarks, the patient is positioned supine with the neck in a slightly extended position for the procedure. The skin is prepared with an antiseptic solution and the patient is draped. The top and both sides of the patient's head and neck are then accessible to the surgeon and the assistants. (Figure 2).

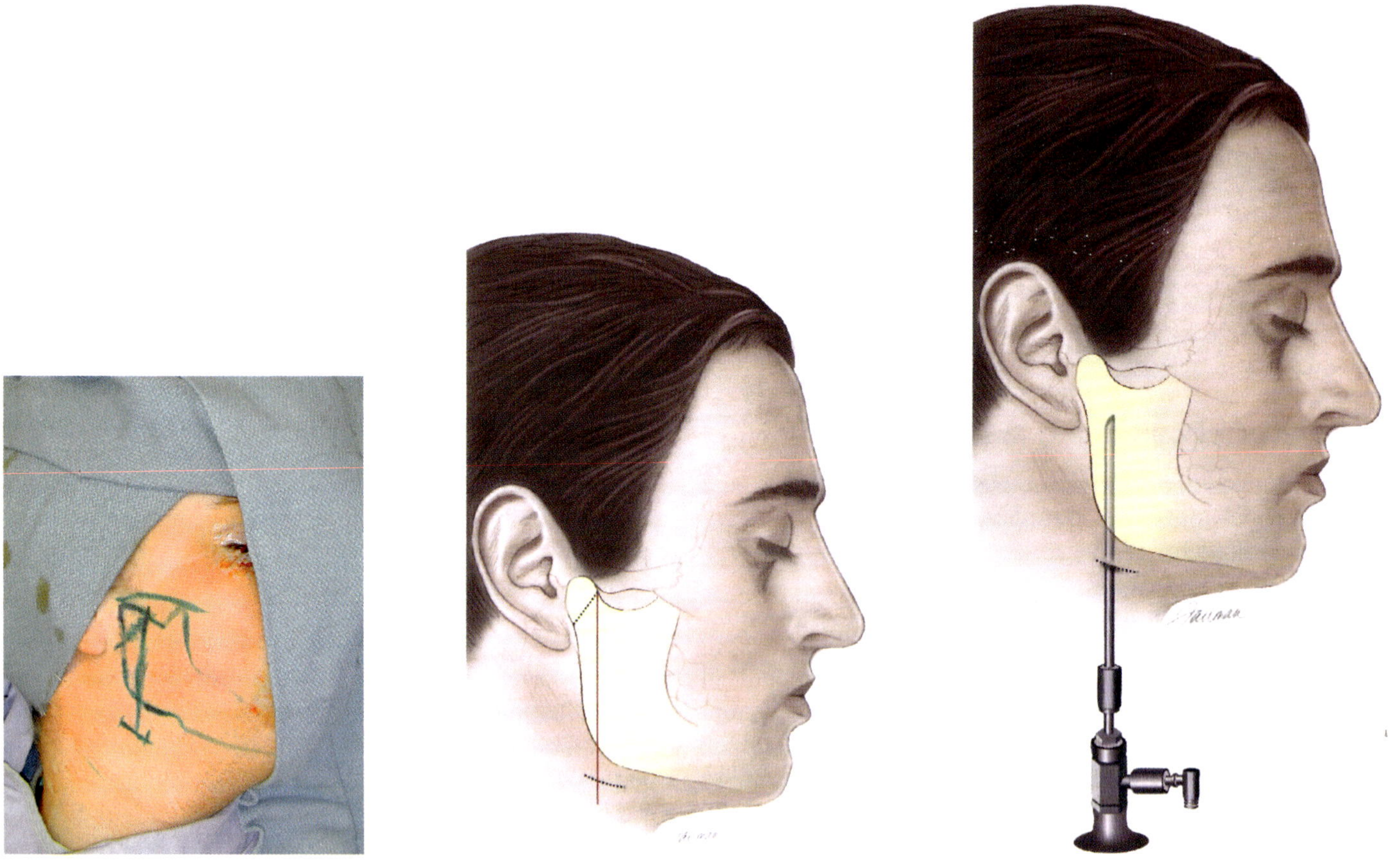

(A) (B) (C)

FIGURE 1 • **Surface Landmarks** Actual patient (A) and schematics (B and C) demonstrating the surface landmarks. The landmarks are palpated and drawn using Malachite Green. The Ink is allowed to dry and the patient is then prepared and draped. The entire ramus/condyle unit and inferior border of the mandible is marked out. The planned osteotomy or area of interest is also marked out. The planned incision is marked in direct line to the area of interest. A 1.5 cm incision is planned parallel to the skins creases, at the angle of the mandible (B). The incision hugs the angle so that the endoscope will be parallel to the posterior border (C). Reprinted with permission from Troulis MJ + Kaban LB: Minimally invasive approach to the mandibular ramus-condyle unit. Endo-Press. 2007 Tuttlingen, Germany.

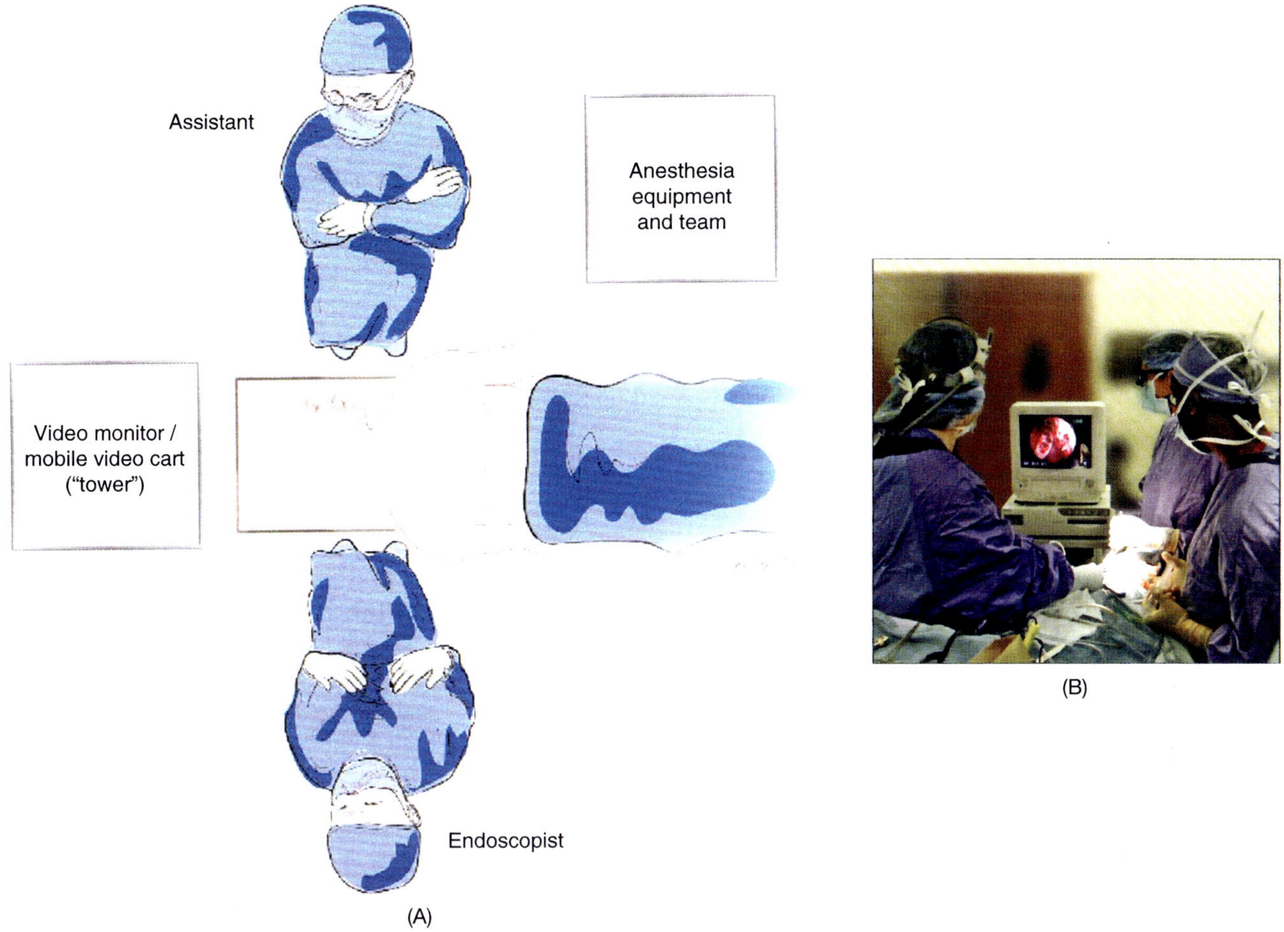

FIGURE 2 • **Positioning** Schematic of the ideal positioning of the endoscopic team relative to the patient. The tower is positioned at 12 o'clock relative to the patient. The endoscopist and assistant are each positioned at either side of the patient (A), Photograph of an endoscopic team at a teaching institute. In this case, both the endoscopist and the first assistant are on the patient's right side. The monitor is a 10 o'clock relative to the patient (B), Close-up photograph showing the endoscopist working on the right ramus/condyle unit. The endoscopist is on the patient's right side with the assistant on the left and the monitor at 12 o'clock. The assistant holds the retractor and the irrigation and follows the procedure on the monitor. The endoscopist supports the scope in the left hand with the suction-elevator in the right hand (see Figure 3C). Reprinted with permission from Troulis MJ + Kaban LB: Minimally invasive approach to the mandibular ramus-condyle unit. Endo-Press. 2007 Tuttlingen, Germany.

The required equipment includes the "tower": camera box, light source and monitor, the endoscope (Hopkin's 2.7mm 30 degree endoscope. (In operating rooms with a hardwired, digital set-up, one monitor is positioned at the head. The remainder of the equipment is at the foot/side).

The next step, prior to making the incision, is to white balance the endoscope. This is done by aiming the scope on a white-non-reflective background and moving the silver button. The blue button is used to focus. (Figure 3) Two suction catheters are set-up, one with a small french suction tip and one with the suction elevator.[15]

Incision and Dissection

A number 15 blade is used to make a 1.5 cm submandibular incision parallel to a neck crease. The dissection is carried sharply to the subcutaneous fat. Using a curved "Jake" hemostat, the tissue is spread (parallel to the facial nerve fibers) down to the platysma layer and small retractors are used to stretch the incision vertically and horizontally. Then, the retractors are placed parallel to the incision and marginal mandibular branch of the facial nerve. The hemostat is used to puncture the platysma lager. The platysma is undermined and opened parallel to the facial nerve. The deep fascia is entered and the masseter muscle identified. (Figure 4)[15]

With the masseter muscle exposed, a needlepoint electrocautery is used to extend down to the bone. The endoscopic elevator is placed in the pocket and the dissection is completed blindly with palpation of the elevator and the landmarks. The endoscope and the endoscopic retractor are inserted into the optical cavity. The retractor (R) is then hooked (h) into the sigmoid notch. If the tissue has not been completely and cleanly elevated, the endoscopic elevator (with a suction port at the tip of the blade) is used to complete the subperiosteal dissection, especially at the region of the sigmoid notch. When all the soft tissue is elevated, the retractor can be positioned into the sigmoid notch under direct (endoscopic) visualization. When the dissection is clean, there is little bleeding and visibility is sharp and clear.[15]

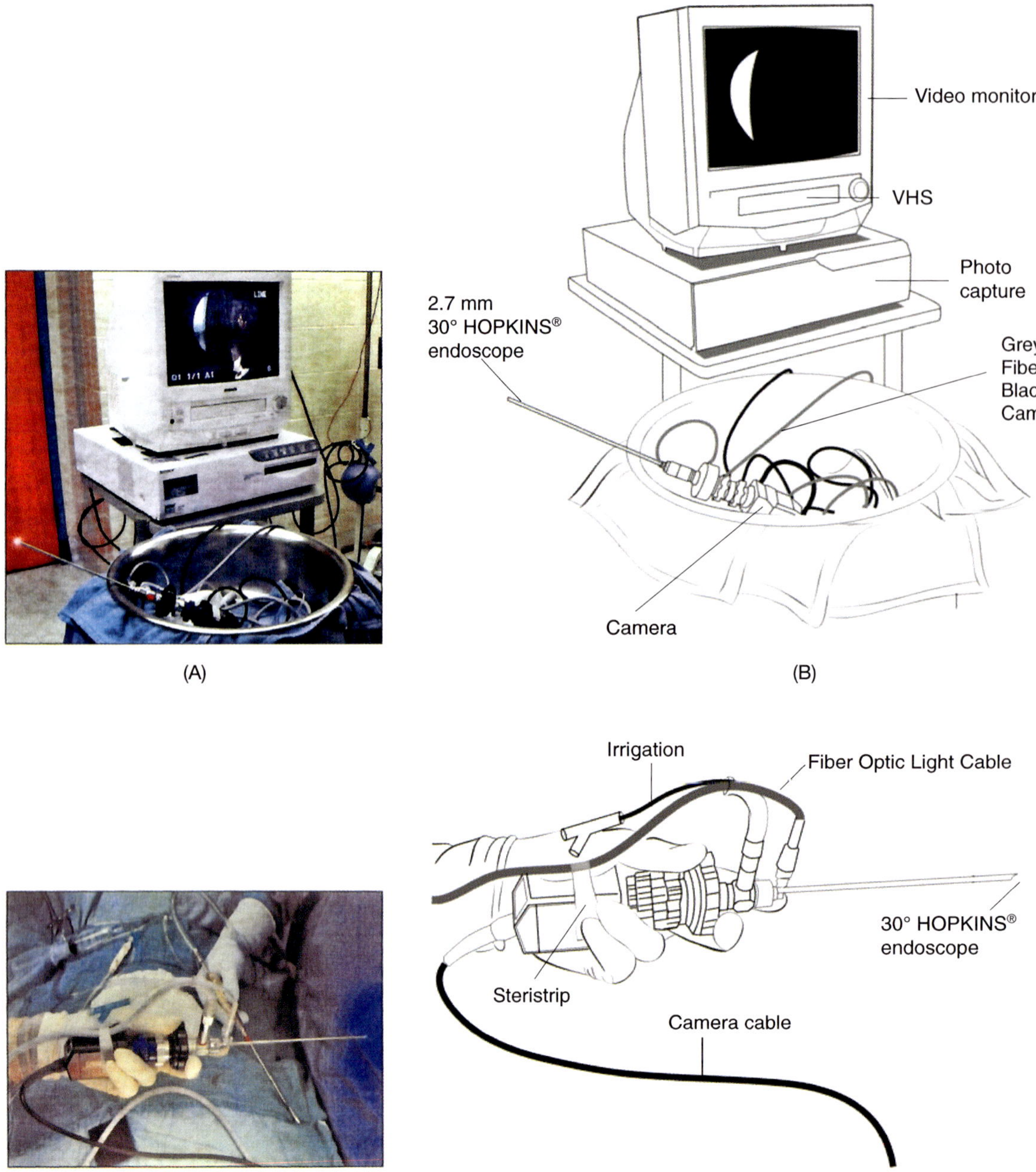

(A) (B) (C) (D)

FIGURE 3 • **Set-up** Photograph (A) and schematic (B) of the tower. Photograph (C) and schematic (D) of the handling of the scope. Note the steristrip helps keep the cables aligned and the scope from rotating. Reprinted with permission from Troulis MJ + Kaban LB: Minimally invasive approach to the mandibular ramus-condyle unit. Endo-Press. 2007 Tuttlingen, Germany.

Endoscopic Dissection and Surgical Anatomy of the Endoscopic Approach

With the endoscopic retractor in the sigmoid notch, the subperiosteal dissection is completed to expose the entire ramus/condyle unit: posterior border, condylar neck, sigmoid notch, coronoid process, and anterior border of the ramus. When performing a condylectomy, the entire condyle is dissected free from the joint capsule, disk attachments and lateral pterygoid muscle (Figure 5).[15]

ENDOSCOPIC ORTHOGNATHIC SURGERY

Background

During the past 10 years there has been great interest in the development of endoscopic techniques for exposure and reconstruction of the maxillofacial skeleton (specifically for subcondylar fracture repair)[13–16] (see Chapter 2, "Endoscopic Treatment of Subcondylar Fractures—Extraoral Approach" and

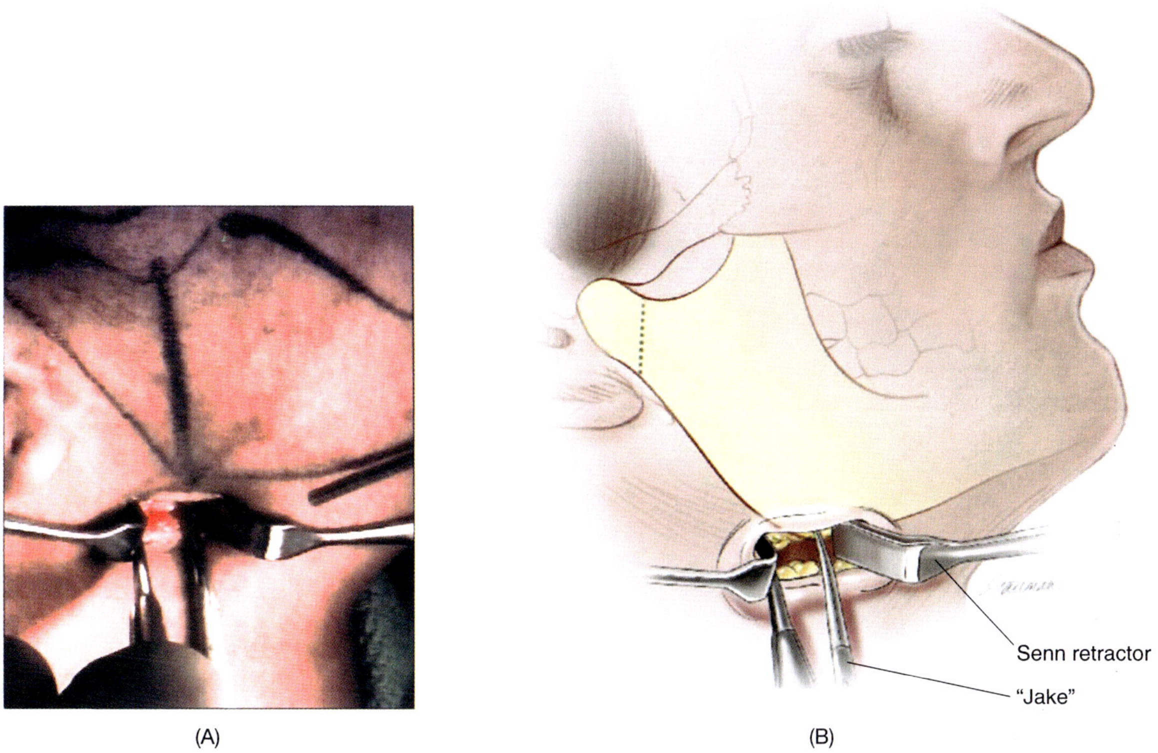

FIGURE 4 • **Incision and Dissection RIGHT ramus/condyle unit** Photograph (A) and schematic (B) of the dissection. Note the positioning of the Senn Retractors, parallel to the marginal mandibular nerve. Reprinted with permission from Troulis MJ + Kaban LB: Minimally invasive approach to the mandibular ramus-condyle unit. Endo-Press. 2007 Tuttlingen, Germany.

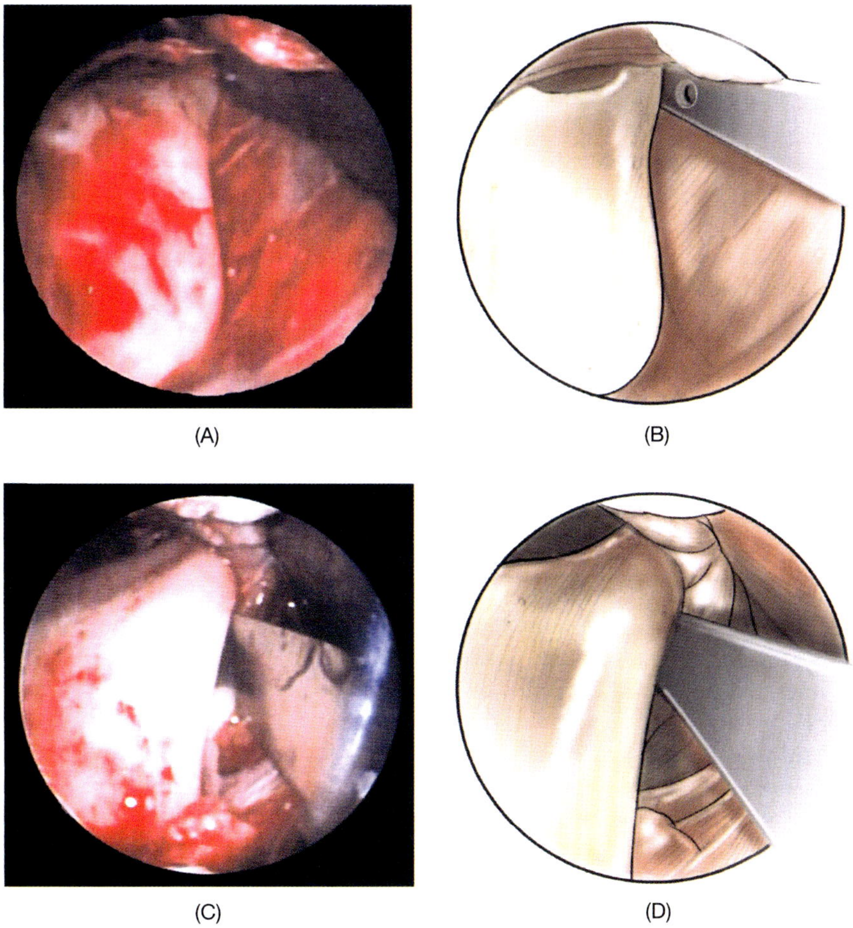

FIGURE 5 • **Endoscopic Landmarks of the LEFT ramus/condyle unit.** Photograph (A) and schematic (B) endoscopic view of the LEFT ramus condyle unit. The entire lateral mandible is well-visualized and the endoscopic landmarks identified: angle, posterior border, and condylar neck. Photograph (C) and schematic (D) endoscopic views. The dissection is proceeding to better visualize the condylar neck. (*continued*)

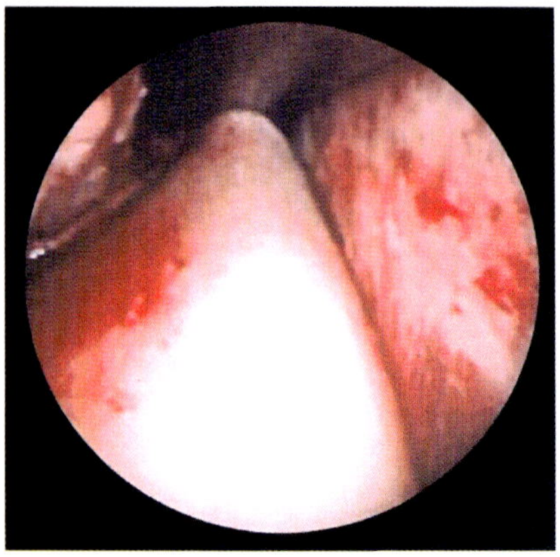

(E)

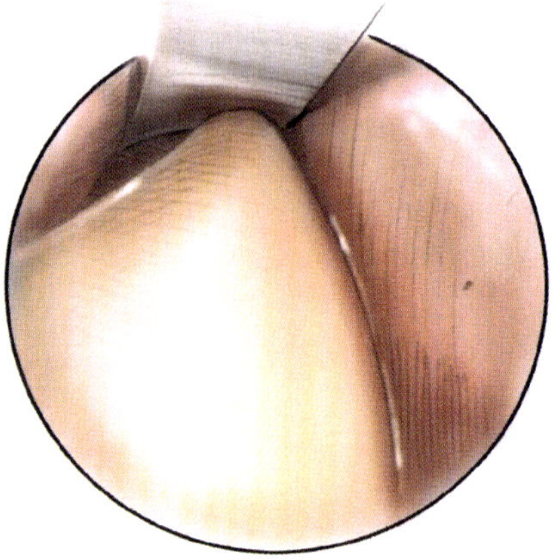

(F)

FIGURE 5 • (*continued from previous page*) Photograph (E) and schematic (F) endoscopic views of the same case. The dissection is carried further, over the top of the condyle. This is only performed for condylectomy cases. Reprinted with permission from Troulis MJ + Kaban LB: Minimally invasive approach to the mandibular ramus-condyle unit. Endo-Press. 2007 Tuttlingen, Germany.

Chapter 3, "Endoscopic Treatment of Subcondylar Fractures—Intraoral Approach").

The benefits of endoscopic surgery include small and remotely placed incisions, hidden scars, and direct visualization of a magnified and illuminated operative field. Minimal tissue dissection and manipulation result in less edema[20] and morbidity.[21–23] Length of hospital stay is shorter[15,21–25] and patients return more quickly to normal activities.[1,21–25]

A small, extraoral submandibular incision for access to the RCU was chosen. In comparison to an intraoral incision, the extraoral approach significantly reduces the extent of manipulation and soft tissue trauma during the dissection and exposure. There is less bleeding and swelling because of the very well-defined nature of the subperiosteal optical cavity. The submandibular approach also allows the surgeon to visualize the operative field "en face," the most comfortable orientation from which to operate, teach and learn. The patient can be placed in maxillomandibular fixation (MMF) prior to fixation without compromising access. All these attributes allow the surgeon to use this approach for a wide variety of procedures on the ramus condyle unit.

Surgical Techniques

Endoscopic Vertical Ramus Osteotomy

For mandibular setback, there are currently two options: (1) Vertical ramus osteotomy accomplished intraorally (IVRO) or extraorally or (2) bilateral sagittal split osteotomy (BSSO). The IVRO offers the distinct advantage of decreased risk of postoperative neurosensory changes in the inferior alveolar nerve. However, it is not possible to rigidly fix the fragments and the patient is placed into maxillomandibular fixation. The BSSO can be accomplished with rigid internal fixation (RIF). However, there is a well-known risk to the inferior alveolar nerve.[24,25] In contrast, the endoscopic vertical ramus osteotomy can be accomplished with minimal risk of neurosensory disturbance and the fragments can be rigidly fixed under direct vision.[15,21] The endoscopic vertical ramus osteotomy with rigid fixation is also an appropriate alternative for patients with well-developed asymptomatic third molars intimately involved with the inferior alveolar nerve. The mandible can be set back, rigid fixation applied, and the third molars left in place.

Anatomic landmarks of the RCU are identified: posterior border, sigmoid notch, coronoid process, anterior border, and posterior body (Figure 6). A curved, long-handled retractor is positioned to maintain the optical cavity (Synthes Maxillofacial, Paoli, PA). Irrigation is performed through the endoscope irrigation port.

With endoscopic visualization, an osteotomy is created from the sigmoid notch to the mandibular angle using a long-shaft reciprocating blade (MicroAire Surgical Instruments, Charlottesville, VA) (Figure 6). The medial pterygoid muscle is partially stripped to allow overlap of the proximal and distal segments. When required, a clamp is used to control the proximal segment. The patient is placed into wire MMF in the desired, preplanned occlusion. Three 12 to 14-mm length, 2.0 mm diameter screws are used to secure the proximal and distal segments into position (through four cortices) (Figure 6). Alternatively, if there is minimal overlap, plate fixation may be used. Screw holes are drilled and screws placed through the incision or with the aid of a percutaneous trocar.[18–21]

Outcomes We reported a retrospective analysis of our first 14 patients (mean age 28 years, nine males and five females) who had endoscopic vertical ramus osteotomies (EVRO) with rigid internal fixation (n = 23 sides).[19] The mean operative time was 37 minutes per side. In all cases, the procedure was successfully executed via a 1.5 cm submandibular incision. The desired, preplanned occlusion was achieved and rigid internal fixation (RIF) was successfully applied through the working port (trochar access was not required). In this early series, no patients required maxillomandibular fixation. Proper positioning of skeletal segments was documented by lateral and posterior-anterior cephalograms and panoramic radiographs immediately postoperatively. The planned occlusal position was confirmed by postoperative clinical examination and by comparison with the articulated surgical models. Nine patients had mandibular prognathism (n = 9) and had a mean setback of 4.7 mm (range, 2–9 mm). The remaining five patients had correction of mandibular asymmetry with asymmetric mandibular setback and RIF. Lateral cephalograms at a mean of 1.7 years postoperatively (range, 6 months to 2 years) documented stability of the bone position (see Figure 6).

To date, 51 patients (19 male and 32 female) have been successfully treated with EVRO and RIF for prognathism or asymmetric prognathism. Complications included marginal mandibular nerve weakness lasting less than 1 week (n = 1 patient), temporary inferior alveolar nerve paresthesia (n = 1), intraoperative bleed secondary to transection of the masseteric artery requiring transfusion of 1 unit of blood (n = 1). The length of hospital stay was less than 1 day for 30 patients and

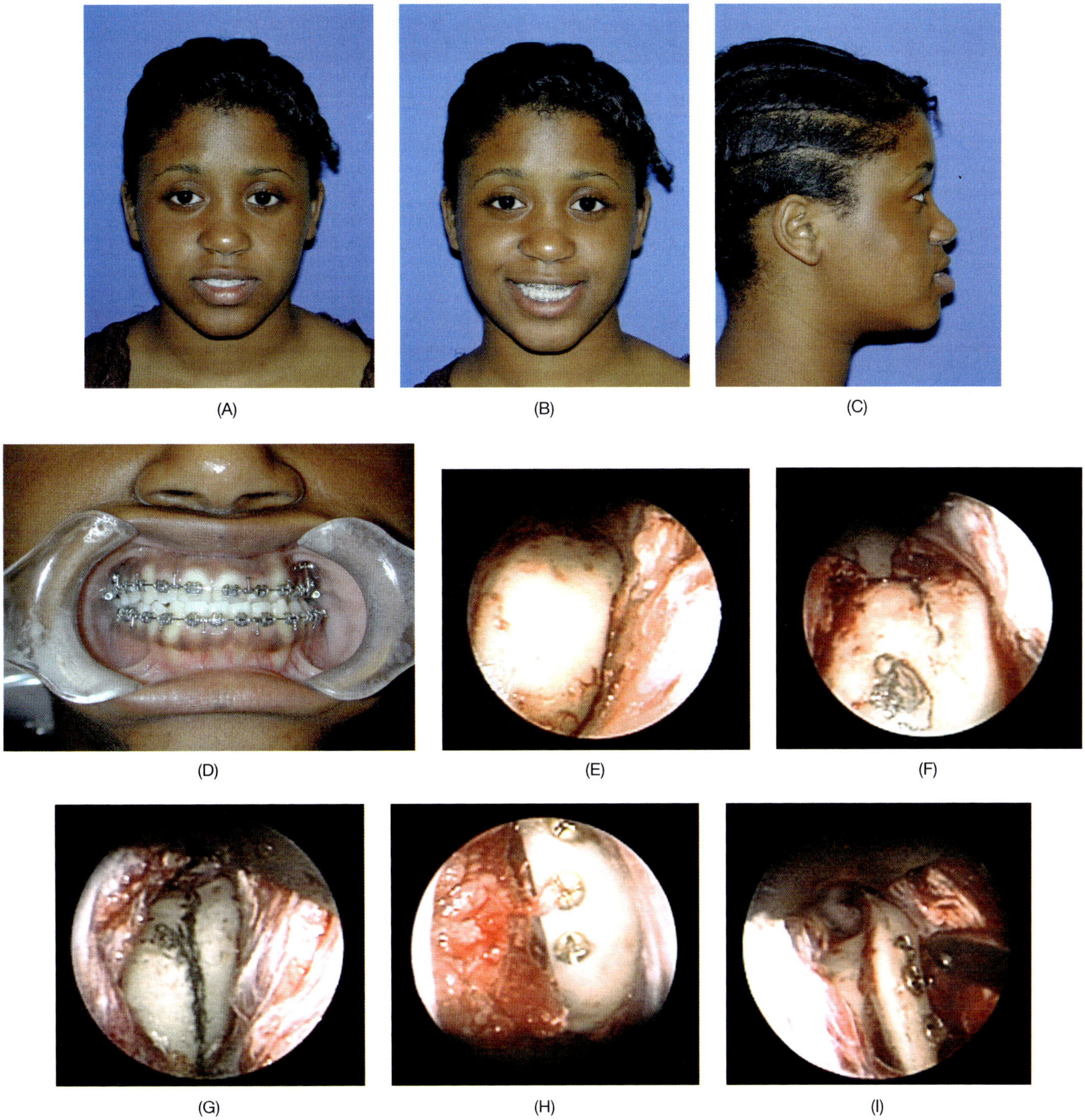

FIGURE 6 • Endoscopic vertical ramus osteotomies. Frontal rest (A), frontal smiling (B), lateral photograph (C), and intraoral view (D) of a 20-year-old female with mandibular prognathism and class III malocclusion. For mandibular setback, the endoscopic vertical ramus osteotomy (EVRO) was chosen to avoid maxillomandibular fixation and to minimize the risk to the inferior alveolar nerve. Endoscopic views of the left ramus condyle unit (E), start of proposed osteotomy seen behind the antilingula, which is seen shaded (F), the proposed osteotomy (G), and completed osteotomy with rigid fixation (H). Bony overlap between the distal and proximal segments (I), (*continued*)

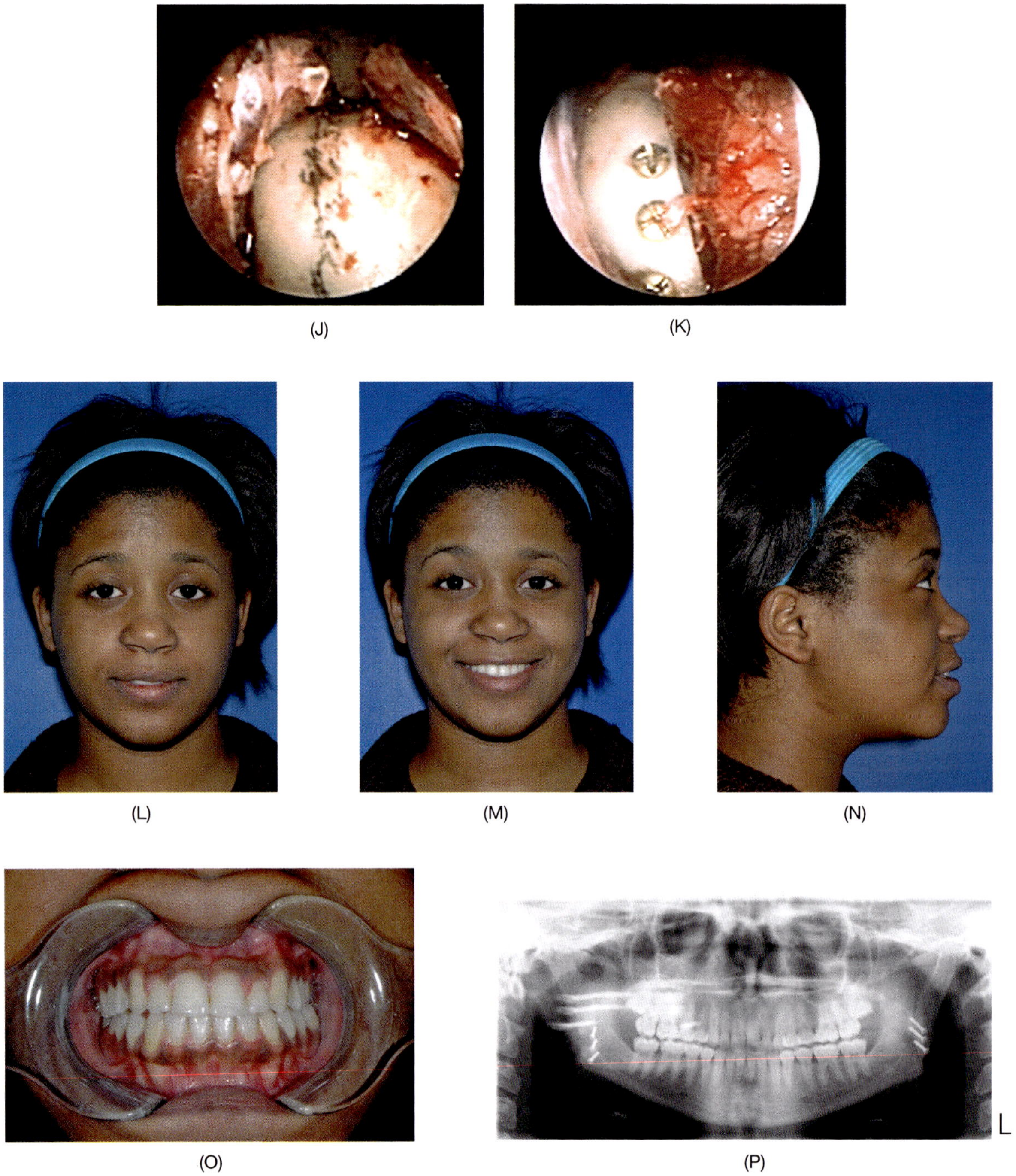

FIGURE 6 • (*continued from previous page*) Endoscopic views of the right ramus condyle unit: the proposed osteotomy is outlined (J), the completed osteotomy with rigid fixation (K). Frontal rest (L), Frontal smiling (M), lateral photograph (N), intraoral view (O), and panoramic radiograph (P) 1 year postoperatively.

2 days for two patients (one for the masseteric bleed and one for concurrent medical problems having nothing to do with the operation). One patient required MMF because we could not achieve acceptable positioning of the proximal fragment with RIF.

Condylectomy and Costochondral Graft

Temporomandibular joint (TMJ) disorders that manifest in progressive alteration of condylar shape and decrease in condylar mass or size include idiopathic condylar resorption or degenerative joint disease. Condylar resorption (condylysis) has been associated with rheumatoid arthritis, systemic lupus erythematosus, intraarticular and systemic steroid usage, trauma, neoplasia, orthodontic treatment, and orthognathic surgery. However, there is frequently no identifiable etiology and hence the term idiopathic condylar resorption.[26] Patients with this condition have decreased posterior face height, retrognathism, and progressive anterior open bite with clockwise rotation of the mandible. The condition is usually bilateral and appears to have a predilection for females in an age range of 15 to 35 years.[26]

The management of condylar resorption remains controversial. Orthognathic surgical correction has been advocated if the resorptive process stabilizes. Condylectomy and reconstruction with either autogenous grafts (eg, costochondral grafts) or alloplastic materials represent other treatment options.[26]

The Massachusetts General Hospital (MGH) protocol for management of patients with active or progressive idiopathic condylysis includes: (1) documentation of the diagnosis by a history of TMJ symptoms and progressive retrognathism, open bite and clockwise rotation of the mandible; (2) documentation of progressive occlusal and condylar changes by evaluation of serial radiographs, photographs, and dental casts; (3) technetium-99m methylene diphosphonate quantitative bone scan to determine activity of the resorptive process; (4) management of TMJ symptoms by splint therapy, physical therapy, muscle relaxants, nonsteroidal anti-inflammatory drugs and other modalities as indicated; (5) longitudinal observation for at least 2 years to document response to treatment and stability of the condition; (6) orthognathic surgical correction for stable patients with a negative bone scan; (7) condylectomy and costochondral graft reconstruction for patients with active/progressive condylar resorption and persistent TMJ symptoms and/or persistent positive bone scan.[26,27]

The standard technique for condylectomy and costochondral graft RCU reconstruction (regardless of diagnosis) involves extended preauricular and submandibular incisions. In contrast, minimally invasive access is achieved with a 1.5 cm submandibular incision located in the skin folds below the mandibular angle (Figures 7 and 8). This approach eliminates risk to the frontal branch of the facial nerve and because of decreased dissection and tissue manipulation reduces the risk of marginal mandibular weakness. This has resulted in excellent patient acceptance of the technique.

Surgical technique The same landmarks and 1.5 cm incision and dissection, as described for the vertical ramus osteotomy, are used. The same care must be taken to draw the surface landmarks on the skin and to do the same meticulous subperiosteal dissection to create the optical cavity and identify the bony landmarks (see Figures 7 and 8). In the case of the condylectomy procedure, the endoscopic view of the RCU allows the condylar head and

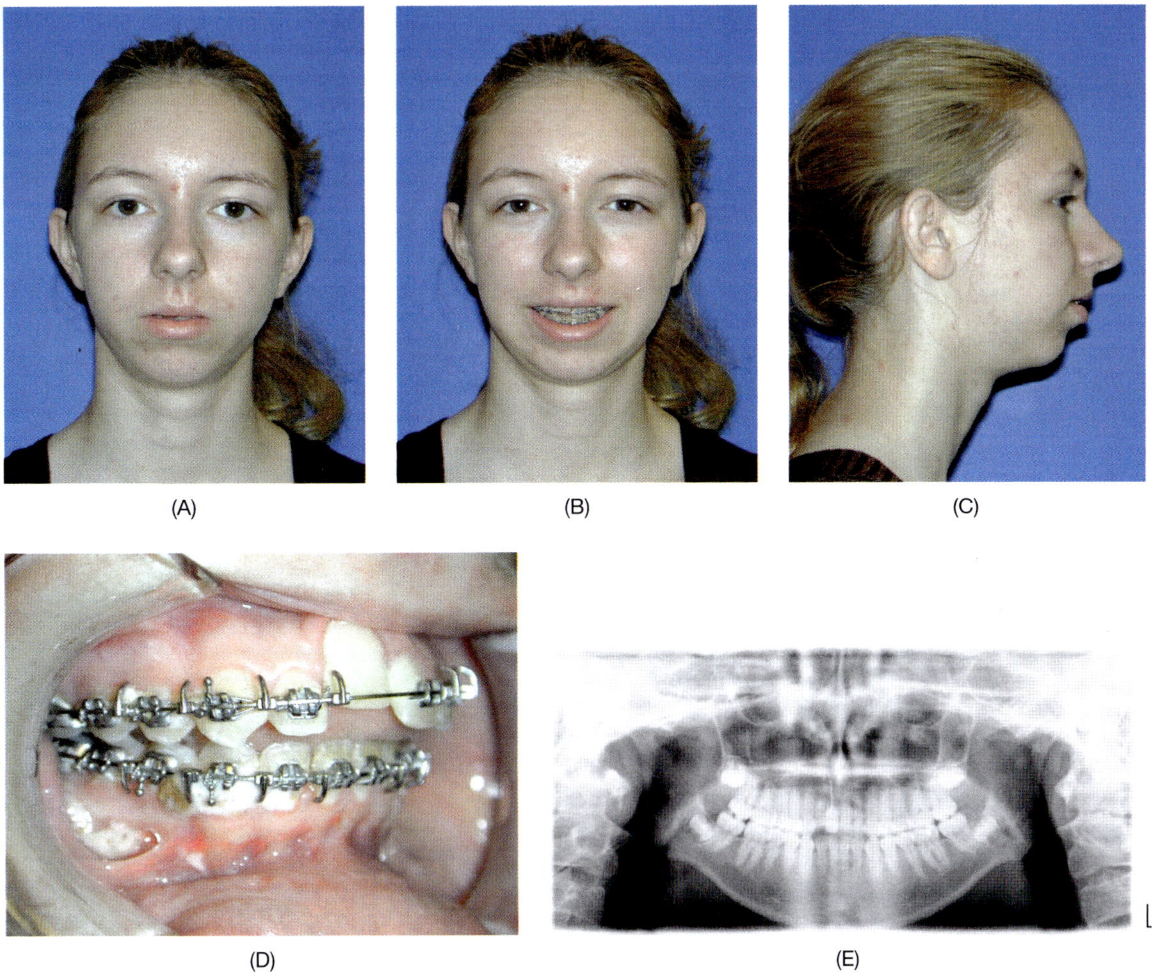

(A) (B) (C) (D) (E)

FIGURE 7 • Endoscopic condylectomy and reconstruction. Frontal rest (A), smiling (B), lateral (C), and intraoral (D) photographs of a 16-year-old female with mandibular retrognathism, clockwise rotation of the mandible, and resultant open bite malocclusion secondary to idiopathic condylar resortion. The panoramic radiograph (E) (*continued*)

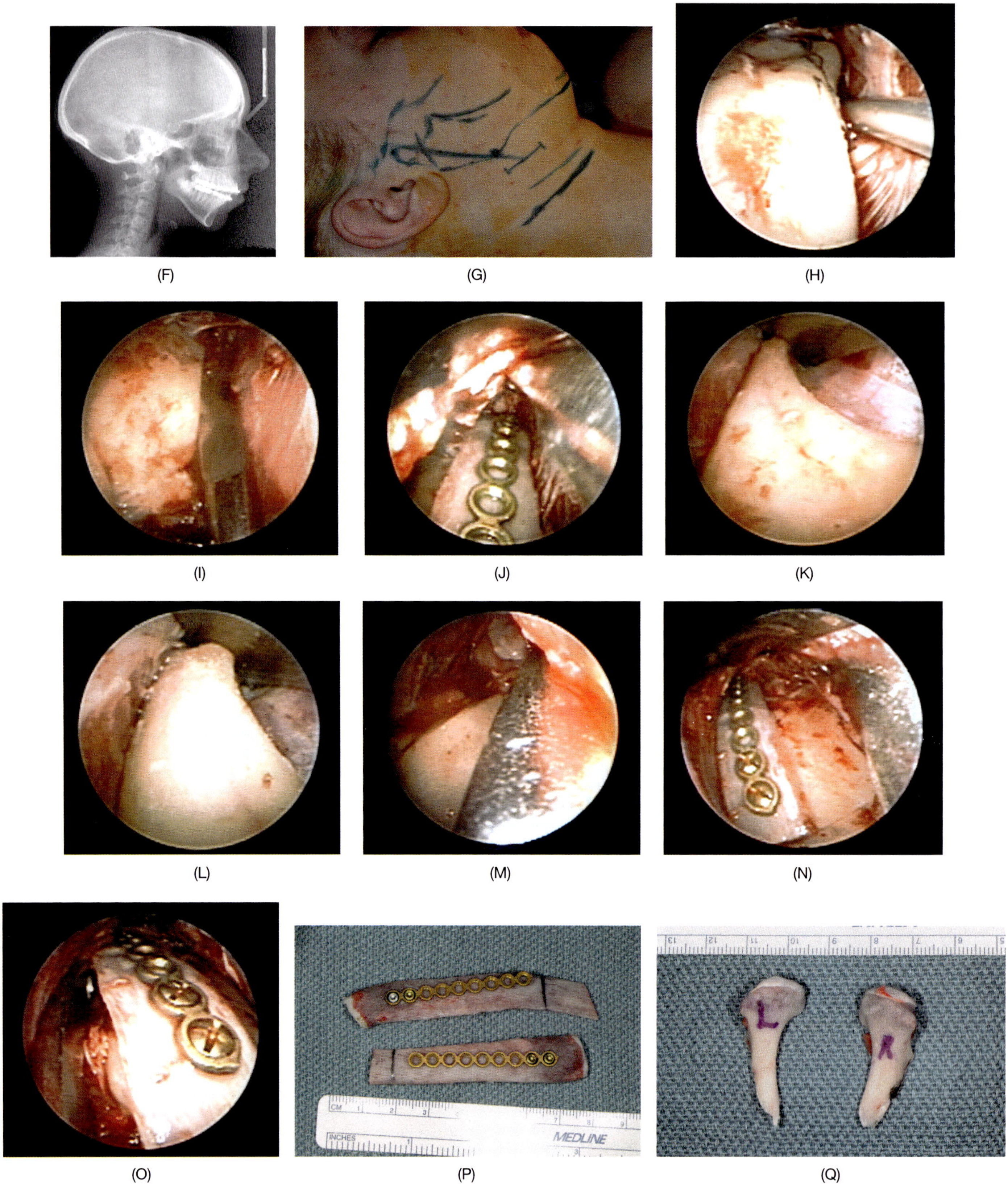

FIGURE 7 • (*continued from previous page*) demonstrates the resorbed condyles and the lateral cephalogram (F) shows the convex profile. The intraoperative lateral view (G) demonstrates the anatomic landmarks. The smaller endoscopic submandibular incision is marked. Endoscopic views of the left condyle: the proposed osteotomy (H) and completed osteotomy and the disk visualized prior to rib placement (I) and the costochondral graft placed against the disk and fixed into position with 2.0 mm screws (J). Endoscopic views of the right condyle: the lateral surface of the ramus and small condyle (K), a close-up view of the well-dissected condyle (L), the completed osteotomy and the disk visualized prior to rib placement (M) and the graft placed against the disk and fixed into position with 2.0 mm screws (N) and view from behind showing the contact between the distal mandible and the rib (O). The costochondral rib grafts with only 1–2 mm of cartilage for the articulating surface (P) condylectomy specimens (Q) and are shown intraoperatively. (*continued*)

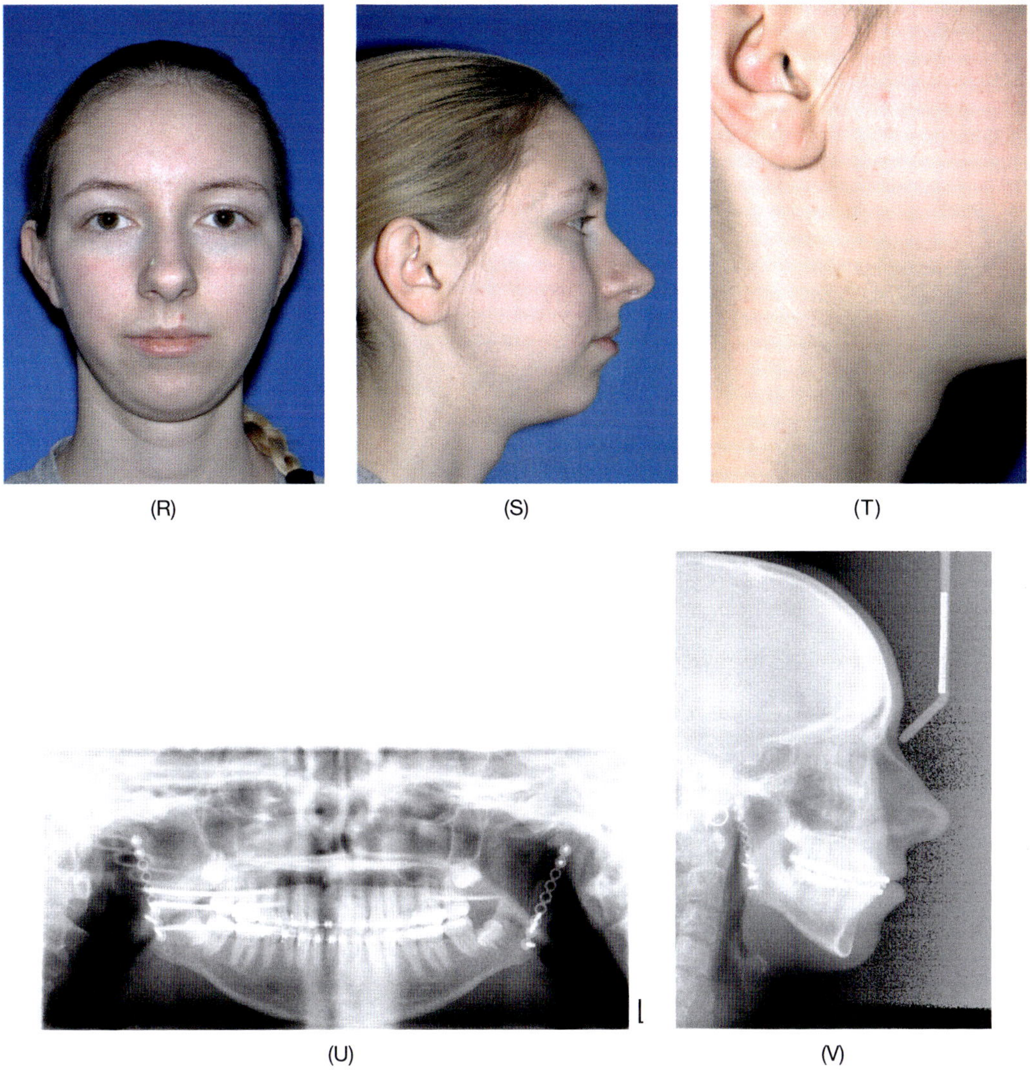

FIGURE 7 • *(continued from previous page)* Frontal (R), lateral photograph (S) and lateral close-up (T) photographs of the patient 2 years after treatment. The panoramic (U) and lateral cephalograph (V) radiographs show the rib position and rigid fixation.

neck to be readily dissected from the surrounding soft tissues. To facilitate the dissection, the RCU is distracted inferiorly with a 26 gauge stainless steel traction wire (placed at the angle). The condylectomy is performed from the sigmoid notch to the posterior border of the mandible (see Figures 7 and 8) using a long-shafted reciprocating blade (MicroAire, Charlottesville, VA). Skeletonization of the condyle is completed and it is atraumatically removed. The undersurface of the articular disk is visualized.[15,21,23] The patient is placed in MMF with a splint producing a 2 to 3 mm posterior open bite on the sides to be reconstructed with a costochondral graft (CCG). This compensates for loss of vertical height that occurs during settling and remodeling of the CCG during healing and remodeling.[26,27]

Costochondral grafts are harvested in the standard manner through an inframammary incision. Then, using the endoscope for direct visualization, the disk is identified and the CCG is placed into the glenoid fossa. The graft is fixed in position using a 2.0 mm titanium miniplate, which acts as a washer, and three to five 12 to 14 mm long, 2.0 mm diameter screws (Synthes Maxillofacial, Paoli, PA). Screw placement is accomplished through the incision or with the aid of a percutaneous trocar.

Outcomes In a retrospective evaluation by Troulis, Williams, and Kaban,[23] 10 females, mean age of 32 years (range, 19–46), underwent successful endoscopic condylectomy and costochondral graft reconstruction of the RCU (n = 17 sides).

Mean operating time was 52 minutes (range, 25–75 minutes) per mandibular side exclusive of the rib harvesting procedure. The average length of stay (LOS) for all patients was 2.5 days (range, 2–4 days). In all cases, the procedure was executed via the 1.5 cm incision. The desired pre-planned occlusion was achieved and rigid internal fixation was successfully applied through the working port (trocar access was not required). All patients were placed in intermaxillary fixation for 10 to 14 days (to allow for initial graft healing).

To date, a total of 30 patients (6 males and 24 females) have had condylectomy and costochondral graft reconstruction for the treatment of idiopathic condylar resorption or degenerative joint

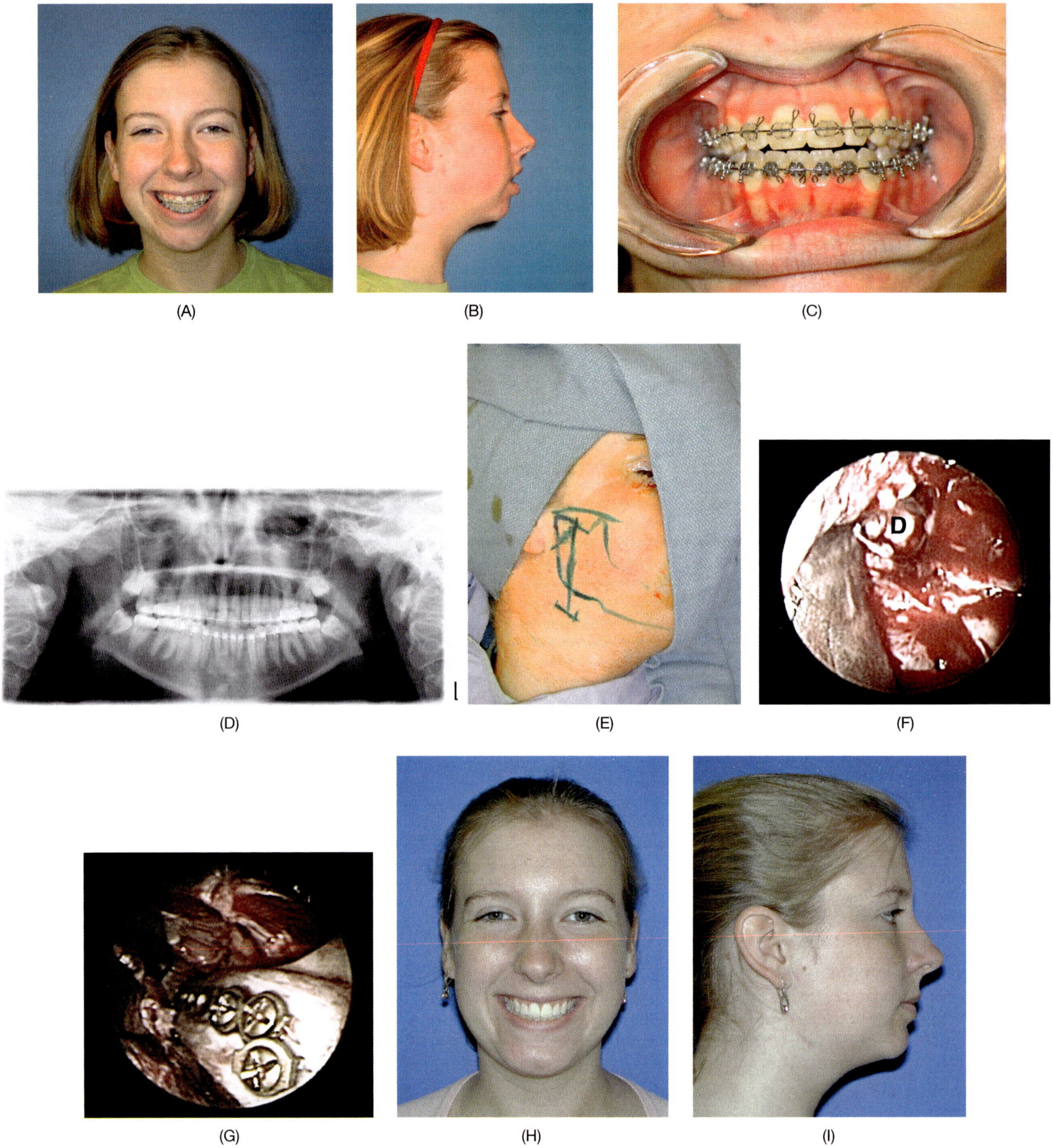

FIGURE 8 • Endoscopic condylectomy and reconstruction. Frontal smiling (A), lateral (B), intraoral (C) photographs of a 17-year-old female with mandibular retrognathism, clockwise rotation of the mandible and resultant open bite malocclusion secondary to idiopathic condylar resorption. The panoramic radiograph (D) demonstrates the resorbed condyles. The intraoperative lateral view (E) demonstrates the planned incision and the surface landmarks. Endoscopic views of the right ramus condyle unit area: the pathologic condyle has been removed and the disk (D) is visualized (F), then the costochondral rib is inserted against the disc and fixed into position (G). Smiling (H), lateral (I), (*continued*)

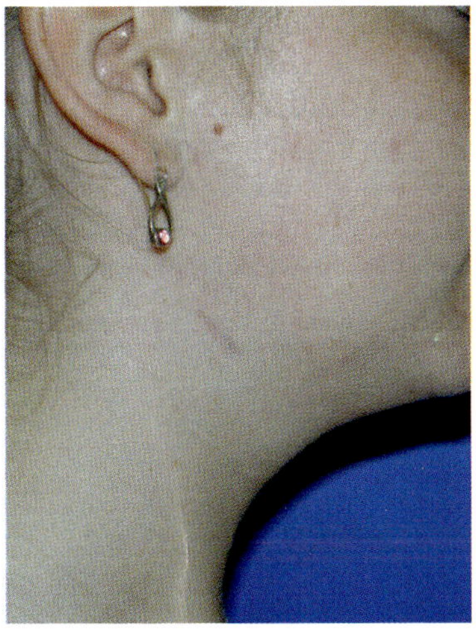

(J)

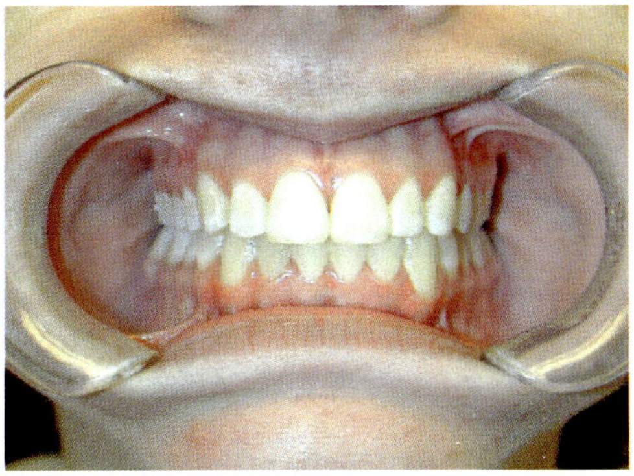

(K)

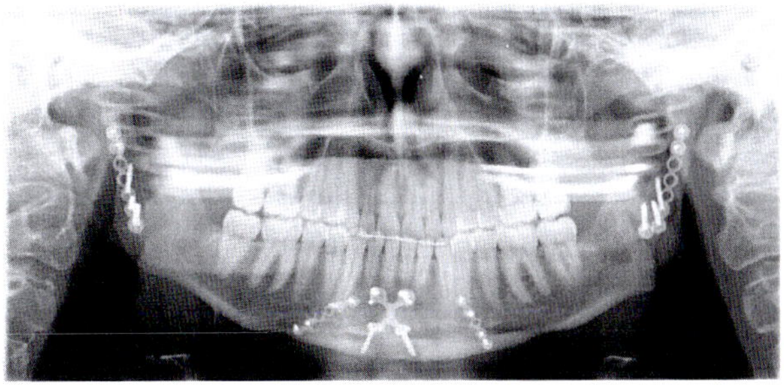

(L)

FIGURE 8 • (*continued from previous page*) close-up lateral of incision site (J) and intraoral (K) photographs as well as panoramic radiograph (L) 2 years postoperatively.

disease. None of the patients suffered any long-term neurologic changes related to the inferior alveolar, lingual or facial nerves. All patients demonstrated good range of motion both postoperatively and at long-term follow-up. Maximal incisal opening returned to preoperative values by 1 year (see Figures 7 and 8).

High Condylectomy and Vertical Ramus Osteotomy

Condylar hyperplasia is the most common acquired deformity of the mandibular condyle. It occurs more frequently in females than males, and it is the result of an abnormal growth center in the condyle. Children with condylar hyperplasia are typically born with a symmetrical jaw. The asymmetry most commonly begins with the onset of the pubertal growth spurt of the mandible.[26] Less frequently, condylar hyperplasia occurs with the first mandibular growth spurt between ages 4 and 7 years.

There are two growth patterns in this disorder documented by localized increase uptake of technetium 99c-methylene diphosphonate in the abnormal condyle.

1. Vertical. Excessive mandibular growth occurs in a predominantly vertical direction. The patients exhibit increased vertical height of the ramus and intraorally, there is an open bite on the affected side. There is no crossbite and little or no deviation of the chin point or dental midline.
2. Rotational. In this pattern, there is not only a vertically long ramus on the affected side, but the mandibular body is convex and the chin point and dental midline are deviated toward the normal side. Intraorally, there is a crossbite on the unaffected side.[24] This pattern is similar in presentation to patients who have mandibular hemi hypertrophy. However, in the letter group, the bone scan demonstrates generalized increase in uptake of the affected condyle, ramus and body.

In patients with vertical pattern condylar hyperplasia, there is a lateral open bite. As dental compensations (ie, supra-eruption of the maxillary teeth) occur, the open bite closes and the maxillary occlusal plane cants downward on the affected side. These compensations are in the dentoalveolar segment and, in contrast to hemifacial microsomia, the alar base and piriform apertures are not canted.[26]

To develop a treatment plan, the surgeon must make the correct diagnosis and determine whether abnormal condylar growth is progressive or stable. Patients in the active phase may be observed until the growth cycle is complete or may be offered surgical excision of the abnormal growth center (high condylectomy) as a growth-arresting procedure.[26]

The standard technique for the high condylectomy is a preauricular incision; the hyperplastic condylar region is usually well-demarcated and easily identified as a soft, marrow-like cap on the condylar head. This is excised (high condylectomy), leaving the normal condylar stump in place. In contrast, minimally invasive access is achieved with the 1.5 cm submandibular incision. A vertical ramus osteotomy is performed and the entire proximal segment is removed. The high condylectomy is performed, the proximal fragment is then replaced and rigidly fixed in a more superior position. The ipsilateral open bite closes as the mandible rotates into its new position (Figure 9).

To date, seven patients, three male and four female, with vertical pattern condylar hyperplasia have been treated by this minimally invasive growth-arresting procedure. In all cases, the vertical ramus osteotomy, high condylectomy and rigid fixation of the ramus was executed via the 1.5 cm incision.

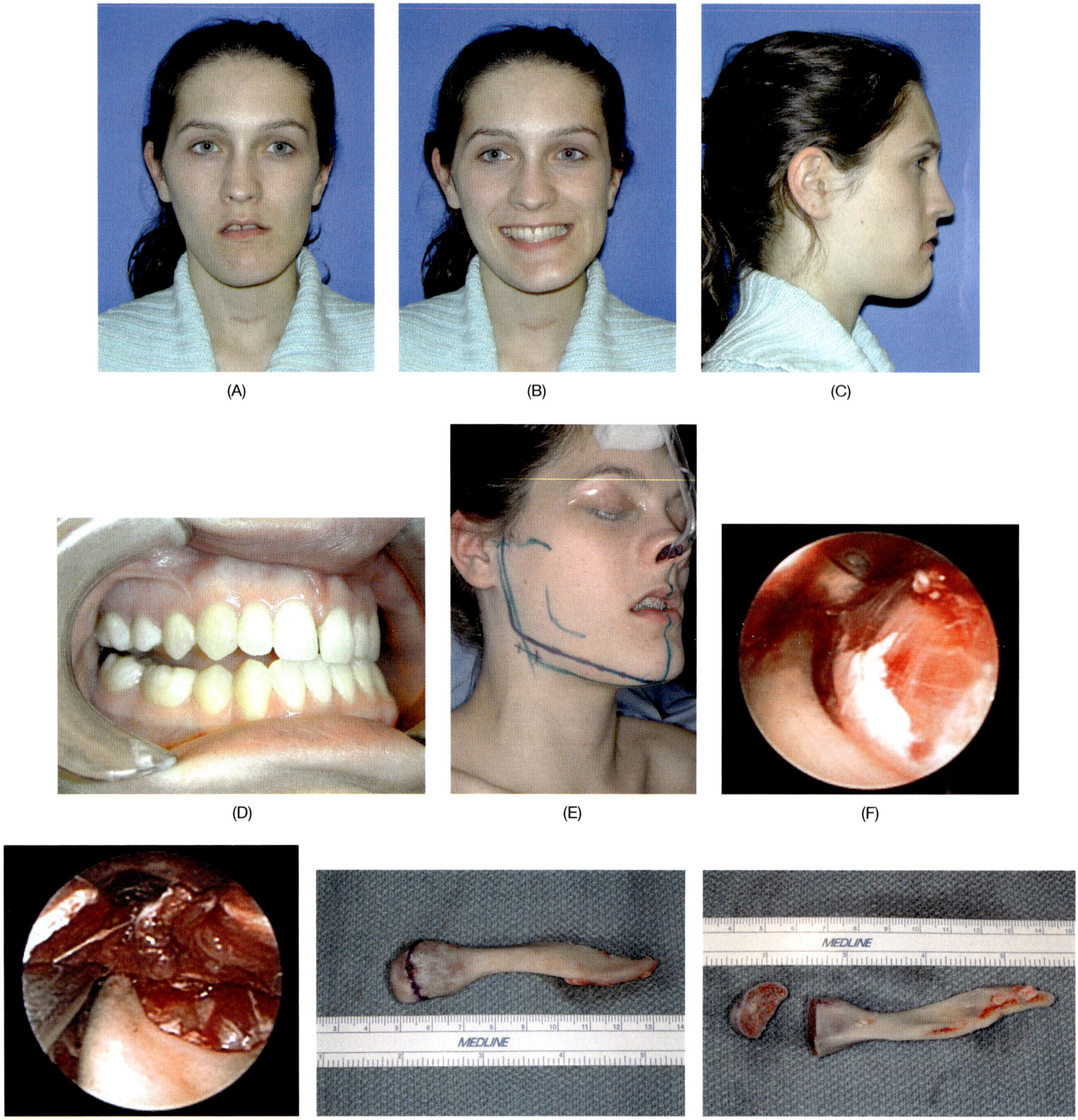

FIGURE 9 • Endoscopic vertical ramus osteotomy and high condylectomy. Frontal rest (A), smiling (B), lateral (C), and intraoral (D) photographs of a 23-year-old female with right mandibular condylar hyperplasia. The intraoperative lateral view (E) demonstrates the planned incision and the surface landmarks. The proposed inferior border ostectomy is also outlined. Endoscopic views of the right ramus condyle unit area: sigmoid notch and condylar neck is well-visualized (F). As the dissection is continued over the condylar head (G), the long neck is appreciated. The intraoperative view of the extracorporal ramus condyle unit (H) with the proposed ostectomy which is completed in (I). *(continued)*

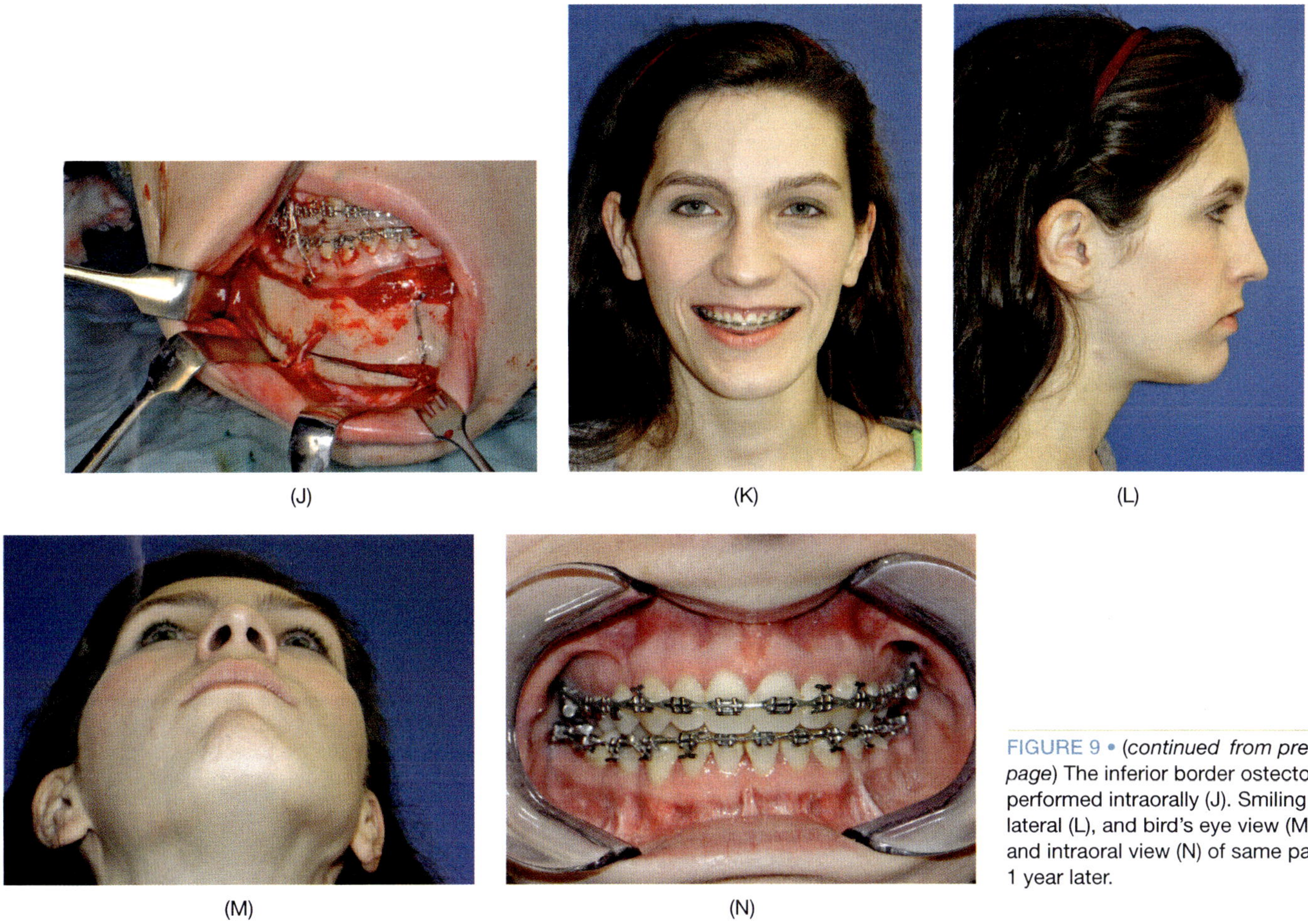

FIGURE 9 • (*continued from previous page*) The inferior border ostectomy is performed intraorally (J). Smiling (K), lateral (L), and bird's eye view (M) and intraoral view (N) of same patient 1 year later.

Results were documented by serial clinical examinations, anterior-posterior and lateral cephalograms, and panoramic radiographs. The desired preplanned occlusion was achieved and rigid internal fixation was successfully applied through the working port (trocar access was not required. None of the patients suffered any long-term neurologic changes related to the inferior alveolar, lingual, or facial nerve. All patients demonstrated good range of motion both postoperatively and at follow-up. Results have been stable at latest follow-up (1 month to 7 years).

Other Procedures

Condylotomy for the treatment of temporomandibular painful click, as described by Hall[28] has also been successfully performed using the endoscopic approach. To date, eight female patients (five left and three right sides) have undergone endoscopic condylotomy. The patient is placed in wire MMF. The previously described landmarks are drawn on the skin, the 1.5 cm incision is made and the dissection carried out. In all cases, the condylotomy was successfully executed. None of the patients had any permanent neurologic changes of the inferior alveolar or facial nerve.

Coronoidectomy is usually performed if the patient has trismus and the coronoid extends superior to the zygomatic arch. The standard approach is to gain exposure of the coronoid process by either an intraoral incision (if the patient can open at least 20 mm) or a bicoronal flap incision.

We have used the approach to the RCU described in this chapter to perform coronoidectomies. To date, eleven patients (n = 17 sides), 3 females and 8 males, with bilateral (n = 7) or unilateral (n = 3) coronoid hyperplasia have undergone endoscopic coronoidectomy. In all cases, the coronoidectomy was executed via the 1.5 cm incision. None of the patients suffered any long-term neurologic changes related to the inferior alveolar or facial nerve. All patients demonstrated good range of motion (>35 mm) both immediately postoperatively and at follow-up.

Biopsies (n = 3) of the subcondylar region or the medial ramus have also been performed using this approach (Figure 10). None of the patients suffered any long-term neurologic changes related to the inferior alveolar or facial nerve. One patient had an intraoperative bleed requiring packing and embolization of the internal maxillary artery (no blood transfusion was required).

CONCLUSIONS

During the past 25 to 30 years, minimally invasive techniques have been developed for gynecologic and urologic surgery and subsequently for general, cardiovascular, otolaryngologic and

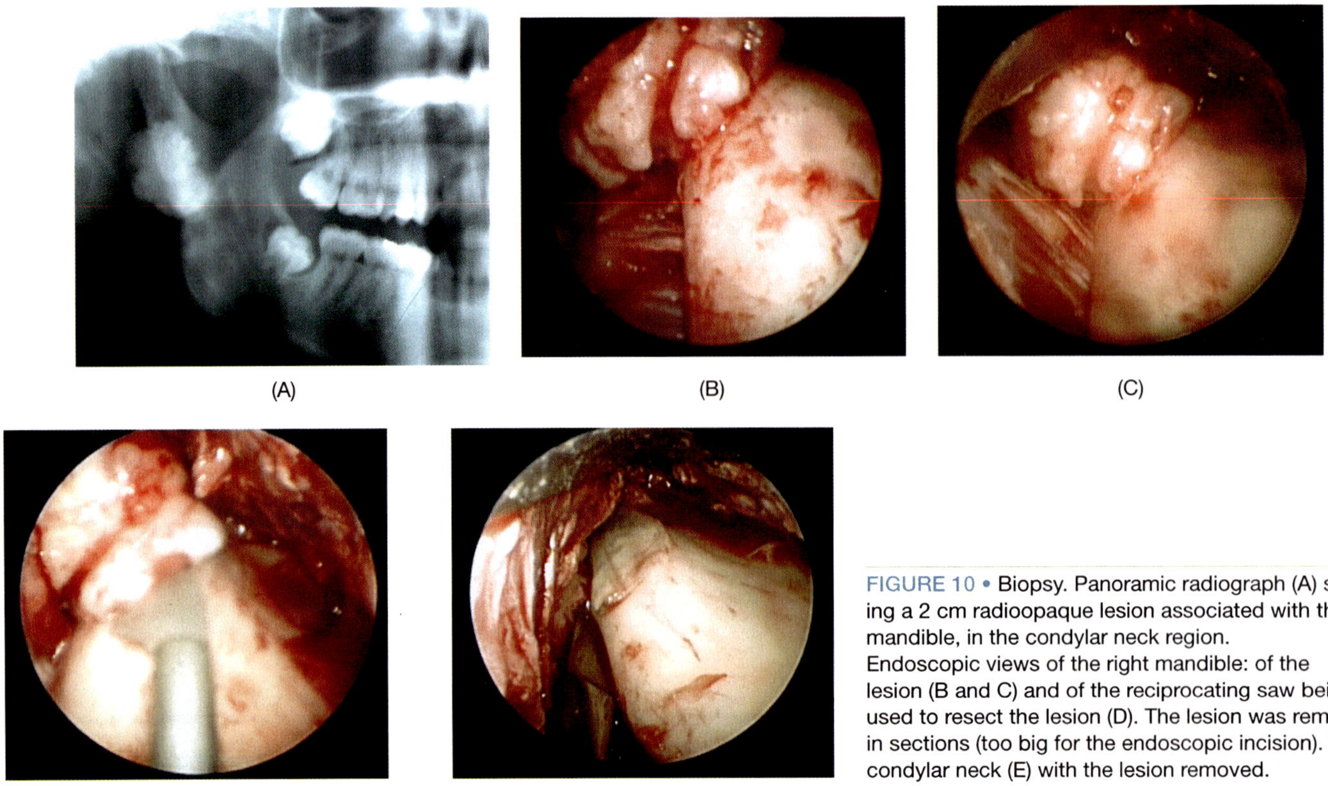

(A) (B) (C) (D) (E)

FIGURE 10 • Biopsy. Panoramic radiograph (A) showing a 2 cm radioopaque lesion associated with the right mandible, in the condylar neck region. Endoscopic views of the right mandible: of the lesion (B and C) and of the reciprocating saw being used to resect the lesion (D). The lesion was removed in sections (too big for the endoscopic incision). The condylar neck (E) with the lesion removed.

facial esthetic surgery. It is only recently, that oral/maxillofacial surgeons have begun to develop minimally invasive (endoscopic) techniques for correction of skeletal deformities[16–19] (see Chapter 2, "Endoscopic Treatment of Subcondylar Fractures—Extraoral Approach," and Chapter 3, "Endoscopic Treatment of Subcondylar Fractures—Intraoral Approach) and diseases of the salivary glands[29] (see Chapter 8, "Sialoendoscopy" and Chapter 9, "Sialoendoscopy—The U.S. Experience").

The benefits of endoscopy include small and remotely placed incisions, acceptable scars and direct visualization of an illuminated and magnified operative field. Minimal dissection and tissue manipulation result in decreased pain and swelling, less overall morbidity, and faster recovery. For these reasons, minimally invasive surgery has gained enthusiastic public acceptance.

References

1. Troulis MJ, Kaban LB. Minimally invasive orthognathic surgery. In: Bell B, Guerro G, eds. Distraction osteogenesis of the facial skeleton. Ontario, Canada: BC Decker Inc; 2007. p. 451–60.
2. Brieger GH. The development of surgery. In: Townsend CM, Beauchamp RD, Evens BM, Mattox KL, eds. Sabiston's textbook of surgery: the biological basis of modern surgical practice. 15th ed. Philadelphia; PA: W.B. Saunders; 1997. p. 1–15.
3. Kaban LB. Biomedical technology revolution: opportunities and challenges for oral and maxillofacial surgeons. Int J Oral Maxillofac Surg 2002;31:1–12.
4. Wolfe SA. Gunther Cohn Stock, MS DDS Father of Maxillary Orthognathic Surgery. J Cranio-Max-Fac Surg 1989;17:331–4.
5. Obwegeser HL: Zur Operationstechnik bei der Progenic und andever unter Kieferanomalien. Dtsch Zahn Mund Kieferheilkid 23:1, 1955
6. Trauner R, Obwegeser H. The surgical correction of mandibular prognathism and retrognathia with consideration of genioplasty. Oral Surg Oral Med Oral Pathol 1957 Jul;10(7) p. 677–89.
7. Trauner R, Obwegeser H. The surgical correction of mandibular prognathism and retrognathia with consideration of genioplasty. Oral Surg Oral Med Oral Pathol 1957 Sep;10(9) p. 899–909.
8. Bell WH. Biologic basis for maxillary osteotomies. Am J Phys Anthropol 1973;38:279–89.
9. Dolman RM, Bentley KC, Head TW, English M. The effect of hypotensive anesthesia on blood loss and operative time during Le Fort osteotomy. J Oral Maxillofac Surg 2000;58:834–9.
10. Samman N, Cheung LK, Tong AC, Tideman H. Blood loss and transfusion requirements in orthognathic surgery. J Oral Surg 1996;54:21–4.
11. Gersema L, Baker K. Use of corticosteriods in oral surgery. J Oral Maxillofac Surg 1992;50:270–7.
12. Huaman ET, Juvet LM, Nastri A, Denman WT, Kaban LB, Dodson TB. Changing patterns of hospital length of stay after orthognathic surgery 2008;66:492–7.
13. Troulis MJ, Ramirez JL, Kaban LB. Endoscopic approach for mandibular orthognathic surgery. Facial Plastic Surg Clin N Am 2006;14:45–50.
14. Lobe TE. In minimal access surgery. Pediatr Rev 1996;17:441–7.
15. Troulis MJ, Kaban LB. Minimally invasive approach to the mandibular ramus condyle unit. Tuttlingen, Germany: Endo-Press, 2007.
16. Hayashi A, Maruyama K, Onishi K, Sawaizumi M. Endoscopic osteosynthesis of zygomatic fractures using minimal-access incisions. Min Invas Ther & Allied Technol 1998;7:53–6.

17. Lauer G, Schmelzeisen R. Endoscope-assisted fixation of mandibular condylar process fractures. J Oral Maxillofacial Surg 1999;57:36–9.
18. Lee C, Mueller RV, Lee K, Mathes J. Endosopcic subcondylar fracture repair: functional, aesthetic and radiographic outcomes. Plast Reconstruc Surg 1998;102:1434–43.
19. Sandler NA, Andreasen KH, Johns FR. The use of endoscopy in the management of subcondylar fractures of the mandible: a cadaver study. Oral Surg Oral Med Oral Path Oral Radiol Endod 1999;88:529–31.
20. Williams WB, Abukawa H, Shuster V, Kaban LB, et al. A comparison of postoperative edema after introral vs. endoscopic mandibular ramus osteotomy. J Oral Maxillofac Surg 2003;61;(8S):61a–2.
21. Troulis MJ, Kaban LB. Endoscopic approach to the ramus condyle unit: clinical applications. J Oral Maxillofac Surg 2001;59:503–9.
22. Troulis MJ, Kaban LB. Endoscopic vertical ramus osteotomy: early clinical results. J Oral Maxillofac Surg 2004;62:824–8.
23. Troulis MJ, Williams BW, Kaban LB. Endoscopic mandibular condylectomy and reconstruction: early clinical results. J Oral Maxillofac Surg 2004;62:460–5.
24. MacIntosh RB. Experience with the sagittal osteotomy of the mandibular ramus: a 13 year review. J Maxillofac Surg 1981;9:151–65.
25. August M, Marchena J, Donady J, Kaban L. Neurosensory deficit and functional impairment after sagittal ramus osteotomy: a long term follow-up study. J Oral Maxillofac Surg 1998;56:1231–5.
26. Kaban LB. Acquired abnormalities of the temporomandibular joint. In Kaban LB and Troulis MJ, eds. Pediatric oral and maxillofacial surgery. Philadelphia, PA: Elsievier, 2004, p. 340–75.
27. Papadaki ME, Tayebaty F, Kaban LB,Troulis MJ. Condylar Resorption. Oral Maxillofac Surg Clin North Am 2007;19:223–34.
28. Werther JR, Hall HD, Gibbs SJ. Disk position before and after modified condylotomy in 80 symptomatic temporomandibular joints. Oral Surg Oral med Oral Pathol Oral Radiol Endod 1995;79:668–79.
29. Nahlieli O, Baruchin AM. Sialoendoscopy: three years' experience as a diagnostic and treatment modality. J Oral Maxillofac Surg 1997;55:912–8.

Endoscopic Treatment of Subcondylar Fractures—Extraoral Approach

2

W. Bradford Williams, DMD, MD / Maria J. Troulis, DDS, MSc / Leonard B. Kaban, DMD, MD

OUTLINE

INTRODUCTION

Mandibular subcondylar fractures have been defined as fractures of the mandible extending from just below the condylar head to the sigmoid notch region.[1–3] It has been estimated that fractures involving the condylar head, neck, and subcondylar region together make up approximately one-third of all mandibular fractures.[4] Not surprisingly, there is a large body of literature dealing with the diagnosis, management, and complications of subcondylar fractures.[5–7] Despite this extensive literature, there is no universally agreed-upon optimal treatment protocol and management of subcondylar fractures remains controversial.

The earliest, and most widely supported treatment, of subcondylar fracture, was closed reduction and immobilization. Beginning in the 1950s through the early 1970s, prior to the development of rigid fixation, closed reduction and immobilization by maxillomandibular fixation (MMF) was the standard of care. Morbidity was low and outcomes were generally satisfactory.[3,5–8] Through a series of animal studies, Dr. R.V. Walker demonstrated that early mobilization of subcondylar fractures in rhesus monkeys produced excellent results in terms of symmetry, occlusion and mandibular mobility. Additionally, this work suggested that early mobilization prevents ankylosis of the temporomandibular joint.[9,10] Closed reduction has remained a common strategy for management because of the surgeon's familiarity with the technique, rapidity of the procedure, and lack of facial scarring or risk for nerve damage. Today, most experienced surgeons manage subcondylar fractures along a wide-ranging continuum, from observation to open reduction with rigid internal fixation (ORIF). Surgeons who favor ORIF argue that treatment time is less and occlusal and temporomandibular joint function are better than in patients treated by closed reduction.[11–16] In addition, patients very much appreciate not having the jaw immobilized with MMF. To address negative patient perceptions of closed reduction and to improve temporomandibular joint function after subcondylar fractures, many surgeons choose to place patients into guiding elastics with close follow-up instead of wire MMF.[9,10,14]

Since the development of rigid fixation there has been an increasing body of literature promoting open reduction for treatment of subcondylar fractures.[3,10] Many authors suggest that open reduction results in improved occlusal stability, facial symmetry, and mandibular range of motion.[11–17] In addition, patients are rehabilitated sooner. For these reasons ORIF is well-accepted by patients.[14,15] Furthermore, open reduction can be employed to correct persistent malalignment of the jaw in patients who have failed closed reduction with any combination of the known complications, which include pain, malocclusion, asymmetry, limited mobility, and gross radiographic abnormalities.[3,18]

TRADITIONAL OPEN TECHNIQUES

Traditional open approaches to subcondylar fractures include submandibular, retromandibular (or rhytidectomy), and preauricular incisions and dissection.[13,19] The most commonly noted drawbacks of open reduction have been the resultant facial scars, incidence of facial nerve injury, bleeding, and infection.[13,15]

Although most facial nerve injuries are temporary, there is a small, but finite number of permanent nerve injuries that are distressing to surgeon and patient alike. The incidence of permanent facial nerve injury with an open approach is low. Ellis, in his review of the literature found a total of 455 patients with subcondylar fractures treated by open approach including submandibular, rhytidectomy, retromandibular, and preauricular incisions. Less than 1% were reported to have permanent facial nerve weakness and only 12% experienced transient facial nerve weakness.[13] In an analysis of his own patient series, 16 of 83 patients treated by open reduction (retromandibular) and 10 of 66 treated by closed reduction experienced transient facial nerve weakness. After 6 weeks, no patients in either group demonstrated facial nerve weakness.

In Ellis' series, facial scarring in patients treated by open reduction via retromandibular approach was also assessed. Only seven patients were found to have wide or hypertrophic scars based on subjective measures. Of those seven, five patients were African American, a population predisposed to hypertrophic scarring.[13]

The submandibular incision places the marginal mandibular branch of the facial nerve at risk, although it protects the larger, more proximal branches of the facial nerve. This incision also results in a relatively esthetic scar that can be hidden within or parallel to the neck creases.

The retromandibular approach provides a well-concealed scar with more direct access to the ramus condyle unit. However, this approach places the facial nerve, as well as the retromandibular vein and auriculotemporal nerve, at greater risk of injury. The preauricular approach may also be well-concealed and may provide close access to the ramus condyle unit. This approach may be complicated by facial nerve injury and an increased risk of Frey syndrome and/or salivary fistula.[13,19]

In order to provide the benefits of open reduction and internal fixation while minimizing the known complications of ORIF, surgeons have turned their attention to minimally invasive approaches to the ramus condyle unit.[20–24] As in other fields of surgery, this effort has included the use of endoscopes, which allow visualization of the ramus condyle unit through smaller, more remote, and less conspicuous incisions. There is an illuminated and magnified operative field easily visualized by both surgeon and assistant. This permits less dissection with less postoperative edema and morbidity for the patient.[22–26] There is a shorter length of hospital stay and most importantly, faster patient return to normal activities (see Chapter 1, "Endoscopic Reconstruction of Ramus Condyle Unit").[14,15,22–26]

Disadvantages of minimally invasive procedures for oral and maxillofacial surgeons treating subcondylar fractures include the expense of required equipment and the time and effort required to learn to operate remotely with visualization on a two-dimensional screen. The surgeon must invest the time to become competent in the new technique.

ENDOSCOPIC OPEN: INTRAORAL AND EXTRAORAL APPROACHES

There have been two approaches to the ramus condyle unit described for treatment of subcondylar fractures: (1) intraoral (see Chapter 3 "Endoscopic Treatment of Subcondylar Fractures—Intraoral Approach") and (2) extraoral.[20,22,24] The intraoral approach has significant limitations, especially for the management of medially displaced and dislocated fractures. In addition, it is only satisfactory for use in specific types of subcondylar fractures. It is unlikely that most surgeons will learn a technique with such limited use (subcondylar fractures) (see Chapter 3, Discussion). In contrast, the extraoral approach can be adapted as a general method to obtain access to the ramus condyle unit for a variety of procedures (see Chapter 1, "Endoscopic Reconstruction of Ramus Condyle Unit"). In this chapter, the extraoral endoscopic approach for treatment of subcondylar fractures is discussed.

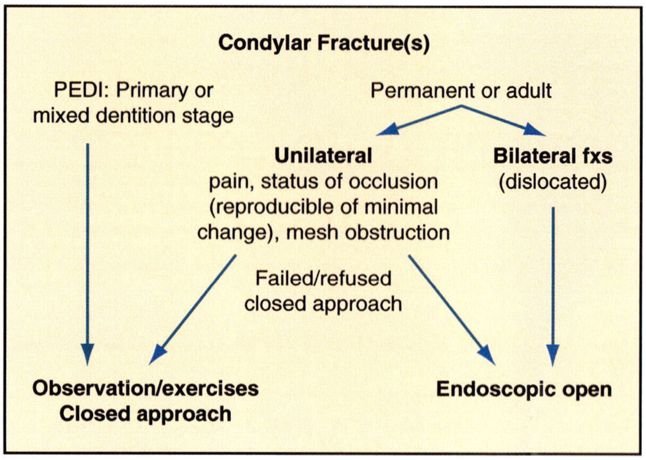

FIGURE 1 • Algorithm for treatment of condylar fractures.

Endoscopic Extraoral Approach

At the Massachusetts General Hospital (MGH), management of subcondylar fractures includes both closed and open reduction on a case-by-case basis. Based on data in the literature and our own clinical experience an algorithm for management of subcondylar fractures has evolved (Figure 1). Management is based on patient age and whether the fracture is unilateral or bilateral.

1. *Age: pediatric versus adult.* The majority of patients who present with subcondylar fractures in primary or mixed dentition are treated by some combination of observation, analgesics dietary restriction, closed reduction and a short period of MMF (<7–10 days) or guiding elastics and exercises. Patients presenting in permanent dentition are separated according to fracture laterality.
2. *Unilateral versus bilateral fractures.* Those adult patients with single, unilateral fractures are treated with closed reduction if a reproducible, premorbid (or minimally changed) occlusion can be achieved without pain or obstruction. Adult patients with bilateral subcondylar fractures are treated with open or closed reduction. Adult patients with bilateral fractures may undergo ORIF of a single side followed by closed reduction of the contralateral side or they may undergo ORIF of both sides. Those patients who fail closed reduction, or refuse the closed approach, are then offered endoscopic open reduction and internal fixation. Failed closed reduction is defined as those patients who, following treatment with closed reduction, experience limited range of mandibular motion (<40 mm) or persistent pain or malocclusion.

The technique used in all minimally invasive ORIF cases at the MGH is the extraoral approach via a 1.5-cm submandibular incision. For reduction of subcondylar fractures, this approach allows the jaw to be placed in maxillomandibular fixation during rigid fixation of the fracture.[20,27]

Techniques

Surface Landmarks

Prior to making the incision and beginning the dissection, the surgeon must accurately and carefully identify the surface landmarks. In addition to the standard surface landmarks described in Chapter 1 ("Endoscopic Reconstruction of Ramus Condyle Unit") the clinically and radiographically identified fracture site is outlined to guide dissection and fixation.

Once the surface landmarks are identified and drawn, the patient is placed into wire maxillomandibular fixation prior to open reduction. Incision and dissection are then carried out as described in Chapter 1 ("Endoscopic Reconstruction of Ramus Condyle Unit").

Subcondylar Fracture

Arch bars are placed, the mandible manipulated into the correct occlusion and wire MMF is applied. With endoscopic visualization, the proximal fragment is identified and the lateral surface stripped of soft tissue (allowing mobilization). A hole and a traction wire are placed at the angle so that the distal mandible can be distracted to help with reduction. The reduction is verified endoscopically by visualizing the alignment at the sigmoid notch and posterior border. The condylar head is verified to be in the joint. Fixation is achieved using a 2.0-mm titanium miniplate. The proximal fragment is screwed to the plate first. This allows the surgeon to remove the plate-holder and to control the proximal fragment. The reduction is confirmed once again and the plate is fixed to the distal fragment.

Medial Displacement

When the proximal fragment is medially displaced, an attempt is made to visualize the fragment and to manipulate it endoscopically into a lateral position. This can usually be accomplished via the extraoral approach because the operator can visualize the medial posterior aspect of the ramus condyle unit. If this is not possible, an endoscopic vertical ramus osteotomy is performed to permit access to the condylar fragment.

Under endoscopic visualization, an osteotomy is made from the sigmoid notch to the mandibular angle using a long-shaft reciprocating blade. The medial pterygoid is stripped to allow removal of the proximal fragment of the vertical ramus. The fractured condyle is then retrieved with a long-handled endoscopic grasper. The two fragments are then rigidly fixated on the table with a 2.0-mm plate and screws reconstructing the ramus condyle unit (RCU). Then, using the endoscope for direct visualization, the disk is identified and the reconstructed RCU is placed into the fossa. Good bony contact between the proximal and distal segments and alignment of the sigmoid notch and the inferior border are verified. The bone is fixed in position using a 2.0-mm titanium miniplates (Synthes Maxillofacial, Paoli, PA). Screw placement is accomplished through the incision or with the aid of a percutaneous trocar.

Wound closure is achieved as described in Chapter 1 ("Endoscopic Reconstruction of Ramus Condyle Unit").

Outcomes

Although standard open and closed treatments for subcondylar fractures are well-documented, there have been few reports on patients treated by endoscopic open reduction internal fixation (EORIF).[20,24,27,29] Currently, retrospective and prospective studies are underway at the MGH. To date, a retrospective evaluation of 14 consecutive patients who underwent endoscopic open reduction and internal fixation of subcondylar fractures (n = 16 sides) at the MGH has been completed. Similar outcomes have been reported in Miloro's series of six consecutive subcondylar fractures treated by extraoral endoscopic ORIF.[27] In these two studies a total of 22 subcondylar fractures treated by extraoral endoscopic open reduction were evaluated. There were no injuries of the marginal mandibular branch of the facial nerve and there were no problems with unacceptable facial scarring. In all cases, the patients were successfully returned to their preinjury occlusion and they had return of normal temporomandibular joint function without pain.[27,29]

In the MGH series specifically, 14 patients, entered into the algorithm (see Figure 1) for subcondylar management, were offered and chose EORIF. This group consisted of 12 adult patients with unilateral fractures and 2 with bilateral fractures for a total of 16 subcondylar fractures. Nine patients were initially treated by EORIF, whereas five patients were treated secondarily after an unacceptable result by closed reduction and maxillomandibular fixation. Eight sides were laterally displaced and eight sides were medially displaced at the level of the fracture. In reference to the fossa, four condyles were nondisplaced and twelve displaced from the fossa. Pre (T_0), postoperative (T_1) and follow-up (T_2) clinical examinations, lateral cephalograms, computed tomography (CT) scans, and panoramic radiographs were used to evaluate the outcomes. Healing of the incision, pain, facial symmetry, facial, inferior alveolar and lingual nerve function, jaw motion, and occlusion were evaluated by history and physical examination. Serial lateral cephalograms were traced to determine the position of the mandible in relation to the maxilla and occlusion. Panoramic radiographs were used to calculate RCU height. A tangent to the condylar head was drawn to represent the true horizontal. The perpendicular distance from this true horizontal to the intersection of tangents to the vertical ramus and the body was recorded as the RCU height.

The ramus condyle unit height on average was increased by 0.65 cm from the preoperative to immediate postoperative period and at latest follow-up ramus condyle unit height was increased by 0.50 cm as compared to the preoperative state. Preoperatively, the ratio of RCU height of fractured to nonfracture side was 0.93 and immediately postoperative was 0.99 and on follow-up was 1.00.

None of the patients required maxillomandibular fixation postoperatively and none experienced facial, lingual, or inferior alveolar nerve injury. All patients were discharged from the hospital in less than 23 hours following completion of the procedure.

Four patients, with medial displacement of the proximal fragment, were treated by EORIF as well as endoscopic vertical ramus osteotomy. Removal of the proximal fragment opened access to the medially displaced proximal fragments. These proximal fragments would not have otherwise been accessible via standard open approach or intraoral endoscopic approach.

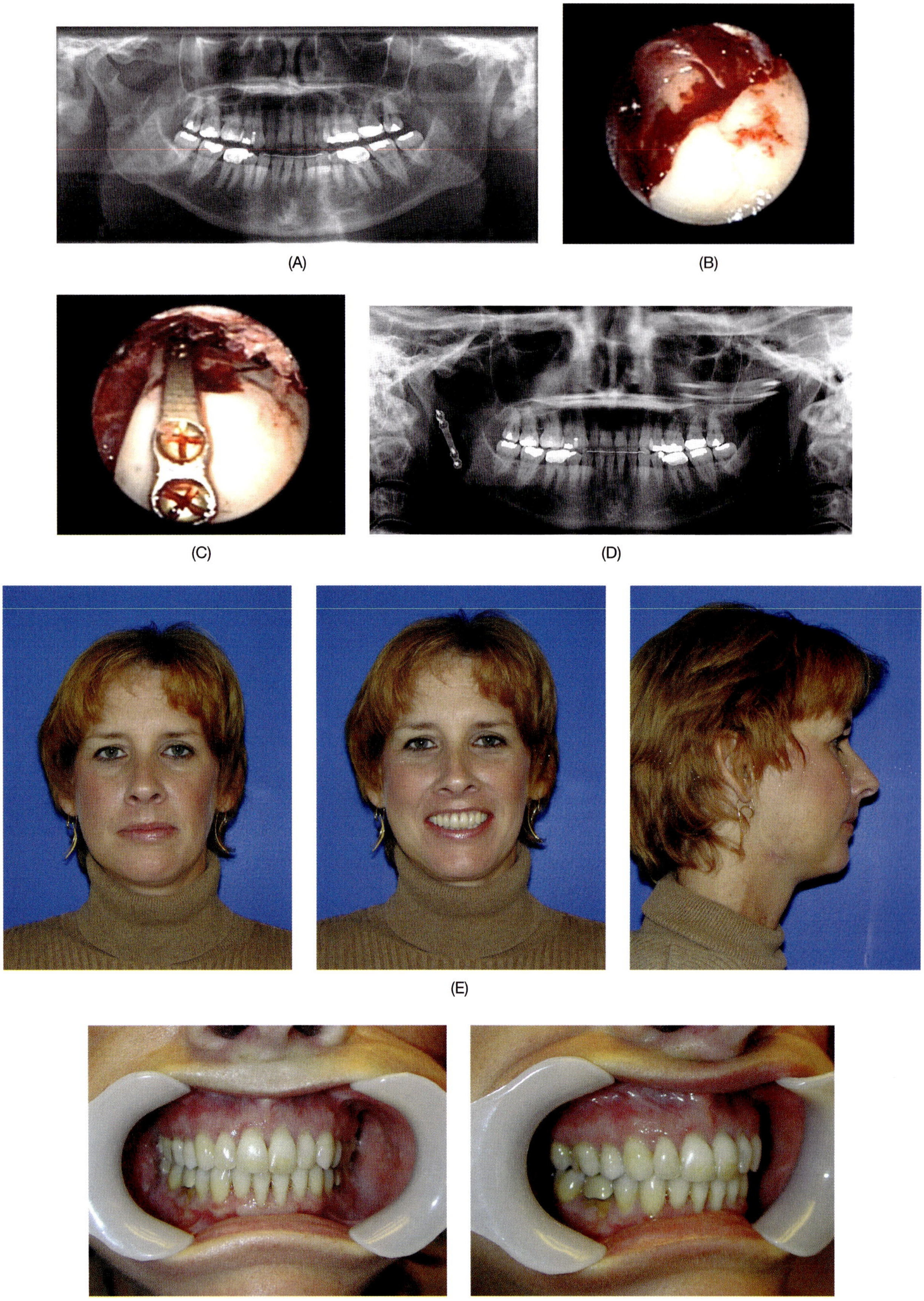

FIGURE 2 • Panoramic radiograph of a right subcondylar fracture (A). Endoscopic view of the same right subcondylar fracture (B). Endoscopic view of the same right subcondylar fracture reduced and rigidly fixed (C). Immediately postoperative panoramic radiograph showing that the RCU has been reconstructed (D). Same patient frontal repose, frontal smiling and lateral view at one month postoperative. Note that the marginal nerve is well functioning and the small access incision is well healed (E). Same patient's occlusion seen at one month postoperatively. There is good interdigitation with no open bite (F).

In an ongoing prospective study with 25 patients currently enrolled, the extraoral endoscopic approach to the RCU, results in a good clinical outcome with minimal morbidity. To date, 2 plate fractures (of no clinical significance) have been observed at late follow-up (Figures 2–4).

CONCLUSIONS

Acknowledging that the benefits of standard techniques for ORIF may not outweigh the potential risks, surgeons have begun to develop minimally invasive endoscopic approaches with the intent of improving patient outcomes.[14,15,20–27] Our experience with the extraoral endoscopic approach to the ramus condyle unit for treatment of subcondylar fracture demonstrates that this approach is technically feasible. It is possible to achieve excellent anatomic reduction and rigid fixation. Following EORIF there is no need for maxillomandibular fixation. The versatility of this procedure also allows the surgeon access to medially displaced fragments that may not be otherwise accessible. The entire operation is completed through a small, 1.5-cm incision that preserves facial nerve function.

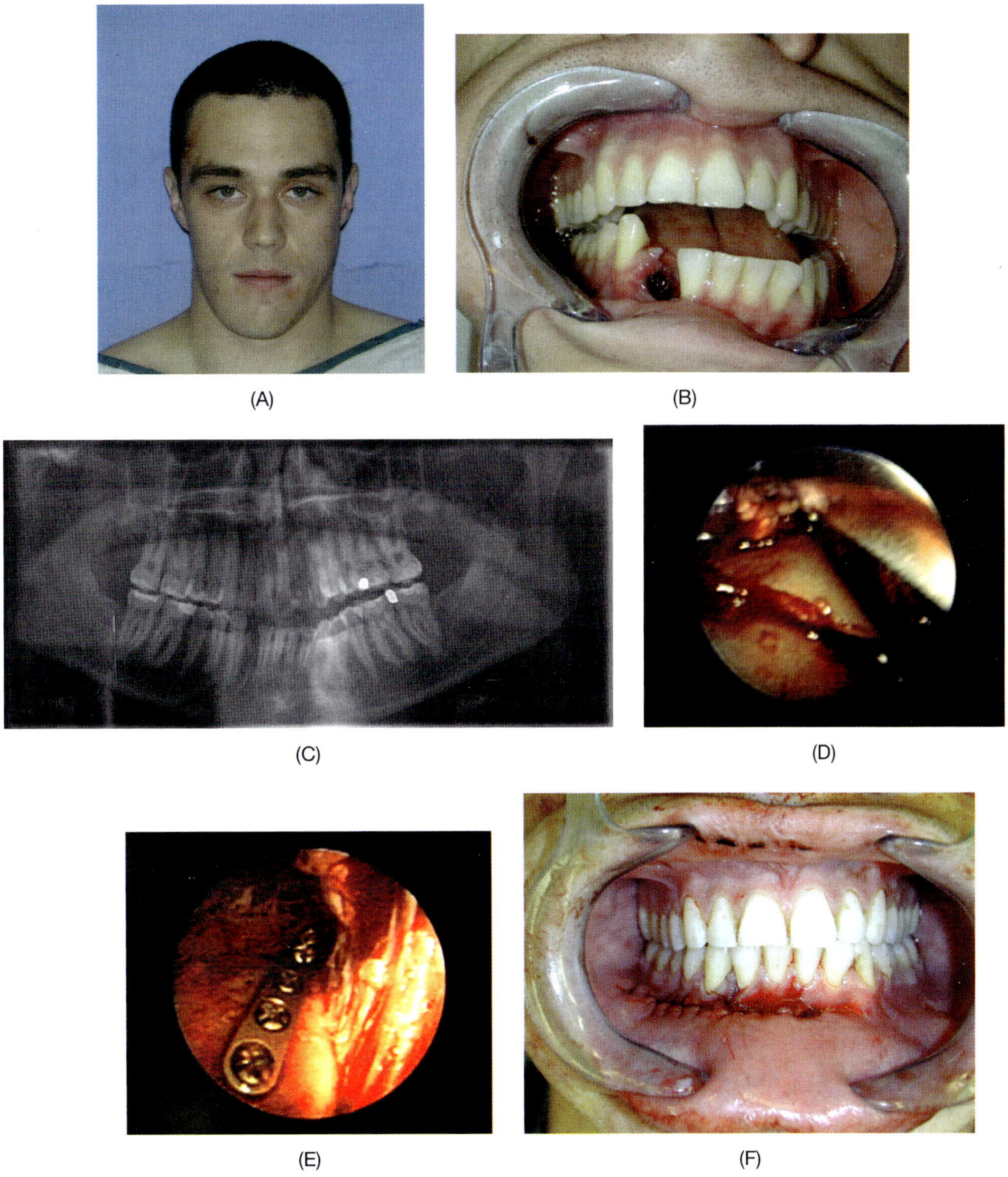

FIGURE 3 • Frontal view of a male involved in a sports injury receiving a blow to the chin (A). Intraoral view shows severe malocclusion including an occlusal step-off (B). Panoramic radiograph shows a left subcondylar fracture, right parasymphyseal fracture (C), Endoscopic view of the same left subcondylar fracture with lateral override (D). Endoscopic view of the same left subcondylar fracture, reduced and rigidly fixed (E). Intraoperative photograph showing good interdigitation of the occlusion. The maxillomandibular fixation is removed (F). *(continued)*

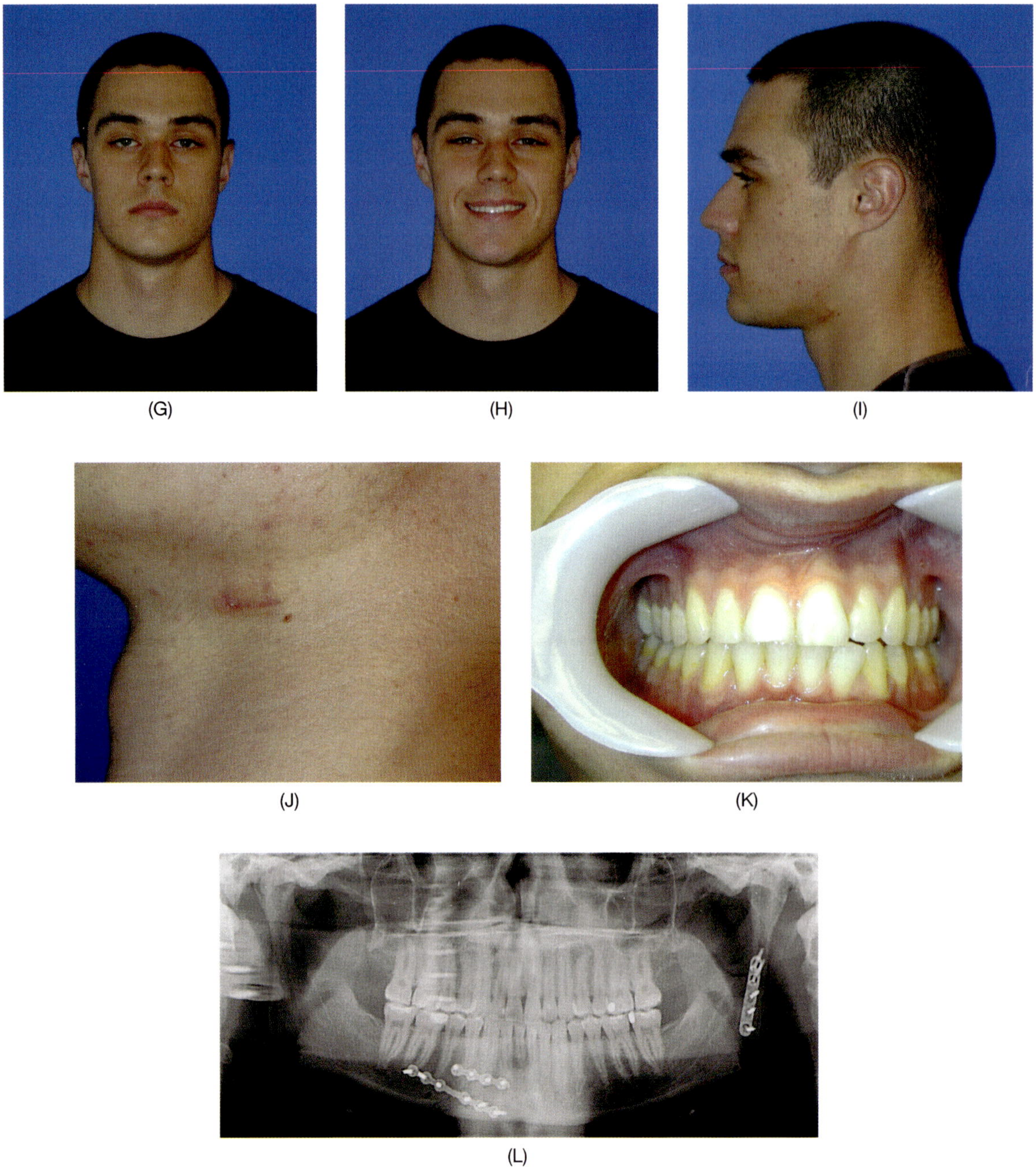

FIGURE 3 • (*continued from previous page*) Same patient frontal repose (G), frontal smiling (H), lateral view (I), close-up of access incision (J), occlusion (K), and panoramic radiograph (L) at four weeks postoperatively.

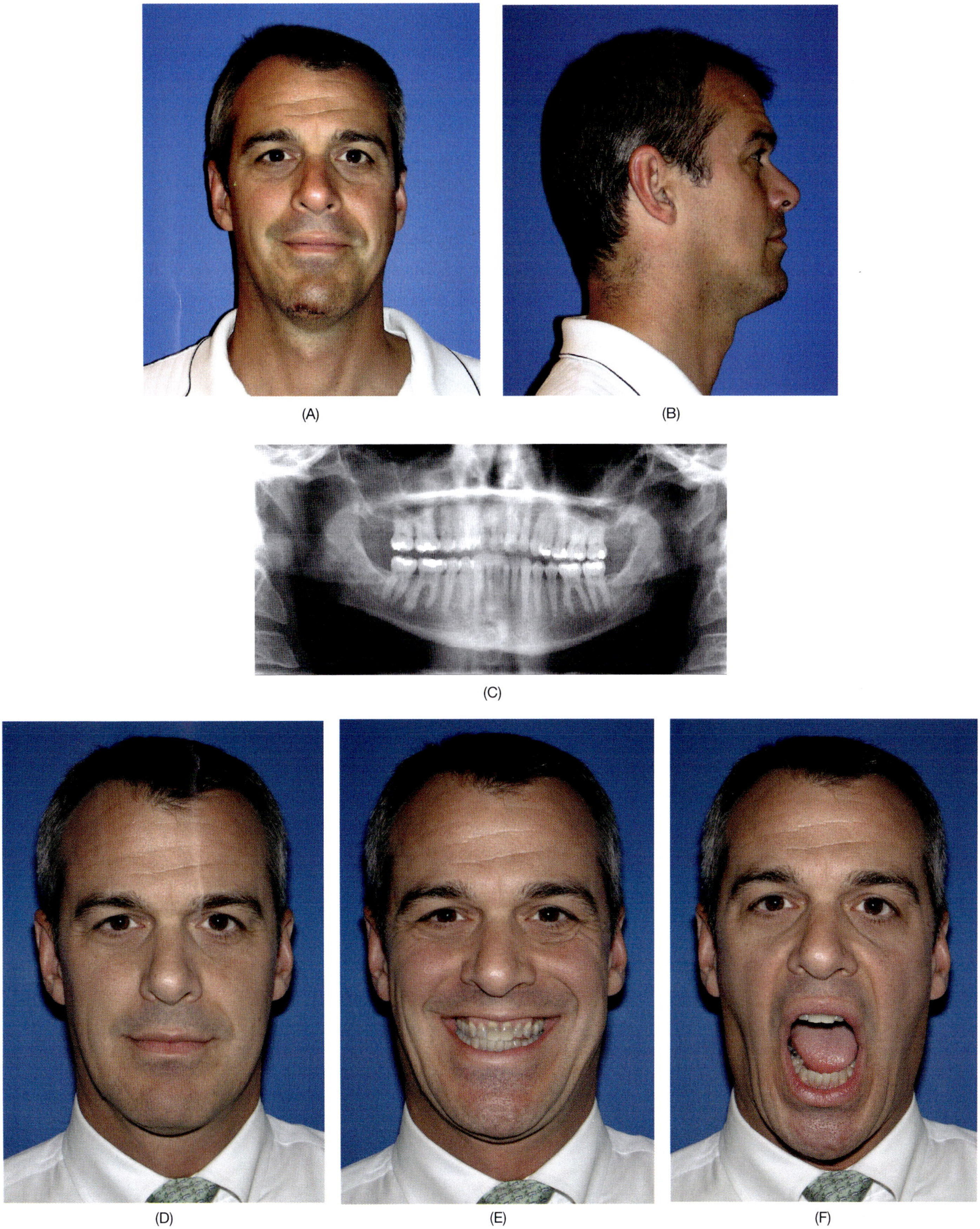

FIGURE 4 • Frontal (A) and profile (B) view of a 44-year-old male who sustained a left subcondylar fracture after a fall from a bicycle. Panoramic radiograph (C) showing the left subcondylar fracture. Frontal repose (D), smiling (E) and opening (F) at 6 months postoperatively showing good facial symmetry, facial nerve function and opening. (*continued*)

Endoscopic Treatment of Subcondylar Fractures—Intraoral Approach

3

Batya Goldwaser, DMD, MD / Robert Kellman, MD

OUTLINE

INTRODUCTION

The traditional intraoral approach for open reduction and rigid internal fixation (ORIF) of subcondylar fractures, first described by Steinhauser in 1964, and later detailed by Jeter et al in 1988, has been modified multiple times.[1,2] However, the general concepts of treatment and strategies for execution are similar. The patient is placed into maxillomandibular fixation (MMF). An intraoral incision through the mucosa, and periosteum extends over the external oblique ridge. After exposing the lateral ramus from sigmoid notch to the mandibular angle, the surgeon can identify the fracture and the proximal fragment. If medially displaced, the surgeon manipulates the condyle laterally while pulling the mandibular angle inferiorly. A trocar is inserted percutaneously in the preauricular region over the fracture site. A hole is drilled through the trocar into the proximal fragment. Tightening of the first screw into the proximal segment occurs through the trocar with the plate held in place intraorally. A second hole is drilled in the distal fragment and the remaining screws placed. After removal of rigid MMF and confirmation of occlusion, the patient uses training elastics, maintains a soft diet, and performs range of motion exercises.[3] Descriptions of minor modifications to this technique and regimen exist.[2]

Although apparently advantageous due to minimal scarring and decreased risk of facial nerve damage, the traditional intraoral technique presents multiple difficulties in execution. Visibility and manipulation are impaired, especially in a medially displaced proximal segment or high condylar neck fracture.[3] Little data exists on the clinical and radiographic outcomes of this method. In one case series describing intraoral ORIF as "technically simple," satisfactory reduction of the condyle occurred in only 80% of patients, and only in nondislocated fractures. Even though range of motion returned to normal, 23% of patients' jaws deviated to the fractured side on maximal opening.[2] These results point to the difficulty in reducing subcondylar fractures using the standard intraoral approach. Most clinicians therefore limit their use and recommendation of this technique. The standard approach to fractures in this region involve the use of preauricular, submandibular, or retromandibular incisions[15] as described in Chapter 2 ("Endoscopic Treatment of Subcondylar Fractures—Extraoral Approach").

ENDOSCOPIC INTRAORAL APPROACH

In 1996, as minimally invasive techniques became more widespread in the treatment of maxillofacial injuries, surgeons first used endoscopy to treat subcondylar fractures. The expected advantages of using an intraoral endoscopic approach instead of the standard intraoral approach included: (1) better visibility intraoperatively because of the illuminated and magnified view; (2) an enhanced view of the posterior border of the ramus enabling a more precise reduction; (3) decreased bleeding due to a less extensive dissection and less invasive procedure; and (4) decreased postoperative morbidity (ie, pain, edema, limitation of mouth opening).[4] These benefits allowed for open reduction by a minimally invasive intraoral approach without the difficulty and complications of the traditional open method.

A 1998 case report by Jacobowitz detailing an intraoral endoscopic technique for ORIF of bilateral subcondylar fractures showed promising results. Using the same technique, we make an intraoral incision in the buccal sulcus of the posterior mandible continued into a subperiosteal dissection of the lateral ramus, posterior border, and angle. Placement of a guarded trocar inserted percutaneously, in the preauricular region perpendicular to the fracture, maintained the optical cavity along with a cheek retraction attachment inserted intraorally. A 4-mm 30-degree endoscope placed though the intraoral incision was rotated toward the fracture to visualize and dissect the proximal segment. The use of elastic instead of rigid MMF enabled manipulation of the posterior mandibular occlusion (Figure 1, 2 and 3). Medially directed force applied with the trocar helped to reduce the proximal segment (Figure 4 and 5). After verification of the reduction on endoscopic view, fixation screws held the reduction in place while a fixation plate secured permanent fracture stabilization (Figure 6, 7, 8 and 9). The patient needed no postoperative MMF. At 8 weeks, the authors confirmed good anatomic reduction, mandibular projection, lower face width, facial height, premorbid occlusion, and a maximal incisal opening of 39 mm. The

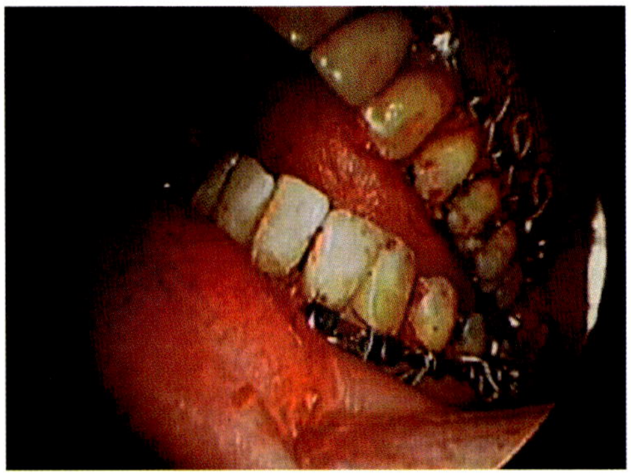

FIGURE 1 • Partial arch bars placed for a left subcondylar fracture. Elastic maxillomandibular fixation is used.

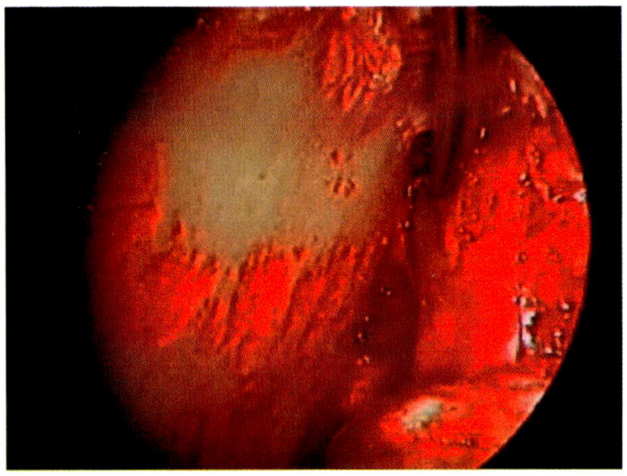

FIGURE 2 • Endoscopic view of a left subcondylar fracture using an intraoral incision, a subperiosteal dissection of the lateral ramus is performed.

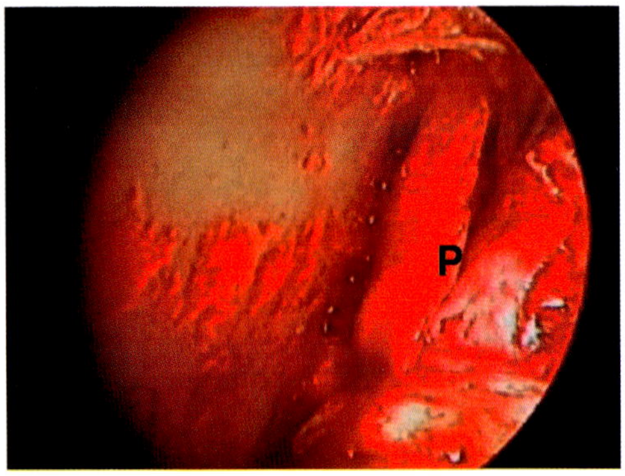

FIGURE 3 • The endscope is used through the intraoral incision toward the fracture to visualize and dissect the proximal (P) segment.

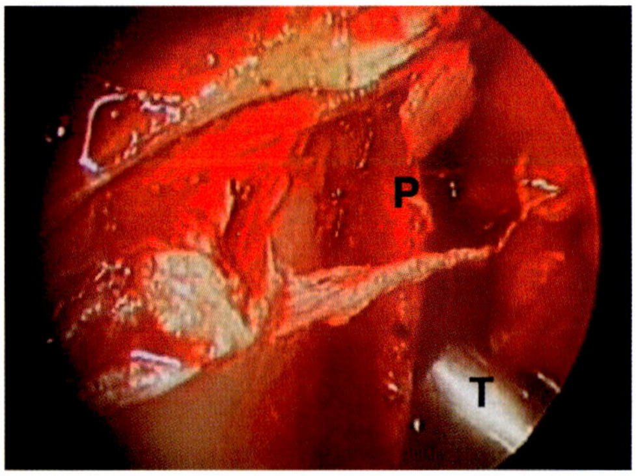

FIGURE 4 • The trocar (T) percutaneously in the left preauricular. The trocar is used to apply medial force on the proximal (P) fragment.

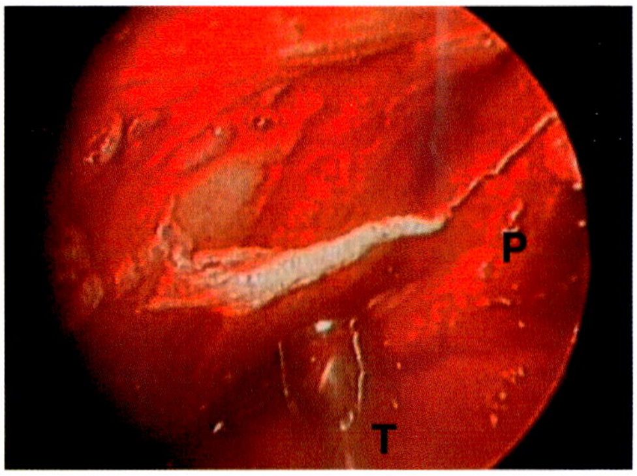

FIGURE 5 • Endoscopic view of same left subcondylar fracture with proximal fragment (P) well-reduced, trocar (T).

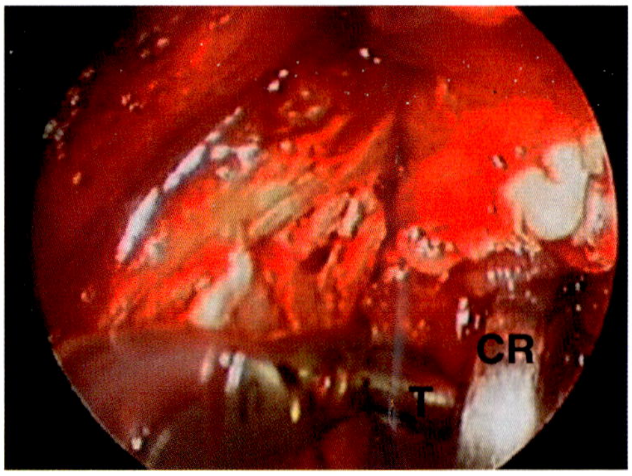

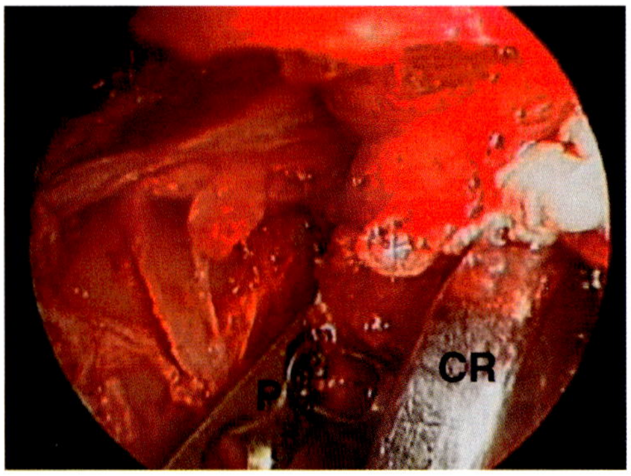

FIGURE 6 • Endoscopic view of same left subcondylar fracture, the plate & endoscope are inserted through the intraoral incision. The trocar + fixation plate are placed through the percutaneous incision (P), (CR) cheek retractor attached to trocar (T).

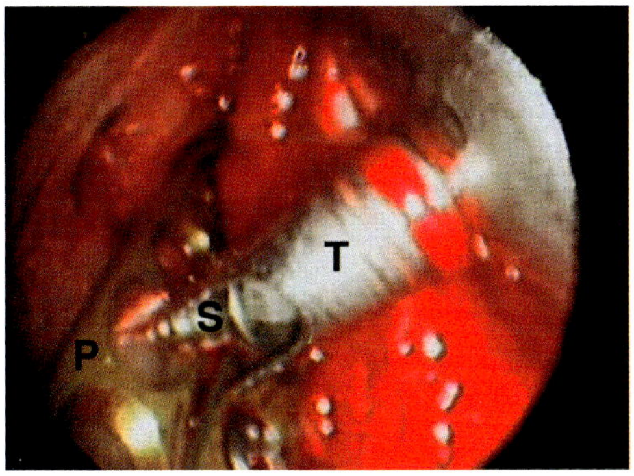

FIGURE 7 • Endoscopic view of same left subcondylar fracture, a second plate is being placed. T = trocar, S = screw, P = plate.

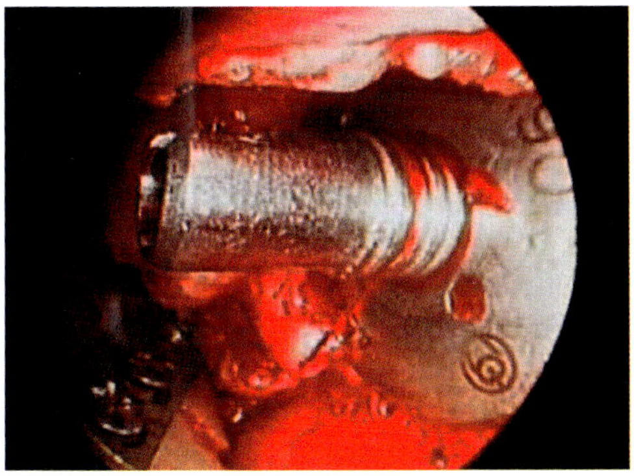

FIGURE 8 • Endoscopic view of the trocar being removed.

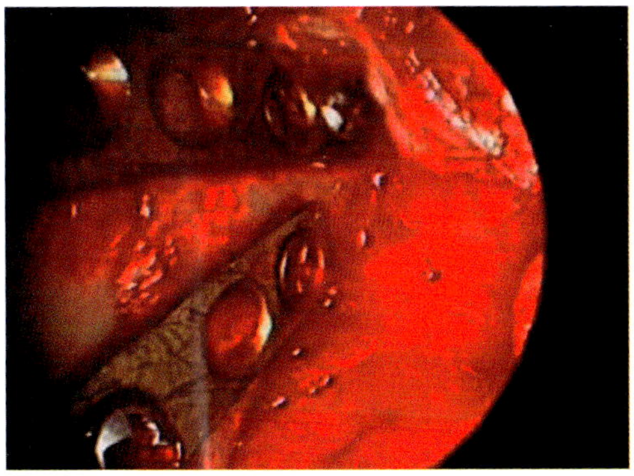

FIGURE 9 • Endoscopic view of same left subcondylar fracture. The 2 fragments are seen to be well reduced and rigidly fixed with 2 plates.

patient experienced no pain with jaw motion. This favorable outcome led to expanded application in a larger cohort of patients that included medial and laterally displaced as well as dislocated condyles.[5,6]

Since 1998, multiple studies using similar techniques for intraoral endoscopic reduction and fixation of subcondylar fractures with various modifications have been published.[7–12] Sandler used two separate trocars—the first along the canthal-tragal line, 20 mm anterior to the tragus, and a second at the site used in TMJ arthroscopy, 10 mm below the first trocar site.[8] Schon et al used angulated drills and screwdrivers, thereby eliminating any transbuccal stab incision.[9] Kellman described the addition of a second 1-cm incision in the submandibular area for easier viewing and fracture reduction. Kellman noted that when the operator became more skilled, the second incision could be eliminated, leaving only a single intraoral endoscopic port and a transbuccal trocar for screw placement. In addition, Kellman used MMF for stabilization during placement of the plate with the first screw, removed it for reduction when the posterior mandible was distracted inferiorly, and then replaced it during insertion of the remaining screws.[10] Chen et al used a bone clamp to distract the distal fragment inferiorly to aid in fracture reduction and when using 1-mm mini-plates, they placed the patient in MMF after reduction, but before fixation. The patients were kept in MMF for 3 to 6 days postoperatively.[11] Pham et al described the placement of two 2.0 mini-plates across the fracture line.[12]

In 2005, a long-term follow-up study of subcondylar fractures treated via intraoral endoscopy was reported. The authors used both 30-degree and 45-degree endoscopes with no intraoperative MMF to allow inferior distraction of the distal segment.[13] They converted all medial override cases to a lateral override position before reduction and fixation. They then used a plate-screw–holding device to place the first screw, together with an AO/ASIF noncompression modified zygoma plate (Synthes CMF, West Chester, PA), into the proximal segment.[13]

REPRESENTATIVE STUDIES

Between 1998 and 2005, the results of four important studies examining the intraoral endoscopic technique for ORIF of subcondylar fractures illustrate the variable outcomes of this approach depending on surgeon experience and type of fracture.

Lee et al, in 1998, presented an early case series of 20 patients with 22 subcondylar fractures treated by intraoral endoscopic ORIF. This series included adults with extracapsular, noncomminuted, and displaced fractures only. The authors studied twenty laterally displaced, two medially displaced, three medially subluxated, and one completely dislocated condylar fracture. They accessed the condyles via an intraoral endoscopic port and a percutaneous trocar insertion site. The average operation lasted 143 minutes. However, in edentulous, patients' OR time averaged 156 minutes because of the inability to put the patient in MMF during the procedure. In addition, when the proximal fragment was medially displaced operating time averaged 269 minutes because of the difficulty in identifying, controlling, and reducing the fracture. Temporary facial nerve palsy that resolved by 6 weeks, the one reported complication, occurred in one of the two medially displaced fractures.

All patients with dentition returned to their premorbid occlusion with a mean maximal incisal opening of 43 mm, ipsilateral excursion of 9.1 mm, and contralateral excursion of 7.9 mm at 8 weeks. All 20 patients expressed satisfaction with the aesthetics of chin projection, jaw line, and symmetry on opening, although one patient deviated slightly. One medial override fracture resulted in imperfect reduction, but normal function. Anatomic reduction occurred in the remaining 21 fractures.[6]

In 2000, Lee et al presented a larger series of 65 patients with various facial fractures including 40 subcondylar fractures reduced and rigidly fixated using intraoral endoscopy. The study included only fractures with clinical instability or displacement on radiograph and excluded intracapsular and comminuted fractures. Thirty-six fractured condyles had lateral override, four had medial override, five had medial subluxation, and three had complete dislocation from the fossa. Operating time averaged 145 minutes.

In 38 of the 40 fractures, the authors achieved anatomic reduction and stable rigid internal fixation. One of these condyles later fractured and needed replating. Comminution and high fracture location in two cases prevented fixation following fracture reduction. In 37 of the 40 fractures, the patients achieved functional recovery with >40 mm maximal incisal opening at 8 weeks postoperatively. A single case of temporary traction palsy of the facial nerve occurred after aggressive retraction of soft tissues.[7]

The data for this technique was further enhanced in 2003, when Kellman reviewed 17 noncomminuted, extracapsular subcondylar fractures treated via intraoral endoscopic approach. Successful reduction and fixation occurred in 10 fractures using this method. However, the difficulty of visualizing and reducing the fractures necessitated the addition of a second, extraoral, submandibular port in seven of the successful cases. Once the surgeon developed more experience with the technique, he used a single intraoral endoscopic port in the remaining three cases. Operative time in these three single port patients ranged from 55 to 60 minutes, significantly shorter than the overall time range of 55 minutes to 6 hours in this series.

In two cases, the attempt at endoscopic access failed and the author converted the surgery to a traditional open approach. In five instances the operation consisted of endoscopic exploration and fracture reduction, but no rigid fixation; all five had medial override and one severe comminution. Due to lack of plating, these five patients needed MMF postoperatively from 2 days to 4 weeks. No cases of facial nerve injury and only one case of prolonged trismus resulted.

The most success occurred late in the series as operator experience and skill increased and operating times shortened. Medially displaced and comminuted fractures resulted in many more complications and longer operating times. The surgeon encountered difficulty in repositioning the medially displaced proximal fragment into a lateral position.[10]

The first long-term follow-up study, published by Schoen et al (2005) examined the 5-year success rate of displaced or dislocated condylar fractures treated with the intraoral endoscopic technique. All patients presented with malocclusion and functional impairment requiring surgical management. Of 62 fractures, 45 were laterally displaced and 17 medially displaced. The surgeons used a single intraoral port and angulated screwdrivers and drills, eliminating the need for a percutaneous trocar. Operating time averaged 65 minutes.

Anatomic reduction, restoration of premorbid occlusion, >40 mm maximal incisal opening, no deviation, and good lateral excursion ultimately occurred in all cases. However, in two patients, hardware was removed some months postoperatively. In another patient, with medial override and dislocation, the proximal fragment could not be retrieved so the fracture healed with 7 days of MMF instead of open reduction and internal fixation. No cases of facial nerve damage occurred. The results of this study illustrate the successful use of the intraoral endoscopic technique even for medial override fractures. The operating time was shorter than previously reported. However, the 65-minute average operating time only included the last 30 cases, indicating the need for increased surgeon experience.[13]

DISCUSSION

Although not comprehensive, the results of the above studies form the basis for most authors' recommendations for the utilization of this technique, its indications, advantages and disadvantages. All endoscopic procedures require specialized training, but the intraoral endoscopic approach to subcondylar fractures demands particular expertise to learn and master.[12] Numerous studies confirm that with practice, surgeons become more adept at this method of fracture reduction. However, the learning curve varies greatly. In some institutions, even surgeons who practice the technique on cadavers, or have previous experience with endoscopy, find this approach challenging.[7,10,13] Some say the difficulty of the surgery correlates with the type of fracture and operator experience alone.[6,8,14] Others believe in an inherent difficulty associated with the intraoral endoscopic approach irrespective of the fracture type.[15] In support of the former view, in one of the earlier studies by Schoen et al in 2002, the authors concluded that the intraoral approach reliably succeeds for dislocated subcondylar fractures or those with lateral override, but that the extraoral method was indicated for those with medial override or comminution.[9] However, their recommendation changed significantly in a follow-up study 3 years later. In 2005, they advocated the transoral endoscopic approach as the method of choice, even in condylar fractures with medial override, with intensive surgeon training in this technique.[13] A thorough review of the literature shows that operating times range from 50 to 269 minutes. If length of operation reliably signifies degree of difficulty, clearly a wide range exists depending on fracture override position, degree of comminution, experience of the surgeon, and whether operating on a cadaver or live person.[6,7,9–11,13,16,17]

The existence of four different access combinations for this procedure underscores the varying difficulty of the technique. The most skilled surgeons, with specialized 90-degree screwdrivers and drills, use a single intraoral port for the endoscope, screws, and plate insertion.[17] A slightly more invasive version includes a preauricular percutaneous stab incision for insertion of the drill and screws through a trocar. This eliminates the requirement for right-angle tools. A third variation includes the use of an intraoral incision and two preauricular trocars, 10 mm apart from each other.[8] Finally, the least advanced modification of this approach combines intra- and extraoral incisions, and a preauricular trocar.[10] Similarly, the type of MMF used during fixation can affect the difficulty of the procedure. If the patient is edentulous and no MMF can be done, the instability may prevent use of this technique.[6] Intraoperative elastics make it somewhat easier by increasing stability while still allowing manipulation of the mandible inferiorly for fracture reduction. Some argue the benefits of wire MMF and believe it can only be utilized with the extraoral approach. In that scenario, a wire attached to the mandibular angle aids distraction of the distal fracture segment inferiorly during reduction.[15] Others have used wire MMF with the intraoral approach.[11] When insufficient stability of the proximal fragment presents, fixation screws temporarily hold the reduced fracture in place until completion of plating.[5] The variation just described reflects the difficulty in quantifying the degree of skill needed to perform reduction and fixation intraorally.

Most authors agree that medial override fractures present the most difficulty in reduction and fixation regardless of the technique used, but especially with the intraoral approach. In medial override, the more aggressive dissection and retraction needed to access the proximal segment increases the incidence of complications, especially temporary facial nerve palsy. However, these types of fractures rarely occur (<10% of all subcondylar fractures) because the lateral pterygoid pulls the condylar head medially and the lateral (temporomandibular) ligament prevents lateral displacement.[6,18,19] Many believe that the low incidence of medial override increases the usefulness and benefits of the intraoral approach in the majority of fractures, and makes it worth learning a new technique. Some even recommend this approach for all subcondylar fractures with indication for open reduction, regardless of override position.[13] In support of this recommendation, the overall rate of facial nerve dysfunction in eight different studies of intraoral endoscopic ORIF of 192 fractures, including medial override, was only 1.04%. Moreover, the few cases of impairment lasted only transiently.[5–11,13,16,17]

INTRAORAL VERSUS EXTRAORAL APPROACH

Despite the clear benefits of using endoscopy for ORIF, the choice of which specific technique to use in treating a particular patient remains controversial. Advocates of the intraoral approach cite numerous benefits including lack of visible scars, negligible risk of facial nerve damage, no cutting of the masseter or parotid gland, and decreased operative time for experienced surgeons. Although not as versatile, the intraoral approach has been used to remove a foreign object from the condyle. The drawbacks include the fundamental challenge of learning this difficult method, especially when one must view the fracture site from a parallel view of the mandible, rather than en face (see Chapter 2 "Endoscopic Treatment of Subcondylar Fractures—Extraoral Approach"). Other negatives include the higher complication rate when treating medial override fractures with such limited access, difficulty of putting the patient into wire MMF when placing screws and plates, and theoretically increased risk of infection from contact with the contaminated oral cavity.[15–17,20,21,22]

Advocates of the extraoral approach use it for dislocated, and especially medial override fractures. The ability to view the mandible face-on ("en face") make this technique less cumbersome for most surgeons. Most patients easily accept the unavoidable 1.5 cm facial scar that accompanies this approach. Clinicians gain more flexibility with this method because of its wide versatility (see Chapter 1 "Endoscopic Reconstruction of Ramus Condyle Unit," and Chapter 2 "Endoscopic Treatment of Subcondylar Fractures—Extraoral Approach"). Although increased compared to 1.04% with intraoral endoscopy, the risk of facial nerve damage with the extraoral approach has been reported as 4.5%, which is significantly lower than the 30% risk with the traditional open approach. Finally, some predict that extraoral endoscopic ORIF will eventually be performed in the outpatient setting with IV sedation only. This may greatly decrease time, cost, and manpower of the procedure which seems unlikely with the intraoral approach.[9,15]

CONCLUSIONS

The pros and cons of various techniques should not distract from the ultimate goals of open reduction: return of premorbid occlusion, pain-free range of motion >35 to 40 mm, good excursions, facial symmetry, and minimal scarring without facial nerve dysfunction.[4] The algorithm must include each patient's individual preferences, fracture characteristics, and surgeon comfort and skill. The extraoral approach makes the procedure easier for the surgeon which may influence outcome, but the intraoral approach offers potentially increased benefit for the patient. The decision must be a joint one between patient and doctor. The most important issue for patients is the function.[23]

References

1. Steinhauser E. Eingriffe am processus articularis auf dem oralen. Weg Dtsch Zahn Z 1964;19:694–7.
2. Lachner J, Clanton JT, Waite PD. Open reduction and internal rigid fixation of subcondylar fractures via an intraoral approach. Oral Surg Oral Med Oral Pathol Oral Rad Endod 1991;71:257–61.
3. Jeter TS, Van Sickels JE, Nishioka GJ. Intraoral open reduction with rigid internal fixation of mandibular subcondylar fractures. J Oral Maxillofac Surg 1988;46:1113–6.
4. Miloro M. Considerations in subcondylar fracture management. Arch Otolaryngol Head Neck Surg 2004;130:1231–2.
5. Jacobovicz J, Lee C, Trabulsy PP. Endoscopic repair of mandibular subcondylar fractures. Plast Reconstr Surg 1998;101:437–41.
6. Lee C, Mueller RV, Lee K, et al. Endoscopic subcondylar fracture repair: functional, aesthetic and radiographic outcomes. Plast Reconstr Surg 1998;102:1434–43.
7. Lee C, Stiebel M, Young DM. Cranial nerve VII region of the traumatized facial skeleton: optimizing fracture repair with the endoscope. J Trauma 2000;48:423–31.
8. Sandler NA. Endoscopic-assisted reduction and fixation of a mandibular subcondylar fracture: report of a case. J Oral Maxillofac Surg 2001;59:1479–82.
9. Schon R, Gutwald R, Schramm A, et al. Endoscopy-assisted open treatment of condylar fractures of the mandible: extraoral vs intraoral approach. Int J Oral Maxillofac Surg 2002;31:237–43.
10. Kellman R. Endoscopically assisted repair of subcondylar fractures of the mandible: an evolving technique. Arch Facial Plast Surg 2003;5:244–50.
11. Chen CT, Lai JP, Tung TC, et al. Endoscopically assisted mandibular subcondylar fracture repair. Plast Reconst Surg 1999;103:60–5.
12. Pham AM, Strong EB. Endoscopic management of facial fractures. Curr Opin Otolaryngol Head Neck Surg 2006;14:234–41.
13. Schon R, Fakler O, Gellrich N, et al. Five-year experience with the transoral endoscopically assisted treatment of displaced condylar mandible fractures. Plast Reconstr Surg 2005;116:44–50.
14. Sandler NA, Andreasen KH, Johns FR. The use of endoscopy in the management of subcondylar fractures of the mandible: a cadaver study. Oral Surg Oral Med Oral Path Oral Radiol Endod 1999;88:529–31.

15. Troulis MJ. Endoscopic open reduction and internal rigid fixation of subcondylar fractures. J Oral Maxillofac Surg 2004;62: 1269–71.
16. Yang WG, Chen CT, Tsay PK, et al. Functional results of unilateral mandibular condylar process fractures after open and closed treatment. J Trauma 2002;52:498–503.
17. Schon R, Schramm A, Gellrich NC, et al. Follow up of condylar fractures of the mandible in 8 patients at 18 months after transoral endoscopic assisted open treatment. J Oral Maxillofac Surg 2003;61:49–54.
18. Vural E. Treatment of adult subcondylar mandibular fractures: closed vs open vs endoscopic approach. Arch Otolaryngol Head Neck Surg 2004;130:1228–30.
19. Zachariades N, Mezitis M, Mourouzis C, et al. Fractures of the mandibular condyle: a review of 466 cases. Literature review, reflections on treatment and proposals. J Craniomaxillofac Surg 2006;34:421–32.
20. Antonyshyn O. In discussion of : Lee C, Mueller RV, Lee K, Mathes SJ. Endoscopic subcondylar fracture repair: functional, aesthetic, and radiographic outcomes. Plast Reconstr Surg 1998; 102:1434.
21. Suarez-Cunqueiro MM, Schon R, Gellrich NC, et al. Endoscopic assistance in the removal of a foreign body in the condylar process. J Craniofac Surg 2004;15:98–101.
22. Toma VS, Mathog RH, Toma RS, et al. Transoral versus extraoral reduction of mandible fractures: a comparison of complication rates and other factors. Otolaryngol Head Neck Surg 2003;128:215–9.
23. Assael LA. Open versus closed reduction of adult mandibular condyle fractures: An alternative interpretation of the evidence. J Oral Maxillofac Surg 2003;61:1333–9.

Frontal Sinus and Zygomatic Complex Fractures

4

Arnulf Baumann, DMD, MD, PhD

OUTLINE

INTRODUCTION

Fractures of the midface and the frontal sinus region may be caused by motor vehicle accidents, sports injuries, and assaults. History and physical examination remain the primary diagnostic modalities. However, axial and coronal computed tomography (CT) scans are invaluable for defining the status of the skeletal anatomy and soft-tissue involvement in the intracranial, orbit, and midface regions.

As recommended by Manson and colleagues[1] and Gruss and colleagues[2] in the 1980s, treatment of midface fractures should include adequate dissection to expose all fracture fragments to facilitate, precise anatomic reduction, rigid fixation, immediate bone grafting if necessary, and definitive soft-tissue management. By following these principles, late aesthetic and functional sequelae of facial fractures can be diminished or eliminated. Internal and external incisions with extensive soft-tissue elevation are necessary to achieve these aims. The coronal incision, with appropriate modifications, is the preferred approach for the upper midface and the craniofacial region. It provides good exposure of the frontonasal, orbital, and zygomatic regions. In severe comminuted orbitozygomatic fractures, the coronal incision should be extended inferiorly into the preauricular region.

Although the coronal incision allows for excellent exposure, disadvantages include risk to the frontal branch of the facial nerve, cosmetic complications such as temporal hollowing, alopecia, and hypertrophic scar.[3,4]

Smaller scars with better visualization are the main advantages of the endoscopic technique which was initially introduced for brow and forehead lift to avoid a coronal incision.[5,6] Decreased soft-tissue manipulation results in less postoperative swelling and pain and a faster overall patient recovery.

The use of an endoscopic technique instead of the coronal incision has not been widely accepted for trauma because of difficulties in maintaining an optical cavity. The soft tissue has to be prevented from draping over the endoscope and obscuring the surgical field. Therefore, special sheaths were developed for endoscopes to support the soft tissue and maintain the optical cavity in facial cosmetic surgery. This is also critical when operating in the midface/zygoma and frontal regions.[7] Experience in endoscopic facial rejuvenation surgery has been subsequently applied to develop endoscopic surgery for zygomatic and frontal sinus fracture repair.[8,9]

FRONTAL SINUS FRACTURES

Surgical Anatomy

The anterior wall of the frontal sinus is uniformly thicker and stronger than the posterior wall and can resist 400 to 1000 kg of force before fracturing. This is twice the force that any of the other facial bones can withstand.[10] High-velocity impact to the anterior wall can also produce significant damage to the posterior wall, resulting in comminuted anterior and posterior wall fractures. Fractures of the posterior table are also associated with dural laceration and cerebrospinal fluid (CSF) leak. The orbital roof and nasofrontal duct are located inferior to the frontal region. The frontonasal duct is located at the most anterior and superior portion of the anterior ethmoid complex and is the only drainage pathway for the frontal sinus into the middle meatus of the nose. Traumatic disruption of this duct can lead to stasis of frontal sinus secretions that may lead to infectious complications.

The frontal sinus is part of the forehead region which extends from the hairline inferiorly to the eyebrows and to the lateral margin of the eyebrows. The bone is covered by five soft-tissue layers (SCALP): skin, subcutaneous fat, galea aponeurotica, which includes the mimic muscles (*m. frontalis, m. corrugator supercilii,* and *m. procerus*), loose connective tissue (subgaleal layer) and periosteum (pericranium). The blood and sensory supply of the forehead and anterior scalp are provided by the supratrochlear and supraorbital vessels and nerves. These structures reach the forehead soft tissue via notches or foramina in the supraorbital rims. The positions of the supraorbital and supratrochlear notches and foramina are variable. However, the supratrochlear notches are always medial to the supraorbital foramina. The bundles pierce the periosteum and muscle and lie in the deep subcutaneous fat layer. These vital structures should be preserved regardless of the soft tissue dissection and exposure technique.

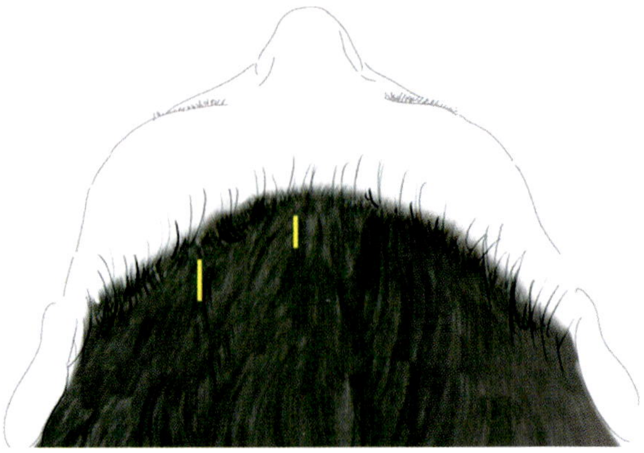

FIGURE 1 • Incision lines (yellow) for an endoscopic approach for a left side frontal sinus fracture repair. These are the visualization and working ports.

Diagnosis

Frontal sinus fractures constitute 5% to 15% of maxillofacial fractures.[10] Motor vehicle crashes are the most common etiology followed by sports injuries. Since a high-velocity impact usually causes these fractures, approximately 75% of the patients have associated orbital, nasal, and/or other midfacial fractures. CT scans in the axial and coronal planes are necessary to assess the involvement of the anterior and/or posterior walls. CT scans also reveal any damage to adjacent structures, such as the frontonasal duct, or associated intracranial abnormalities.

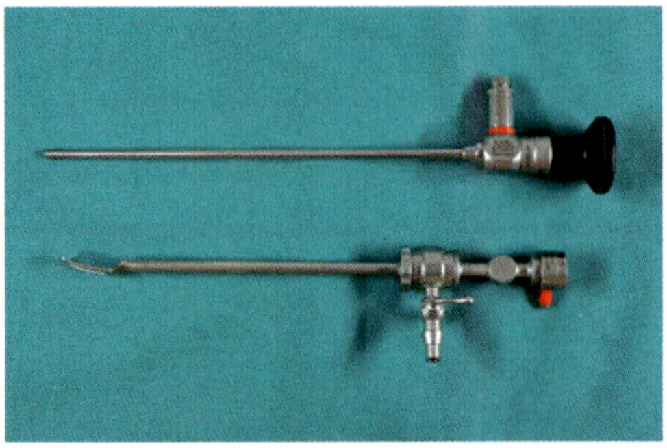

FIGURE 2 • Endoscope (above) and sheathed retractor (below).

Treatment Modalities

Treatment planning for frontal sinus fracture repair depends on which of the frontal sinus walls are involved (anterior and/or posterior). If the fracture involves only the anterior wall, without the nasofrontal duct region, the treatment is an anatomic reduction and stabilization of the fracture with titanium microplates or resorbable plates. If there is an additional frontonasal duct injury or posterior wall injury with dural tear or CSF leakage, the sinus mucosa should be removed and the sinus cavity obliterated before the anterior frontal sinus wall is reconstructed.[10]

In isolated anterior wall fractures removal of the sinus mucosa and sinus obliteration are not necessary. The aim in isolated anterior wall fractures is to restore normal forehead contour for aesthetic reasons. This procedure can be done by a coronal incision or by a less-invasive endoscopic approach.[9,11]

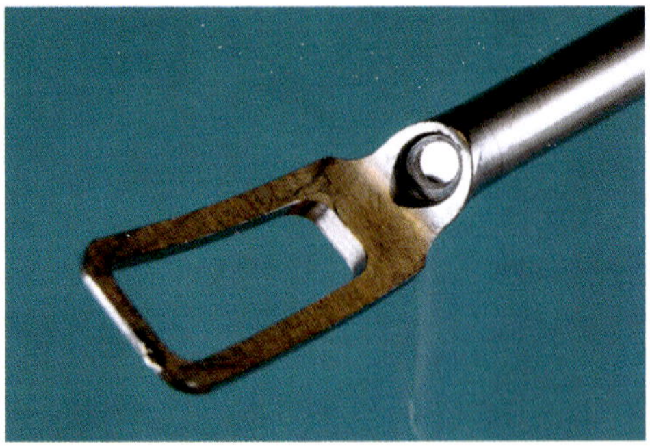

FIGURE 3 • Endoscope locked in the retractor tube.

Endoscopic Procedure for Frontal Sinus Fracture Repair

The instruments and the technique for frontal sinus fracture repair are similar to that for endoscopic forehead lift. Two hairline incisions are made after injection of local anesthesia (1% lidocaine and 1:200000 epinephrine). The incisions are 1 cm to 2 cm long and located 2 cm to 3 cm behind the hair line (Figure 1). In patients with a prominent forehead or a receding hair line, the incisions should be closer to the hair line, to allow better visualization around the curvature of the forehead. Two separate ports are used for access and instrumentation: visualization port for the endoscope and working ports for the instruments.

A 4 mm diameter, 30-degree angle Hopkins II endoscope (Karl Storz, Tuttlingen, Germany) with an external sheath for retraction of the soft tissue is used for this operation. The sheathed retractor has an overhang over the lens and a locking metal tube for the endoscope, which allows simultaneous movement of the retractor and the endoscope during the dissection of the optical cavity (Figures 2 and 3). The endoscope is linked to a 250 watt to 300 watt xenon light source (Olympus CLV-U20) and to a digital camera for magnification of the operation field (Stryker Endoscopy, 888i). The surgery is visualized on the monitor (Sony Trinitron Colour video monitor) and the procedure can be documented by a digital video recorder (Sony, DVCAM).

A blind subperiosteal dissection is performed through the two incisions for the first 3 cm to 4 cm. Then, the endoscope is inserted, the visualized dissections are completed and the two

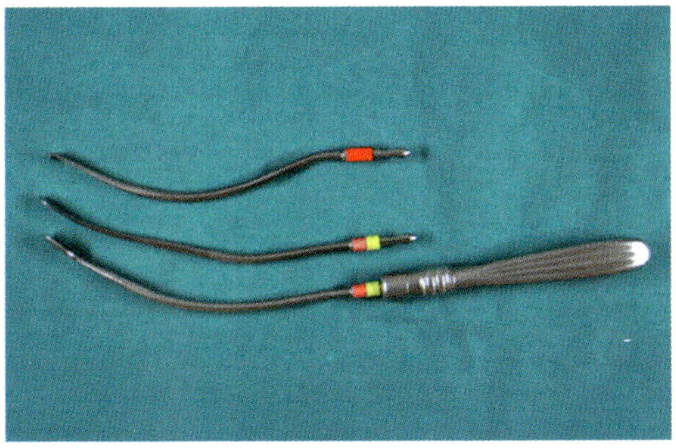

FIGURE 4 • Curved dissectors with different blade size.

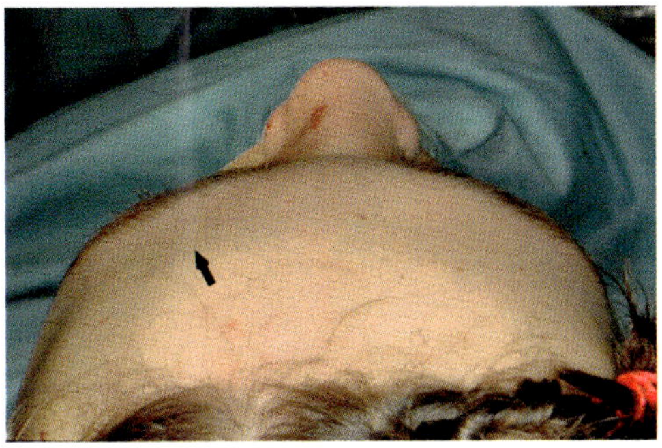

(A)

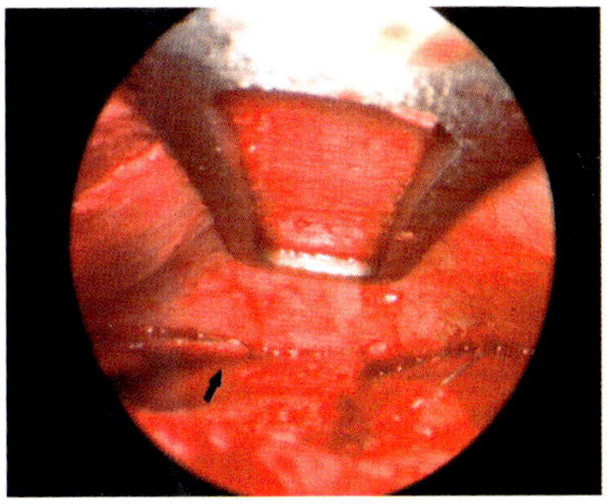

(B)

FIGURE 5 • *A*, Clinical situation of a depressed left frontal sinus fracture (black arrow); *B*, Endoscopic view of the depressed left frontal sinus fracture (black arrow).

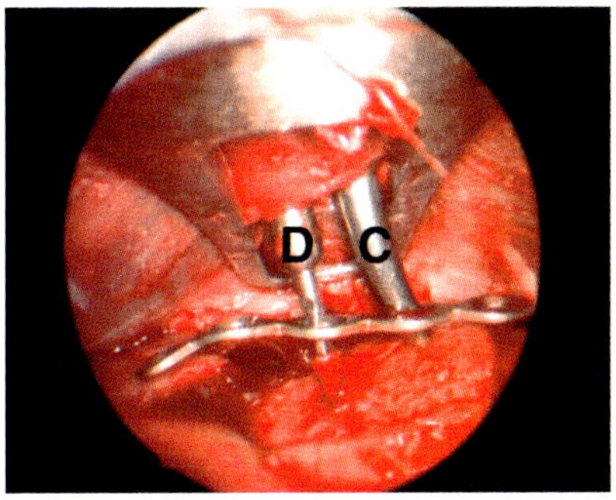

FIGURE 6 • Miniplate positioned and held by a clamp (C). The drill (D) is also inserted percutaneously by a stab incision in the eyebrow.

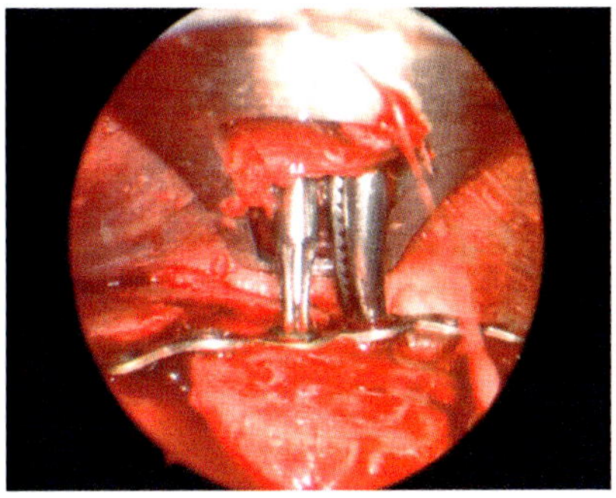

FIGURE 7 • The screw fixes the bone to the miniplate.

tunnels are connected with curved endoscopic dissectors (Figure 4). The tissue is elevated subperiosteally to the level of the supraorbital rim, creating the optical cavity and the fracture endoscopically visualized (Figure 5). The supratrochlear and supraorbital vessels and nerve bundles are identified and preserved. Once the fracture is visualized, the bone fragments are elevated and repositioned through the working port. Often it is necessary to approach the fragments directly through a stab incision either in the eyebrow or through a preexisting forehead laceration (Figure 6). Sometimes the fragments will remain reduced after gentle elevation of the bone. If so, no additional osteosynthesis is necessary. Otherwise, the fragments are fixated with miniplates or resorbable polylactidmesh. The miniplate is inserted via one of the hairline incisions and the screws inserted percutaneously via direct stab incisions (Figure 7). If there are small irregularities in the bony contour after repositioning of the fracture, these are camouflaged by an Ethisorb Dura Patch (Ethicon, GmbH, Norderstedt, Germany).

In some cases the fracture segments are unstable (usually due to severe comminutions) and cannot be adequately plated under endoscopic access. In these patients, the irregularities of the bone are camouflaged or the procedure is converted to a conventional open approach via coronal incision.

Discussion

Since the first case report of endoscopic repair of frontal sinus fracture by Graham and Spring,[9] there have been several published papers describing results and complications of frontal sinus repair by the endoscopic approach. Shumrick[12] has published the largest patient series of endoscopic anterior sinus wall fracture repair. He documented successful repair of anterior sinus wall fractures in 12 of 19 patients using the endoscopic technique. In the other 7 patients the procedure had to be converted to the standard coronal approach, due to extensive fracture comminution. Shumrick[12] concluded that the endoscopic technique is appropriate in fractures with several large segments without extensive comminution. These clinical findings were previously described in the cadaver study by Strong and colleages.[13] They generated frontal sinus fractures in 11 cadavers and compared frontal fracture reduction and fixation with miniplates to fracture camouflage with bone cement. Reduction was most difficult in comminuted fractures which were only partially reduced and fixation was incomplete. Fracture camouflage with bone cement via a coronal incision was easier to perform and offered better results. However, using bone cement (Hydroxyapatite

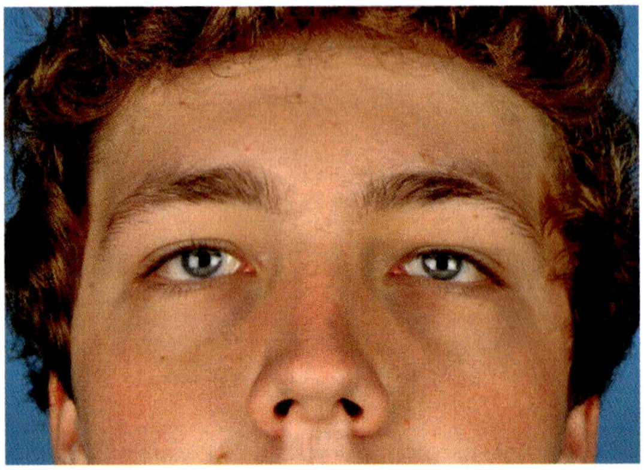

(A)

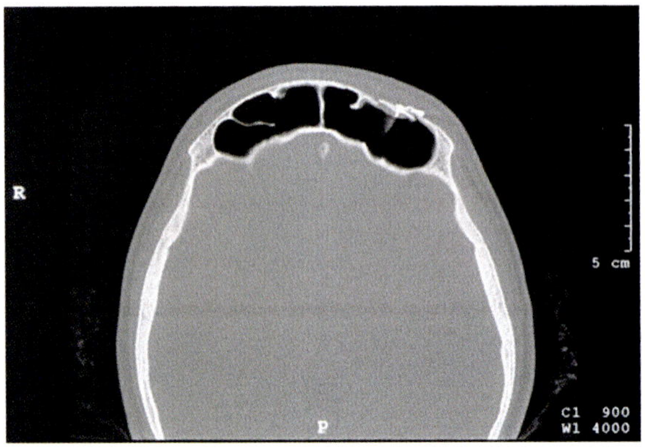

(B)

FIGURE 8 • *A*, Patient 3 months after endoscopic repair of a left frontal sinus fracture. Screws were inserted by stab incisions in the eye brow. In the same procedure a left orbital floor fracture was repaired via a transconjunctival incision. *B*, Postoperative CT scan. The left anterior sinus wall fracture was fixed by miniplate osteosynthesis.

cement) for camouflage in anterior wall reconstruction in the acute reconstruction phase may result in infection of the frontal sinus.[14] Hydroxyapatite cement appears to be an ideal material for elective delayed reconstruction.[14] We found that a combination of endoscopic reduction and adjuvant camouflage with a resorbable Ethisorb Dura Patch results in an adequate aesthetic result in most cases (Figure 8).

Chen and colleagues[15] reported a group of seven patients with good anatomic reduction of the anterior frontal sinus wall by removing all fragments through stab incisions. The fragments were assembled and fixed with microplates on a side table and reinserted as a free graft after the sinus mucosa was removed. There was no comment of bone resorption during the follow-up.

Most anterior wall fractures are nondisplaced or only minimally displaced and need no surgical treatment.[16] The possible aesthetic deformity, in mildly to moderately displaced anterior frontal sinus fractures, cannot be easily determined during the early posttraumatic period due to edema. Therefore, an alternative treatment option would be fracture camouflage after complete resolution of the edema, which takes several months. In these cases a porous polyethylene implant (Medpor-Porex Surgical, Newnan, GA) can be inserted and fixed with screws using endoscopic procedure.[17] The hairline incision for this procedure has to be larger than the standard endoscopic incision to accommodate implant insertion.

Conclusion

Reconstruction of the forehead region after an anterior frontal sinus wall fracture is usually an aesthetic procedure and can be executed via a coronal or endoscopic approach. The advantages of the endoscopic approach are small scars, avoidance of alopecia, decrease risk of numbness in the scalp region, and less postoperative swelling. However there is a steep learning curve associated with endoscopic procedures. The success of endoscopic repair in frontal sinus fracture depends on careful patient selection. Endoscopic repair offers an additional treatment option that should be considered for frontal sinus fracture repair.

ZYGOMATIC COMPLEX FRACTURES

Zygomatic complex fractures are commonly occurring injuries of the facial skeleton. Complexity of the fracture is related to the velocity and force of the offending impact. Low-velocity blows result in less displacement and more linear fractures than high-velocity impact. High-velocity injuries predispose to segmentation, comminution, and significant displacement of the bone fragments.

Surgical Anatomy

The zygoma, because of its prominent position and contour, is highly susceptible to injury. The anterior portion of the bone, which forms the projection of the cheek, articulates with the frontal, maxillary, and sphenoid bones and supports the inferior lateral wall of the orbit. The posterior portion of the zygoma is a thin but solid convex bone that articulates with the temporal bone forming the zygomatic arch.

In a zygomatic complex fracture its four articulations with the adjacent craniofacial skeleton may be disrupted. These are the zygomaticofrontal, zygomaticosphenoid, zygomaticomaxillary, and zygomaticotemporal sutures. The zygomatic arch connects the midface with the skull base and maintains the forward projection of the malar prominence. It determines the vertical and transverse dimensions of the face. The zygomatic arch is the site of origin of the masseter muscle and insertion of the temporalis fascia. Displacement of this bone is a result of masseter muscle pull. In complex midface and orbitozygomatic fractures, the zygomatic arch is used as a landmark for reconstruction of the face.[2]

The usual approach for open reduction and rigid fixation of the zygomatic arch is a coronal incision. In making this incision and completing the dissection, the course of the ramus frontalis of the facial nerve must be considered. The nerve proceeds over the zygomatic arch approximately 2 cm anterior to the tragus and within 1 cm or less of the frontal branch of the superficial temporal artery beneath the temporoparietal fascia.[3] An understanding of the three important fascial layers in the temporal region is necessary to prevent facial nerve injury: temporoparietal fascia, and the superficial and deep layers of the deep temporal fascia. The deep temporal fascia overlies the temporalis muscle and the fascia splits in the temporal line in two parts at the level of the superior orbital rim. Between the two parts, the superficial and the deep layer of temporal fascia is a loose areolar plane. This plane is avascular and contains the superficial temporal fat pad. Dissection in this plane to the zygomatic arch prevents injury of the frontal branch of the facial nerve. These anatomic correlations in the zygomatic arch region have to be considered when using an open approach for the zygomatic arch.

Diagnosis

Assessment of zygomatic complex fractures includes history, physical examination and CT scans in axial, coronal, and sagittal planes. Detailed 3-dimensional skeletal anatomy of the fracture demonstrated on the CT scan, allows for development of an appropriate treatment plan.

Treatment Modalities

The extent of exposure required in the management of zygomatic fractures depends on the displacement of the bone at the suture lines and status of the zygomatic arch.[18] If the zygomatic arch is minimally displaced and the projection is maintained, the zygoma can be reduced and stabilized via an eyebrow or supratarsal fold incision, a transconjunctival or lower eyelid incision and/or an intraoral upper buccal sulcus incision.

Displaced zygomatic fractures in combination with a displaced zygomatic arch should be additionally stabilized via zygomatic arch reconstruction. This is advisable to restore the anteroposterior projection in the arch region.[18] The standard exposure of the zygomatic arch for open reduction is a coronal incision. As noted above, careful anatomic dissection to prevent weakness or permanent paralysis of the frontal ramus of the facial nerve is important. The coronal approach is associated with other sequelae such as a large scar, hollowing of the temporal region, hypoesthesia in the incision line, and possible alopecia.[3] There is also more bleeding and edema because of interruption of venous and lymphatic pathways by the extended soft-tissue dissection of the forehead and temporal regions. These sequelae can be minimized or avoided by using an endoscopic approach for this procedure.[8,19]

Endoscopic Procedure for Zygomatic Complex Fracture Repair with Zygomatic Arch Reconstruction

Lateral eye brow or supra-tarsal fold, transconjunctival or subciliary, and intraoral upper buccal sulcus incisions in some combination are used to expose the orbitozygomatic complex. If the zygomatic arch is reconstructed via an endoscopic approach, it is useful to map the course of the frontal ramus of the facial nerve before starting the operation (according to the measurement of Schmidt and colleagues).[20] This helps for orientation if transcutaneous screw fixation of the arch is necessary (Figure 9).

The endoscopic equipment is the same as for frontal sinus fracture repair. The zygomatic arch is approached endoscopically by the upper buccal sulcus incision and two temporal incisions (Figure 10). After infiltration of local anaesthesia (1% Lidocaine and 1:200000 Epinephrine), the 2 cm temporal incision is made behind the temporal hairline and extended deep to the temporal fascia (Figure 11). Blunt dissection is carried out to the temporal line of fusion, where the deep temporal fascia splits into the superficial and deep layers. The fascia is incised with an electric knife (Figure 12). Dissection is continued

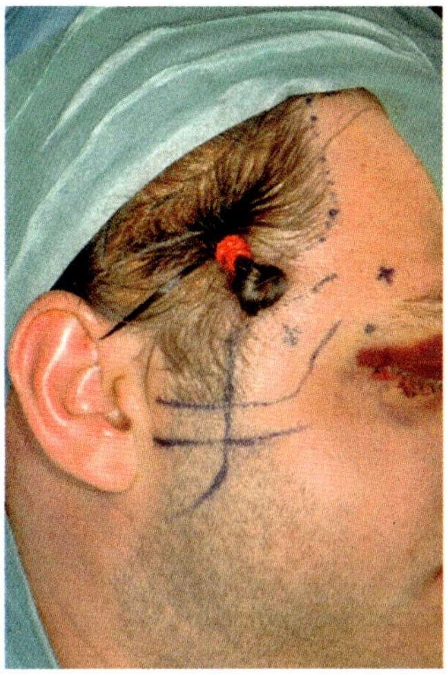

FIGURE 9 • Preoperative marking of the frontal branch of the facial nerve according to the landmarks.[20] (Right zygomatic complex fracture)

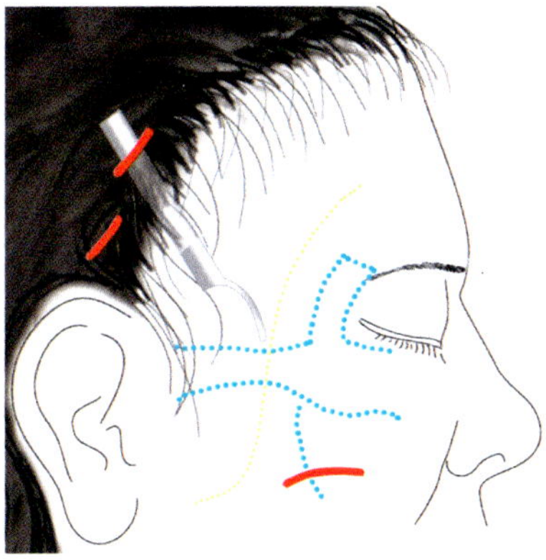

FIGURE 10 • Schematic endoscopic approach to the right zygomatic arch with intraoral and hairline incision. The course of the frontal branch of the facial nerve is marked (yellow).

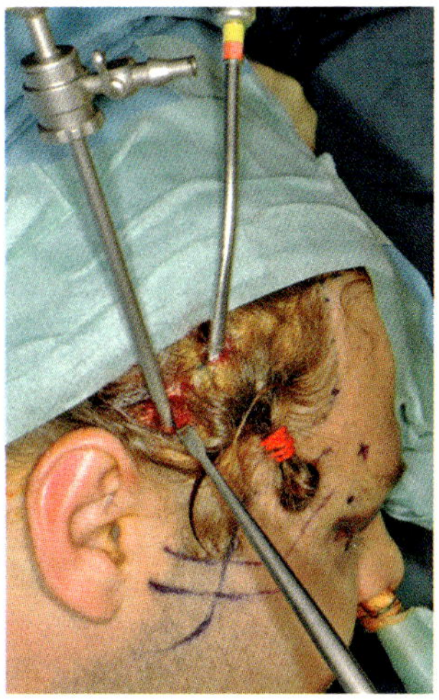

FIGURE 11 • Endoscope inserted through a temporal incision (left). The curved dissector is inserted from above.

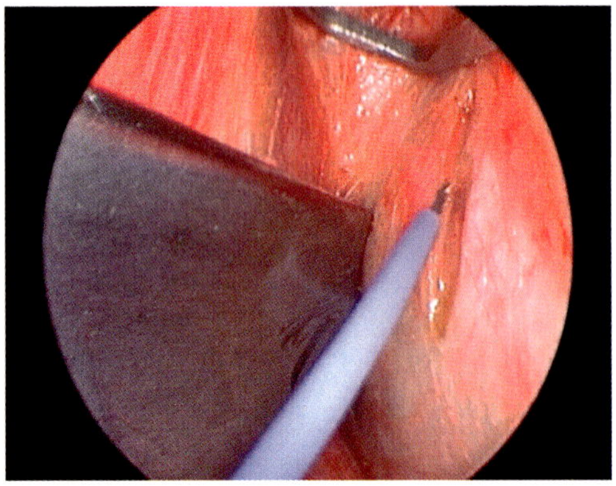

FIGURE 12 • Endoscopic view: Incision of the fascia in the temporal fusion line with an electric knife.

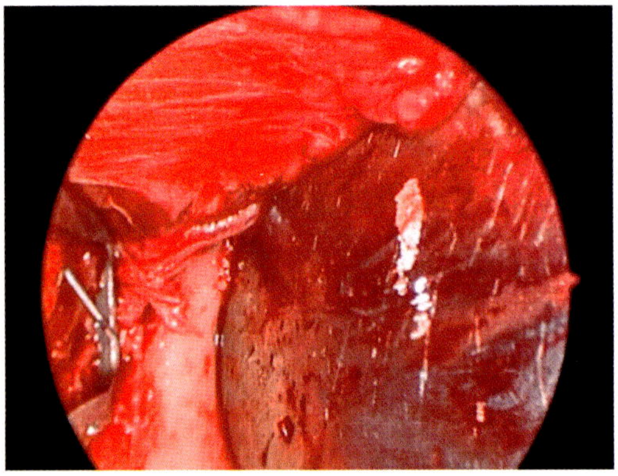

FIGURE 13 • The zygomatic arch is exposed between the endoscopic retractor (left) and the spatula (right).

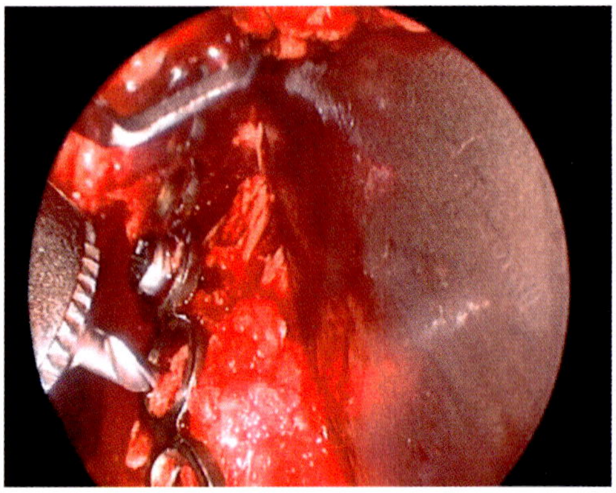

FIGURE 14 • Endoscopic view: Drilling for screw placement. One screw is already inserted.

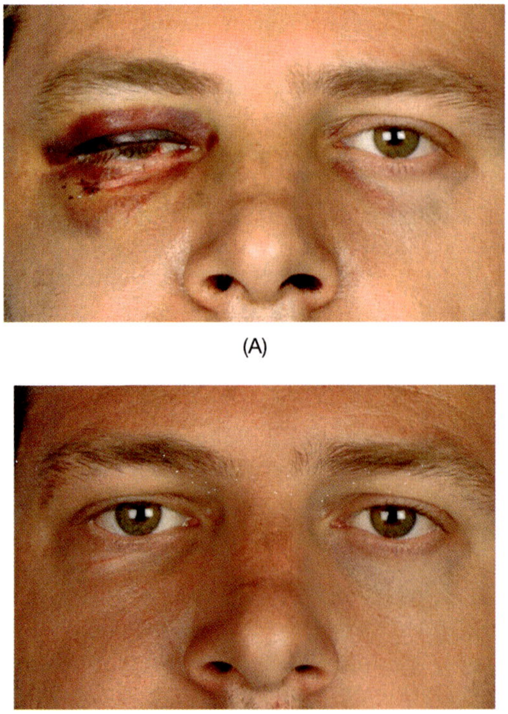

(A)

(B)

FIGURE 15 • *A*, Patient preoperative with a right orbitozygomatic fracture and an impressed zygomatic arch fracture. He had additionally a right infraorbital skin laceration. *B*, Patient 3 months later after fracture repair via lateral brow incision, transconjunctival and intraoral upper buccal sulcus incision. The zygomatic arch was repaired via an endoscopic approach.

to the deep temporal fascia. The superficial temporal fat pad is adherent to the superficial fascia and not to the deep temporal fascia. By using this approach the middle temporal artery, which runs in the fat pad, is preserved and there is no bleeding during the endoscopic dissection. The superficial fascia with the fat pad can be reflected with the endoscopic sheath retractor and the temporalis muscle is reflected by a retractor if necessary. The periosteum of the zygomatic arch is incised and the arch is dissected in an anterior direction (Figure 13). The zygomatic arch is also dissected endoscopically via the intraoral upper buccal sulcus approach. The tunnels are connected and the entire zygomatic arch is exposed. Reduction and stabilization of the fractured arch is done by miniplate osteosynthesis. The screws are inserted percutaneously under endoscopic visualization (Figure 14). The zygoma is fixed as usual by miniplates. Orbital walls are reconstructed if necessary (see Chapter 5, "Orbital Floor Fractures"). The incision in the superficial temporal fascia and the temporal incisions in the hairline are sutured closed (Figures 15, 16, and 17).

Discussion

The zygomatic arch is a surgical landmark for anatomic restoration of facial width and malar projection after a complex, comminuted zygomatic complex fracture.[2] Restoration of the arch can be successfully achieved with an endoscopic access. Lee and colleagues[21] reported of a series of 25 patients in whom the endoscopic approach was used for zygomatic arch repair. They used a preauricular incision at the anterior margin of the helical crus with a 2 cm extension superiorly above the auricle in the hairline. This incision is visible but miniplate placement can be easily accomplished with this technique. In this study group, 8 of the 25 patients had temporary paralysis of the frontal branch of the facial nerve. This was likely caused by a traction injury. Facial weakness recovered by the tenth postoperative week.[21]

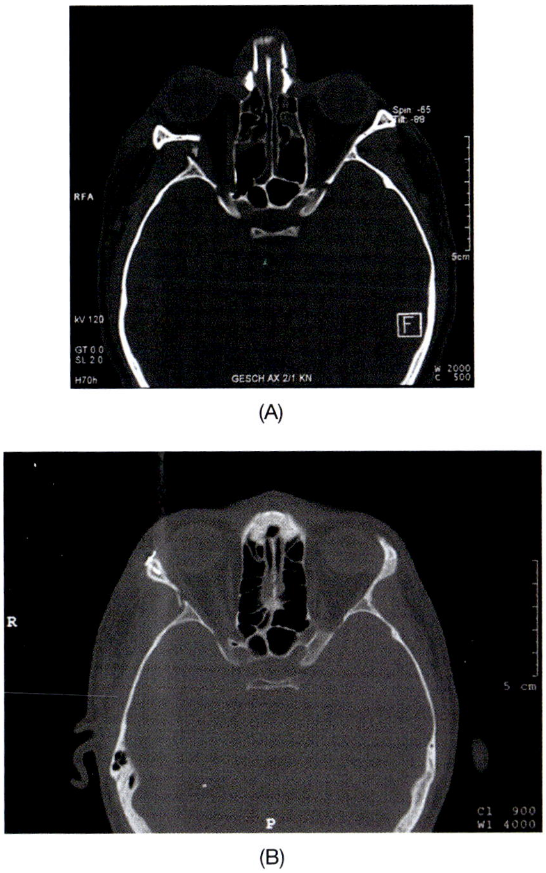

FIGURE 16 • CT scans preoperative (A) and postoperative (B) at the level of the lateral orbital wall.

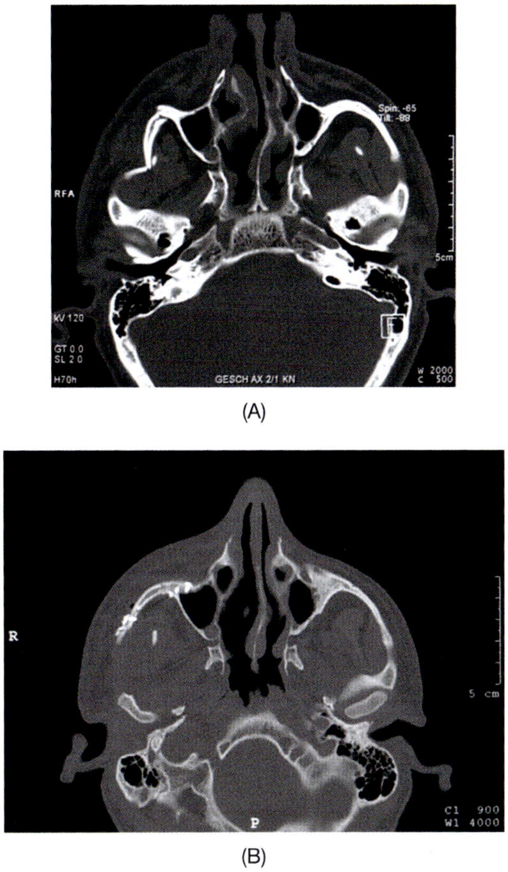

FIGURE 17 • CT scans preoperative (A) and postoperative (B) at the level of the zygomatic arch.

Chen and colleagues[22] used a temporal endoscopic approach to treat 15 patients with isolated zygomatic arch and zygomatic complex fractures. The dissection plane was superficial to the deep temporal fascia and 2 of the 15 patients developed a transient frontal nerve palsy.

The risk of facial nerve damage may be avoided by subfascial dissection.[23] The temporal muscle is exposed directly with the plane of dissection on the muscle. With this technique, no weakness of the frontal branch of the facial nerve was observed in 25 patients. In comparison, 2/17 patients had temporary weaknesses of the frontal branch nerve with the superficial plane dissection.

Fixation of the zygomatic arch fragments on a side table after endoscopic removing them may not produce a better outcome.[24] Arch necrosis and bone resorption has been noted with this approach.

If elevation of the zygomatic arch to the correct anatomic position cannot be achieved by the standard indirect (Gillies) technique, the endoscopic approach offers a method to restore the zygomatic arch which is less invasive than the coronal incision. As with endoscopic techniques in general, postoperative swelling is decreased because of the minimal tissue manipulation and dissection (Figure 18).

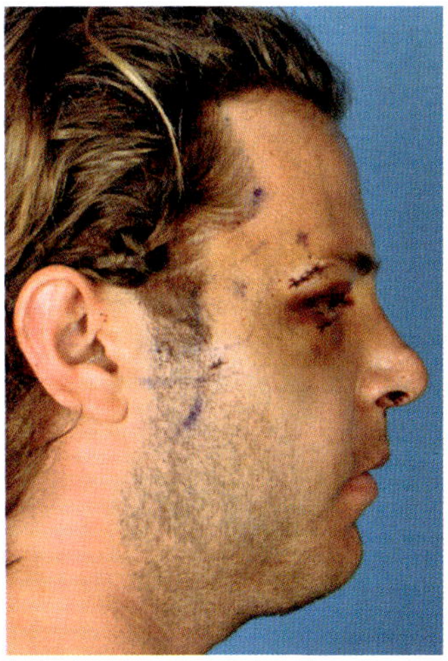

FIGURE 18 • Patient in Figure 15 at the first postoperative day.

Conclusion

The endoscopic technique offers a minimally invasive alternative to stabilize and fixate a fractured, displaced and comminuted zygomatic arch, either as an isolated injury or in the setting of displaced orbitozygomatic fractures. The temporal incision with a deep dissection to the deep temporal fascia reduces the risk of facial nerve (frontal branch) palsy. Further improvement of instrumentation will result in increasing utilization of this technique.

References

1. Manson PN, Crawley WA, Yaremchuk MJ, et al. Midface fractures: advantages of immediate extended open reduction and bone grafting. Plast Reconstr Surg 1985;76:1–12.
2. Gruss JS, Van Wyck L, Phillips JH, Antonyshym O. The importance of the zygomatic arch in complex midfacial fracture repair and correction of posttraumatic orbitozygomatic deformities. Plast Reconstr Surg 1990;85:878–90.
3. Frodel JL, Marentette LJ. The coronal approach: anatomic and technical considerations and morbidity. Arch Otolaryngol Head Neck Surg 1993;119:201–7.
4. Burm JS, Oh SJ. Prevention and treatment of wide scar and alopecia in the scalp: wedge excision and double relaxation suture. Plast Reconstr Surg 1999;103:1143–9.
5. Isse NG. Endoscopic facial rejuvenation: endoforehead the functional lift. Case reports. Aesth Plast Surg 1994;18:21–9.
6. Ramirez OM. Endoscopic techniques in facial rejuvenation: an overview. Part I. Aesth Plast Surg 1994;18:141–7.
7. Isse NG. Endoscopic facial rejuvenation. Clin Plast Surg 1997;24:213–31.
8. Kobayashi S, Sakai Y, Yamada A, Ohmori K. Approaching the zygoma with an endoscope. J Craniofac Surg 1995;6:519–24.
9. Graham III HD, Spring P. Endoscopic repair of frontal sinus fracture: case report. J Craniomaxillofac Trauma 1996;2:52–5.
10. Yavuzer R, Sari A, Kelly CP, et al. Management of frontal sinus fractures. Plast Reconstr Surg 2005;115:79e–93e.
11. Lappert PW, Lee JW. Treatment of an isolated outer table frontal sinus fracture using endoscopic reduction and fixation. Plast Reconstr Surg 1998;102:1642–5.
12. Shumrick KA. Endoscopic management of frontal sinus fractures. Facial Plast Surg Clin North Am 2006;14:31–5.
13. Strong EB, Buchalter GM, Moulthrop TH. Endoscopic repair of isolated anterior table frontal sinus fractures. Arch Facial Plast Surg 2003;5:514–21.
14. Reddi SP, Stevens MR, Kline SN, Villanueva P. Hydroxyapatite cement in craniofacial trauma surgery: indications and early experience. J Craniomaxillofac Trauma 1999;5:7–12.
15. Chen DJ, Chen CT, Chen YR, Feng GM. Endoscopically assisted repair of frontal sinus fracture. J Trauma 2003;55:378–82.
16. Gossman DG, Archer SM, Arosarena O. Management of frontal sinus fractures: a review of 96 cases. Laryngoscope 2006;116:1357–62.
17. Kim KK, Mueller R, Huang F, Strong EB. Endoscopic repair of anterior table: frontal sinus fractures with a medpor implant. Otolaryngol Head Neck Surg 2007;136:568–72.
18. Gruss JS, Phillips JH. Rigid Fixation of zygomatic fractures. In: Yaremchuk MJ, Gruss JS, Manson PN, eds. Rigid fixation of the craniomaxillofacial skeleton. Boston: Butterworth-Heinemann; 1992. pp. 263–82.
19. Lee CH, Lee C, Trabulsy PP, et al. A cadaveric and clinical evaluation of endoscopically assisted zygomatic fracture repair. Plast Reconstr Surg 1998;101:333–45.
20. Schmidt BL, Pogrel MA, Hakim-Faal Z. The course of the temporal branch of the facial nerve in the periorbital region. J Oral Maxillofac Surg 2001;59:178–84.
21. Lee C, Stiebel M, Young DM. Cranial nerve VII region of the traumatized facial skeleton: optimizing fracture repair with the endoscope. J Trauma 2000;48:423–32.
22. Chen CT, Lai JP, Chen YR, et al. Application of endoscope in zygomatic fracture repair. Br J Plast Surg 2000;53:100–5.
23. Lee SS, Lin SD, Chiu YT, et al. Deep dissection plane for endoscopic-assisted comminuted malar fracture repair. Ann Plast Surg 2002;49:452–9.
24. Krimmel M, Cornelius CP, Reinert S. Endoscopically assisted zygomatic fracture reduction and osteosynthesis revisited. Int J Oral Maxillofac Surg 2002;31:485–8.

Orbital Floor Fractures

5

Michael Miloro, DMD, MD

OUTLINE

- Introduction
- Indications
- Technical Considerations
- Discussion
- Conclusions

INTRODUCTION

Surgical innovation and technological development should be guided by clinical necessity, and ideally provide improvement in our current understanding of disease processes, as well as surgical management. Although endoscopic diagnosis and treatment of the craniomaxillofacial region has expanded tremendously over the past decade,[1] the application to the area of orbital injuries is perhaps most justified, due to the potential morbidity of current treatment philosophies and methodologies. Orbital fractures may involve any of the bony components, including the medial and lateral walls, roof, and floor, yet due to its inherent weakness, orbital floor fractures are most common.[2] The orbital rim is composed of dense compact bone that does not fracture easily. Although the orbital wall thickness varies, and the medial wall is usually thin, the intricate network of ethmoidal air cells and bony trabeculae provide support to the medial orbital wall, yet medial wall fractures do occur often.[3] Orbital fractures may present with a range of signs and symptoms, including periorbital edema and ecchymosis, chemosis, ptosis (which can make further examination difficult), lacerations, decreased visual acuity, visual field defects, canalicular disruption, decreased extraocular muscle movement (entrapment or paresis), diplopia (in certain fields of gaze), ophthalmoplegia, abnormal pupillary response to light and accommodation, subconjunctival hemorrhage, proptosis, enophthalmos, increased intraocular pressure (retrobulbar hemorrhage), corneal abrasions, retinal disruption, lens disruption, hyphema, and possibly paresthesia of the supraorbital, infraorbital, and supratrochlear nerves.

Following a complete ophthalmologic examination, the imaging study of choice is a computed tomographic (CT) scan, or 3D CT when available, to help accurately delineate the precise patterns of fracture.[4] Plain films may be useful (e.g., Water's view) to determine herniation of orbital contents into the maxillary sinus, but a non-contrast CT scan with fine axial and coronal sections, and three-dimensional reconstruction if available, provides improved detail and resolution. Once a diagnosis has been established, the need for an operation must be determined based upon the degree/magnitude of bony displacement (usually greater than 50% floor defect), presence of enophthalmos, extraocular muscle entrapment, and diplopia. A treatment plan is developed and the management of orbital floor fractures has traditionally been performed via a variety of incisions,[5] including a transconjunctival incision,[6] with or without a lateral canthotomy with inferior cantholysis. Other surgical approaches include the subciliary incision with a higher rate of complications than the transconjunctival approach;[7] and the retroseptal, infraorbital, transcaruncular, and orbital rim approaches have been employed as well based upon surgeon preference. The overall complication rate using any of these skin or conjunctival incisions varies between 1.2% and 42.5% (with a mean complication rate in the 5% range).[8–10] Potential complications include unacceptable unesthetic facial scarring, prolonged lower lid edema, lower lid retraction (Figure 1), ectropion, entropion, iatrogenic canalicular injury, iatrogenic globe injury, damage to the levator or extraocular muscles and periorbita, and possible granuloma formation along the incision line. The use of endoscopy for orbital floor repair may provide an excellent option for treatment

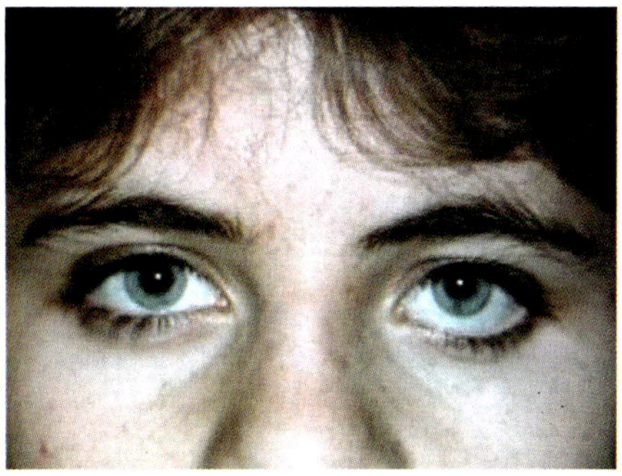

FIGURE 1 • Left lower lid retraction following a transconjuctival incision.

that obviates the need for periorbital incisions and significantly decreases postoperative morbidity. In addition, the endoscopic technique can allow visualization of the entire orbital vault, especially the posterior compartment, which is usually difficult with direct visualization due to prolapsed orbital contents, edema, and bleeding.

The use of endoscopy for orbital floor repair is not a new concept, and it was described in 1901 using a modified cystoscope.[11] The transantral approach to the orbital floor was described in 1972 using indirect illumination, manual fracture reduction, and short-term sinus packing to provide support for the comminuted orbital floor.[12] With the introduction of functional endoscopic sinus surgery in 1985,[13] the connection was made clear to apply this technology to the midface for other applications, such as fractures. Throughout the past two decades there were several descriptions of endonasal approaches to the orbit; however, these provide only limited access and, therefore, these techniques were not adopted by most surgeons due to inherent technical difficulties.[14–17] As a result, many surgeons have employed the Caldwell-Luc approach to the orbit through the maxillary sinus with endoscopic assistance in order to achieve wider surgical access and improved visualization of the orbital floor (and medial orbital wall in some cases) and to provide optic nerve decompression, when indicated.[18–23] In a comparative cadaveric study of endoscopic transantral versus transconjunctival repair of orbital floor fractures, there were no statistically significant differences between postrepair CT volumetric analysis. Furthermore, this study confirmed the efficacy of endoscopic-assisted repair of orbital floor fractures.[24] This technique has several advantages including the use of a simple standard transmucosal approach to the anterior wall of the maxillary sinus; the use of a standard Caldwell-Luc antrostomy, which may be modified in size to accommodate materials needed for floor repair (eg, Medpor or titanium mesh); the development of a large "optical cavity" (ie, the entire maxillary sinus) for access and visualization; minimal bleeding following the use of local topical hemostatic measures and drying agents; demucosalization of the sinus lining for visualization of all fracture segments; and effective control of the periorbital tissues and fat. In addition, this approach provides unparalleled visualization of the posterior shelf of the orbital floor, including the orbital foramen, if necessary; access to the medial orbital wall is possible as well in the case of fracture; ease of surgical placement of floor replacement material, if necessary; lack of a skin or transconjunctival incision and avoidance of resultant morbidity (eg, ectropion); significantly decreased postoperative edema when compared to transfacial approaches; and, overall better surgical access, exposure, visualization, and treatment of the orbital floor fracture.

INDICATIONS

In the management of orbital floor fractures, the general indications are significant enophthalmos, diplopia within normal gaze, limitation of extraocular muscle movement (entrapment) (Figure 2), and a large floor defect (>50%), and these apply to all patients, including those who are considered for an endoscopic-assisted repair technique. A CT scan is useful for assisting in the diagnosis and planning definitive treatment (Figures 3, 4, and 5).

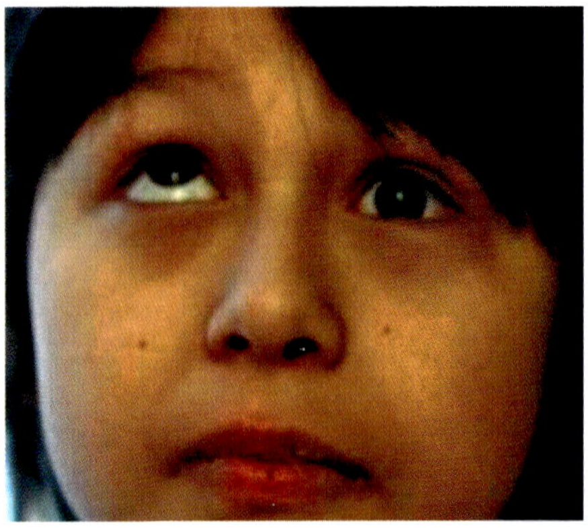

FIGURE 2 • Left inferior rectus muscle entrapment upon upward gaze due to a left orbital floor fracture.

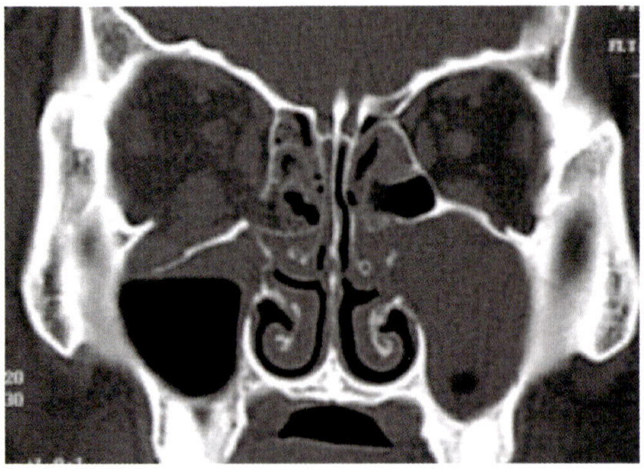

FIGURE 3 • Coronal CT scan showing a large right orbital floor fracture with a medial hinge component.

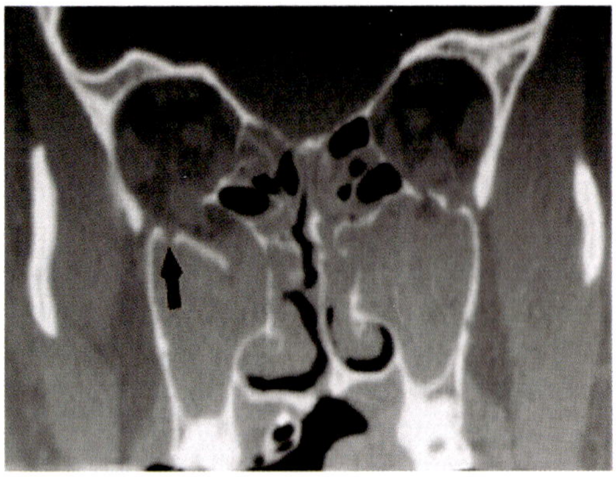

FIGURE 4 • Coronal CT scan showing a large right orbital floor fracture with a lateral hinge component, and a nondisplaced left orbital floor fracture (Note: impingement of bone on right inferior rectus muscle arrow).

The CT scan can reveal many variations of orbital floor fractures, but these can generally be classified into trap door (or hinge-type) fractures, or complete (comminuted) fractures. There may be medial and/or lateral components to the fracture, based upon the location with regards to the infraorbital canal. In general, blowout fractures of the orbital floor, with either a medial or lateral hinge component, or a complete comminuted

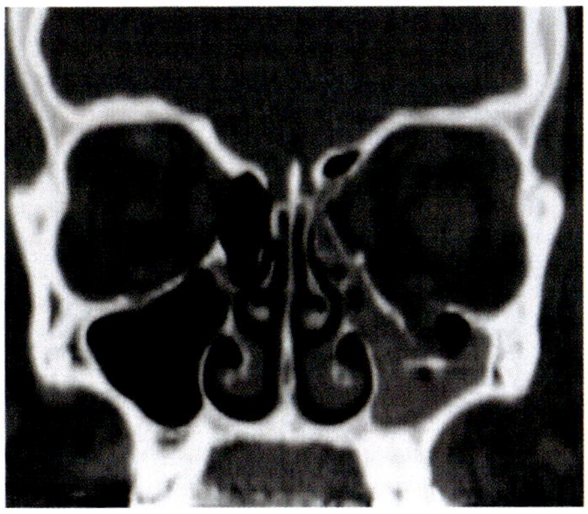

FIGURE 5 • Coronal CT scan showing a comminuted left medial blowout fracture and a left medial orbital wall fracture.

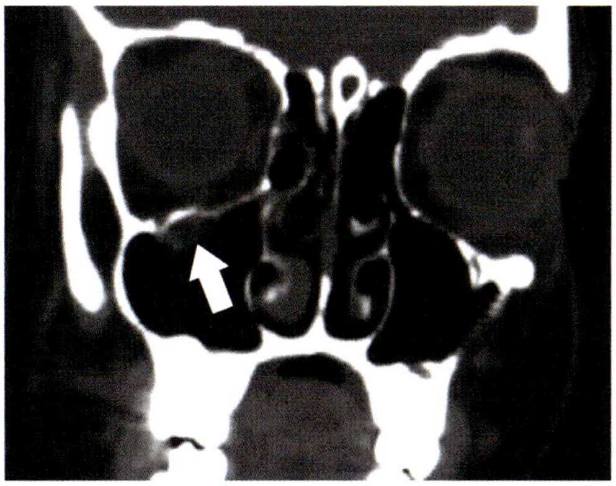

FIGURE 7 • Minimally displaced right orbital floor fracture.

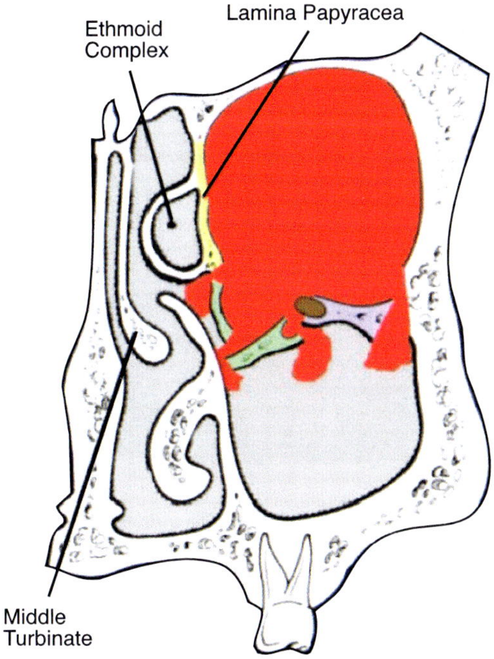

FIGURE 6 • The components of an orbital floor fracture based upon the position of the infraorbital canal. There is a medial wall fracture (yellow), and a medial (green) and lateral (blue) orbital floor fracture (courtesy of R. Mueller).

blowout fracture, are more amenable to endoscopic repair, than is an isolated medial orbital wall fracture, due to difficulty with access and visualization (Figure 6). Considerations favoring the endoscopic approach for orbital floor fracture repair include: need for early operation before swelling resolves; isolated orbital floor fractures; large orbital floor defect. Of course, the endoscopic-assisted approach should only be used by surgeons who have obtained the proper training and who have prior experience with the technique (ie, the learning curve has been overcome). Regarding potential contraindications to endoscopic repair of orbital floor fractures, the following may apply: lack of indication for open reduction, in general; no ocular signs or symptoms; minimal fracture displacement, but the presence of another indication for open repair (eg, entrapment). In this case, an endoscopic approach from below would require removal of a significant amount of nondisplaced bony orbital floor, which may be overly aggressive and not prudent, therefore a standard open approach (e.g. via a transconjunctival incision) may be preferable (Figure 7); and, possibly active sinus disease or infection.

TECHNICAL CONSIDERATIONS

A standard surgical technique must be employed in order to avoid patient morbidity and allow acceptable operative times to complete the surgical procedure. In one of the first reports of using a transantral technique to assist in the diagnosis of orbital floor fractures, an "unobscured" fracture was seen in only one of seven orbital floor fractures.[25] This resulted from surgeon inexperience, variation in surgical technique, the use of different endoscopes with varying fields of view angulations, as well as a lack of control of sinus mucosa, bleeding ecchymosis, and fat herniation. In another series of orbital floor fractures,[26] the floor replacement material (either Medpor, high molecular weight porous polyethylene, or titanium mesh) was placed on the "sinus side" of the orbital floor using a zero-degree endoscope. The authors reported only one out of nine cases of postoperative enophthalmos, despite the lack of restoration of orbital volume using their technique.[27] Also, the use of a zero-degree endoscope makes positioning of the instruments during surgery difficult and awkward. The authors did describe the need for a 2.0 $cm^2 \times$ 1.5 cm^2 antrostomy window in an attempt to standardize the technique. Finally, the authors recognized the difficulties in medial wall fracture management using endoscopic techniques, as they had noted in a previous publication.[28]

The current technique used for endoscopic-assisted repair of orbital floor fractures has been developed and reported previously by Strong.[21,29] Initial studies included 16 cadaveric dissections, as well as 10 clinical observations, which led to the development of the diagnosis of the fracture patterns, and the endoscopic approach to treatment. Although the technical developments were still being revised, the majority of clinical postoperative CT scans were rated as excellent or good, and only one was rated as poor. The patient was reoperated through a standard open transconjunctival approach.

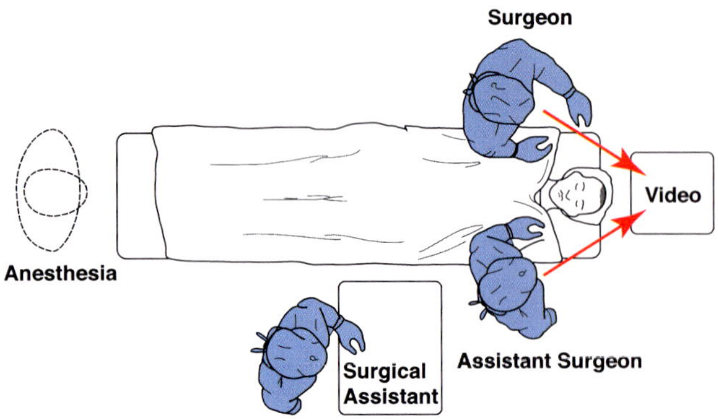

FIGURE 8 • Surgical team positioning for endoscopic surgery.

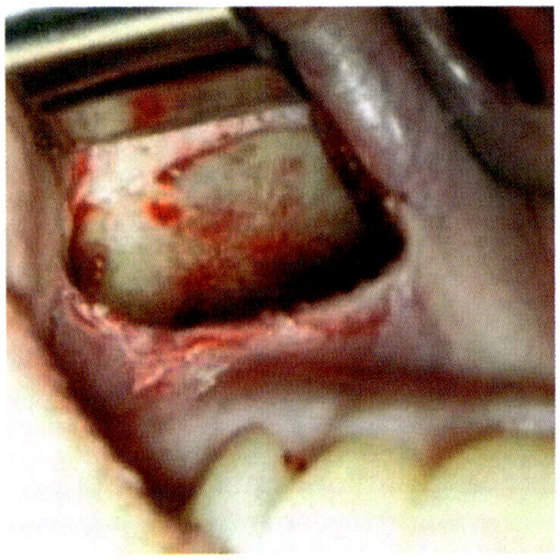

FIGURE 9 • Right maxillary buccal sulcus incision with exposure of the anterior wall of the maxillary sinus.

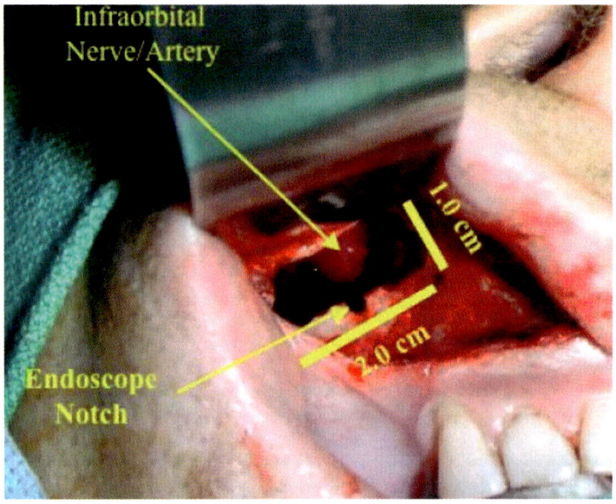

FIGURE 10 • Caldwell-Luc osteotomy for endoscopic orbital floor access (1.0 cm × 2.0 cm) (courtesy of R. Mueller).

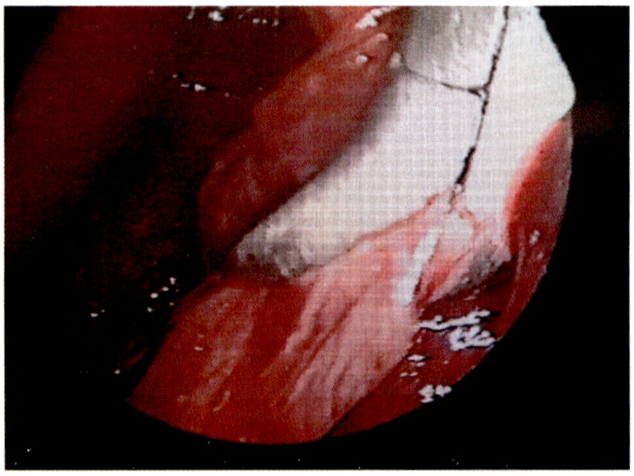

FIGURE 11 • Demucosalization of the maxillary sinus membrane.

The surgery begins with the patient in the supine position under general anesthesia, with the surgeon on the right side of the table and the video monitor at the head of the table (Figure 8). As opposed to a periorbital approach, the antral endoscopic approach does not require the use of a corneal shield and, therefore, the chance of an iatrogenic corneal abrasion is eliminated. An oral antimicrobial rinse, such as chlorhexidine 0.012%, may be used and a throat pack is placed. Local anesthetic with vasoconstrictor is used to infiltrate the area of the posterior superior alveolar and middle superior alveolar nerves. A 3 cm incision is made in the maxillary buccal sulcus region approximately 3 mm to 5 mm above the mucogingival junction, from approximately the first molar to the lateral incisor region on the side of the orbital fracture. The periosteum is elevated to expose the anterior wall of the maxillary sinus to enable visualization and protection of the infraorbital neurovascular bundle (Figure 9). On occasion, there may be a fracture of this maxillary bone, and the antrostomy should be designed to accommodate for this by incorporating the fracture segment into the osteotomy design as much as possible, in order to avoid creating an overly large window. A standard approach to enter the maxillary sinus in the canine fossa region, avoiding the root of the maxillary canine, should be performed using osteotomes and Kerrison rongeurs for bone removal. A window 1.0 cm in height by 2.0 cm in width is created (Figure 10). The width of the osteotomy may need to be widened later for the introduction of a floor implant, if necessary. The osteotomy window may or may not be replaced following the surgical procedure so it should be stored in saline. Also, in some cases, the window may be used as an orbital floor graft if the defect is small. Finally, a small "endoscope notch" may be created at the inferior aspect of the window to provide stability for the endoscope in order to minimize movement during surgery.

Upon entry into the maxillary sinus, a decongestant should be used, such as oxymetazoline on cotton pledgets, and left for several minutes. Next, a 4-mm 30-degree Hopkins endoscope (Karl Storz, Carver City, CA) is used to inspect the maxillary sinus (optical cavity). An advantage of having the maxillary sinus as the optical cavity is that any bleeding will settle to the back of the sinus and not obscure visualization, and this could be continuously suctioned using a curved Frazier tip suction catheter. A critical component of the procedure is to next demucosalize the entire maxillary sinus, taking care to remove all remnants of the sinus mucoperiosteum (Schneiderian membrane), avoiding injury in the region of the ostium (Figure 11). Once this is completed, a full, unobstructed view of the orbital fracture is possible: location and size of the fracture, degree of fat herniation,

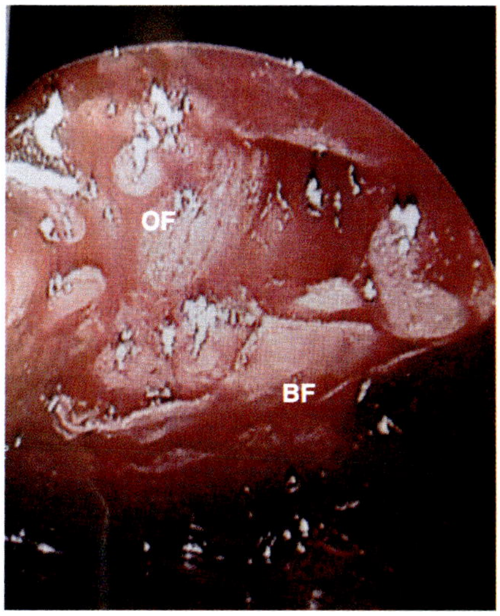

FIGURE 12 • Endoscopic view of the left orbit (OF) from below showing comminuted bone segments posteriorly (BF) and herniated periorbital tissues anteriorly.

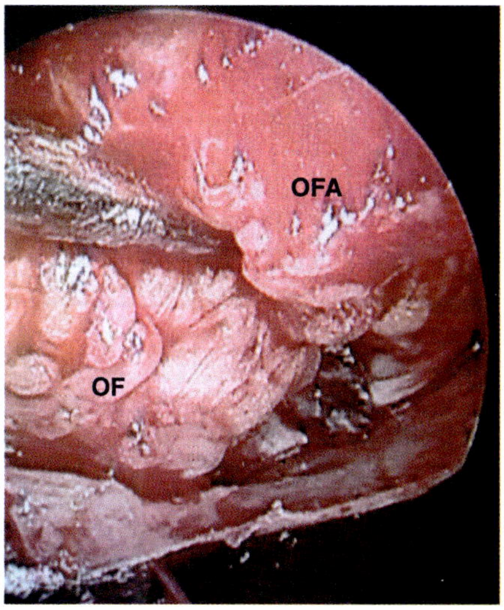

FIGURE 13 • Endoscopic view of the left orbit showing removal of large posterior bone fragment from the orbital floor (OF) with prolapsed orbital fat anteriorly (OFA).

bony comminution, and muscle entrapment. Gentle pressure on the globe will allow visualization of periorbital fat herniation and floor defects (Figure 12). In the case of "trap door" fractures of either the medial or lateral portion of the floor, it may be possible to ensure that all periorbital contents are reduced back into the orbit and, if the fracture segment is large enough, it may be used to stabilize the reduction through interfragmentary friction, or an implant may be used, on the "sinus side," to stabilize the reduced bony floor. In many cases, the floor is fractured with comminuted bone fragments that will not allow this to be done. In these cases, cautious and meticulous removal of all, even small, comminuted bone segments is performed endoscopically using long, narrow straight and curved instruments as necessary to accomplish this critical step in the surgery (Figure 13). Once all bony fragments have been meticulously removed, a retractor

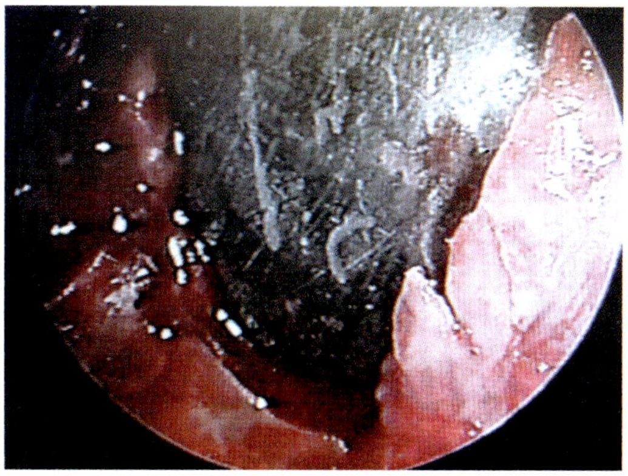

FIGURE 14 • Endoscopic view of the left orbit following demucosalization and bone removal, with retraction of the periorbital tissues into the orbital cavity, showing the residual floor defect requiring an implant (Note: bottom of picture shows the posterior orbital shelf).

(eg, malleable retractor) can be used to help elevate all of the periorbital contents, fat, infraorbital neurovascular bundle, and muscles, back into the orbital cavity (Figure 14). Of course the bony floor defect will be much larger than previously seen on CT scan, and an orbital floor replacement material should be chosen that has enough inherent strength to support the orbital contents, and that may be retrievable, if necessary. A high molecular weight polyethylene material seems to be the near ideal synthetic material (eg, Medpor, Porex Surgical, Newnan, GA, USA), because titanium mesh is very difficult to remove following soft tissue ingrowth, and the irregular edges of the mesh following trimming, may lead to iatrogenic injury. An autogenous bone graft may also be considered (eg, calvarial or costochondral split grafts). In general, the porous polyethelene implant is 0.85 to 1.0 mm in thickness to allow for some flexibility in placement. This implant is also available with titanium mesh sandwiched in between two layers of porous polyethylene to provide structural rigidity as well as the ability to contour the material, while avoiding the problems with titanium mesh, as discussed. The implant should provide "overcorrection" of orbital volume to prevent postoperative enophthalmos due to loss of orbital volume. The implant is trimmed so that it is 1 mm to 2 mm larger in all dimensions than the floor defect so that it will obtain stabilization from the bony edges and provide support for the orbital contents and maintain globe position. A "spacer" such as a suture package can be used to determine the size of the defect, and then used as a template for trimming the implant to the appropriate size. The implant is inserted through the antrostomy window, which may require enlargement to accommodate the implant. The implant is gently manipulated into position, and the malleable retractor is removed, in an "exchange-like" maneuver. Care is taken to prevent strangulation of periorbital fat, application of pressure on the infraorbital nerve, or the introduction of bony fragments into the orbit. Once the implant is positioned ideally (Figure 15), gentle pressure is applied to the globe to ensure stability of the implant, and a forced duction test is performed to rule out entrapment of the inferior rectus muscle. The implant is held in place by the weight of the periorbita and globe and is stabilized circumferentially by the remaining bony edges of the orbital floor

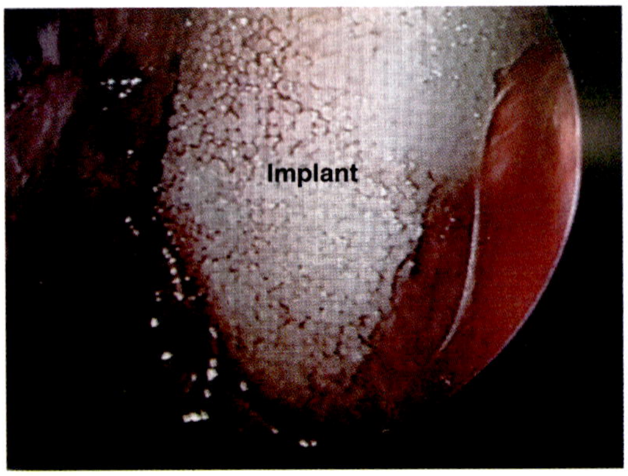

FIGURE 15 • Endoscopic view of the left orbit showing the in situ position of the implant (high molecular weight polyethylene) replacing the majority of the orbital floor defect.

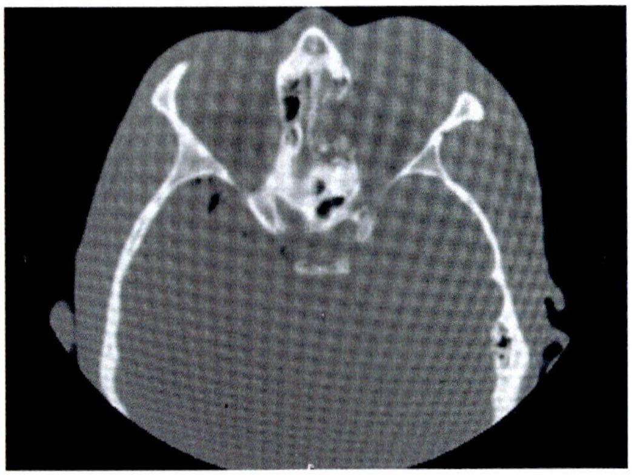

FIGURE 17 • Isolated left medial orbital wall fracture.

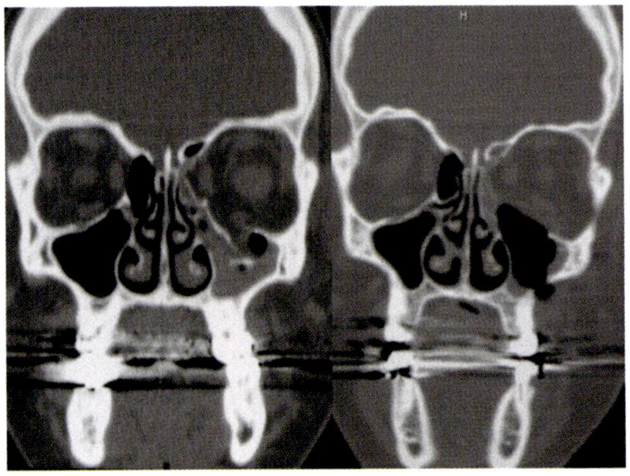

FIGURE 16 • Preoperative and postoperative CT scans before and after endoscopic-assisted repair of a left medial orbital floor fracture via a Calwell-Luc approach (note the defect in the anterior wall of the sinus in the postoperative CT), using a (radiolucent) porous polyethylene implant (note the slight overcorrection of orbital volume to prevent enophthalmos).

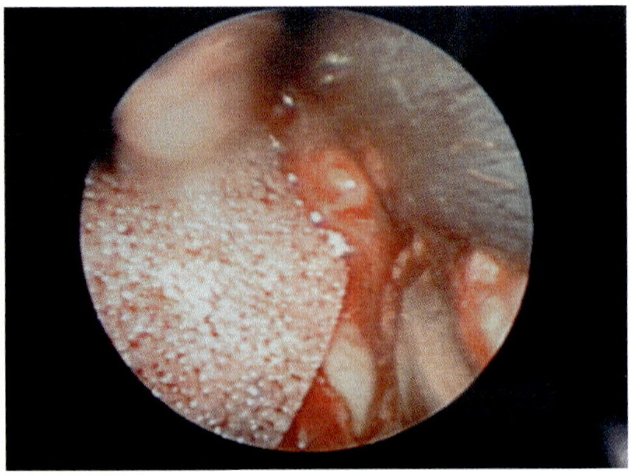

FIGURE 18 • Endoscopic view of porous polyethelene implant placed via a transcaruncular incision to repair a left medial orbital wall fracture.

and medial and lateral orbital walls. No fixation devices (plates and/or screws) are generally used, because they are, for the most part, unnecessary, and also to decrease the number of foreign bodies introduced into the orbit that may become mobile or displaced in the future, resulting in inflammatory reactions necessitating removal. Next, the maxillary ostium may be inspected to ensure integrity and patency. A temporary nasal antrostomy is rarely required for sinus drainage. The wound is irrigated and suctioned, and in most cases, the bony window of the antrostomy is not replaced, but the mucosa is closed with resorbable sutures. A forced duction test is performed at the conclusion of the surgical procedure. Frequent postoperative patient assessments, including visual acuity checks, are necessary, and a CT scan is desirable to assess the surgical result as well (Figure 16).

As mentioned previously, pure medial orbital wall fractures are difficult to manage endoscopically, but may be approached through a Caldwell-Luc, intranasal, or transcaruncular approach (Figures 17, 18, and 19). The reason for the surgical difficulty is not only the location of the fracture, but also the nature of the thin ethmoid bone and inherent problems with manipulating the bony segments. Based upon the literature[30] and anecdotal reports, perhaps a medial transcaruncular incision is better suited to approach these fractures than a Caldwell-Luc approach. An endoscope may be used for assistance via the transcaruncular approach as well, and the use of a smaller diameter endoscope such as a 2.7 mm with either a zero- or 30-degree field of view may be appropriate in individual cases.

A variety of possible complications do exist including inability to complete the surgical procedure perhaps due to inexperience, damage to teeth, infraorbital paresthesia (usually transient), postoperative maxillary sinusitis due to various reasons including preexisting sinus disease, ostium obstruction, or delayed mucoperiosteal regeneration, implant displacement, and introduction of bone fragments into the orbit due to possibly poor visualization and lack of recognition of the presence of the bone fragment. The other risks usually associated with orbital floor fracture repair, such as lower lid retraction, scarring, and postoperative edema, are certainly minimized or eliminated with this endoscopic-assisted approach. If revision survey is required, it is best accomplished through a standard transconjunctival or subciliary approach.

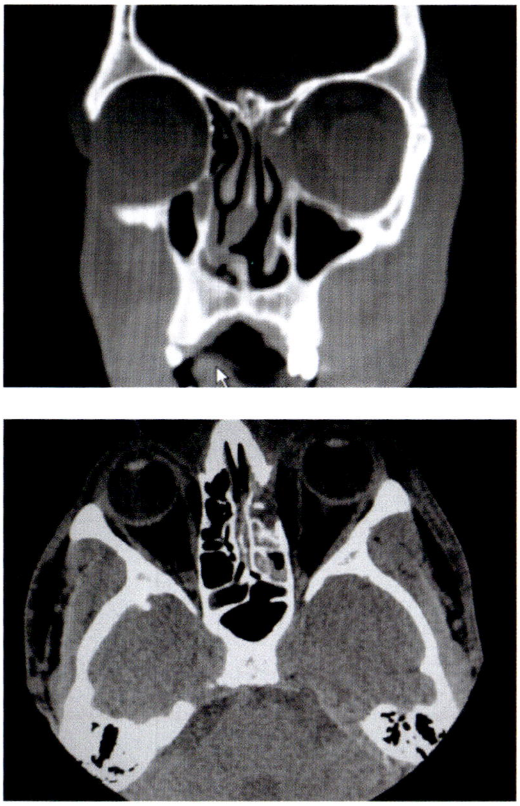

FIGURE 19 • Postoperative CT scans (coronal and axial) following repair of left medial wall fracture with (radiolucent) porous polyethylene.

DISCUSSION

The utilization of endoscopic assistance for repair of orbital floor fractures is well established in the literature and in clinical practice. The technology and instrumentation can be applied to numerous other periorbital surgical procedures making it very versatile for optic nerve decompression (eg, for Grave's ophthalmopathy),[31] as well as for surgical access to pituitary and skull base tumors. Additionally, the endoscope may be used adjunctively during standard approaches to other facial fractures, including zygomatico-maxillary complex fractures (see chapter 4, "Frontal Sinus and Zygomatic Complex Fractures"), in order to confirm reduction of fractures that would typically not be exposed for assessment (eg, sphenozygomatic suture). Of course any new technology requires training and experience and overcoming the issues of a "learning curve" in clinical practice. Many of the steps involved in the transantral approach to the orbital floor are routine for most oral and maxillofacial surgeons, otorhinolaryngologists, and plastic surgeons and, therefore, the largest hurdle may be the development of a familiarity with using indirect visualization through the endoscope in this area.

The advantages of an endoscopic-assisted approach to the orbital floor are apparent, and the ability to operate early without waiting for edema to resolve in order to perform a standard transconjunctival or subciliary approach from a technical standpoint, makes endoscopy very attractive. Especially for the pediatric patient with a fixed globe, or any patient with entrapment and/or diplopia, a surgical time delay could result in devastating consequences. It is clear that traditional approaches to treat orbital floor fractures carry potentially significant morbidity including lower lid malposition, especially with more extensive fractures requiring more extensive soft tissue dissection and globe manipulation. The minimal periorbital soft-tissue dissection and mild globe manipulation results in a postoperative patient with minimal, if any, increased edema beyond that which was present presurgically. From a surgical standpoint, the ability to clearly visualize the posterior aspects of the orbital floor is a major advantage over traditional approaches where the posterior orbital cannot be visualized at all due to an obscured visual field by periorbital tissues as well as the hesitancy to cause iatrogenic injury to vital structures in close proximity.

CONCLUSIONS

Endoscopic-assisted orbital floor fracture repair is an efficacious technique that avoids the morbidity associated with traditional approaches to orbital fractures. Although the application of endoscopy to many areas in head and neck and general surgery are still expanding, and some applications are eliminated because endoscopy offers little advantage, its use in the orbit has become a mainstay of therapy because it allows meticulous surgical manipulation in and around vital structures, and avoids long-term morbidity of other surgical techniques. The success of this technique is highly dependent upon a well-trained surgeon, and as more surgical residents become exposed to the technique, the wider the potential application will become in the future, as an option in the treatment algorithm for management of orbital floor fractures.

References

1. Schon R, Gellrich NC, Schmelzeisen R. Frontiers in maxillofacial endoscopic surgery. Atlas Oral Maxillofac Surg Clin N Am 2003;11:209–38.
2. Jones DE, Evans JN. Blow-out fractures of the orbit: an investigation into their anatomic basis. J Laryngol Otol 1967;81: 1109–20.
3. Burm JS, Chung CH, Oh SJ. Pure orbital blowout fracture: new concepts and importance of medial orbital blowout fracture. Plast Reconstr Surg 1999;103:1839–49.
4. Sinn DP, Karas ND. Imaging in orbital trauma. Oral Maxillofac Surg Clin N Am 1993;5:431–42.
5. Werther JR. Cutaneous approaches to the lower lid and orbit. J Oral Maxillofac Surg 1998;56:60–5.
6. Converse JM, Firmin F, Wood-Smith D, et al. The conjunctival approach in orbital fractures. Plast Reconstr Surg 1973;52: 656–7.
7. Patel PC, Sobota BT, Patel NM, et al. Comparison of transconjunctival versus subciliary approaches for orbital fractures: a review of 60 cases. J Craniomaxillofac Trauma 1998;4:17–21.
8. Mullins JB, Holds JB, Branham GH, et al. Complications of the transconjunctival approach. A review of 400 cases. Arch Otolaryngol Head Neck Surg 1997;123:385–8.

9. Appling WD, Patrinely JR, Salzer TA. Transconjunctival approach vs subciliary skin-muscle flap approach for orbital fracture repair. Arch Otolaryngol Head Neck Surg 1993;119:1000–7.
10. Lorenz HP, Longaker MT, Kawamoto HK. Primary and secondary orbit surgery: The transconjunctival approach. Plast Reconstr Surg 1999;103:1124–8.
11. Hirschmann A. Uber endoskopic der nuse und deren nebemhohler. Eine neue untersuchungsmetnode. Arch Otolaryngol Rhinol (Berlin) 1901;14:195–7.
12. Walter WL. Early surgical repair of blowout fracture of the orbital floor by using the transantral approach. South Med J 1972;65:1229–43.
13. Kennedy D. Functional endoscopic sinus surgery. Technique. Arch Otolaryngol 1985;111:643–9.
14. Yamaguchi N, Arai S, Mitani H, et. al. Endoscopic endonasal technique of the blowout fracture of the medial orbital wall. Operative Techniques in Otolaryngology Head and Neck Surgery 1991;2: 269–74.
15. Michel O. Isolated medial orbital wall fractures: results of minimally invasive endoscopically controlled endonasal surgical technique. Laryngorhinootologie 1993;72:450–4.
16. Chen CT, Chen YR. Endoscopic orbital surgery. Atlas of Oral Maxillofacial Surg Clin NA 2003;11:179–208.
17. Rhee JS, Chen CT. Endoscopic approach to medial orbital wall fractures. Facial Plast Surg Clinics of North Am 2006;14:17–23.
18. Saunders CJ, Whetzel TP, Stokes RB, et al. Transantral endoscopic orbital floor exploration: a cadaver and clinical study. Plast Reconstr Surg 1997;100:575–81.
19. Woog JJ, Hartstein ME, Gliklich R. Paranasal sinus endoscopy and orbital fracture repair. Arch Ophthalmol 1998;116:688–91.
20. Otori N, Haruna S, Moriyam H. Endosocpic endonasal or transmaxillary repair of orbital floor fracture: a study of 88 patients treated in our department. Acta Otolaryngol 2003;123: 718–23.
21. Strong EB, Kim KK, Diaz RC. Endosopic approach to orbital blowout fracture repair. Otolaryngol Head Neck Surg 2004;131: 683–95.
22. Farwell DG, Strong B. Endoscopic repair of orbital floor fractures. Facial Plast Surg Clinics of North Am 2006;14:11–16.
23. Pletcher SD, Sindwani R, Metson R. Endoscopic orbital and optic nerve decompression. Otolaryngol Clin North Am 2006;39: 943–58.
24. Wallace TD, Moore CC, Bromwich MA, et al. Endoscopic repair of orbital floor fractures: computed tomographic analysis using a cadaveric model. J Otolaryngol 2006;35:1–7.
25. Sandler NA. The use of maxillary sinus endoscopy in the diagnosis of orbital floor fractures. J Oral Maxillofacial Surg 1999;57: 399–403.
26. Chen C, Chen Y. Endoscopically assisted repair of orbital floor fractures. Plast Reconstr Surg 2001;108:2011–8.
27. Miloro M. Commentary: Chen C, Chen Y. Endoscopically assisted repair of orbital floor fractures. Plast Reconstr Surg 2001; 108:2011–6, Arch Facial Plast Surg 2002;124:4.
28. Chen C, Chen , Tung TC, et al. Endoscopically assisted reconstruction of orbital medial wall fractures. Plast Reconstr Surg 1999;103: 714–20.
29. Strong EB. Endoscopic repair of orbital blow-out fractures. Facial Plast Surg 2004;20:223–30.
30. Rhee JS, Chen CT. Endoscopic approach to medial orbital wall fractures. Facial Plast Surg Clinics of North Am 2006;14: 17–23.
31. Lund VJ, Larkin G, Fells P, et al. Orbital decompression for thyroid eye disease: a comparison of external and endoscopic techniques. J Layngol Otol 1997;111:1051–5.

Endoscopic Surgery of the Paranasal Sinuses

6

Seth J. Isaacs, MD / Robert M. Kellman, MD

OUTLINE

INTRODUCTION

In the early 1900s, Hirschmann described the first endoscopic evaluation of the nose and paranasal sinuses. However, this was not widely accepted due to inadequate visualization and instrumentation. The Caldwell-Luc procedure remained the treatment of choice for chronic sinusitis. In the 1960s, Hopkins developed rod telescopic lenses, which greatly increased the light intensity and improved the overall resolution. With the development of endoscopes, Messerklinger was able to describe the anatomy of the paranasal sinuses and the pathophysiology of diseases affecting these regions. In the 1980s, Stamberger and Kennedy applied these principles and developed endoscopic surgical techniques.[1] Over the last 30 years, functional endoscopic surgery has become the standard of care for the treatment of inflammatory diseases involving the paranasal sinuses. Its role in the management of sinonasal and skull-base neoplasms continues to expand due to advances in equipment, imaging modalities, and navigation software.

Functional endoscopic sinus surgery is predicated on the notion that chronic sinus disease can be reversed if the sources of obstruction at the outlet of a sinus can be removed. The mucosa of the obstructed sinuses can return to normal function if proper aeration and mucociliary clearance are re-established. A thorough understanding of the complex anatomy of the paranasal sinuses and proper identification of surgical landmarks are paramount for safe and successful endoscopic surgery.

SURGICAL ANATOMY

A fundamental principle of sinonasal function is having open drainage pathways between the paranasal sinuses and the nose, making the anatomy important to understand. Along the lateral nasal wall are the inferior, middle, and superior turbinates with their associated meatuses. The inferior turbinate is the largest of the turbinates and can be easily identified on anterior rhinoscopy extending posteriorly along the floor of the nasal cavity. The nasolacrimal duct drains into the inferior meatus approximately 1 cm from the anterior edge of the inferior turbinate. The middle turbinate attaches to the skull base anteriorly (Figure 1). As it courses posteriorly it comes to lie in the coronal plane before attaching to the lamina papyracea or medial orbital wall. This horizontal portion of the middle turbinate is the ground lamella and separates the drainage pathways of the anterior and posterior ethmoid air cells. The maxillary sinus, frontal sinus, and anterior ethmoid air cells drain into the middle meatus. The middle turbinate may have variable

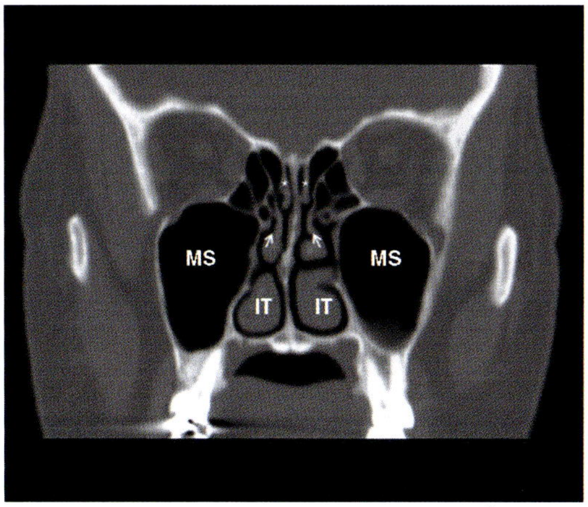

FIGURE 1 • A coronal CT image through the midportion of the maxillary sinus (MS) demonstrating the position of the inferior, middle, and superior turbinates within the nasal cavity. IT, inferior turbinate; arrow, middle turbinate; asterisk, superior turbinate.

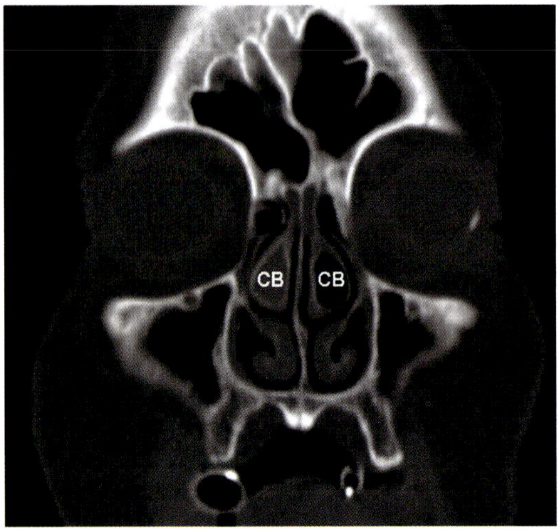

FIGURE 2 • A coronal CT depicting bilateral concha bullosa (CB). There is fluid within the aerated right middle turbinate indicating a similarity to a functioning paranasal sinus. With significant enlargement of these structures, the middle meatus can become obstructed.

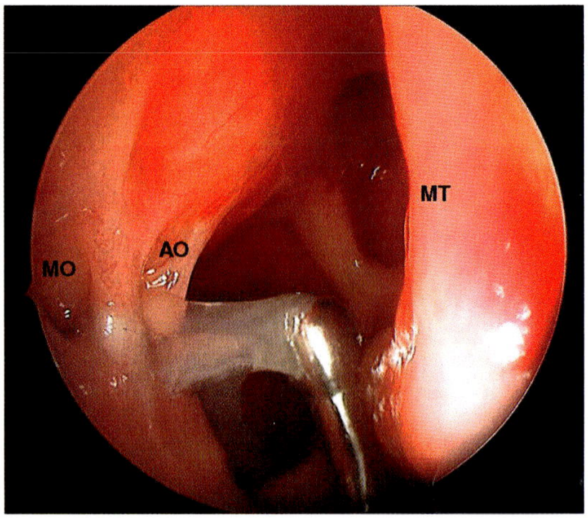

FIGURE 4 • Endoscopic view of the middle meatus demonstrates the presence of both the natural maxillary ostium (MO) and accessory ostium (AO). The presence of an accessory ostium may lead to recirculation of mucus from the accessory ostium to the natural ostium. MT, middle turbinate.

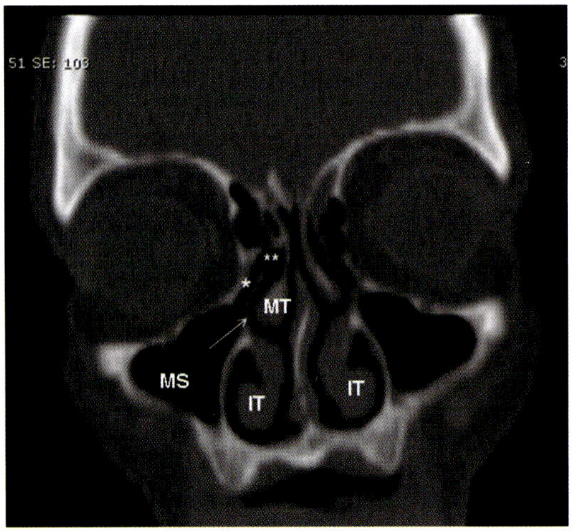

FIGURE 3 • A coronal CT image through the middle meatus. The single asterisk marks the infundibulum into which the maxillary, anterior ethmoid, and frontal sinuses drain. The double asterisk identifies the hiatus semilunaris, which is the space between the uncinate process (arrow) anteriorly and ethmoid bulla posteriorly. MT, middle turbinate; IT, inferior turbinate; MS, maxillary sinus.

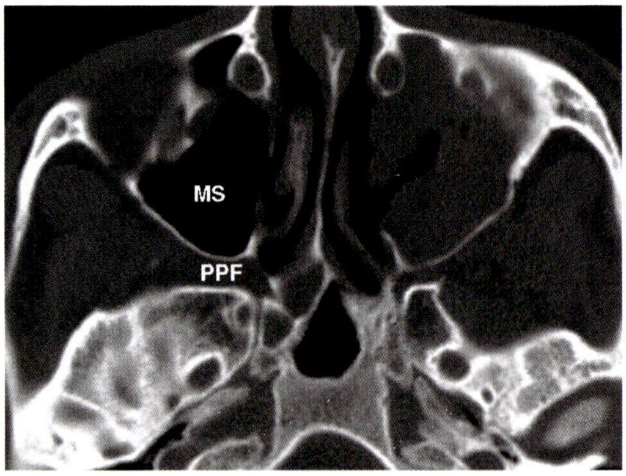

FIGURE 5 • Axial CT image depicts the position of the pterygopalatine fossa (PPF) behind the posterior wall of the maxillary sinus (MS). This important space houses many important neurovascular structures, including the internal maxillary artery.

morphology. It can have a paradoxical curvature, which may obstruct the middle meatus. Additionally, it may become aerated and form a concha bullosa. Behaving as a functional sinus, it can become a site of sinus disease or when sufficiently enlarged serve as a source of obstruction (Figure 2). The superior turbinate lies posterior and superior to the middle turbinate. Within the superior meatus or sphenoethmoid recess, drain the posterior ethmoid air cells and sphenoid sinus.

The ostiomeatal complex is an important structure that plays a role in the pathophysiology of maxillary, ethmoid, and frontal sinus disease, by providing a common outflow tract within the middle meatus. The uncinate process is a sickle-shaped bone that lies in close proximity to the frontal recess anterosuperiorly. As it courses posteroinferiorly, it rotates from a parasagittal plane to lie horizontally where it attaches to the lacrimal bone laterally and the inferior turbinate. Lateral to the uncinate process is a three-dimensional (3D) space known as the infundibulum. Between the posterior free edge of the uncinate process and the anterior wall of the ethmoid bulla is the hiatus semilunaris (Figure 3). The superior attachment of the uncinate process is variable and determines the course of the frontal sinus outflow tract.[2]

Posterior to the uncinate process along the lateral nasal wall are the anterior and posterior fontanels, which form the natural ostia and accessory ostia of the maxillary sinus, respectively. An accessory ostium is present in 10 percent to 40 percent of the population (Figure 4).[3] The maxillary sinus is pyramidally shaped with its base oriented medially. The boundaries of the maxillary sinus are the orbital floor superiorly, zygoma laterally, lateral nasal wall medially, hard palate and alveolus inferiorly, and the pterygopalatine fossa posteriorly.[4,5] Important neurovascular structures in this region include the infraorbital nerve, which lies along the roof of the maxillary sinus, and the internal maxillary artery within the anterior compartment of the pterygopalatine fossa (Figure 5).[6]

The anatomy of the ethmoid region is the most variable and directly affects the drainage pathways of the other paranasal sinuses. The anterior and posterior ethmoids are separated by the basal lamella, which is

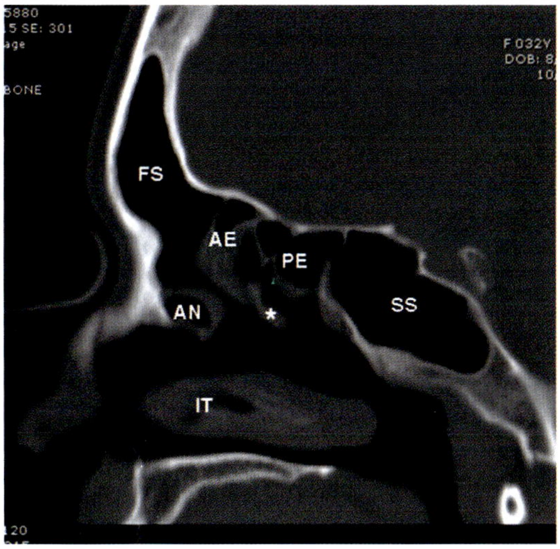

FIGURE 6 • Sagittal CT image depicts the anatomy of the ethmoid and frontal sinuses. The basal lamella (asterisk) separates the anterior ethmoids (AE) from the posterior ethmoids (PE). The agger nasi cell (AN) is the anterior-most ethmoid air cell. The anatomy of the ethmoid air cells affects the course and patency of the frontal sinus (FS) outflow. IT, inferior turbinate; SS, sphenoid sinus.

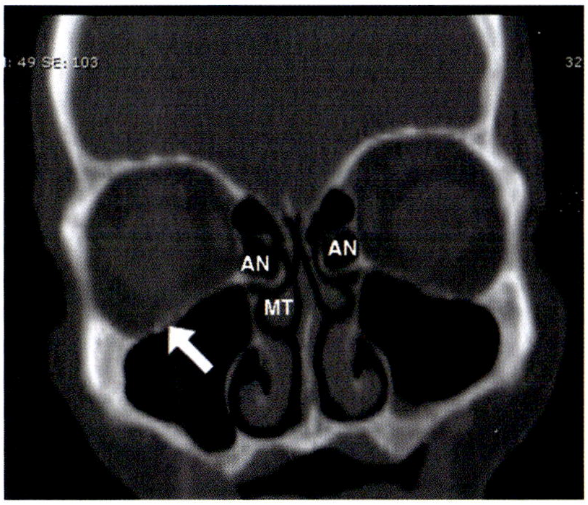

FIGURE 7 • Coronal CT image shows the position of the agger nasi cell superior to the anterior-most aspect of the middle turbinate.

the horizontal portion of the middle turbinate. The agger nasi cell is the most anterior of the ethmoid air cells and is located along the lateral nasal wall anterior to the middle turbinate. Its posterior wall forms the anterior border of the frontal sinus outflow tract (Figure 6). A single agger nasi cell is only one variation of the anatomy within the frontal recess region (Figure 7). Frontal ethmoid cells are located superior to the agger nasi and are classified based on the number of cells and their proximity to the frontal sinus. The most common classification was described by Kuhn in 1995 (Table 1).[7] These ethmoid cells influence the course, dimension, and patency of the frontal recess and outflow tract (Figure 8).

A more consistent structure within the anterior ethmoids is the ethmoid bulla. It lies directly posterior to the free edge of the uncinate process and forms the posterior extent of the hiatus semilunaris. When the anterior face of this cell extends to the skull base, it forms the posterior border of the frontal recess. Otherwise the space superior to the ethmoid bulla is known as the suprabullar recess. Posterior to the bulla is the basal lamella of the middle turbinate. These structures are separated by a space known as the retrobullar recess. The natural ostium of the ethmoid bulla is within the medial aspect of the retrobullar space. The Haller cell is an infraorbital ethmoid cell located along the inferior and medial aspect of the orbit (Figure 9). If large enough, this structure can create an obstruction of the maxillary ostium.

TABLE 1 Kuhn Classification of Frontal Ethmoid Cells

Type 1	Single cell above the agger nasi cell
Type 2	Tier of two or more cells above the agger nasi cell
Type 3	Single cell pneumatizing into the frontal sinus
Type 4	Isolated cell within the frontal sinus

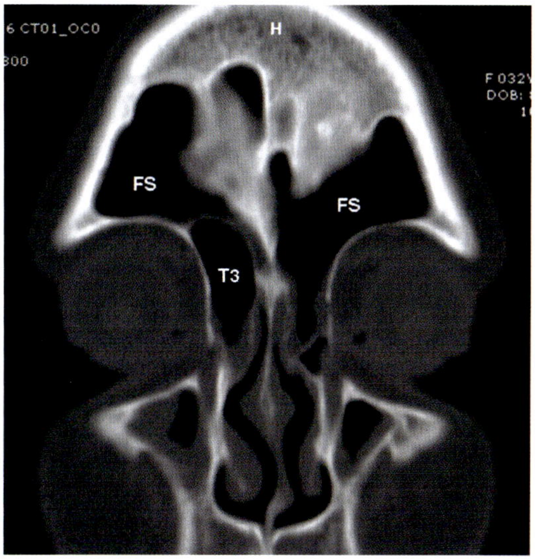

FIGURE 8 • Coronal CT image demonstrates a Type 3 frontal ethmoid cell (T3), which extends into the frontal sinus (FS).

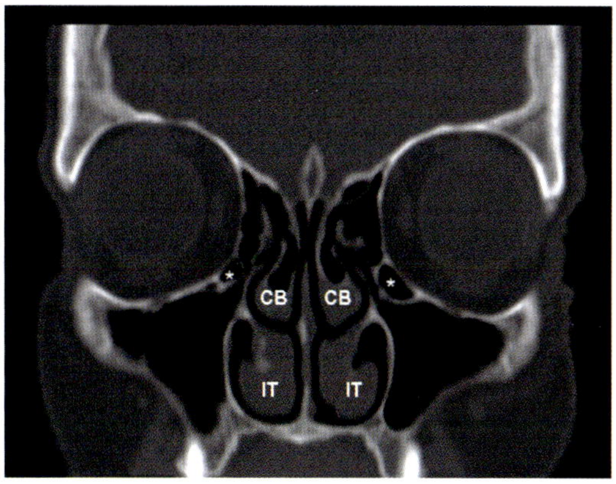

FIGURE 9 • Coronal CT image depicts a patient with bilateral Haller cells (asterisks). If sufficiently enlarged, these structures may obstruct the maxillary ostia. The patient also has bilateral concha bullosa (CB). IT, inferior turbinate.

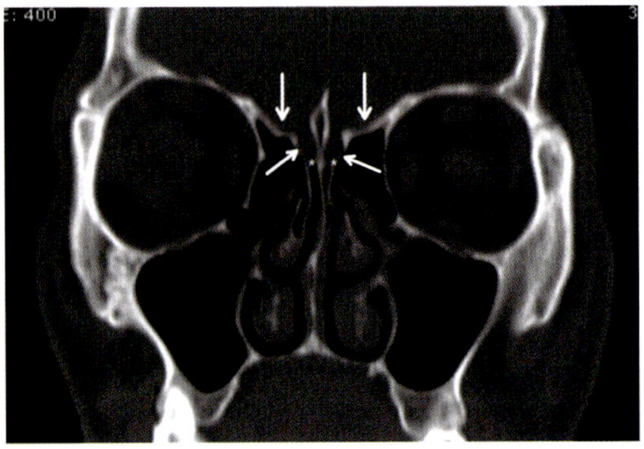

FIGURE 10 • Coronal CT image depicting the anterior skull base. Between the cribriform plate (asterisk) and the fovea ethmoidalis (thick arrow) is the lateral lamella (thin arrow). This portion of the skull base is the thinnest and the most common site of an iatrogenic CSF leak.

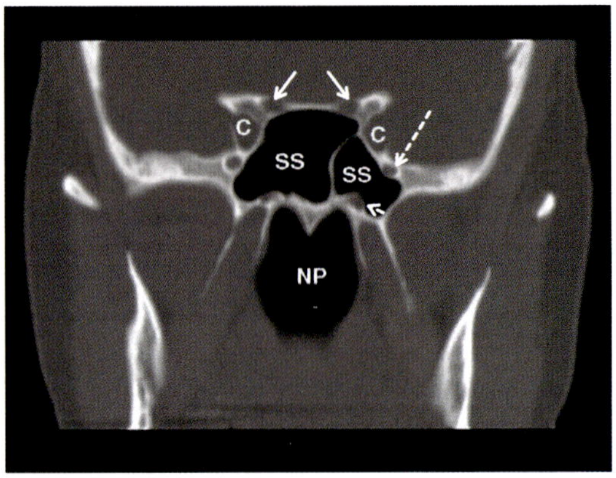

FIGURE 12 • A coronal CT image through the sphenoid sinuses (SS). They are bordered by several important neurovascular structures including the optic nerve (solid arrow), internal carotid artery (C), foramen rotundum (dashed arrow), and vidian canal (arrow head). Within the foramen rotundum is the second branch of the trigeminal nerve. The vidian canal houses the vidian nerve, a parasympathetic branch of the facial nerve.

TABLE 2 Keros Classification of the Lateral Lamella

Type 1	Less than 3 mm
Type 2	Between 4 mm and 7 mm
Type 3	Between 8 mm and 16 mm

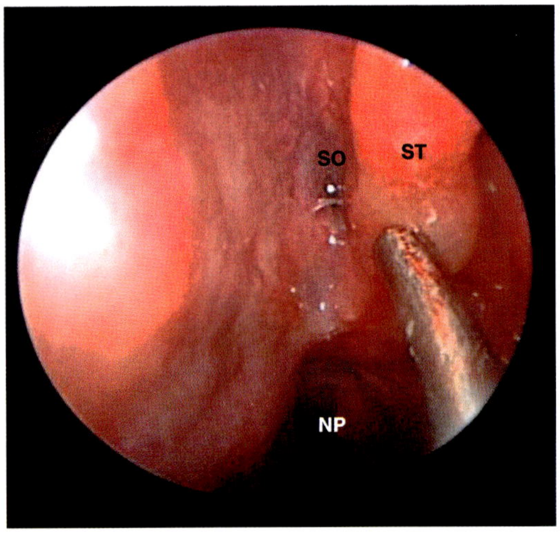

FIGURE 11 • Endoscopic view of the posterior nasal cavity demonstrates the position of the sphenoid ostium (SO) medial to the superior turbinate (ST). The sphenoid sinus lies superior to the choana and nasopharynx (NP).

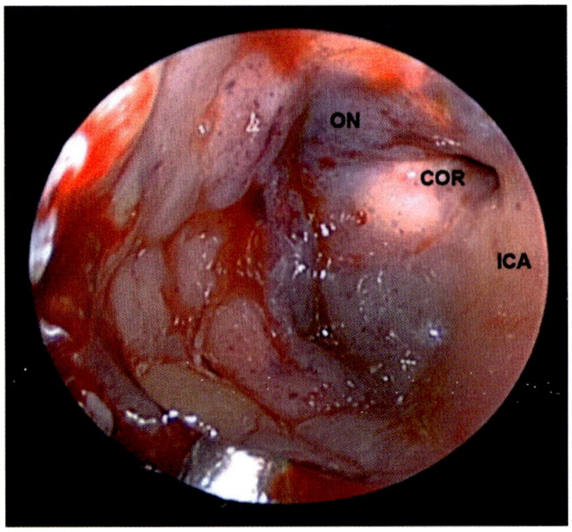

FIGURE 13 • An endoscopic view into the left sphenoid sinus shows the anatomy of the lateral sphenoid wall. Between the optic nerve (ON) and the internal carotid artery (ICA) is a space named the carotid-optic recess (COR).

The posterior ethmoid cells extend from the basal lamella anteriorly to the anterior face of the sphenoid sinus posteriorly. They are variable in number and shape. Pre-operative planning with CT imaging in both coronal and sagittal planes aids in the proper identification of these cells (Figure 6).

The anterior skull base forms the superior limit of the posterior ethmoids and slopes inferiorly as it extends posteriorly. It is critical to adequately evaluate the skull base prior to surgery and clearly identify it during the ethmoidectomy. The anterior skull base is composed of the cribriform plate at the midline and the fovea ethmoidalis laterally. Between these two structures is the lateral lamella, which is the thinnest portion of the skull base and most common site of an iatrogenic CSF leak (Figure 10). The length of lateral lamella is variable and can be described based on the Keros classification (Table 2).[8]

The paired sphenoid sinuses are the most posterior of the paranasal sinuses (Figure 6). The sphenoid sinus ostium is located medial to the superior turbinate in 83 percent of the population and drains into the sphenoethmoid recess (Figure 11).[9] The sphenoid sinus is bordered by several important neurovascular structures, including the optic nerve, internal carotid artery, and the cavernous sinus (Figures 12 and 13). The roof of the sphenoid sinus forms the floor of the sella, which houses the pituitary gland. The endoscopic hypophysectomy is based on this relationship. Onodi cells are extensions of the posterior ethmoids that have pneumatized into the superolateral aspect of the sphenoid sinus (Figure 14).

The frontal sinuses are contained within the frontal bone and are bordered by the cranial vault posteriorly and the orbits inferolaterally. The frontal recess is a 3D potential space that connects the narrow frontal ostium to the middle meatus.[10] The shape, course, and patency

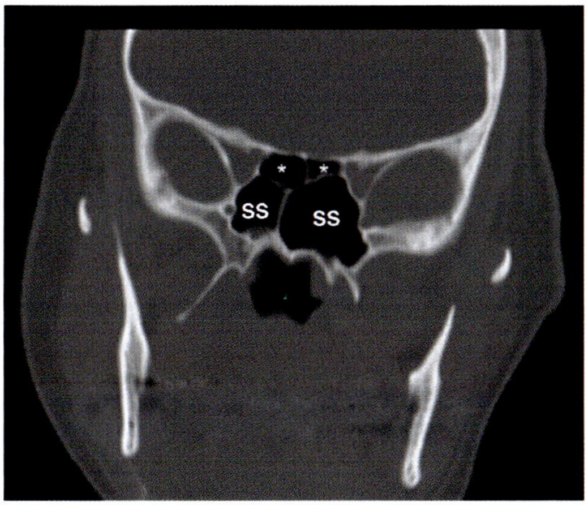

FIGURE 14 • The coronal CT image of a patient with bilateral Onodi cells (asterisks), which are posterior ethmoid cells that pneumatize along the posterior and superior surface of the sphenoid sinus (SS).

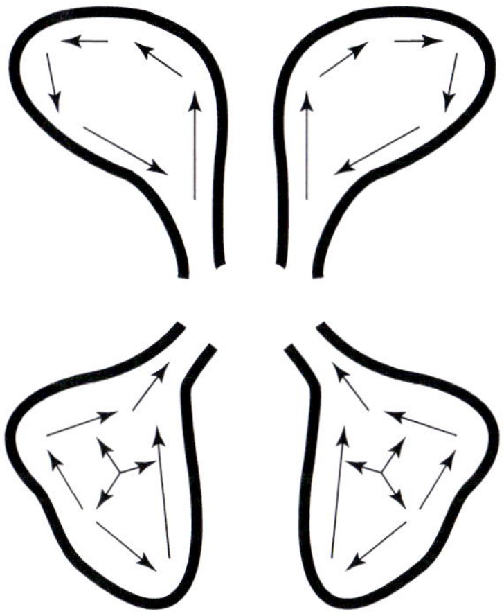

FIGURE 15 • The mucociliary clearance pattern of the paranasal sinuses is an important mechanism in the pathophysiology of sinus disease. The frontal sinuses drain in a counterclockwise direction beginning along the medial wall. The maxillary sinuses drain from all directions toward the natural ostium. The presence of an accessory ostium may result in recirculation.

of the frontal recess depend upon surrounding structures.[2] These include the agger nasi and frontal ethmoid cells anteriorly, and the bullar cells posteriorly (Figure 6).[11] Variable superior attachment of the uncinate process to either the skull base or lateral nasal wall also determines whether the frontal sinus will drain laterally into the infundibulum or medially into the middle meatus, respectively.

PATHOPHYSIOLOGY

The concept of the ostiomeatal unit is based on the specific patterns of mucociliary clearance through the natural ostia of the maxillary and frontal sinuses (Figure 15). The presence of accessory ostia and even surgically created ostia can lead to recirculation within the affected sinus and a persistence of symptoms (Figure 4). Inflammation and mucosal hypertrophy within the ostiomeatal unit can lead to mucociliary obstruction and recurrent infections of the paranasal sinuses (Figure 16).

In 2003, the Rhinosinusitis Task Force sought to characterize the spectrum of inflammatory disease that affects the paranasal sinuses.[12]

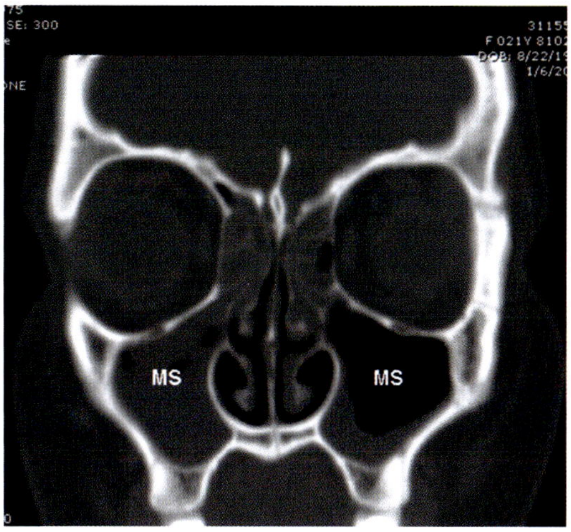

FIGURE 16 • Coronal CT image shows significant mucosal inflammation within both maxillary and ethmoid sinuses. The right maxillary sinus (MS) has more advanced disease.

TABLE 3 Classification of Rhinosinusitis Based on Duration of Symptoms

Acute	Up to 4 weeks
Subacute	4 to 12 weeks
Chronic	Greater than 12 weeks
Recurrent acute	4 or more episodes per year with return to baseline between
Acute exacerbation of chronic disease	Worsening of chronic rhinosinusitis with return to baseline

Acute and chronic rhinosinusitis were characterized based on duration of symptoms (Table 3). Chronic rhinosinusitis refractory to medical management remains the most common indication for functional endoscopic sinus surgery. The etiology of the inflammation associated with this disease process is multifactorial. Host factors include allergy, immunodeficiency, congenital, anatomical, neoplastic, and mucociliary dysfunction. Environmental factors include microorganisms, inhaled irritants, and trauma. The purpose of surgery is to remove the anatomical obstruction of the sinus drainage pathways and restore mucociliary clearance. The patient must be informed that there may be other factors that predispose him or her to persistent or recurrent disease.

Advances in endoscopic surgery have expanded its use beyond inflammatory disease of the sinonasal tract toward a growing role in skull-base surgery. Applications of endoscopic skull base surgery have included transsphenoidal approaches to the pituitary gland, repair of CSF leaks, and resection of clival lesions.[13] Endoscopic resection of benign and malignant neoplasms of the sinonasal tract is replacing open procedures and reducing the associated morbidity.[14,15]

Absolute indications for sinus surgery, either endoscopic or open approaches, include chronic rhinosinusitis refractory to medical treatment, orbital cellulitis, subperiosteal abscess, mucoceles, CSF leaks, and invasive or allergic fungal sinusitis. Relative contraindications for endoscopic sinus surgery include meningitis, intracranial abscesses, recurrent frontal sinusitis with ostial stenosis, lateral frontal disease,

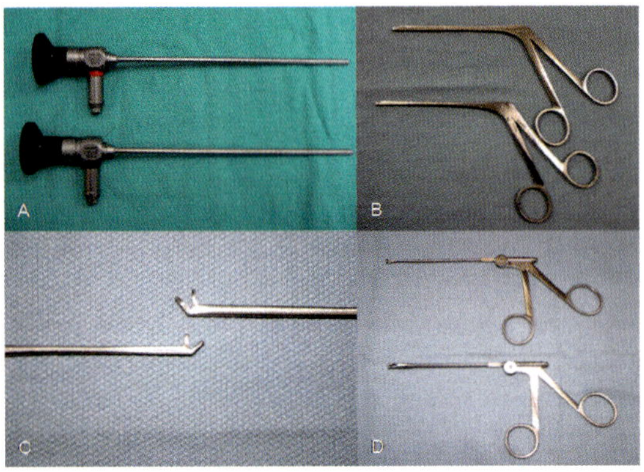

FIGURE 17 • A, Straight and angled telescopes; B, Straight through-cutting forceps; C, 45-degree through-cutting instruments; D, back-biting forceps.

malignant neoplasms with significant orbital or skull-base erosion, and smoking.

INSTRUMENTATION

Adequate visualization and precision instrumentation are paramount for safe and successful endoscopic sinus surgery. Angled telescopes (0, 30, 45, and 70 degrees) are necessary tools to provide an assessment of narrow recesses including the frontal sinus outflow tract and lateral extent of the maxillary sinus (Figure 17, A). This permits a broader field of dissection and improved capability to completely eradicate the disease process. Over the last 10 years, powered microdebriders have become necessary instruments for endoscopic sinus surgery.[16] They are effective at precisely removing diseased tissue with the preservation of normal surrounding mucosa. This prevents potential scarring and stenosis along the drainage pathways of the sinuses. However, these instruments can cause a significant amount of damage along the orbit and skull base. Through-cutting instruments remain invaluable tools for focused resection of diseased mucosa (Figure 17, B–D). Advances in self-irrigating drills and curved through-cutting instruments have expanded the role of endoscopic frontal sinusotomies as an alternative to frontal sinus obliteration for the treatment of chronic frontal sinusitis.

Endoscope cleaners prevent the accumulation of blood and mucus along the lens. This effectively improves visibility and overall safety of sinus surgery. In addition, it allows the surgery to progress at a more efficient pace and reduces anesthesia time. Image guidance systems have improved the accuracy of revision endoscopic sinus surgery by augmenting the identification of altered intranasal landmarks. The addition of MRI and CT fusion images has assisted intraoperative decision making during the endoscopic resection of paranasal sinus and skull-base neoplasms.

RADIOGRAPHIC ANALYSIS

Computed tomography (CT) has replaced plain x-ray films as the primary imaging modality of the paranasal sinuses. They aid in the diagnosis of sinus pathology and are essential for pre-operative planning.[17] Most institutional protocols use a multislice helical CT scanner to produce images in an axial plane from 0.5-mm to 3-mm intervals. Coronal reconstructions are preferred because they closely represent the sinonasal anatomy encountered during endoscopic surgery. Recent studies have demonstrated that sagittal imaging of the paranasal sinuses provides additional information that increases the examiner's confidence and alters surgical decision making.[18,19] Triplanar imaging is essential for the conceptualization of 3D sinus anatomy from two-dimensional CT data.

During the pre-operative evaluation, Kennedy describes seven anatomic regions that are essential for preventing surgical complications. They include the (1) anterior skull base, specifically the lateral lamella; (2) medial orbital wall; (3) anterior ethmoid artery and its proximity to the skull base; (4) height of the posterior ethmoid cells; (5) Haller cells and accessory maxillary ostia; (6) sphenoid sinus and surrounding neurovascular structures; and (7) frontal sinus outflow tract.[1]

Magnetic resonance imaging (MRI) is not a routine part of the pre-operative assessment because it does not provide adequate bony definition and overestimates the degree of mucosal thickening. MRI scans are essential for the evaluation of intranasal and skull-base neoplasms. They help to differentiate tumor from retained mucus in the sinuses and to identify involvement of the orbit or dura.

PATIENT ASSESSMENT

The patient evaluation begins with a thorough history including the characteristics, onset, and duration of symptoms associated with rhinosinusitis (Table 4).[12] An assessment of prior treatment regimens is important. A major indication for endoscopic sinus surgery is chronic rhinosinusitis that is refractory to maximal medical management including systemic steroids, long-term antibiotics, and nasal preparations. The medical history should include allergies, immunodeficiency disorders, and any associated pulmonary diseases, including COPD, asthma, and bronchitis. Prior sinus or nasal surgeries should be documented.

Diagnostic nasal endoscopy in the office is performed in a systematic manner with a 30-degree endoscope. Kennedy describes a three-pass method.[1] During the first pass, the endoscope is maneuvered along the nasal floor posteriorly toward the nasopharynx, which allows visualization of the septum, inferior turbinate, and nasopharynx. The second pass is made between the middle and inferior turbinates medially into the sphenoethmoidal recess. The sphenoid ostium can be identified. The final pass involves the middle meatus with visualization of the ostiomeatal complex, infundibulum, hiatus semilunaris, and ethmoid bulla. The extent of disease in the nasal cavity will determine which structures can be identified on routine endoscopy.

The most important aspect of the pre-operative assessment is patient education. In the majority of presentations, functional endoscopic sinus surgery is an elective procedure. Both the risks and benefits of the

TABLE 4 Symptoms of Rhinosinusitis

MAJOR FACTORS	MINOR FACTORS
Facial pain/pressure	Headache
Nasal obstruction	Fever
Purulent drainage	Halitosis
PND (Postnasal discharge)	Fatigue
Anosmia/hyposmia	Dental pain
Purulence in nasal cavity	Cough
Fever (acute)	Ear pressure

TABLE 5 Complications Associated with Endoscopic Sinus Surgery

MAJOR	MINOR
CSF leak	Adhesions
Orbital hematoma	Epistaxis
Diplopia	Hyposmia
Blindness	Periorbital ecchymosis
Meningitis	Periorbital emphysema
Nasolacrimal duct injury	Dental injury
Anosmia	
Carotid injury	

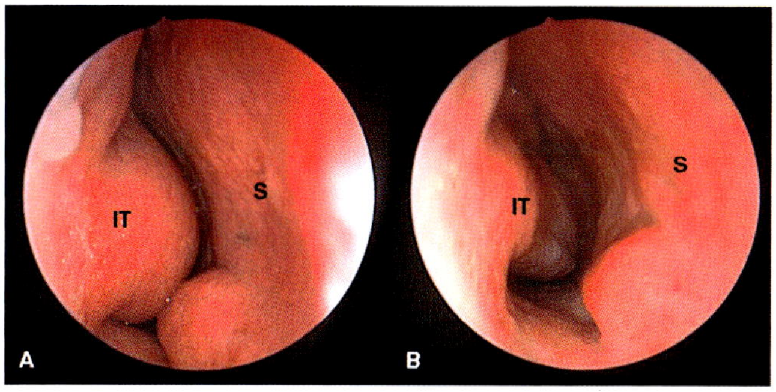

FIGURE 18 • Endoscopic view of the anterior nasal cavity. A, Inferior turbinate (IT) and septum (S) prior to decongestion. B, post-decongestion.

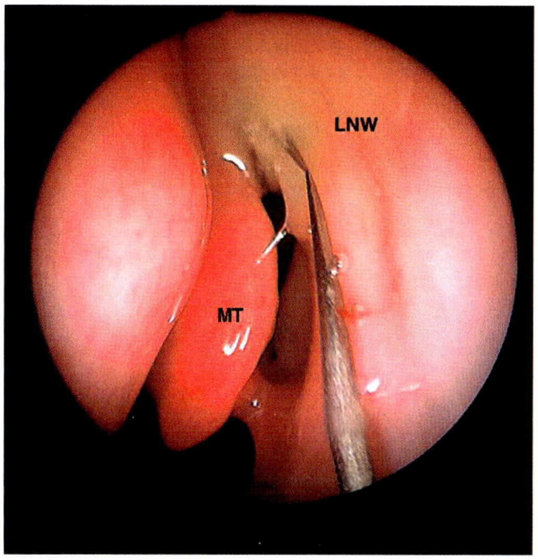

FIGURE 19 • Application of local anesthesia along the lateral nasal wall (LNW) via endoscopic guidance. MT, middle turbinate.

surgery must be clearly explained. Chronic rhinosinusitis is an inflammatory process of multiple etiologies. Surgery addresses only the anatomical causes of sinonasal obstruction, and therefore is not an absolute cure. The most serious complications of endoscopic sinus surgery include injuries to the brain and orbit secondary to their intimate anatomical relationship to the paranasal sinuses (Table 5).

ANESTHESIA

Functional endoscopic sinus surgery can be performed under either local anesthesia with sedation or general anesthesia. Proponents of local argue that identification of intracranial or intraorbital injuries is quicker with the patient awake. As endoscopic procedures have become more extensive, general anesthesia has become the more accepted method. Topical decongestion and infiltration of local anesthetics are still performed prior to surgery, however. Typically, the intranasal decongestion is begun in the pre-operative holding area with the instillation of oxymetolazine one hour prior to surgery and repeated at 15-minute intervals (Figure 18). Once the patient is in the operating room, cotton pledgets with topical adrenaline (1:10,000) are placed along the inferior turbinate and into the middle meatus.

Infiltration of local anesthesia can be performed either directly with a headlight or endoscopically. One percent lidocaine with 1:100,000 epinephrine is injected 1 cm anterior to the superior attachment of the middle turbinate, in the region of the anterior ethmoid artery (Figure 19). A sphenopalatine block can be performed either intranasally or intraorally via the greater palatine artery.[20] The intranasal injection is made with a spinal needle at the junction of the basal lamella and lateral nasal wall. The greater palatine block is performed by infiltrating the anesthetic within the greater palatine foramen medial to the first maxillary molar. It is important to bend the needle at 2 cm, as this length will prevent entry into the infraorbital foramen through the pterygopalatine fossa.[20]

ENDOSCOPIC TECHNIQUE

The keys to successful endoscopic sinus surgery include complete removal of diseased tissue and creation of adequate sinus ventilation. It is important that all osteitic bone be thoroughly removed while at the same time preserving sinonasal mucosa. Ultimately, the type of pathology and degree of disease will determine the extent of surgery.

After adequate decongestion and local anesthesia, the procedure is begun by identifying the middle turbinate and uncinate process with a 0-degree endoscope. The middle turbinate can be gently medialized with a Freer elevator. Care must be taken not to destabilize the middle turbinate and fracture the anterior skull base with this maneuver. The uncinate process and ethmoid bulla should be clearly visualized at this point. A backbiter can be placed behind the posterior free edge of the uncinate process into the infundibulum. Beginning at the most inferior extent, the uncinate process can be removed in a superior direction with a combination of the backbiter and microdebrider. The anterior extent of this resection is the hard lacrimal bone. This will avoid injury to the nasolacrimal duct and prevent post-operative epiphora.

Removal of the uncinate process allows visualization of the infundibulum. A 30-degree endoscope angled laterally is used to identify the maxillary ostia (Figure 20). The natural ostium is oval shaped and usually located behind the inferior 1/3 of the uncinate process. A curved suction can be used to dilate the ostium. The maxillary antrostomy can be enlarged inferiorly and posteriorly using through-cutting instruments and the microdebrider (Figure 21). If an accessory ostium is present, it is typically located more posteriorly. It is important to connect the accessory ostium to the natural ostium in order to prevent recirculation (Figure 4).

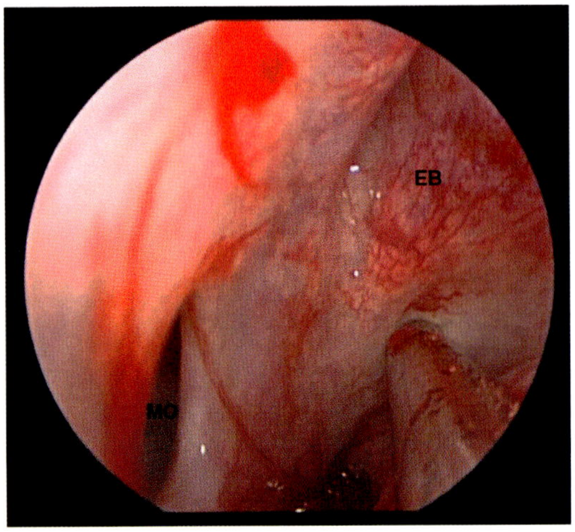

FIGURE 20 • Endoscopic view of into the right middle meatus and infundibulum after removal of the uncinate process. The natural ostium of the maxillary sinus (MO) can be visualized along the lateral nasal wall. The ethmoid bulla (EB) determines the posterior extent of the infundibulum.

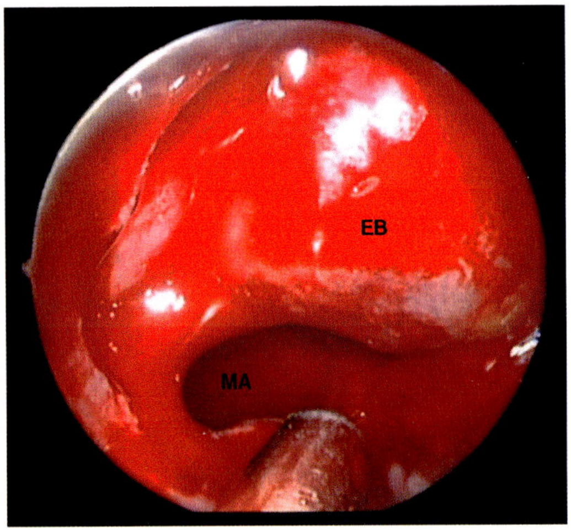

FIGURE 21 • Endoscopic view of the right middle meatus after creation of a large maxillary antrostomy (MA). The ethmoid bulla (EB) is still intact.

The most prominent structure within the middle meatus is the ethmoid bulla. This cell can be entered medially and inferiorly using a sinus seeker or curette. The lamina papyracea must be identified in order to prevent entrance into the orbit. This is the lateral boundary of the dissection during an ethmoidectomy. The basal lamella serves as the posterior limit of the anterior ethmoidectomy (Figure 22). The middle turbinate should be preserved in most cases. One important exception is a concha bullosa, or aerated middle turbinate. In this circumstance, a sickle knife may be used to penetrate into the anteroinferior aspect of the turbinate. Then through cutting instruments or a microdebrider can be used to resect the lateral half of the turbinate. Maintaining the skull base attachment of the middle turbinate prevents destabilization and lateralization of the turbinate, which would result in obstruction of the ostiomeatal unit.

The posterior ethmoidectomy proceeds by penetrating through the medial and inferior aspect of the basal lamella with a small Frazier tip suction. Dissection within this region should remain low until the skull base and anterior wall of sphenoid sinus are identified. Superior

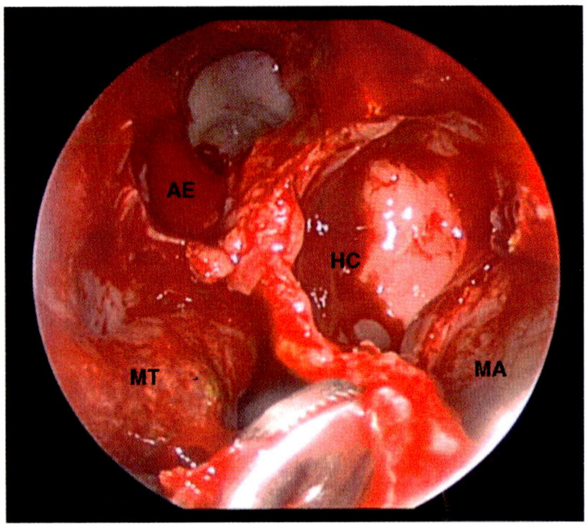

FIGURE 22 • Endoscopic view of the left middle meatus showing a large maxillary antrostomy (MA) and an anterior ethmoidectomy (AE). A Haller cell (HC) was resected in order to provide access to the area. MT, middle turbinate.

partitions may be removed from a posterior to anterior direction using 45-degree through-cutting forceps. Care must be taken not to interrupt the anterior ethmoid neurovascular bundle along the anterior skull base. A complete ethmoidectomy is accomplished when the point of intersection of the sphenoid sinus, skull base, and lamina papyracea is identified.

The ostium to the sphenoid sinus can be identified after resecting the lower 1/3 of the superior turbinate. For isolated sphenoid disease, the ostium can be approached medial to the superior turbinate, approximately 1 cm above the entrance to the choana. A sphenoid sinusotomy is created by enlarging the natural ostium with through-cutting punches and forceps (Figure 13).

The shape, course, and patency of the frontal recess depend on surrounding ethmoid cells. Therefore, successful frontal sinusotomy requires adequate resection of these structures. Agger nasi and frontal ethmoid cells define the anterior limit of the frontal recess. A frontal sinusotomy begins with resection of these cells if present (Figure 23). A frontal sinus seeker can be passed gently through the frontal ostium into the sinus. A microdebrider can be used conservatively to enlarge the ostium (Figure 24). Mucosal preservation is paramount in order to prevent stenosis within the frontal recess and ostium.

POST-OPERATIVE CARE

The post-operative management after endoscopic sinus surgery is variable and dependent on the surgeon's preferences. Important aspects to consider include packing, antibiotics, steroids, irrigation, and debridement.

Post-operative nasal packing is a controversial topic. Materials range from Telfa® and Silastic® sheets to resorbable packs including MeroGel®, Seprafilm®, and Floseal®. Proponents of nasal packing within the middle meatus cite state the importance of maintaining the middle turbinate in the medialized position and preventing synechiae to the lateral nasal wall.[21] However, recent studies have demonstrated that there is no significant effect of packing on the incidence of adhesions within the middle meatus during the early post-operative period.[22]

In-office debridement is becoming a standard component of the post-operative management. The rationale for debridement includes removing crusts that may serve as a source of infection and synechiae formation. Several randomized control trials have demonstrated that

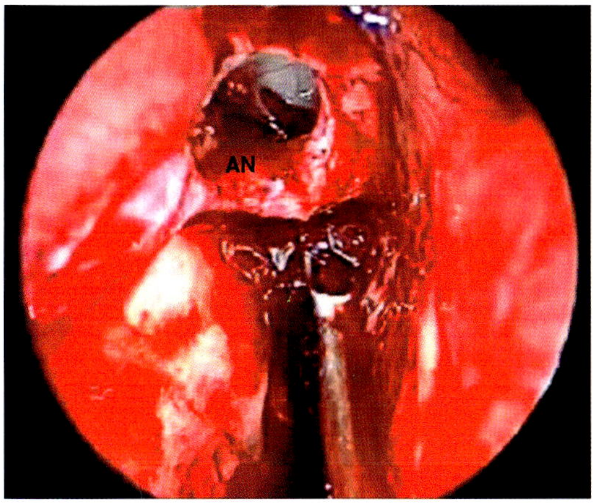

FIGURE 23 • Endoscopic view of the right frontal recess region. The agger nasi cell (AN) is being resected with a curette.

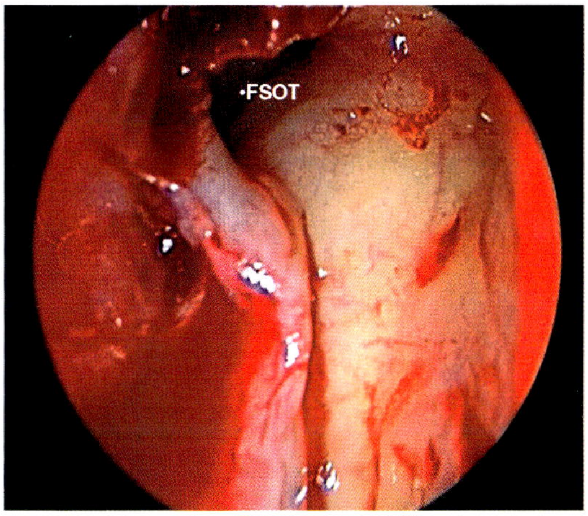

FIGURE 24 • Endoscopic view of the superior middle meatus. The agger nasi cell was already removed with a curette. FSOT, frontal sinus outflow tract.

aggressive debridement reduces the formation of synechiae and increases the rate at which the patient's nasal symptoms improve.[23]

The benefit of post-operative antibiotics and steroids include reduced mucosal infection and inflammation as well as improved wound healing. The side effects of steroids, however, must be considered, especially immune suppression and cushingoid features. Nasal irrigation with hypertonic saline has proven to be an important component of the post-operative management after endoscopic sinus surgery. The primary function of irrigation is mechanical debridement, effectively removing clots and crusting that would otherwise serve as conduits for synechiae formation. Secondarily, the hypertonicity improves the mucociliary clearance of regenerating mucosa.[24] Currently, several commercially available irrigation systems exist.

CONCLUSION

Endoscopic surgery of the paranasal sinuses has become the standard of care for the management of chronic rhinosinusitis refractory to medical management. A thorough understanding of sinonasal anatomy is paramount for safe and successful surgery. Effective endoscopic sinus surgery is predicated upon the removal of anatomic obstruction in order to restore the natural drainage and ventilation of the sinuses. Despite the multifactorial etiology of chronic rhinosinusitis, proper surgical technique permits improvement of symptoms in nearly 98 percent of patients over the long term.[25] Through advances in instrumentation, imaging, and neuronavigation, the role of endoscopic sinus surgery is expanding beyond inflammatory disease of the paranasal sinuses toward approaches to the skull base.

References

1. Bailey Bea. *Head & Neck Surgery — Otolaryngology*. Philadelphia: Lippincott Williams & Wilkins, 2006:2826.
2. Ercan I, Cakir BO, Sayin I, Basak M, Turgut S. Relationship between the superior attachment type of uncinate process and presence of agger nasi cell: a computer-assisted anatomic study. *Otolaryngol Head Neck Surg* 2006;134:1010–1014.
3. May M, Sobol SM, Korzec K. The location of the maxillary os and its importance to the endoscopic sinus surgeon. *Laryngoscope* 1990;100:1037–1042.
4. Erdogan N, Unur E, Baykara M. CT anatomy of pterygopalatine fossa and its communications: a pictorial review. *Comput Med Imaging Graph* 2003;27:481–487.
5. Daniels DL, Mark LP, Ulmer JL et al. Osseous anatomy of the pterygopalatine fossa. *Am J Neuroradiol* 1998;19:1423–1432.
6. Choi J, Park HS. The clinical anatomy of the maxillary artery in the pterygopalatine fossa. *J Oral Maxillofac Surg* 2003;61:72–8.
7. Bent JP, 3rd, Spears RA, Kuhn FA, Stewart SM. Combined endoscopic intranasal and external frontal sinusotomy. *Am J Rhinol* 1997;11:349–354.
8. Keros P. Contribution to the Study of the Anterior Cranial Cavity. *Rad Med Fak Zagrebu* 1963;11:225–244.
9. Kim HU, Kim SS, Kang SS, Chung IH, Lee JG, Yoon JH. Surgical anatomy of the natural ostium of the sphenoid sinus. *Laryngoscope* 2001;111:1599–602.
10. Daniels DL, Mafee MF, Smith MM et al. The frontal sinus drainage pathway and related structures. *Am J Neuroradiol* 2003;24: 1618–27.
11. Wormald PJ. The agger nasi cell: the key to understanding the anatomy of the frontal recess. *Otolaryngol Head Neck Surg* 2003;129:497–507.
12. Benninger MS, Ferguson BJ, Hadley JA et al. Adult chronic rhinosinusitis: definitions, diagnosis, epidemiology, and pathophysiology. *Otolaryngol Head Neck Surg* 2003;129:S1–32.
13. Rodziewicz GS, Kelley RT, Kellman RM, Smith MV. Transnasal endoscopic surgery of the pituitary gland: technical note. *Neurosurgery* 1996;39:189–192.
14. Busquets JM, Hwang PH. Endoscopic resection of sinonasal inverted papilloma: a meta-analysis. *Otolaryngol Head Neck Surg* 2006;134:476–82.
15. Schlosser RJ, Bolger WE. Nasal cerebrospinal fluid leaks: critical review and surgical considerations. *Laryngoscope* 2004;114:255–65.
16. Hackman TG, Ferguson BJ. Powered instrumentation and tissue effects in the nose and paranasal sinuses. *Curr Opin Otolaryngol Head Neck Surg* 2005;13:22–6.
17. Turgut S, Ercan I, Sayin I, Basak M. The relationship between frontal sinusitis and localization of the frontal sinus outflow tract: a computer-assisted anatomical and clinical study. *Arch Otolaryngol Head Neck Surg* 2005;131:518–22.
18. Kanowitz SJ, Shatzkes DR, Pramanik BK, Babb JS, Jacobs JB, Lebowitz RA. Utility of sagittal reformatted computerized

tomographic images in the evaluation of the frontal sinus outflow tract. *Am J Rhinol* 2005;19:159–65.

19. Kew J, Rees GL, Close D, Sdralis T, Sebben RA, Wormald PJ. Multiplanar reconstructed computed tomography images improves depiction and understanding of the anatomy of the frontal sinus and recess. *Am J Rhinol* 2002;16:119–23.
20. Douglas R, Wormald PJ. Pterygopalatine fossa infiltration through the greater palatine foramen: where to bend the needle. *Laryngoscope* 2006;116:1255–57.
21. Chandra RK, Kern RC. Advantages and disadvantages of topical packing in endoscopic sinus surgery. *Curr Opin Otolaryngol Head Neck Surg* 2004;12:21–6.
22. Wormald PJ, Boustred RN, Le T, Hawke L, Sacks R. A prospective single-blind randomized controlled study of use of hyaluronic acid nasal packs in patients after endoscopic sinus surgery. *Am J Rhinol* 2006;20:7–10.
23. Bugten V, Nordgard S, Steinsvag S. The effects of debridement after endoscopic sinus surgery. *The Laryngoscope* 2006;116:2037–2043.
24. Talbot AR, Herr TM, Parsons DS. Mucociliary clearance and buffered hypertonic saline solution. *Laryngoscope* 1997;107:500–03.
25. Senior BA, Kennedy DW, Tanabodee J, Kroger H, Hassab M, Lanza D. Long-term results of functional endoscopic sinus surgery. *Laryngoscope* 1998;108:151–57.

The Temporomandibular Joint

7

King Kim, DMD / Joseph P. McCain, DMD

OUTLINE

BASIC ANATOMY

The human temporomandibular joint (TMJ) is very complex and is like no other joint in the body. It is the articulation between the squamous part of the temporal bone and the mandibular condyle. It is also a ginglimo-diarthroidal joint, meaning it has two types of synovial joints with both rotational and translational properties.[1] The primary load bearing areas of the TMJ include the lateral portion of the squamous part of the temporal bone, mandibular condyle, and articular disc.

The squamous portion of the temporal bone has an avascular articular surface lined with fibrous connective tissue and consists of three distinct areas: anterior, the articular eminence and articular tubercle; intermediate, the glenoid fossa; and posterior, the tympanic plate becoming the postglenoid tubercle. The articular eminence is the bony prominence of the zygomatic process. The articular tubercle provides attachment for the capsule and the lateral temporomandibular ligament. The glenoid fossa is an important structure which separates the joint from the middle cranial fossa. The tympanic plate is located just anterior to the external auditory meatus. The postglenoid tubercle comprises the posterior part of the glenoid fossa and provides attachment for the capsule as well as the retrodiscal tissues.[2]

The articular disc is positioned between the squamous portion of the temporal bone and the mandibular condyle. This structure is primarily avascular, and is composed of dense fibrous connective tissue and fibrocartilage. The disc can be divided into three bands (Figure 1): (1) anterior, thick and located just anterior to the condyle in the closed mouth position; (2) intermediate, thin and sits along the articular eminence with the mouth closed; and (3) posterior, thick and lies superior to the condyle with the mouth in closed position.[6] In addition, the disc is attached by medial and lateral collateral ligaments that anchor the disc to the condyle medially and laterally. Anteriorly, the disc is adhered to the capsule and the superior head of the lateral pterygoid muscle, which allows the disc to rotate over the condylar head in an anteroposterior direction. Posteriorly, the disc blends with the bilaminar zone, which is contiguous with the TMJ capsule.[2]

The mandibular condyle articulates with the articular disc. The articular surface of the condyle is avascular fibrous connective tissue and the primary load-bearing area is on the lateral aspect. The condyle measures 10 mm anteroposteriorly and about 20 mm mediolaterally.[2]

Important structurally, because of its involvement with TMJ internal derangement and patient symptoms, is the bilaminar zone, or posterior attachment complex (Figure 1). This structure lies posterior to the articular disc and contains within it the retrodiscal pad. The other structures in the bilaminar zone are the superior lamina and inferior lamina. They connect the posterior part of the disc to the postglenoid tubercle and condyle, respectively. The retrodiscal pad is a highly vascular and neural structure made up of collagen and elastic fibers. Any pathologic functioning of the condyle against it will produce profound inflammation and pain for the patient. This abnormal situation is often the reason minimally invasive TMJ procedures are necessary.

The TMJ capsule is composed of fibrous connective tissue and encases the entire articular surface of the temporal bone and condyle. It is lined by a vascular synovial membrane, and also contains nociceptors that are responsible for eliciting pain when the joint is inflammed (synovitis). The attachments of the TMJ capsule include: (1) superior, the rim of temporal articular surfaces, (2) inferior the condylar neck, (3) medial it blends with the medial collateral ligament, (4) lateral it blends with lateral collateral ligament, (5) anterior it blends with superior head of lateral pterygoid muscle, and (6) posterior it blends with retrodiscal pad.[2]

The blood supply and innervation of the TMJ are relatively simple. Its arterial blood supply comes from the superficial temporal artery, which is one of the terminal branches of the external carotid artery. Also, blood supply comes from the deep auricular and anterior tympanic arteries, which branch off of the maxillary artery. Venous drainage comes from the superficial temporal and maxillary veins. Sensory innervation of the TMJ is via the auriculotemporal, masseteric, and posterior deep temporal nerves, which are all branches of the mandibular division of the trigeminal nerve.[2]

MOUTH OPENING

The TMJ can be divided into two compartments: superior and inferior. Both compartments contain endothelial cells that produce synovial fluid, which helps lubricate and nourish the TMJ. The superior compartment is the area between the squamous part of the temporal bone and articular disc. It occupies approximately 1.2 mL of volume and is the area where translation of the TMJ occurs. The inferior compartment is situated between the articular disc and mandibular condyle, with a volume of 0.9 mL and provides a place for rotational movement of the TMJ.[2]

At the beginning of mouth opening, up to approximately 20 mm, the condyle rotates in the inferior compartment on the fibrocartilaginous disc, mediated by the medial and lateral collateral ligaments. After 20 mm, translation in the superior compartment takes over, and the condyle/disc unit slides downward and forward along the posterior slope of the articular eminence up to maximum mouth opening, approximately 50 mm. This is a dynamic system, constantly changing according to the body's demands.

TMJ PATHOLOGY

As noted previously, the TMJ is a synovial joint subject to pathology similar to that of other synovial joints. A basic difference, however, is that fibrocartilage rather than hyaline cartilage lines the articular surface. Osteoarthritis may develop primarily or as a result of a traumatically dislocated articular disc. Important structural anatomy of the TMJ is demonstrated in Figure 1.

TMJ pathology often results from parafunctional habits that heavily load the articular tissues beyond their functional capacity. At the point when the functional limit is exceeded, breakdown of TMJ tissues begins, collagen fibers fragment, and the collagen network loses its stiffness. When this occurs, the proteoglycan-water matrix swells and flows out of the joint space, which softens the articular surfaces. This is known as chondromalacia. If excessive loading ensues, the damage can become irreversible. Fibrillation within the joint can occur, which can lead to roughening of the articular surfaces. This roughening can change the frictional character of the articular surfaces and lead to adherence of the articular structures. Prolonged adherence changes the mechanical stability of the condyle and disc relationship during rotation and translation. This strains the surrounding TMJ ligaments, which ultimately leads to disc displacement.

The displaced disc causes pain and mechanical locking of the mandible. Synovial adhesions often accompany herniation of the disc into the anterior recess. Central perforations of the disc occur in primary osteoarthritis. Perforation of the posterior ligament occurs in chronic dislocated discs secondary to micro- or macro-trauma. A surgical cascade is advocated to manage the disease process. Evidence-based research has shown that restoration of normal anatomy or removal of the diseased perforated disc is not always needed to obtain a return to relatively pain-free function.

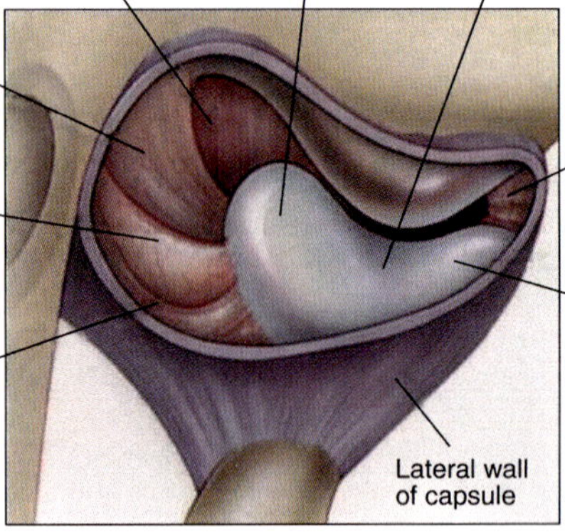

FIGURE 1 • Diagram showing arthroscopic anatomy of left TMJ. Taken from Joseph P. McCain. Principles and Practice of Temporomandibular Joint Arthroscopy, Philadelphia; Mosby, 1996; used with permission from Elsevier.

TMJ ANTERIOR REPOSITIONING APPLIANCE

In the 1970s, Farrar first introduced the anterior repositioning appliance to treat patients who had anteriorly displaced discs.[3] The thinking was to re-establish a more normal disc to condyle relationship to relieve the patients' pain and clicking symptoms. The appliance essentially provides an occlusal relationship that maintains the mandible in the forward position. Originally, the appliance was to be worn for 24 hours for 3 to 6 months. Initially results with this device were promising. Patients' painful clicking joints were reduced and normal disc to condyle relationships were reestablished.[1]

However, patients could not wear this device permanently, so the next step in therapy was unclear. Dental procedures needed to be performed to establish the occlusion with the mandible in the forward, therapeutic position. A study performed with a 2.5 year follow-up time revealed that successful elimination of both pain and joint symptoms was attained in only 28% of patients.[4] Other studies reported similar findings.[5,6] These long-term investigations reveal that the anterior repositioning appliances are not as effective as initially thought.

Recent studies have shown that anterior repositioning appliances do not actually recapture the disc onto the condyle.[7–9] Rather, when the condyle returns into the fossa, it articulates against adapted retrodiscal tissues. If these tissues have adequately adapted, function can continue without pain. Although the newly remodeled TMJ appears to be more normal, there will be a painless click-with movement, because the disc remains displaced.

INDICATIONS FOR MINIMALLY INVASIVE SURGERY

In general, patients who undergo consultation for TMJ internal derangement should undergo nonsurgical therapy first. This includes teaching the patient about limiting eccentric jaw movements to eliminate any pathologic habits, having the patient undergo Boering therapy that is educating the patient of their problem, and instructing them to eat a softer diet and to perform nonclenching techniques. Also, these patients should begin a regimen of oral anti-inflammatory medications and wear a soft orthotic device to either eliminate a bruxing habit or to soften the load placed on the temporomandibular joints during bruxism. The chronic bruxer can also benefit from an oral muscle relaxant to minimize the load as well. The exceptions to this are those patients who present for consultation who have already undergone orthotic therapy with little or no positive results.

After 2 to 4 months of non-operative therapy, the patient is re-evaluated. Sometimes the problem will be resolved with these measures and no additional treatment is warranted. However, there are many who are refractory to this conservative approach, and who desperately wish for resolution of their problem. This is where a discussion with the patient about minimally invasive TMJ surgery is indicated.

The patient is made aware that arthroscopic TMJ surgery may not alleviate the symptoms and that open joint surgery may be required. Additionally, the patient is explained the possibility of seventh nerve deficits secondary to extraoral punctures, malocclusion following surgery, possible scar formation from the punctures, as well as the possibility of pain, bleeding, swelling, and postoperative infection.

The next step is making a decision on the type of arthroscopic procedure the patient will receive. This obviously must be determined based on the diagnosis made. For example, a patient with a clicking joint, intermittent pain and 35 mm maximal incisal opening (MIO), with a Wilkes' stage II (Table 1) may benefit from a lysis of adhesions with lavage of the superior joint space to relieve the symptoms. In contrast, a patient who is in extreme pain with an anteriorly dislocated disc, which is nonreducing on MRI, an MIO of 20 mm, with a preliminary Wilkes' stage IV (Table 1), may benefit from a surgical arthroscopy procedure consisting of anteriorly releasing the disc with posterior repositioning, and fixating the disc to the condylar head.

PERIOPERATIVE CONSIDERATIONS

When the patient is ready for the operation, it is important to wrap the patient's head prior to entering the operating room to keep hair out of the operative field. When the head is wrapped securely, it allows the procedure to be clean and less cumbersome. In addition, the possibility of infection is minimized.

The procedures for TMJ arthroscopy are best performed under general anesthesia via nasal intubation. It is crucial that the assistant is able to manipulate the mandible and close the jaw into occlusion without the endotracheal tube getting in the way. If, however, the surgeon and patient elect to do a procedure such as arthroscopic arthrocentesis in an office setting under local anesthesia and light sedation, the patient must be able to follow commands of moving the jaw open and closed and into excursions during the procedure.[12] A diagram of patient positioning in the operating room setting can be seen in Figure 2, and arthroscopic set-up can be seen in Figure 3.

TABLE 1 Clinical and radiologic criteria for Wilkes' staging of TMJ internal derangement[10,11]

STAGE	CLINICAL FINDINGS	RADIOLOGIC FINDINGS
I	No significant mechanical symptoms, no pain or limitation of motion.	Slight forward displacement and good anatomic contour of disc.
II	First few episodes of pain, occasional joint tenderness and related temporal headaches, increase in intensity of clicking, joint sounds later in opening movement, beginning transient subluxations or joint locking.	Slight forward displacement and beginning anatomic deformity of disc, slight thickening of posterior edge of disc.
III	Multiple episodes of pain, joint tenderness, temporal headaches, locking, closed locks, restriction of motion, difficulty (pain) with function.	Anterior displacement with significant anatomic deformity/prolapse of disc, moderate to marked thickening of posterior edge of disc, no hard-tissue changes.
IV	Characterized by chronicity with variable and episodic pain, headaches, variable restriction of motion, and undulating course.	Increase in severity over intermediate stage, early to moderate degenerative remodeling hard-tissue changes.
V	Crepitus on examination, scraping, grating, grinding symptoms, variable and episodic pain, chronic restriction of motion, difficulty with function.	Gross anatomic deformity of disc and hard tissue, essentially degenerative arthritic changes, osteophytic deformity, subcortical cystic formation.

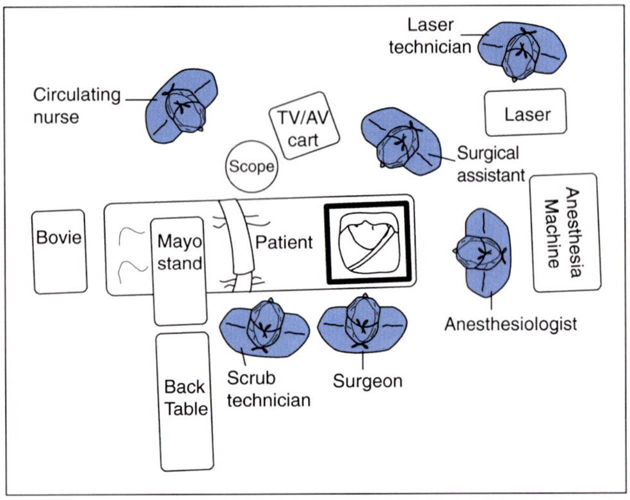

FIGURE 2 • Schematic showing patient positioning in the operating room. Note: surgeon positioned on the side of the joint to be operated. Taken from Joseph P. McCain. Principles and Practice of Temporomandibular Joint Arthroscopy, Philadelphia; Mosby; used with permission from Elsevier.

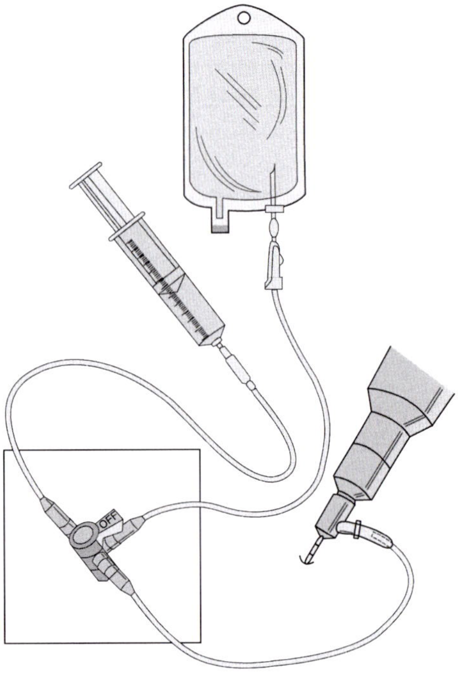

FIGURE 3 • Schematic of TMJ arthroscopic setup. Taken from Joseph P. McCain. Principles and Practice of Temporomandibular Joint Arthroscopy, Philadelphia; Mosby; used with permission from Elsevier.

Perioperative medication considerations are crucial to attaining gratifying results with TMJ arthroscopy. The use of prophylactic antibiotics to prevent postoperative infection is controversial. The orthopedic literature in regards to arthroscopy is scant with good prospective studies, because the incidence for infection following arthroscopic procedures is so low. The American Academy of Orthopedic Surgeons has not elicited an advisory statement regarding the issue.[13] However, literature does exist essentially stating that antibiotic prophylaxis for arthroscopic procedures is not indicated for healthy patients.[14] It is the opinion of these authors that antibiotic usage is indicated for those prone to infection. These include patients with diabetes, compromised immunity, or skin disorders. When antibiotics are indicated, 1 gram of cefazolin given intravenously 1 hour prior to the procedure provides good coverage to prophylactically manage these patients.[13] Intravenous administration of corticosteroids prior to the procedure is an important consideration to prevent postoperative edema. Many times after TMJ arthroscopic procedures it is vital to initiate mandibular motion and opening to prevent arthrofibrosis, hemarthrosis, and train the muscles of mastication to stretch.[15] Edema interferes with this ability to achieve the goal of improved mandibular functioning. Because of this, it is advisable to administer postoperative dosing regimens of corticosteroids in a tapered fashion.

As discussed earlier, the importance of instructing the patient to function the mandible postoperatively cannot be overemphasized. Patients are given these directions for jaw hinge exercises the day of or the day following surgery. These exercises are classified as stage I and stage II. Stage I exercises consist of taking the tongue, rolling it as far back on the palate as possible, and opening as wide as possible without disengaging the tongue. The second of these exercises is to move the jaw in lateral and protrusive excursions as far as possible. Both sets of exercises are to be performed for 20 repetitions, four times per day. Stage II exercises involve the pry bar, painting of teeth, and rubber tubing. The pry bar is placing the thumb on the upper incisors, the middle finger on the lower incisors, and prying to maximum incisal opening. The painting of teeth involves taking the tip of the tongue and starting at the buccal surface of the most posterior molar, painting all of the teeth from right to left and back again. Lastly, a piece of rubber tubing is placed between the canines on one side and moving into lateral excursions while lightly biting on the rubber tubing. Each of these exercises is performed for 20 repetitions, four times per day.

Most patients are instructed to follow stage II exercise, but stage I exercises are indicated for patients under certain circumstances. For example, patients who undergo semirigid diskopexy repair with the suture technique to treat anterior disc dislocation could very easily relapse back into the anterior position if translation is initiated too early. The same is true for those patients who receive posterior retrodiscal scarification procedures to treat slightly dislocated discs. Also, in those patients treated for condylar subluxation who undergo posterior retrodiscal release need to undergo opening exercises in reverse after surgery. This is to say that they need to prevent translating their condyles to allow the disc to settle into correct position. In all three of these instances, the patient is only gradually worked into full function due to the instability of the disc position, and potential for relapse.

SURGICAL CASCADE

Regardless of the pathology the recommended surgical cascade is as follows:

1. Arthroscopic arthrocentesis
2. Lysis and lavage with debridement
3. Diskopexy
4. Contracture for mandibular dislocation

A diagrammatic view of the relevant TMJ anatomy as it relates to puncture can be seen in Figures 4 and 5.

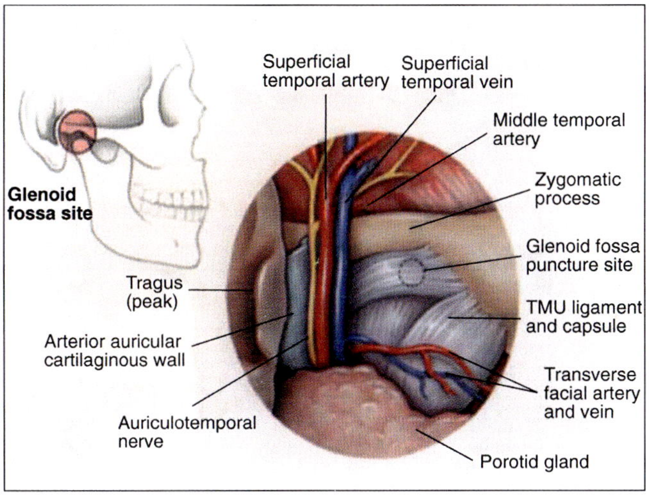

FIGURE 4 • Diagram showing anatomy relevant to the puncture technique, for the left TMJ. Taken from Joseph P. McCain. Principles and Practice of Temporomandibular Joint Arthroscopy, Philadelphia; Mosby; used with permission from Elsevier.

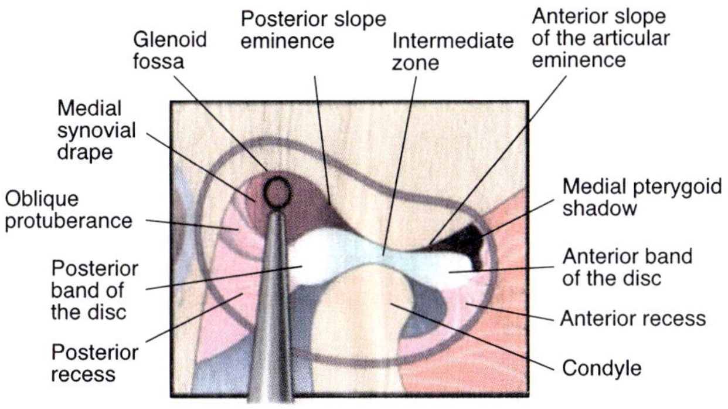

FIGURE 5 • Diagram showing the anatomy and puncture location, for the left TMJ. Taken from Joseph P. McCain. Principles and Practice of Temporomandibular Joint Arthroscopy, Philadelphia; Mosby; used with permission from Elsevier.

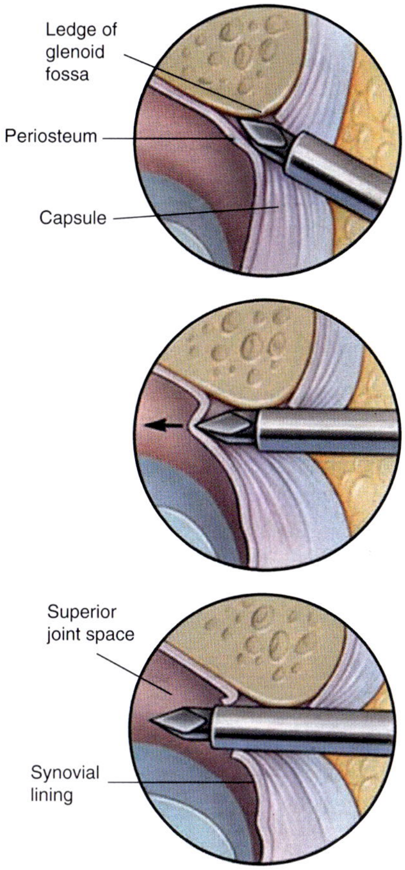

FIGURE 6 • Sequential diagrams for the single puncture technique, for the left TMJ. Taken from Joseph P. McCain. Principles and Practice of Temporomandibular Joint Arthroscopy, Philadelphia; Mosby; used with permission from Elsevier.

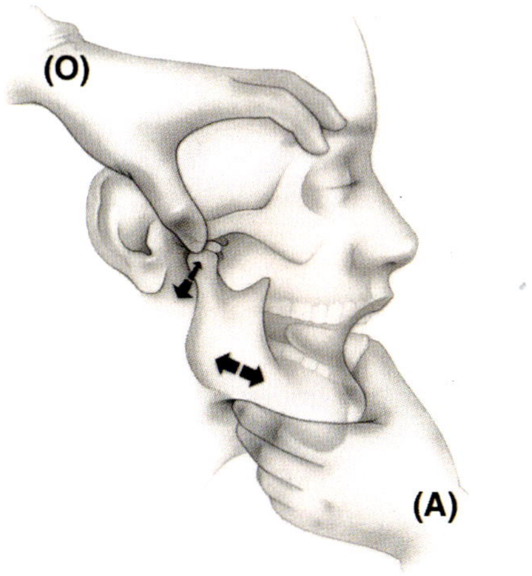

FIGURE 7 • Schematic showing operator (O) and assistant (A) collaborating to locate correct puncture location, for the left TMJ. Taken from Joseph P. McCain. Principles and Practice of Temporomandibular Joint Arthroscopy, Philadelphia; Mosby; used with permission from Elsevier.

Arthroscopic Arthrocentesis

This procedure essentially involves the single puncture technique. The single puncture technique can be seen in Figure 6.

The thumb is used to palpate the maximum concavity of the glenoid fossa at the junction between the periosteum and synovial capsule (Figure 7). The bony landmark is the maximum concavity of the inferior aspect of the zygomatic and temporal bones. To facilitate precision of the technique, this puncture should be done with a sharp trocar. Throughout the puncture into the superior joint space, the trocar and cannula are advanced (Figure 8) deeper into tissues utilizing a twisting and rotational motion to prevent injury to the facial nerve and to provide better control of the sharp instrument. After puncturing skin, the trocar is used to scrape off soft tissue attachments subperiosteally at the maximum concavity of the glenoid fossa. The trocar and cannula are then redirected medially to enter the superior joint space. A depth of 20 to 25 mm indicates a safe puncture. Any more and the risk of perforating through the medial synovial drape increases (Figure 9). It is important that the scope enter the joint as superolateral as possible for best visualization. Once inside the joint, the lavage of the joint ensues, and irrigating needle is placed for outflow. The 22 gauge irrigating needle can be inserted 5 mm anterior and 5 mm inferior to the first puncture (Figure 10A and B). It is important to

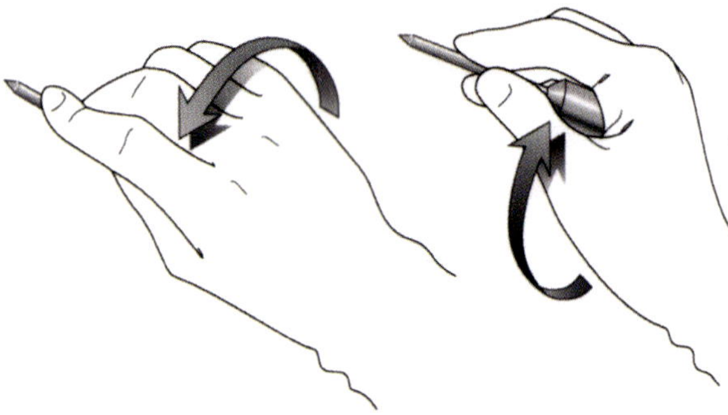

FIGURE 8 • Schematic representation of the rotational hand movement for puncturing the superior joint space. Taken from Joseph P. McCain. Principles and Practice of Temporomandibular Joint Arthroscopy, Philadelphia; Mosby; used with permission from Elsevier.

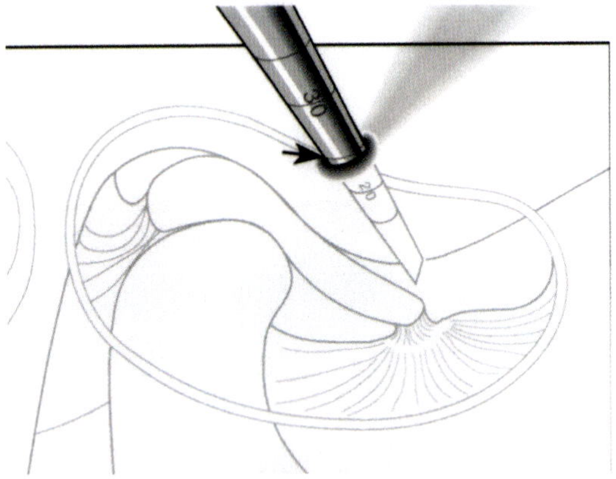

FIGURE 9 • Diagrammatic representation showing the depth of the puncture with the trocar. Taken from Joseph P. McCain. Principles and Practice of Temporomandibular Joint Arthroscopy, Philadelphia; Mosby; used with permission from Elsevier.

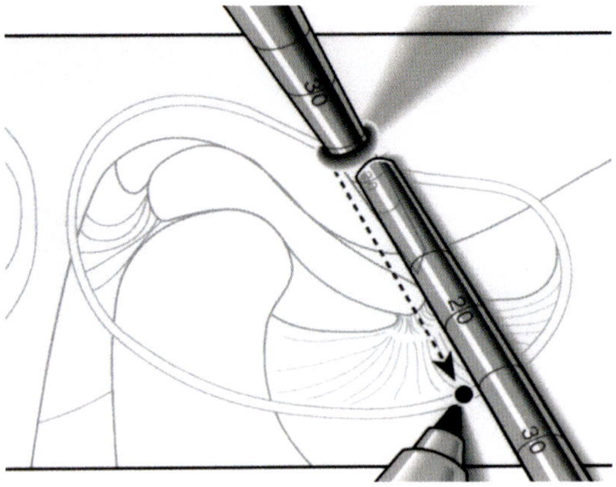

FIGURE 10A • Diagrammatic representation of how to measure the second puncture point, for the left TMJ. Taken from Joseph P. McCain. Principles and Practice of Temporomandibular Joint Arthroscopy, Philadelphia; Mosby; used with permission from Elsevier.

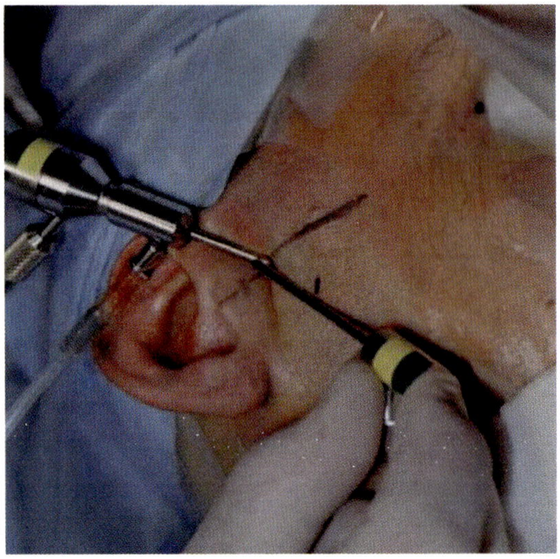

FIGURE 10B • Intraoperative photograph of a patient's left TMJ, the surgeon is measuring the second puncture point. Taken from Joseph P. McCain. Principles and Practice of Temporomandibular Joint Arthroscopy, Philadelphia; Mosby; used with permission from Elsevier.

insufflate with 2 to 3 mL of fluid prior to placing the irrigating needle to prevent collapse of the joint space.

Preliminary results from a 1 year ongoing investigation at Nova Southeastern University/Broward General Medical Center, Department of Oral and Maxillofacial Surgery evaluated 72 patients with internal derangements ranging from Wilkes' stages II to V. After being refractory to conservative therapy, all 72 patients underwent arthroscopic arthrocentesis in the office setting. After up to 1 year follow-up, 11 patients needed additional surgery consisting of arthroscopic diskopexy and 1 needed an open procedure. This coincides with 16.6% who needed re-operation. Sixty-seven percent of these re-operations were on Wilkes' stage III patients. Seventy-five percent of the re-operations occurred between 2 to 3 months after the first procedure. The results of this investigation are indeed promising for this procedure and provides validity to the importance of lavaging a pathologically inflamed joint.

Lysis and Lavage with Debridement

This procedure involves the double puncture technique. The fossa portal is created as described by the single puncture technique. Once the arthroscopic cannula with arthroscope is inserted, the articular eminence portal is then created by triangulation (Figure 11). To perform this technique, the most superior, anterior, and lateral aspect of the articular eminence is targeted. The lavage needle is removed, and a 2.0 trocar and cannula is introduced into the superior joint space. The second cannula inserted is known as the working cannula (Figure 12 and 13). The working cannula is used to accommodate instrumentation such as holmium lasers, shavers, graspers, straight probes, hook probes, as well as certain medicaments if they are needed (Figure 14). The size of the working cannula can be increased as needed.

Diskopexy

This procedure, which is the most complex, involves three to four punctures into the superior joint space. For rigid diskopexy, three punctures are utilized. In addition to the arthroscopic

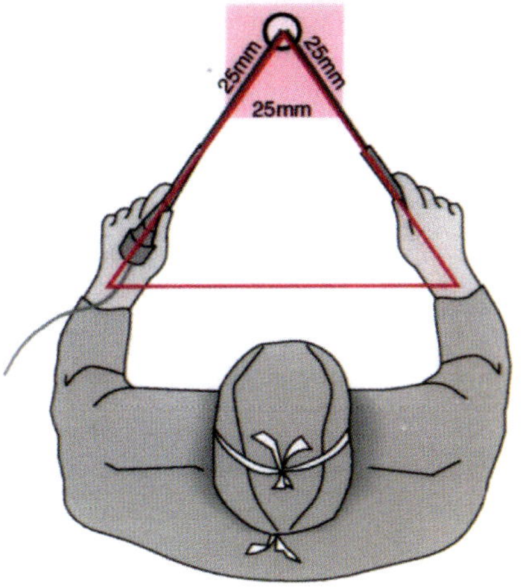

FIGURE 11 • Schematic demonstrating the triangulation concept. Taken from Joseph P. McCain. Principles and Practice of Temporomandibular Joint Arthroscopy, Philadelphia; Mosby; used with permission from Elsevier.

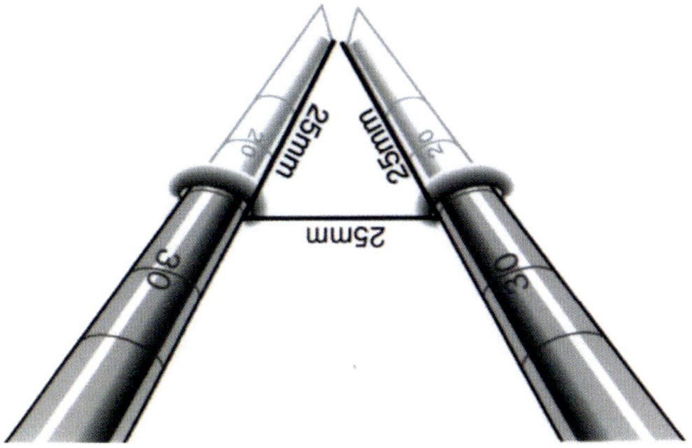

FIGURE 12 • Magnified schematic of the triangulation concept. Taken from Joseph P. McCain. Principles and Practice of Temporomandibular Joint Arthroscopy, Philadelphia; Mosby; used with permission from Elsevier.

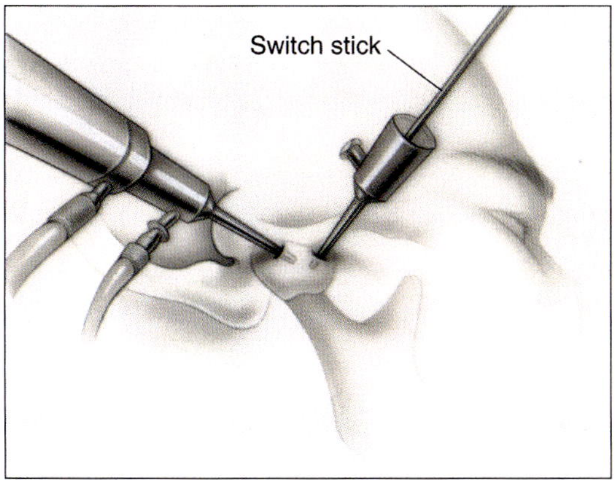

FIGURE 13 • Schematic representation of triangulation on a patient. Taken from Joseph P. McCain. Principles and Practice of Temporomandibular Joint Arthroscopy, Philadelphia; Mosby; used with permission from Elsevier.

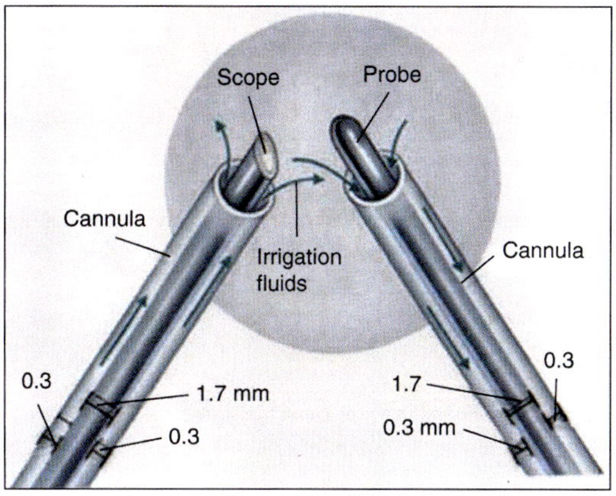

FIGURE 14 • Schematic showing the instrumentation used through both cannulas. Taken from Joseph P. McCain. Principles and Practice of Temporomandibular Joint Arthroscopy, Philadelphia; Mosby; used with permission from Elsevier.

cannula and the working cannula as previously described, the third puncture involves creation of the fixation portal for insertion of the fixation cannula. For semi-rigid diskopexy, four punctures are utilized. The arthroscopic cannula, working cannula, and fixation portal which accommodates two needles: one anterior and one posterior.

For either rigid or semirigid diskopexy, an anterior release must be performed prior to fixation. The anterior release is best performed through the working cannula with a holmium laser. The disc-synovial crease is delineated and scored with a hook probe, and the laser is used to cut along scored area, making sure to detach the superior head of the lateral pterygoid from the discal tissues. After adequate release is performed, the straight probe is utilized to reduce the disc posteriorly, while the mandible is in the forward position. The retrodiscal tissues at this time can be ablated with bipolar cautery or the holmium laser, as needed as a means to scar the tissues and keep the disc reduced.

For the rigid diskopexy procedure, the fixation portal is created while a straight probe in the working cannula reduces the disc posteriorly. The trocar and cannula targets the posterior lateral disc attachment. A 22 gauge needle is used to sound the area predicted for puncture. While the mandible is in the forward position, the fixation portal is created 20 mm inferior to the fossa portal.

Rigid fixation of the disc to the condylar head can now be performed through the fixation cannula which is approximately 3.0 to 3.5 mm in diameter. Fixation is done with either resorbable screws or nails, and can be done with one, two, or three depending on the amount of access that is available. Postoperative rehabilitation consists of immediate functioning with stage II exercises, as disc position is secure. The correct position of the three cannulas can be seen in Figure 15, whereas pictures of the technique can be seen in Figures 16, 17, and 18.

For semirigid fixation, or suture diskopexy, the fixation portal is created the same way as in the rigid fixation technique. The anterior release can be seen sequentially in Figure 19. Through the fixation cannula, a vector is aimed at the posterolateral attachment. A straight meniscus mender is then punctured under the reduction cannula to the condylar head then superiorly

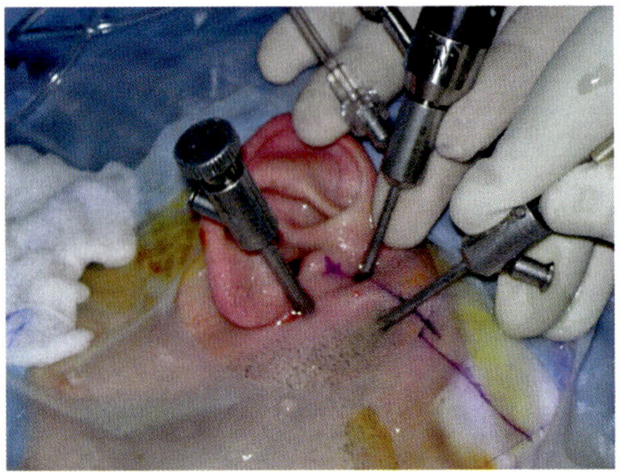

FIGURE 15 • Clinical picture of the arthroscopic cannula, working cannula, and fixation cannula of the patient's left TMJ.

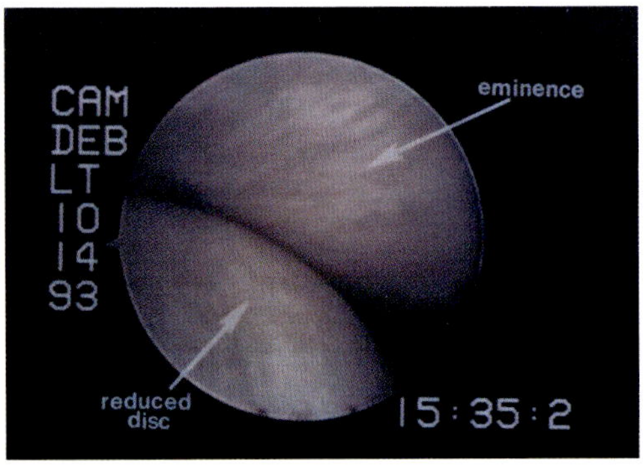

FIGURE 16 • Arthroscopic picture of the left TMJ showing the disc in reduced position.

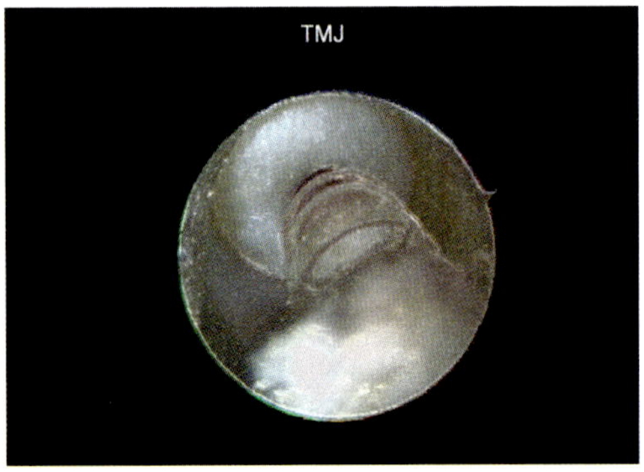

FIGURE 17 • Arthroscopic picture of the fixation technique.

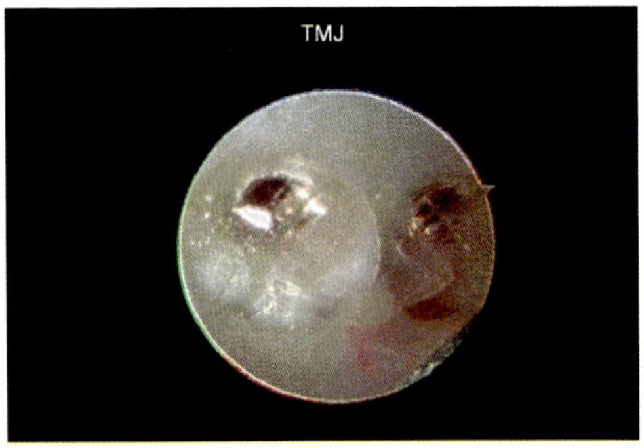

FIGURE 18 • Arthroscopic picture of the resorbable screw fixation in place.

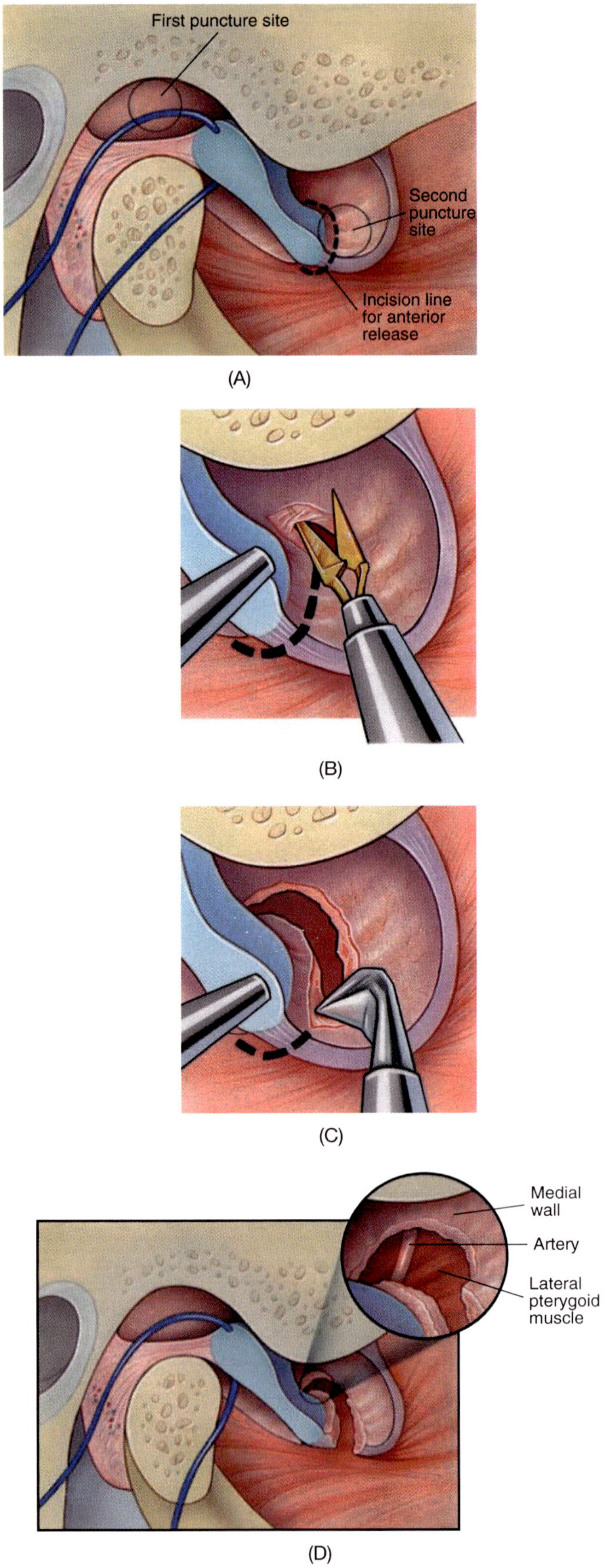

FIGURE 19 • Outlining the location of the anterior release: **A.** Diagram, **B.** The instrument cutting the tissue, **C.** the cutting is continued, **D.** The anterior release is completed. Figures 19A and 19B were taken from Joseph P. McCain. Principles and Practice of Temporomandibular Joint Arthroscopy, Philadelphia; Mosby; used with permission from Elsevier.

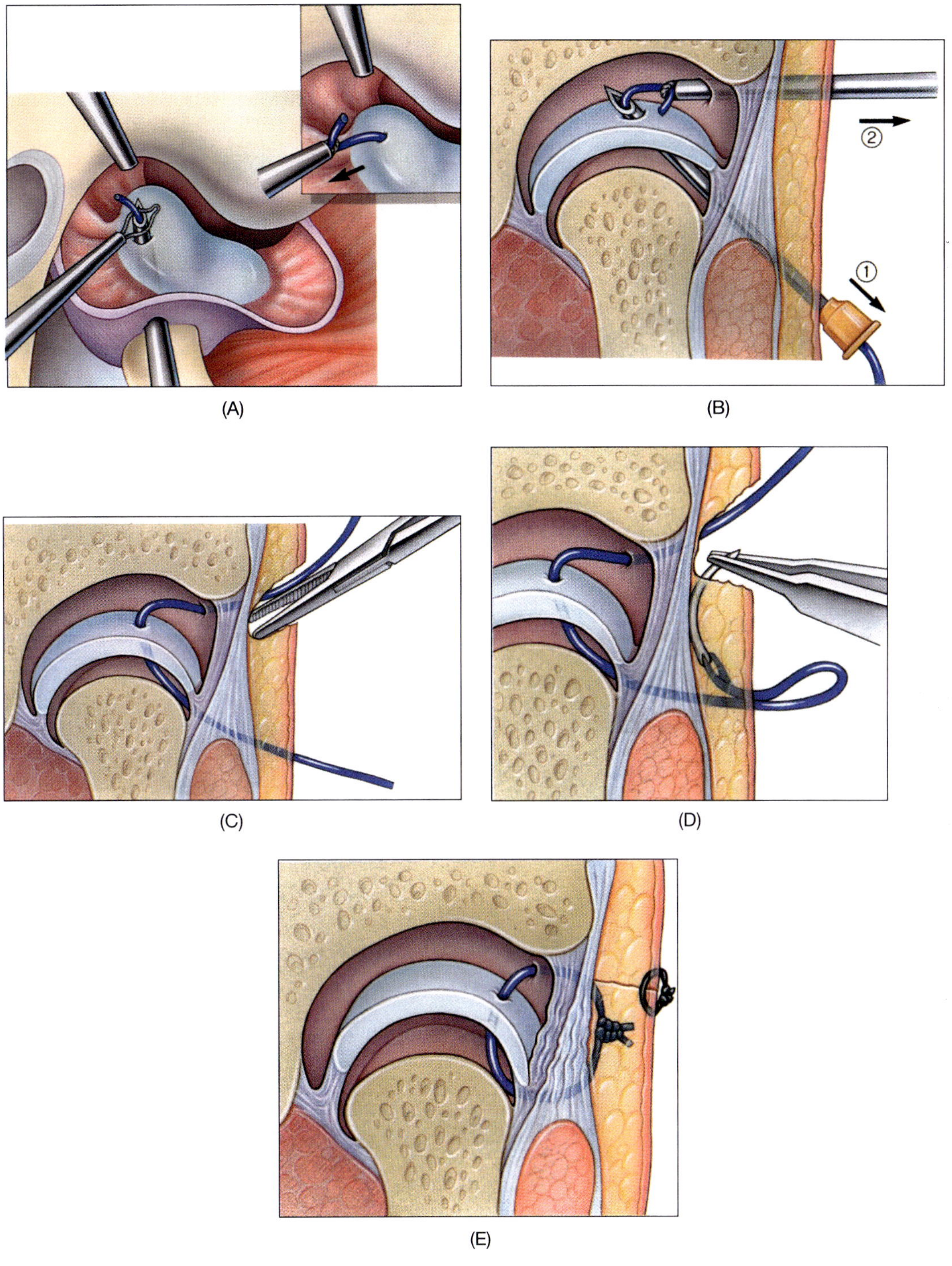

FIGURE 20 • Diagram of left TMJ showing the suture discopexy technique. **A.** Insertion of suture (O-PDS), percutaneously to the disc. **B.** The snare catches the suture posteriorly. **C.** The suture is pulled through the skin. **D.** A skin incision is made the suture passed through. **E.** A surgeons knot is tied to the capsule and disc. Taken from Joseph P. McCain. Principles and Practice of Temporomandibular Joint Arthroscopy, Philadelphia; Mosby; used with permission from Elsevier.

through the posterior lateral portion of the disc. A second meniscus mender is then puncture through skin and is used to catch the suture (Figure 20). A 0-PDS suture is then passed anteriorly and a snare is used to catch it posteriorly (Figure 20B). The location of the suture when it is pulled through skin is parallel and inferior to the apex of the tragal cartilage (Figure 20C). A small skin incision is made after the suture is passed through (Figure 20D), and a surgeon's knot is tied to capsule and disc (Figures 20E). Postoperative rehabilitation for these patients is slow and requires only limited amount of jaw mobility initially, gradually increasing to stage II exercises and beyond. (Figure 20 & 21)

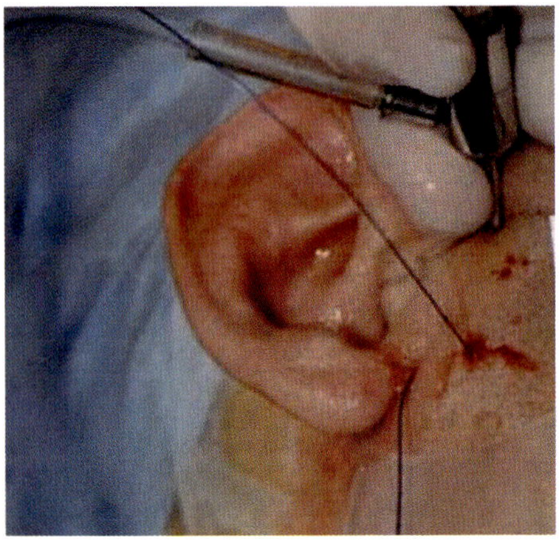

FIGURE 21 • Photograph of patient's left TMJ, undergoing suture discoplexy. Taken from Joseph P. McCain. Principles and Practice of Temporomandibular Joint Arthroscopy, Philadelphia; Mosby; used with permission from Elsevier.

CONCLUSIONS

Overall, TMJ arthroscopy is a safe procedure. It is minimally invasive, and can be performed in an outpatient setting. Complications do, however, exist. These include facial nerve trauma, arthrofibrosis, infection, perforation into the ear canal or tympanic membrane, and subdural or epidural hematoma from perforation into the middle cranial fossa. Management of these complications is outside the scope of this chapter, however, with good surgical technique and attention to detail, as well as adequate and procedure selective postoperative rehabilitation, these complication rates can be drastically reduced.

References

1. Okeson JP, Wright AL. Management of temporomandibular disorders and occlusion. Vol. 1. 4th ed. St. Louis: Mosby, 1998, p. 180–5.
2. Norton NS, Netter FH. Netter's head and neck anatomy for dentistry. Vol. 1. Pennsylvania: WB Sanders, 2007. p. 256–65.
3. Farrar WB. Differentiation of temporomandibular joint dysfunction to simplify treatment. J Prosthet Dent 1972;28: 629–36.
4. Okeson JP. Long-term treatment of disc-interference disorders of the temporomandibular joint with anterior repositioning occlusal splints. J Pros Dentistry 1988;60:611–16.
5. Moloney F, Howard JA: Internal derangements of the temporomandibular joint. III. Anterior repositioning splint therapy. Aus Dentist J 1986;31:30–9.
6. Lundh H, Westesson PL, Kopp S. A three-year follow-up of patients with reciprocal temporomandibular joint clicking. Oral Surg Oral Med Oral Pathology 1987;63:530–3.
7. Kirk WS Jr. Magnetic resonance imaging and tomographic evaluation of occlusal appliance treatment for advanced internal derangement of the temporomandibular joint. J Oral Maxillofacial Surg 1991;49:9–12.
8. Choi BH, Yoo JH, Lee WY. Comparison of magnetic resonance imaging before and after nonsurgical treatment of closed lock. Oral Surg Oral Med Oral Pathology 1994;78:301–5.
9. Chen CW, Boulton JL, Gage JP. Effects of splint therapy in TMJ dysfunction: a study using magnetic resonance imaging. Aus Dentistry J 1995; 40:71–8.
10. Smolka W, Iizuka T. Arthroscopic lysis and lavage in different stages of internal derangement of the temporomandibular joint: correlation of preoperative staging to arthroscopic findings and treatment outcome. J Oral Maxillofac Surg 2005;63:471–8.
11. Wilkes CH. Internal derangements of the temporomandibular joint: Pathological variations. Arch Otolaryngol Head Neck Surg 1989;115:469.
12. Kim K, McCain JP, Rangarajan S, et al. Office-based temporomandibular joint arthroscopy: its use in the treatment of internal derangement. J Oral Maxillofac 2008;66(supl):112–3.
13. Kurzweil PR. Antibiotic prophylaxis for arthroscopic surgery. Arthroscopy. 2006;22:452.
14. Bert J, Giannini D, Nace L. Antibiotic prophylaxis for arthroscopy of the knee: is it necessary? Arthroscopy Journal Arthro Related Surg 2007;23:4.
15. Ohnishi M. Arthroscopy of the temporomandibular joint. Jap Stomatol Sot 1975;42:207.

Interventional Sialoendoscopy

8

Oded Nahlieli, DMD

OUTLINE

INTRODUCTION

Obstructive sialadenitis, with or without sialolithiasis, is the most common inflammatory disorder of the major salivary glands. Satisfactory treatment depends on the establishment of a precise diagnosis and, in the case of sialolithiasis, accurate location of the obstruction. Until recently, many of these glands were removed under general anesthesia.

Sialolithiasis accounts for 50% of major salivary gland disease.[1,2] The submandibular gland is the most prone to sialolithiasis representing approximately 80% of all cases. Of the remainder, 19% are in the parotid and 1% in the sublingual glands. Sialolithiasis is most often found in adults but it may occur in children.[3] Salivary stones may vary in size, shape, texture, and consistency. They may occur as a single stone or as multiple stones and usually occur unilaterally (95%) but rarely (5% of cases) can be bilateral. It has been reported that sialolithiasis was present in approximately 1% of dissected cadavers.[4]

In Great Britain, it has been estimated that there are 57 cases of symptomatic sialolithiasis per million people per annum admitted to the hospital. This represents a total of 3420 patients each year.[1] Assuming an equal incidence in Europe or the United States, with populations in the 300 million range, approximately 17,100 patients per annum might require hospitalization for sialolithiasis and sialadenitis in these geographic areas. This data does not include patients who are treated on an outpatient basis.

Sialolithiasis is more common in males[5] and the peak incidence is between the ages of 30 and 60 years.[5] Stones grow by sequential mineral deposition and range in size from 0.1 mm to 30 mm.[6] Patients typically present with a painful swelling of the gland at meal times.[7] In the past decade, the introduction of salivary gland endoscopy has provided an accurate means of diagnosing and locating intraductal obstructions and a minimally invasive surgical treatment.[8–20]

In 1997, Nahlieli reported[11] on his experience with the use of a rigid miniature endoscope to perform sialoendoscopes on 46 major salivary glands. Iro and Zenk[22] reported a similar experience in 2000. To date, more than 1000 patients with sialolithiasis and sialadenitis have been managed successfully by endoscopic techniques at Barzilai Medical Center, Ashkelon, Israel. The goal of this chapter is to describe our experience with diagnosis, endoscopic management, outcomes and complications.

INDICATIONS FOR SIALOENDOSCOPY

1. Inspection of salivary ducts to identify and localize stones.
2. Inspection of salivary ducts for diagnostic purposes in patients with recurrent episodes of salivary gland swelling without an obvious cause of obstruction.

3. Removal of deeply located small to medium (<5 mm) stones in the middle and posterior portion of Wharton's or Stenson's ducts.
4. Exploration of the duct system for secondary stones, stone particles, or debris after removal or mechanical fragmentation of calculi or lithotripsy.
5. Identification and treatment of strictures and kinks in the submandibular and parotid ducts.
6. Treatment of recurrent nonobstructive submandibular and parotid sialadenitis by endoscopic inspection, dilatation and copious irrigation.

CONTRAINDICATION FOR SIALOENDOSCOPY

The only absolute contraindication for sialoendoscopy is acute sialadenitis. Acute infection should be eliminated by a course of antibiotics before undertaking endoscopy.

PRE-ENDOSCOPY ASSESSMENT

Following clinical evaluation, plain radiographs (panoramic, occlusal, and occlusal oblique) are obtained. Computed tomographic (CT) scans are always useful to identify stones and are particularly indicated for parotid stones in the mid to posterior duct. It may be helpful to map the ductal system by sialography or ultrasound prior to sialoendoscopy. These radiologic studies may demonstrate variations in duct diameter such as dilatations or strictures. Sialography may establish whether or not the duct can be dilated under pressure.

INTRODUCTION OF THE SIALOENDOSCOPE

There are four methods to introduce the sialoendoscope into the duct lumen (Figures 1, 2, 3, and 4):

1. Dilatation of the duct orifice with dilators and lacrimal probes.
2. Papillotomy of the duct orifice (commonly required) performed with a CO_2 laser, a fine scissors or a number 11 scalpel blade.
3. Incision of the floor of mouth mucosa to identify and enter the anterior portion of the submandibular duct. The submandibular duct can be opened safely anywhere along its length. The parotid duct is prone to stricture and the duct should not be opened over more than 1 mm or 2 mm.

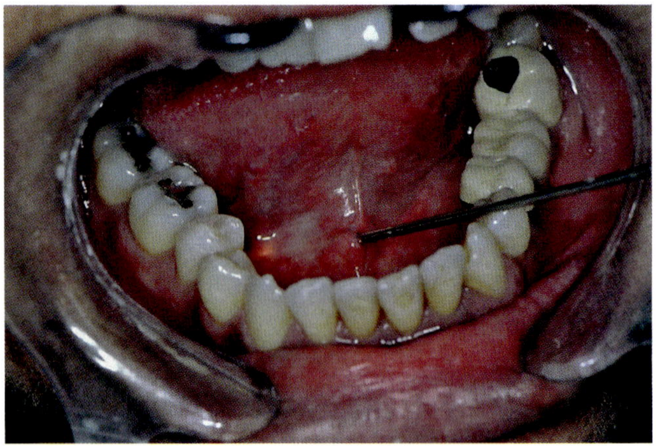

FIGURE 1 • Introduction of the diagnostic sialoendoscope in the natural orifice of Wharton's duct.

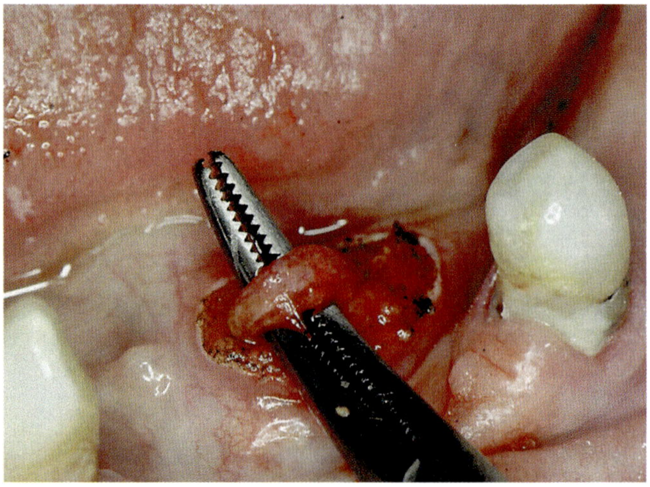

FIGURE 2 • Exploration technique—the anterior part of the duct is exposed.

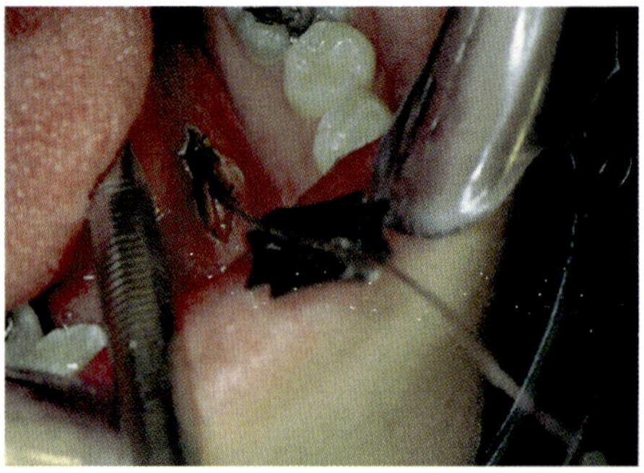

FIGURE 3A • Following identification, the Wharton's duct is incised; note the lacrimal probe is immediately inserted to identify the duct lumen. This is an important technical point as the lumen is easily lost during the procedure.

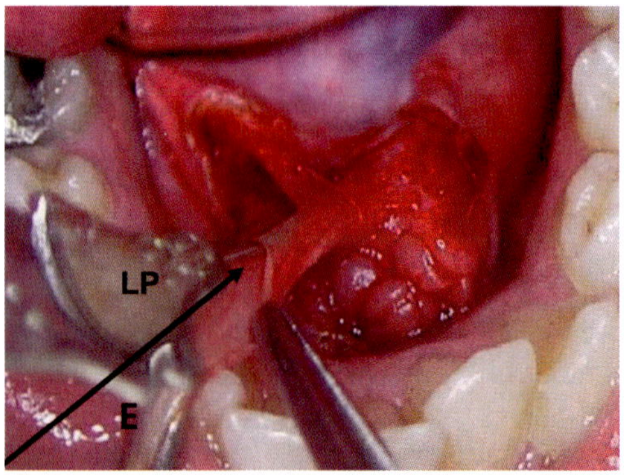

FIGURE 3B • Exploration of the Wharton's duct, note the diameter of the duct sufficient for insertion of the surgical endoscope for interventional sialoendoscopy. LP = lacrimal probe is in the duct for correct location. E = the position of the endoscope for accurate insertion.

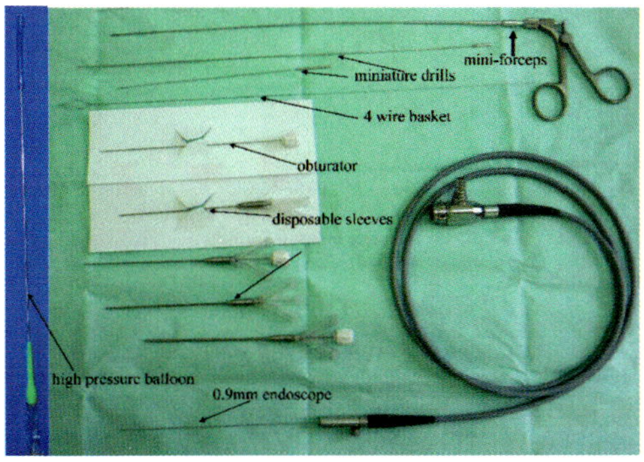

FIGURE 3C • Sialoendoscopy set-up.

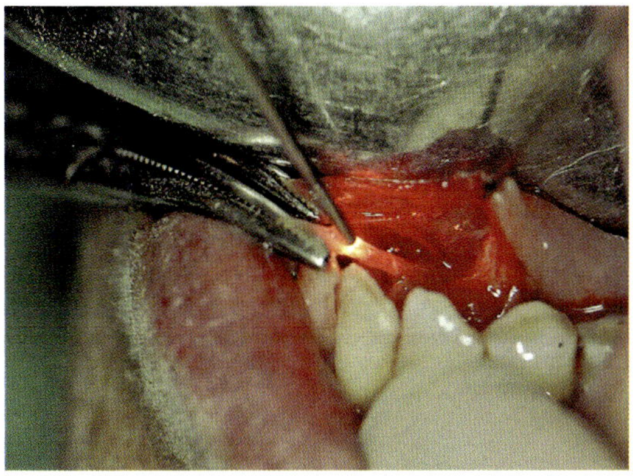

FIGURE 4A • Introducing the diagnostic sialoendoscope through a "ductal cutdown" procedure.

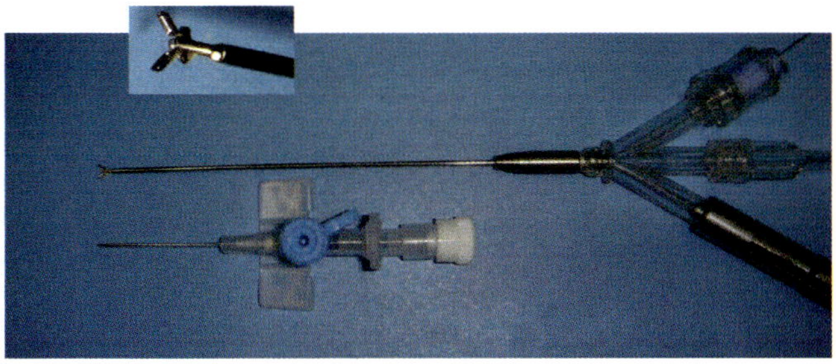

FIGURE 4B • The Nahlieli surgical endoscope (Karl Storz, Germany). seen with the grasping torcep in the surgical or working port and camera + light attached for visualization.

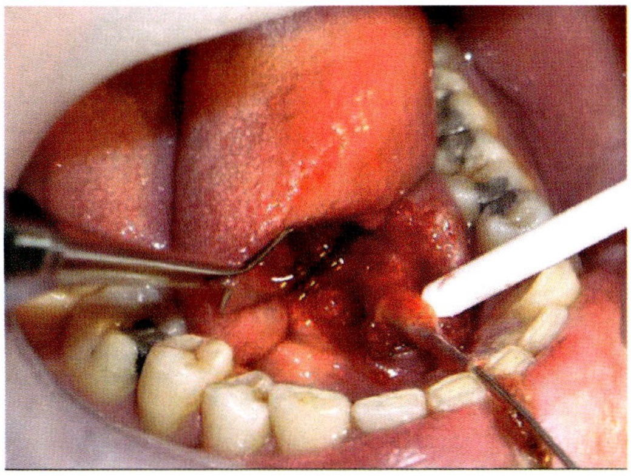

FIGURE 4C • Introducing the surgical endoscope into the ductal lumen following exploration of the duct.

4. Direct insertion after sialolithotomy. The endoscope can be inserted through the opening made for intraoral stone extraction.

IRRIGATION DURING SIALOENDOSCOPY

Irrigation, to maintain the ductal opening, is a crucial component of any sialoendoscopic procedure. The lumen must be filled with isotonic saline to allow adequate visualization manipulation and guidance of the endoscope. The duct system is sensitive to lumenal pressure (much like the kidney) and may constrict in response to high-pressure irrigation. Before irrigating, therefore, 5 mL to 10 mL of 2% lidocaine without epinephrine should be injected into the duct. Irrigation can be accomplished by means of a 500 mL bag of normal saline on a self-filling syringe or simply a 20 mL syringe connected by a short (60 cm) tube to the endoscope. The syringe should not be connected directly to the endoscope because pressing the plunger moves the endoscope and distorts the image.

ENDOSCOPIC SIALOLITHOTOMY TECHNIQUES

Intraoral Techniques

Intraductal Sialolithotomy

The intraductal approach to sialolithotomy is a true endoscopic technique. When a sialolith is encountered, its diameter is estimated using the caliber of the endoscope as a reference. There are four options for endoscopic removal through the duct (<5 mm stones): (1) the calculus may be grasped by forceps; (2) the stone may be grasped by a wire basket; (3) a balloon may be passed over the stone, inflated, and then the stone retrieved by withdrawing the balloon and delivering the stone in front of it; (4) the calculus can be crushed by a mini forceps and then removed in fragments. Alternatively, a stone can be fragmented with a laser (intracorporeal lithotripsy), or by a combination of techniques such as intracorporeal/Extracorporeal Shock Wave Lithotripsy (ESWL). After fragmentation, wire baskets and grasping forceps are used to remove the pieces and debris. When possible, the calculus should be removed in one piece (Figures 5, 6, 7, and 8).

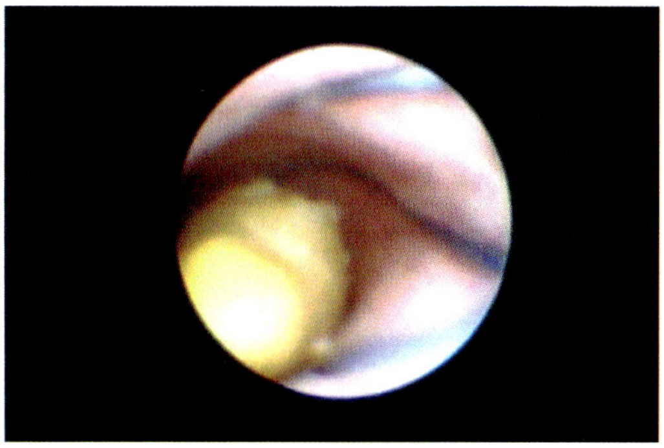

FIGURE 5 • The wire basket (blue) is used to retrieve the sialolith (yellow).

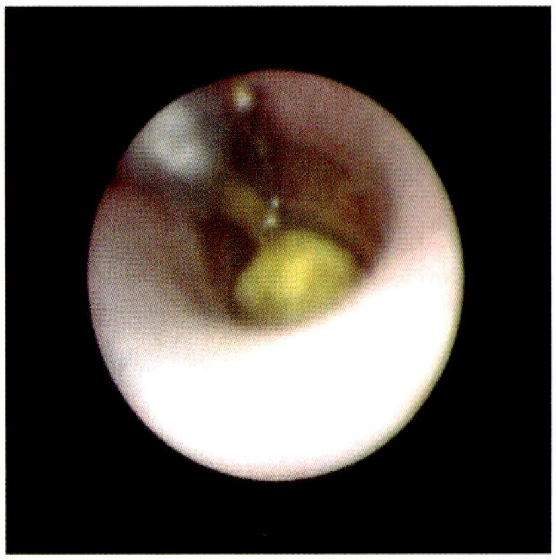

(A)

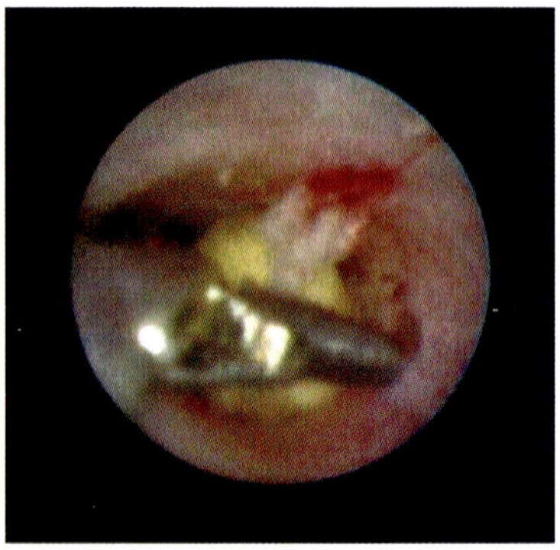

(B)

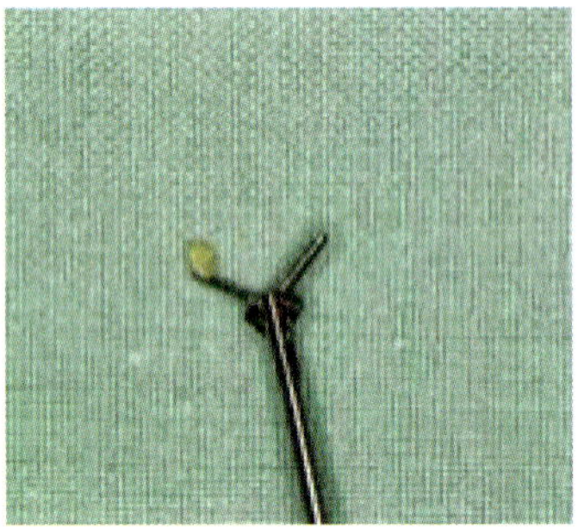

(C)

FIGURE 6A, B, C • Miniature forceps can be used to retrieve a stone.

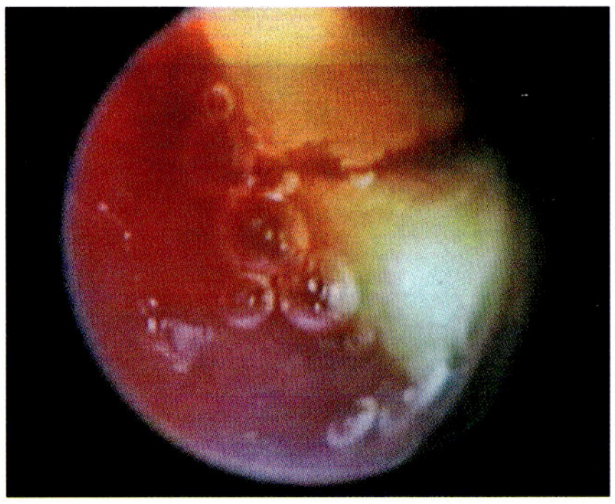

FIGURE 7A • Intracorporeal shock wave lithotripsy with electrohydraulic lithotripter-fragmentation renders the stone into two pieces.

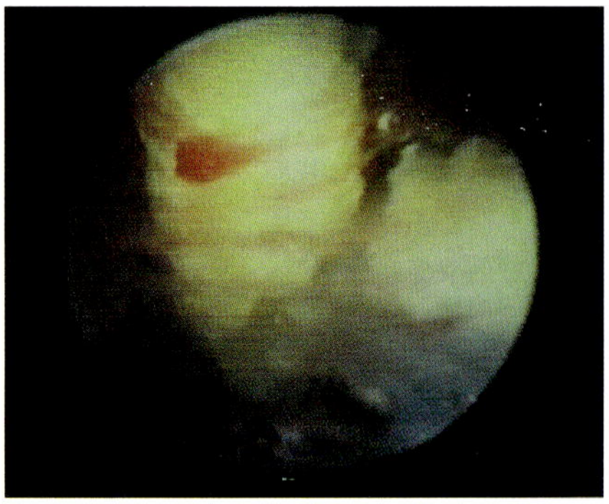

FIGURE 7B • Continued lithotripsy results in fragmentation into multiple pieces.

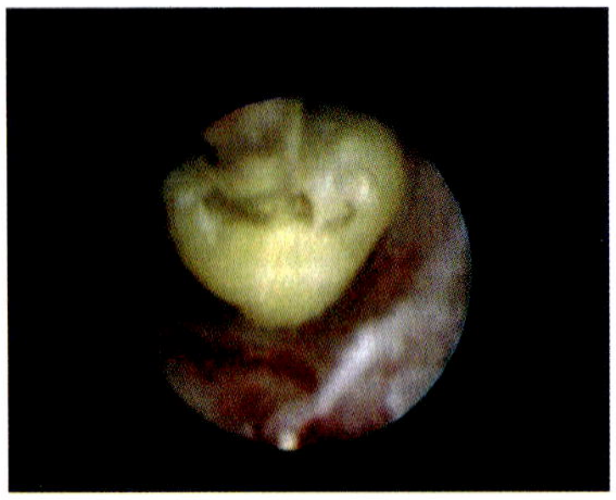

FIGURE 8 • Fragmentation of submandibular hilar stone with intracorporeal erbium laser probe. The stone is reduced to a dust material.

Extraductal Sialolithotomy

This approach is used when attempts to extract a stone through the duct have failed. Other indications for extraductal sialolithotomy include a calculus measuring more than 5 mm and a narrow duct (Figures 9 and 10).

The endoscope is inserted and the duct explored to accurately locate the stone. The duct system is irrigated for dilatation and to loosen the calculus from its ductal attachment. Then, a lacrimal duct probe is inserted into the duct until it contacts the stone. Alternatively, the endoscope can be advanced to contact the stone or a basket can be used to capture and immobilize the stone. The mucosa is incised over the duct using a CO_2 laser or with a number 11 scalpel blade. The probe, or endoscope or basket is used as a guide when making the incision. The duct is dissected and isolated from the surrounding tissue. Using upward digital pressure from the submandibular triangle, the gland is manipulated toward the mouth. The duct is incised over the stone, which is then dissected free of the duct and removed. Then, endoscopic exploration for removal of additional calculi and lavage is carried out. A stent is placed (Sialostent, Sialotechnology, Ashkelon, Israel) or a 19-gauge Silastic angiocatheter for 2 to 4 weeks (Figure 11). If the stone is close to the skin surface, percutaneous (extraoral) retrieval can be performed (Figures 12 and 13).

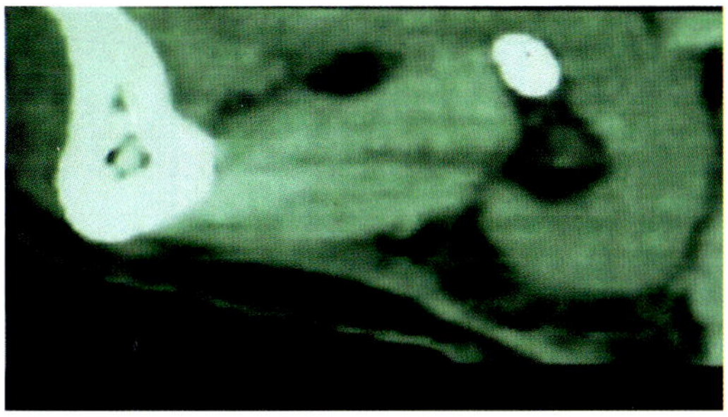

FIGURE 9A • CT of sialolith located in the submandibular hilum not accessible through intraductal approach.

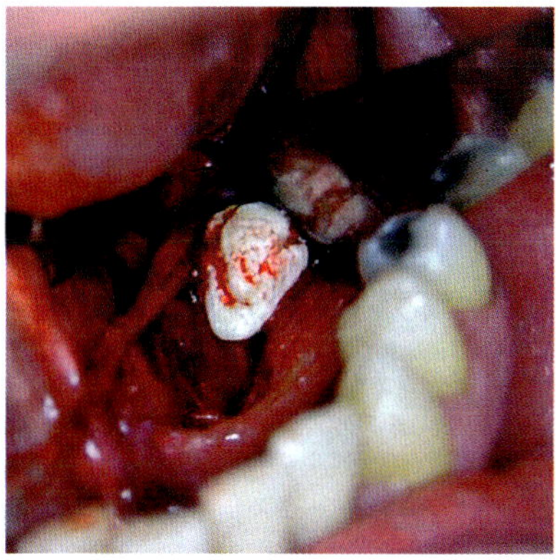

FIGURE 9B • "Stretching technique" pulling the duct and stone forward, to allow stone removal (seen in 9A).

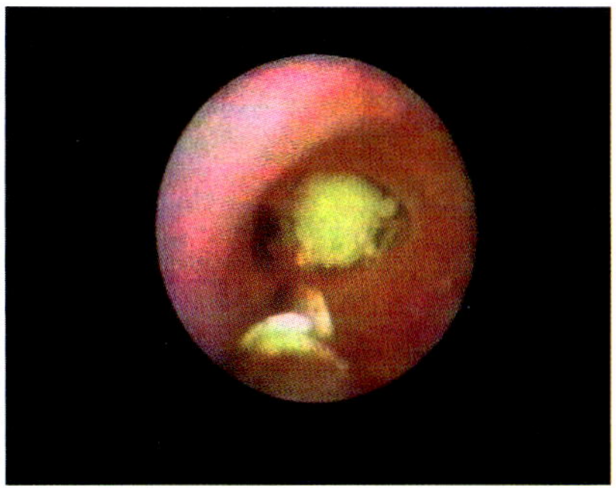

(A)

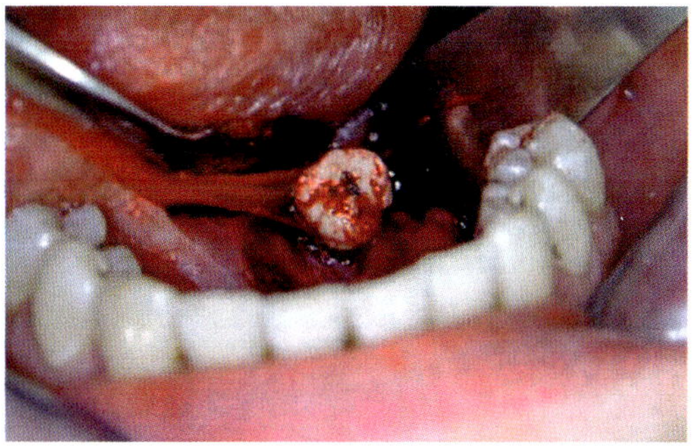

(B)

FIGURE 10 • Following the removal of stone from the posterior part of the submandibular duct via the stretching technique (**A**), endoscopic exploration revealed an additional calculus in the hilum that was endoscopically retrieved (**B**).

Extracorporeal Shock Wave Lithotripsy (ESWL)

ESWL is a well-known treatment for kidney stones. When applied for the treatment of sialolithiasis, the success rate is much lower and hence the technique has not been as widely applied. By generating shock waves that hit the stone, small fragments of calculus can be washed out of the gland. In 1989, Iro[22] reported the use of ESWL for salivary stones. The success rate for stone clearance in the submandibular gland is low (about 30%) but results are better in the parotid (about 60%).[23, 24] Success is also size dependent and stones over 8 mm have less than 10% chance of clearance. The morbidity following ESWL is low. In 2004, Nahlieli[25,26] developed a miniature ESWL (Sialotechnology, Ashkelon, Israel) and used it with sialoendoscopy. (Figures 14 and 15). The indications for ESWL are: (1) to remove small

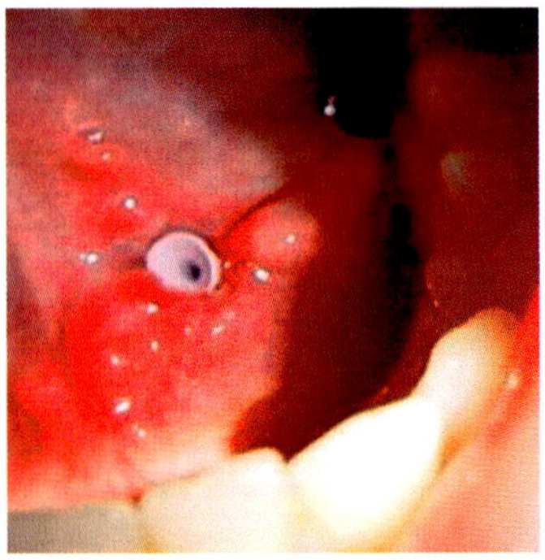

FIGURE 11A • Temporary polymeric sialo stent inside Wharton's duct after sialoendoscopic surgery.

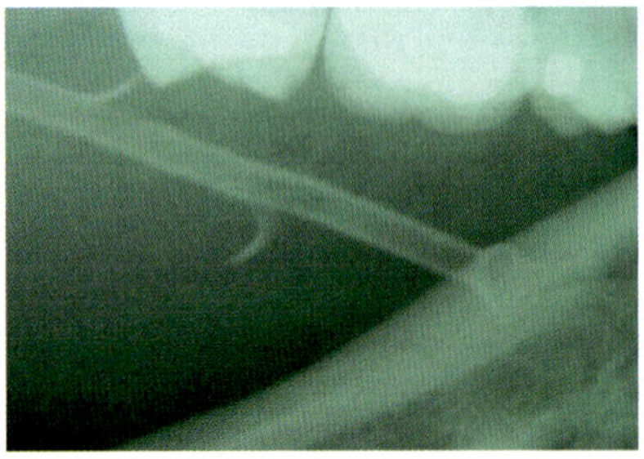

FIGURE 11B • X-ray image (oblique) of the same temporary sialo stent showing the two flaps that hold the stent in place.

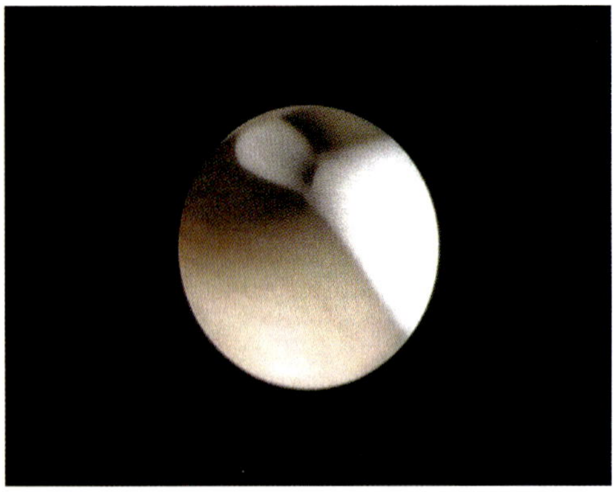

FIGURE 11C • Endoscopic view of the polymeric sialo stent inside the Wharton's duct.

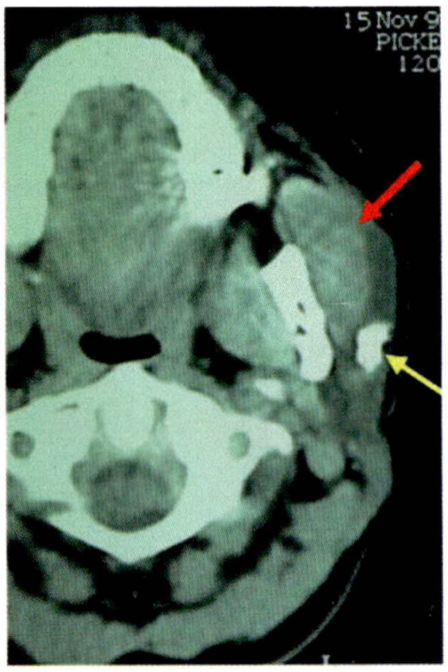

FIGURE 12A • Example of an impacted parotid stones (yellow arrow) which cannot be retrieved via an intraoral approach: stricture inside Stensen's duct (red arrow) anterior to the stone prevented an intraoral approach.

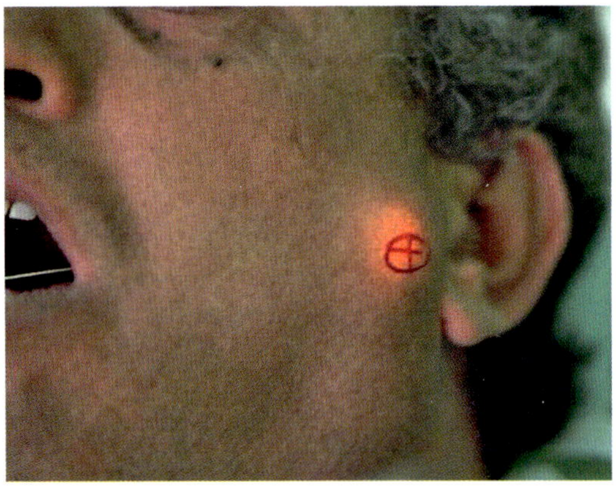

FIGURE 12B • Extraoral sialoendoscopic technique for removal of an impacted stone: Endoscopic identification making use of the transillumination effect.

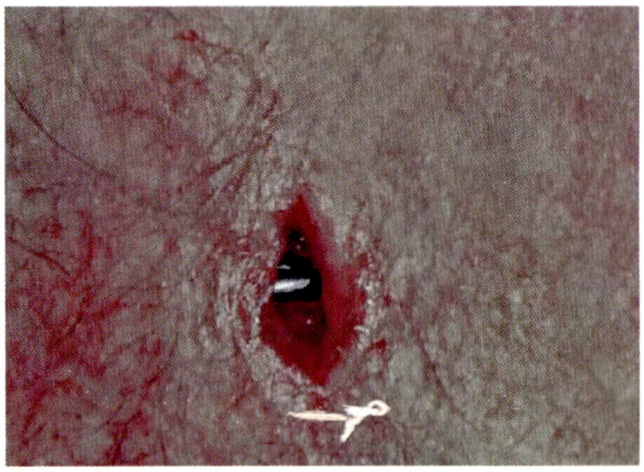

FIGURE 12C • Placement of an intraductal stent.

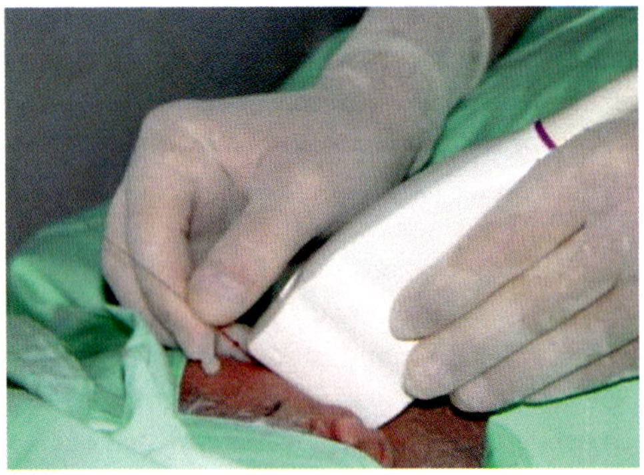

FIGURE 13A • Ultrasound identification of the stone location by means of a wire guide (biopsy marker).

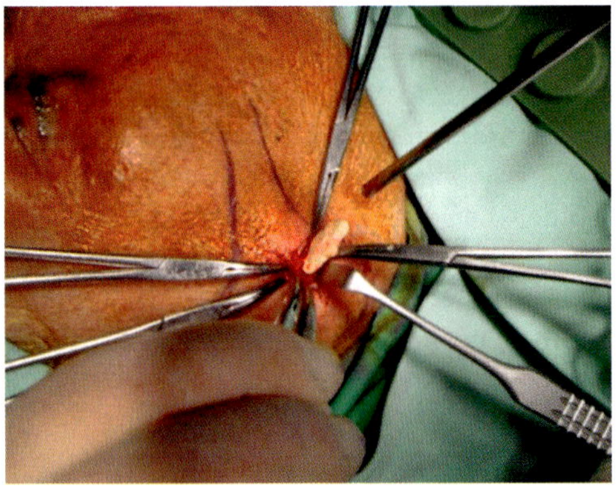

FIGURE 13B • Extraction of the stone, percutaneously.

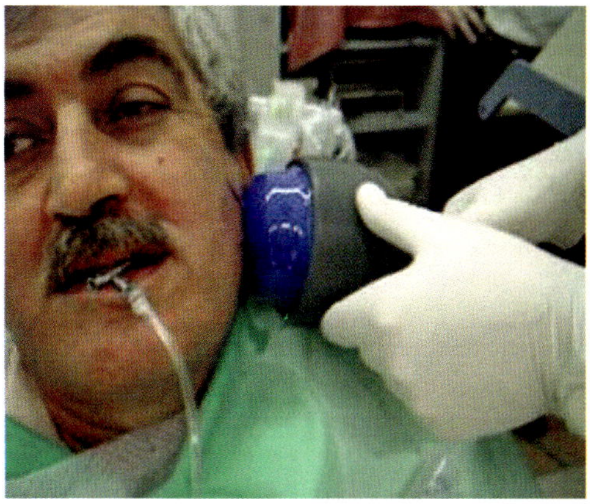

FIGURE 14A • ESWL of parotid stone using the sialoendoscope, for localization.

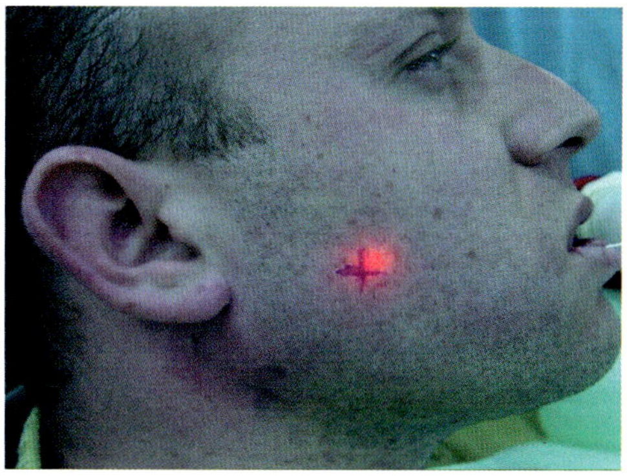

FIGURE 15A • Demonstration of the ESWL treatment stages of the posterior part of the gland. The yellow arrows mark the original location of the stone. The red arrow marks the location of the stone after the second session of ESWL (every session, 1500 shockwaves). The localization of the stone performed with the sialoendoscope and the transillumination effect.

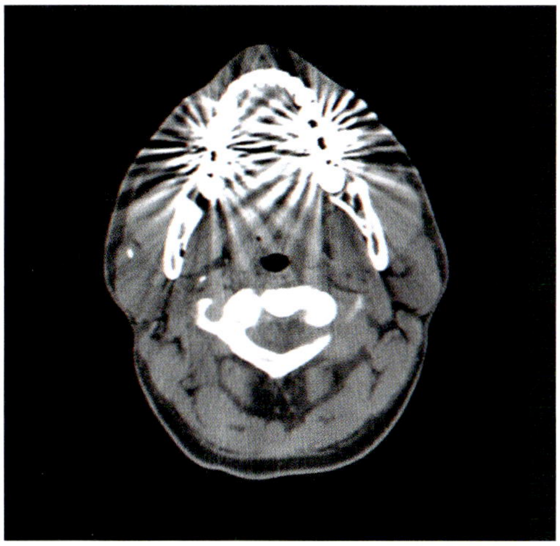

FIGURE 15B • CT scan of the same patient after the first session of lithotripsy.

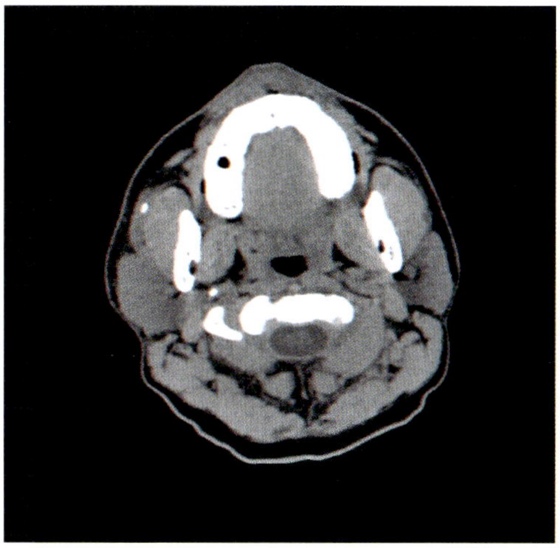

FIGURE 15C • CT scan of the same patient after the second session of lithotripsy.

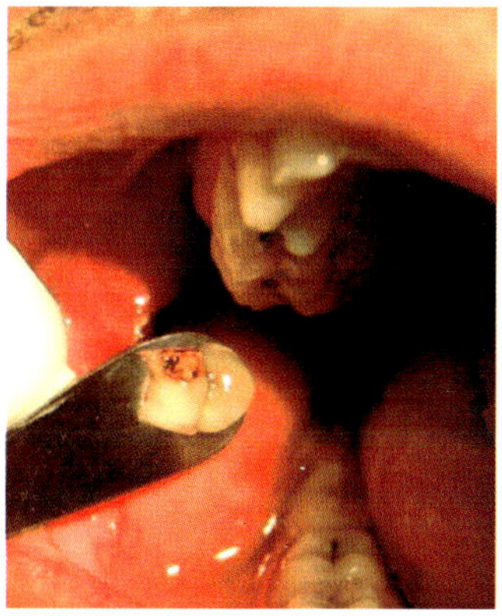

FIGURE 15D • Removal of the stone from the Stensen's duct orifice, with simple sialolithotomy after the third session of lithotripsy.

parotid stones (<5 mm) in conjunction with dilatation and lavage; (2) to fragment and reduce the size of large stones (>5 mm); (3) to treat deep intraparenchymal stones; (4) to treat stones after failure of endoscopic or endoscopic assisted techniques; and (5) to treat medically compromised patients who are not surgical candidates.

DIAGNOSIS AND TREATMENT OF STRICTURES AND KINKS

Strictures and kinks in the salivary ducts are the second-most frequent cause of obstruction after sialoliths.[17] Strictures can be identified by both sialography and sialoendoscopy. In contrast, kinks of the duct are best visualized by sialography.

Buckenham and Brown[27,28] pioneered the use of balloon dilatation of ductal strictures under fluoroscopic guidance. More recently, sialoendoscopic techniques for ductal dilatation have been introduced. This allows the surgeon to accomplish dilatation using less complicated instrumentation and with the ability to see the duct directly via the endoscope[17,29] (Figure 16).

Dilatation Techniques

Endoscopic ductal dilatation can be accomplished by saline hydrostatic pressure balloons or by the use of a miniature forceps expansion maneuver.

Initially, an attempt can be made to dilate the stricture by increasing the pressure of irrigation while advancing the endoscope. If saline pressure irrigation fails to induce dilatation, the next step is to insert a balloon (high- and low-pressure balloons exist) that can be inflated with air or water. The duct is allowed to remain dilated for 5 minutes and the maneuver is repeated 3 times. This is sufficient to dilate most strictures. (Figures 17 and 18)

Grasping forceps can be used to dilate strictures. The forceps is inserted to just beyond the stricture and then opened and slowly withdrawn along the walls of the stricture. Following dilatation, the duct is flushed with 100 mg of hydrocortisone solution.

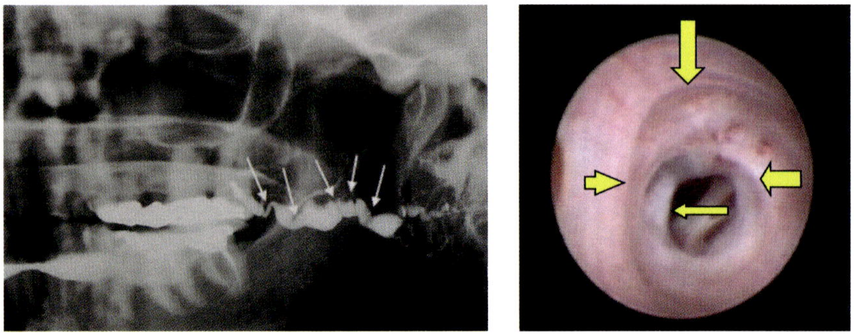

(A) (B)

FIGURE 16 • (**A**) Sialographic demonstration of strictures. (**B**) Endoscopic images of multiple strictures in Stensen's duct in a patient with chronic recurrent parotitis (arrows—strictures).

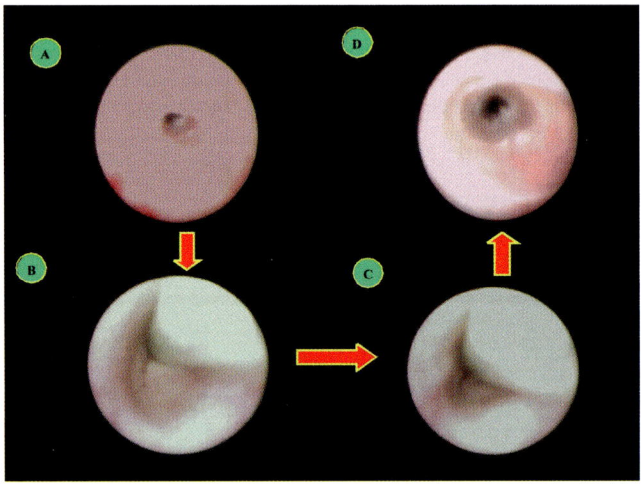

FIGURE 17 • Step-by-step dilatation stricture. **A**, Endoscopic demonstration of stricture in the Stensen's duct; **B,C,** intraoperative dilatation with low-pressure balloon; **D**, immediate postoperative dilation view.

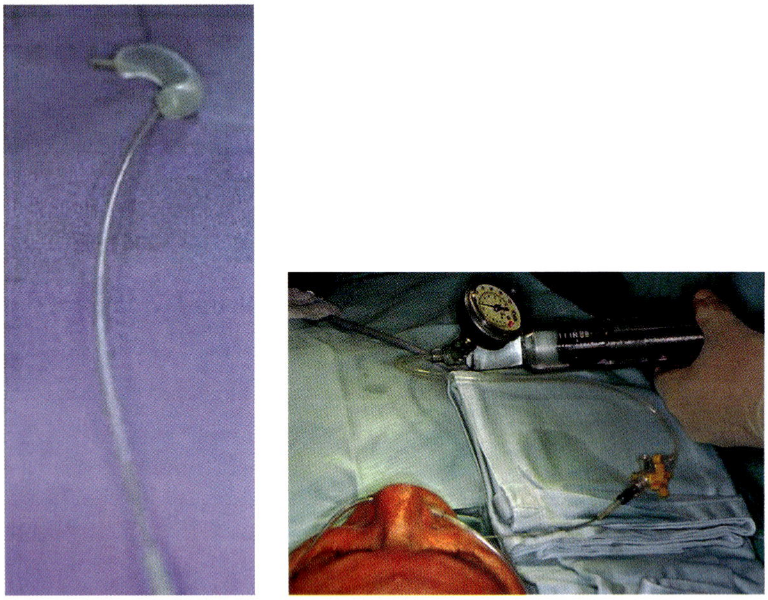

(A) (B)

FIGURE 18 • (**A**) Low-pressure balloon with long shaft. (**B**) High-pressure balloon—the inflation of the balloon with dilatation pump.

When a kink or bend in the duct is present, the anti-kink procedure (Figure 19) is performed. When possible a stent is inserted into the dilated duct, for 2 to 4 weeks, to prevent recurrence. All patients are treated postoperatively with amoxicillin 1.5 g per day for 7 days.

Nahieli et al. have treated 111 strictures endoscopically: Parotid gland ($n = 71, 64\%$); and submandibular gland ($n = 40$, 36%). The success rate for the treatment was 81%. In 15% of the cases, more than 1 procedure was required and in 4%, the dilatation procedure failed.

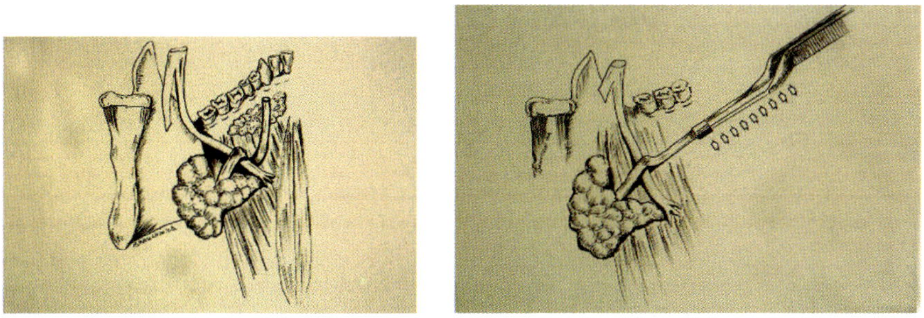

(A) (B)

FIGURE 19 • (A) Kink formation in Wharton's duct inside the hilum region. (B) The anti-kink procedure for stretching the duct to enlarge the acute angle.

TREATMENT OF SIALADENITIS WITH NO MECHANICAL OR ANATOMIC OBSTRUCTION

Submandibular Sialadenitis

In patients who have been treated for sialadenitis without evidence of obstruction, thorough sialoendoscopic-guided lavage of the main and secondary ducts with at least 60 cc of normal saline is a very effective strategy to prevent recurrences. Adhesions may be responsible for stasis and poor flow which then result in recurrent submandibular sialadenitis.

Chronic Recurrent Parotitis[30]

Chronic recurrent parotitis (CRP) is a nonspecific sialadenitis characterized by unilateral or bilateral pre-auricular swelling. The patient may present with multiple episodes of swelling and remission, or persistent swelling of the involved gland. The etiology of this condition may be multifactorial, but commonly there is poor salivary flow with a predisposition to stasis and infection. Sialography demonstrates ductal and parenchymal anatomy, pathologic findings, and gland function. Chronic parotitis starts with punctate sialectasis and dilatation of the duct. These dilatations and strictures create a "sausage-like appearance" as observed on sialographic imaging. Endoscopic-guided examination of the ductal system in CRP reveals multiple mucus plugs, fibrosis of the duct lining, strictures, and dilatations.

The goal of endoscopic treatment is to dilate the multiple strictures. The endoscopic techniques are the same as that previously described in the section on strictures.

ENDOSCOPIC OBSERVATIONS AND TREATMENT IN CLINICAL PRACTICE

During the past 13 years, sialoendoscopy has been performed on 1078 glands with symptoms of obstructive disease at Barzilai Medical Center. There were 545 males and 533 females, with an age range of 2 to 96 years. There were 722 submandibular, 347 parotid, and 9 sublingual glands examined. All patients underwent preoperative and postoperative screening including routine radiography, sialography, and ultrasound. Postoperative examination was routinely performed 1 month following the procedure. The longest follow-up was 40 months postendoscopy. The majority of procedures were performed under local anesthesia on an outpatient basis. The duration of the procedures ranged from 30 to 90 minutes.

The success rate for parotid sialolithotomy was 86% and for submandibular gland sialolithotomy 89%. Immediate failures (introduction of the miniature endoscope failed or proved not feasible) occurred in 1.1%; intraoperative failures (inability to accomplish any of the endoscopic stone retrieval techniques) in 4.4%; and late failures in 4.7% of patients. Complications included: temporary lingual nerve paresthesia in 0.4%, postoperative infection in 1.6%, postoperative bleeding in 0.5%, traumatic ranula in 0.7%, and ductal strictures in 2.5% of patients.

A number of micro-anatomic and pathophysiologic phenomena have been encountered in the course of sialoendoscopic procedures:

1. Sphincter-like mechanisms.
2. Sublingual duct opening (Bartholin's duct). This opening was noted in the anterior portion of Wharton's duct, between 0 mm and 5 mm posterior to the papilla.
3. Changes in the ductal system: a matted appearance, ecchymosis and hypovascularity in chronic sialadenitis or in the presence of long-standing calculi; a shiny appearance of the ductal lining visible mucosal vascularity healthy glands or in patients with short-term obstruction (Figure 20A & B).

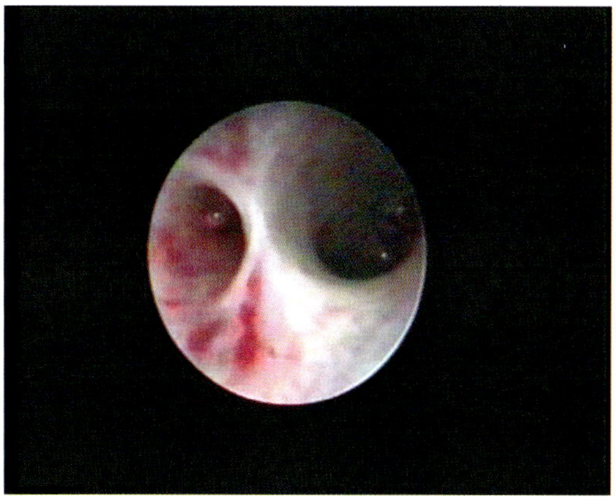

FIGURE 20A • Endoscopic appearance of a healthy submandibular hilum.

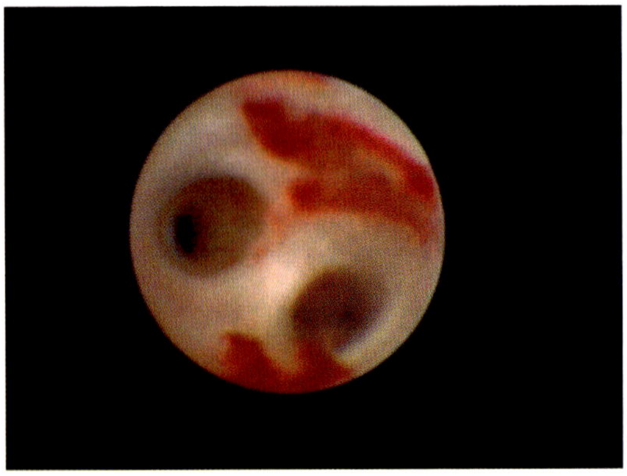

FIGURE 20B • Endoscopic appearance of chronic sialadenitis of submandibular hilum (the lining mucosa of the ductal system had a matted appearance, ecchymosis, and a small number of blood vessels).

4. Peculiar connections between calculi and the ductal wall were observed in the submandibular and parotid glands.
5. Intraparenchymal stones.
6. Foreign bodies: ten foreign bodies in the ductal system were identified—four in the parotid duct and six in the submandibular duct. Of the ten foreign bodies, four were hair shafts and three were parts of a plant that were washed out by irrigation. Five of the ten foreign bodies were associated with calculi, three were found in children. Formation of a sialolith around a hair shaft was observed in two cases.
7. Ductal strictures and kinks: malformations of salivary ducts were detected in 148 cases—37 kinks (29 submandibular and 8 parotid) and 111 strictures (40 submandibular and 71 parotid).
8. Pelvis-like formation: an anatomic malformation in the submandibular hilum was identified to be a pelvis-like configuration of the duct (a basin-like structure) instead of a bifurcation or trifurcation. This pelvis-like formation may have caused an obstructive phenomena.
9. Intraductal evagination: an evagination of the duct was identified in a 10-year-old child who had two sialoliths. The sialoliths were identified in Wharton's duct. During stone extraction, the evagination was noted. It obstructed the ductal lumen and, was probably the cause for the formation of calculi.

CONSENSUS OF SIALOLITHIASIS TREATMENT

In 2006, surgeons from three major centers dealing extensively with minimally invasive treatment for salivary gland sialolithiasis gathered to write a consensus report. The participating centers were the Department of Oral and Maxillofacial Surgery, Barzilai Medical Center, Ashkelon, Israel; Erlangen–Nürenburg University, Germany, and the Department of Oral and Maxillofacial Surgery, Guy's Hospital, London, England. In this chapter we present the algorithms (Algorithms 1 and 2) for selecting the appropriate sialoendoscopic treatment of parotid and submandibular stones based on this consensus conference.

CONCLUSION

Sialoendoscopy is a promising new minimally invasive strategy for diagnosis and treatment of most inflammatory conditions of the major salivary glands. Sialoendoscopy and associated techniques may be performed on an outpatient basis, utilizing local anesthesia. The complication rate is extremely low. Sialoendoscopy appears to be the future solution for the management of complex inflammatory salivary gland pathology. As

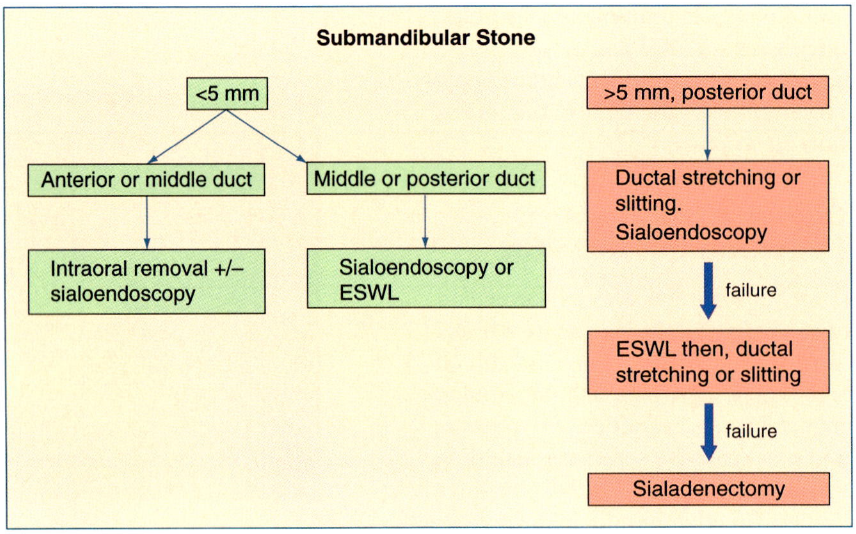

ALGORITHM 1 • Selecting submandibular sialoendoscopy techniques.

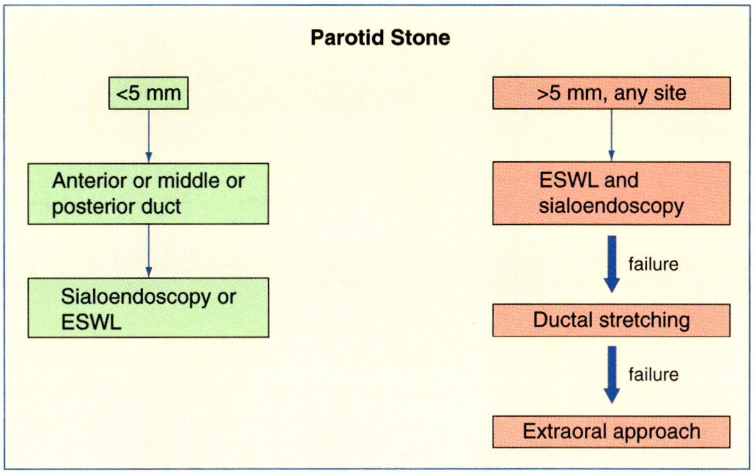

ALGORITHM 2 • Selecting parotid sialoendoscopy techniques.

more surgeons become involved with sialoendoscopy (see Chapter 9, "Sialoendoscopy: the United States Experience"), more findings and innovations will be forthcoming, adding to its effectiveness.

References

1. Escudier MP, McGurk M. Symptomatic sialoadenitis and sialolithiasis in the English population, an estimate of the cost of hospital treatment. Br Dental J 1999;186:463–6.
2. Epker BN. Obstructive and inflammatory diseases of the major salivary glands. Oral Surg Oral Med Oral Pathol 1972;33:2–27.
3. Nahlieli O, Eliav E, Hasson O, et al. Pediatric sialolithiasis. Oral Surg Oral Med Oral Pathol 2000;90:709–12.
4. Rauch S, Gorlin RJ. Diseases of the salivary glands. In Gorlin RJ, Goldman HM, eds. Thoma's oral pathology (6th ed), St. Louis, MO: Mosby; 1970.
5. Lustmann J, Regev E, Melamed Y. Sialolithiasis. A survey on 245 patients and a review of the literature. Int J Oral Maxillofac Surg 1990;19:135–8.
6. Frame JW, Smith AJ. Large calculi of the submandibular salivary glands. J Oral Maxillofac Surg 1986;15:769.
7. Hardy KJ. Submandibular calculus disease at the Royal Melbourne Hospital 1954–1963. Med J Aus 1966;16:670–1.
8. Konigsberger R, Feyh J, Goetz A, et al. Endoscopic controlled laser lithotripsy in the treatment of sialolithiasis. Laryngorhinootologie 1990;69;322–3.
9. Katz PH. Endoscopie des glands salivires. (Endoscopy of the salivary glands) Ann Radiol (Paris) 1991;34:110–3.
10. Nahlieli O, Neder A, Baruchin AM. Salivary gland endoscopy—a new technique for diagnosis and treatment of sialolithiasis. J Oral Maxillofac Surg 1994;52:1240–2.
11. Nahlieli O, Baruchin AM. Sialoendoscopy—three years experience as a diagnostic and treatment modality. J Oral Maxillofac Surg 1997;55:912–8.
12. Nahlieli O, Baruchin AM. Endoscopic technique for the diagnosis and treatment of obstructive salivary gland diseases. J Oral Maxillofac Surg 1999;57:1394–401.
13. Hasson O, Nahlieli O. Endoscopia das glandulas salivares (Sialoendoscopia): nova echnical para remocao da sialolitiase. Revista Da APCD 1998;52:277–9.
14. Nahlieli O, Shacham R, Shlesinger M, Eliav E. Juvenile recurrent parotitis: a new method of diagnosis and treatment. Pediatrics 2004;114:9–12.
15. Nahlieli O, Baruchin AM. Long term experience with endoscopic diagnosis and treatment of salivary gland inflammatory diseases. Laryngoscope 2000:110;988–93.
16. Marchal F, Becker M, Dulguerov P, Lehmann W. Interventional sialoendoscopy. Laryngoscope 2000;110:318–20.
17. Nahlieli O, Shacham R, Yoffe B, Eliav E. Diagnosis and treatment of stricture and kinks in salivary gland ducts. J Oral Maxillofac Surg 2001;59;484–90.
18. Nahlieli O. Development and application of microsalivary gland endoscopy. In MacGurk M, Renehan A, eds. *Controversies in the management of salivary gland disease*. Oxford: Oxford University Press; 2001:274–83.
19. Nahlieli O, London D, Zagury A, Eliav E. Combined approach to impacted parotid stones. J Oral Maxillofac Surg 2002:60: 1418–23.
20. Nahlieli O, Shacham R, Bar T, Eliav E. Endoscopic mechanical retrieval of sialolithiasis. Oral Surg Oral Med Oral Pathol 2003: 95:396–402.
21. Koch M, Zenk J, Bozzato A, Bumm K, Iro H. Sialoscopy in cases of unclear swelling of the major salivary glands. Otolaryngol Head Neck Surg 2005;133:863–8.
22. Iro H, Benzel W, Zenk J, et al. Minimally invasive treatment of sialolithiasis using extracorporeal shock waves. HNO 1993:41: 311–6.
23. Escudier M, Katz P, Capacccio P. Extracorporeal shock wave lithotripsy for salivary calculi: results and future role in the management of salivary calculi. In controversies in the management of salivary gland disease. Oxford: Oxford University Press; 2001:257–70.
24. Brown EJ. Minimally invasive techniques for the treatment of benign salivary gland obstruction. Cardiovasc Interven Radiol 2002;25;345–51.

25. Nahlieli O, Shacham R, Zaguri A. Combined external lithotripsy and endoscopic techniques for advanced sialolithiasis cases. J Oral Maxillofac Surg (accepted for publication, February 2010).
26. Nahlieli O. Sialolithiasis and other obstructions: classical approaches to sialoendoscopy. In Witt, ed. Salivary gland diseases and disorders. New York: Thieme; 2006: 79–93.
27. Buckenham TM, Page JE, Jeddy T. Technical report: Interventional sialography—balloon dilatation of the Stensen's duct stricture using digital subtraction sialography. Clin Radiol 1992;45:34.
28. Brown AL, Shepherd D, Buckenham TM. Per oral balloon sialoplasty: results in treatment of salivary duct stenosis. Cardiovasc Intervent Radiol 1997;20:337.
29. Nahlieli O, Hecht-Nakar L, Nazarian Y, Turner MD. Sialoendoscopy: a new approach to salivary gland obstructive pathology. J Am Dent Assoc 2006;137:1394–400.
30. Nahlieli O, Bar T, Shacham R, et al. Management of chronic recurrent parotitis: current therapy. J Oral Maxillofac Surg 2004;62;1150–5.

9 Sialoendoscopy: The United States Experience

Maria Papadaki, DDS, MD / Joseph McCain, DDS / Leonard B. Kaban, DMD, MD / Maria J. Troulis, DDS, MSc

OUTLINE

INTRODUCTION

Sialoendoscopy has recently gained popularity as a minimally invasive alternative to ablative surgery for obstructive salivary duct disease resulting from stones and strictures. Dilatation of the duct is usually accomplished through the natural orifice and an endoscope is used for visualization, navigation, and treatment of the duct system. Sialoendoscopy may be diagnostic or therapeutic.[1] The purpose of this chapter is to discuss the experience with diagnostic and therapeutic sialoendoscopy at two American Centers—Massachusetts General Hospital, Boston, Massachusetts, and Nova Southeastern University/Baptist Hospital, Miami, Florida.

DIAGNOSTIC SIALOENDOSCOPY

As described in Chapter 8, Interventional Sialoendoscopy, obstructive sialadenitis of the parotid and submandibular glands is a frequently encountered clinical problem. Sialolithiasis, the most common cause of sialadenitis, has been shown to be present in 1.2% of the population in postmortem studies.[2,3] In the United Kingdom, Escudier and McGurk estimated the incidence of sialolithiasis to be one case per 15,000 people per year.[4,5] Traditional diagnostic techniques such as radiographs, sialography, ultrasound, and computed tomographic (CT) scans may fail to identify the cause and location of the obstruction. Sialography is used as the standard imaging method for the diagnosis of sialolithiasis in most institutions in Europe.[1] However, the clinician must consider the risks of allergy to the contrast solution and the possibility of displacing the stone(s) deeper into the duct system. In addition, the radiation dose for sialography is equivalent to approximately 20 chest radiographs.[1,6] In the United States sialography is being replaced by non-contrast CT scan for diagnosis of sialolithiasis.

Diagnostic sialoendoscopy is superior to conventional techniques for identifying the location and etiology of obstruction. In addition to stones, other sources of obstruction such as strictures, ductal polyps, kinks, foreign bodies, adhesions, mucous plugs and anatomic variants, which are often missed by standard imaging, may be identified during sialoendoscopy.[1,7] Koch and colleagues reported that 21 of 103 patients (20.3%) who underwent diagnostic sialoendoscopy for parotid or submandibular gland swelling had sialolithiasis. Strictures, ductal polyps, kinks, adhesions, etc. were present in 56.3% of the patients.[8]

Additional advantages of diagnostic sialoendoscopy include the ability to: (1) identify small stones (calcified or radiolucent), (2) differentiate stones from projecting ductal polyps and mucous plugs, and (3) visualize and dilate the proximal portion of the duct.

INTERVENTIONAL SIALOENDOSCOPY

Introduction and Background

Standard treatment for sialolithiasis has been to remove stones through intraoral incisions when the stones are located in the distal two thirds of Wharton's or Stensen's ducts. When sialoliths are present in the parenchyma of the gland, the proximal portion of the duct, or when the patient has a history of recurrent sialadenitis, conventional management has consisted of excision of the involved salivary gland. These procedures involve a skin incision and the potential for complications such as facial nerve injury. The latter has been reported to be as high as 30%.[9]

Treatment of obstructive sialadenitis with sialoendoscopy is minimally invasive, does not involve extraoral incisions, and avoids removal of the affected salivary gland. Inflamed salivary glands have been shown to gain functionality even after chronic sialadenitis when the obstructive cause is successfully managed.[9] Sialoendoscopy also eliminates the risk of facial nerve injury and patients more readily accept sialoendoscopy versus gland excision.

Interventional sialoendoscopy allows the surgeon to remove stones located near the hilum of the gland, results in less postoperative pain and swelling and fewer complications (infection, nerve injury, hematoma) when compared to traditional treatments.[10–13] In addition, length of hospital stay is reduced and patients may return to work more quickly.

Sialoendoscopy has been reported in large numbers of patients, with a high degree of success.[7,10,11,14–16] Nahlieli has treated over 1000 patients (in a 13-year time period), with an 86% and 89% success rate for the submandibular and parotid gland respectively[12,13,16] (see Chapter 8, "Interventional Sialoendoscopy").

Marchal and colleagues[10] evaluated and treated 129 consecutive patients with suspected ductal disorders of the submandibular gland. Diagnostic sialoendoscopy was used for classifying ductal lesions such as sialolithiasis, stenosis, sialodochitis, or polyps. Interventional sialoendoscopy was used to treat these disorders. The outcome variable was the endoscopic clearing of the ductal system and resolution of symptoms. Diagnostic sialoendoscopy was possible in 131 of 135 glands (97%). Interventional sialoendoscopy was attempted in 110 cases with a success rate of 82%. General anesthesia was used in 12% of cases. Sialoliths were removed in one piece in 70% of 110 cases, fragmentation was required in 26%, and submandibular gland resection was performed in 4% of patients. Minor complications were encountered in 10% of cases.

Katz[11] reported a retrospective case series of 1773 patients with sialolithiasis who underwent sialoendoscopy and/or intracorporeal laser lithotripsy or electromagnetic extracorporeal shockwave lithotripsy. Sialoendoscopy was performed in 1105 cases. Stones were eliminated endoscopically in 96% of these patients. Lithotripsy completely destroyed the stones in 63% of 668 cases; in 35%, stone fragmentation with spontaneous expulsion or ancillary endoscopic removal was achieved. There were no major complications such as nerve or tooth damage. Early in the endoscopic series, one patient required surgical removal of a lodged wire-basket snare. Postoperative edema, bleeding, or infection were temporary and treated as necessary. Four percent of the endoscopic cases and 2% of the lithotripsy cases ultimately required gland excision to remove the stone.

PATIENT SELECTION AND PREOPERATIVE EVALUATION

Theoretically, every patient with documented sialolithiasis or recurrent sialadenitis could be a candidate for sialoendoscopy. Location and size of stones should be taken into consideration before attempting sialoendoscopy. *Stones located at the distal* portion of the duct, close to the ostium, can safely be removed through a mucosal incision. In these cases, however, endoscopic exploration of the duct system to dilate and lavage the duct system and to identify residual calculi should follow stone removal. Stones deep in the gland parenchyma cannot be accessed and removed with the current endoscopes.

The presence of suppurative sialadenitis is a contraindication for sialoendoscopy. Acute infection should be controlled before any intervention. Surgeons should be aware of the limitations of the technique and inform candidate patients accordingly. Risks of the technique include: inability to find the duct, inability to navigate the duct, inability to find the stone or fragment, retrieve, or dislodge it. Complications include infection, damage to lingual or facial nerve, prolonged intubation, equipment failure, and need for gland excision. Patients should understand the possibility of failure to relieve their symptoms with sialoendoscopy and should be required to sign an easily understandable consent form.

Patients and Methods

Patients with sialadenitis have a plain radiographs and a CT scan prior to interventional sialoendoscopy. We do not use sialography. In the presence of sialolithiasis it is very helpful to estimate the size, number, and location of stones prior to sialoendoscopy. If preoperative imaging studies fail to show any obstructive cause, diagnostic sialoendoscopy is attempted.

Recently, we conducted a retrospective study of 94 patients with submandibular ($n = 77$ glands) or parotid ($n = 17$ glands) sialadenitis who underwent sialoendoscopy at Nova Southeastern University/Baptist Hospital, Miami ($n = 52$ patients) or at the Massachusetts General Hospital, Boston ($n = 42$ patients).[17] Patients were included who had obstructive sialadenitis secondary to stones, strictures, or mucous plugs. Patients with neoplasms, purulent infection, duct rupture, or who were medically unfit for general anesthesia were excluded from the study. The diagnosis of sialolithiasis was confirmed by clinical examination, plain radiographs (Panorex, occlusal), and spiral CT scans.

All procedures were performed under general anesthesia on an outpatient basis. Dilatation of the duct through the natural orifice was accomplished with lacrimal or salivary dilators (Karl Storz, Tuttlingen, Germany). A small papillotomy incision was performed in five cases to facilitate access to the duct. A polyethylene tube or an angiocatheter (14 G or 16 G) was inserted into the duct and secured with a suture to protect the duct and maintain access during sialoendoscopy in 39 cases. Three endoscopes of varying sizes were used: the *Karl Storz Nahlieli Sialendoscope* ($n = 44$), a 1 mm semirigid endoscope used with a 1.3 mm diagnostic or 2.3 mm working sheath; the *Karl Storz Marchal Sialendoscope* ($n = 44$), a 1.3 mm semirigid endoscope , this includes the working port (KarlStorz, Tuttlingen, Germany); and the *Millennium Endoscope Sialoview* ($n = 34$), a 1.1 mm diameter endoscope that includes the working port (Millennium Devices Inc, NY).

The scope was navigated through the duct system under continuous irrigation with isotonic saline. Irrigation is essential for visibility because it maintains duct dilatation as the scope is advanced. Stones encountered were either removed or fragmented depending on their size. Small stones (<5 mm) were retrieved using a basket through the working port of the scope (Figure 1). Fragmentation of large stones was achieved with intracorporeal lithotripsy (Calcustat, Karl Storz, Tuttlingen,

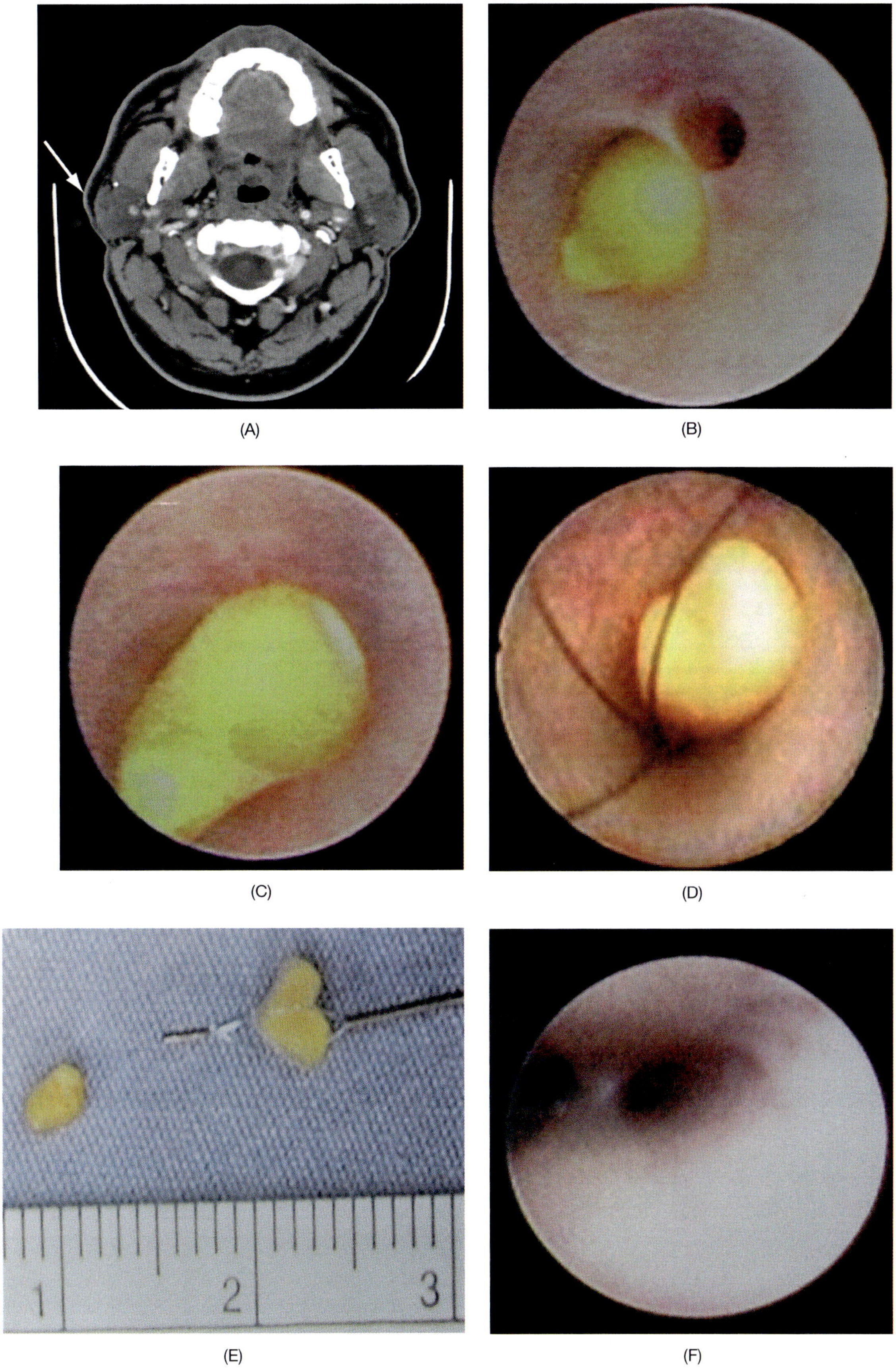

FIGURE 1 • **A**, CT scan of male patient 52 years old demonstrating one stone in the right parotid duct of (*arrow*). **B** and **C**, Sialoendoscopy reveals the stone at a duct bifurcation, wedged in a small branch. Using a basket the stone was pulled forward into the main duct. **D**, Stone engagement into the basket. **E**, Due to the fragility of the stone, it was separated into two pieces while it was removed. **F**, endoscopic view of patent salivary duct.

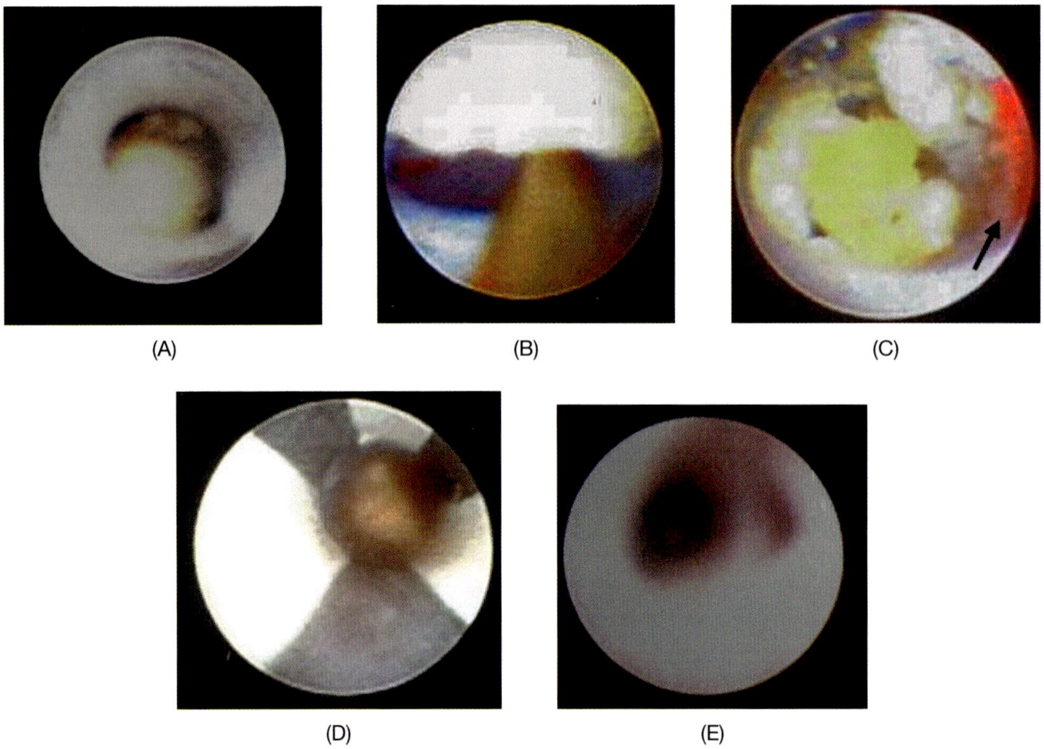

FIGURE 2A • Eleven-year-old female patient with one 5 mm stone at the hilum of the left Wharton's duct. **A**, Endoscopic visualization demonstrates the stone to be wedged in the duct wall. **B**, Lithotriptor probe (yellow) in contact with the stone. **C**, Fragmentation of the stone was achieved. Note small amount of blood caused by minimal injury of the duct wall during *lithotripsy* (arrow). **D**, A stone particle trapped in the angiocatheter-stent. **E**, Patent salivary duct after removal of stone fragments with grasper and basket. Reprinted with permission from Papadaki et al. JOMFS 66:954–962, 2008

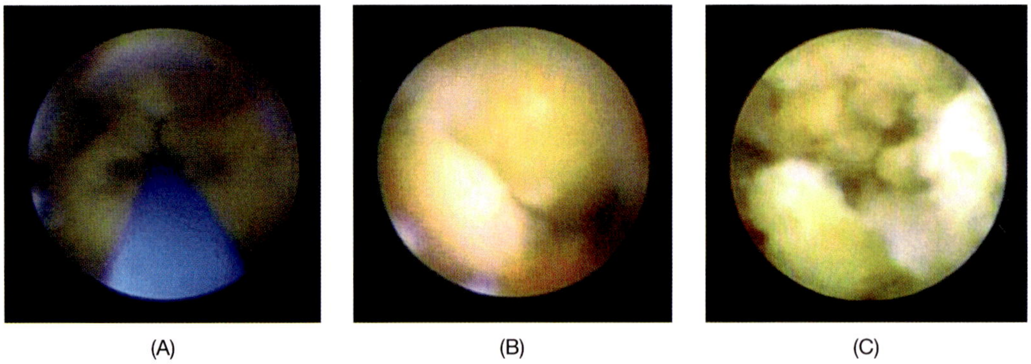

FIGURE 2B • Thirty-year-old male patient with a 10 mm stone. **A**. (Blue) *Laser* (Holmium) against stone repeated a number of times until stone is **B**. fractured, and **C**. fragmented.

Germany) (Figure 2). Strictures and mucous plugs were managed by duct dilatation accomplished with endoscope passage and copious irrigation.

An intraductal stent was maintained for a period of 2 to 4 weeks to prevent postoperative strictures or adhesions, to maintain duct diameter in areas of stricture and to allow residual fragments of calculi to be washed out.

Patients received antibiotics, usually penicillin, for 7 days postoperatively. They were instructed to frequently massage the affected gland, drink lots of water, and use sugar-free lemon drop candies.

Salivary duct dilatation and navigation were accomplished in 91 of 94 patients. In three cases, the duct was not found. Sialoliths were encountered in 73 of 91 patients and strictures or mucus plugs in 18 of 91 patients.

Fragmentation or retrieval of 77 stones was accomplished in 62 of 73 patients (84.93%). Three patients had two stones; three had three stones and two had four stones in the same duct system (Figure 3). Thirty-four patients underwent stone removal in one piece with the basket or a grasper under endoscopic view. In 18 of 62 patients, stones were fragmented with lithotripsy or laser and the particles were subsequently retrieved with the basket or washed out with irrigation. In 5 of 62 patients, stones were fragmented but not all stone particles were removed intraoperatively. These patients experienced extrusion of stone pieces for 1 to 2 months postendoscopy and were totally stone free at the last follow-up (6 months to 4 years).

Finally, in 5 of 62 patients, both fragmentation and stone engagement into the basket failed. In these patients, the stones were removed through intramucosal dissection under the guidance of the endoscope light and the monitor. While the stones were visualized on the monitor, the endoscope light was transilluminating through the oral mucosa pointing to the stone. A mucosal incision was made over the stone and using blunt dissection, the scope light was identified, followed and the stone approached and removed. The angiocatheter that had been previously inserted into the duct was then advanced through the proximal duct segment and sutured in place (Figure 4). In all of these five patients, the stones were wedged in the duct wall and located in the proximal half of Wharton's duct.

In the 11 of 73 patients, where stone removal or fragmentation ($n = 9$) was not possible or stones were not visualized ($n = 2$) although demonstrated on preoperative CT scan, 5 patients underwent subsequent salivary gland removal; three are asymptomatic and do not desire further intervention, one underwent stone retrieval through a papillary incision and subsequent exploratory sialoendoscopy. Two patients are planned to undergo sialoendoscopy again. The average diameter of the sialolith in the failed cases was 1.4 cm. Overall, stone fragmentation was tried in 37 patients and was successful in 18.

Strictures and mucous plugs were managed in 18 of 18 patients. In total, 81 of 91 patients with obstructive sialadenitis who

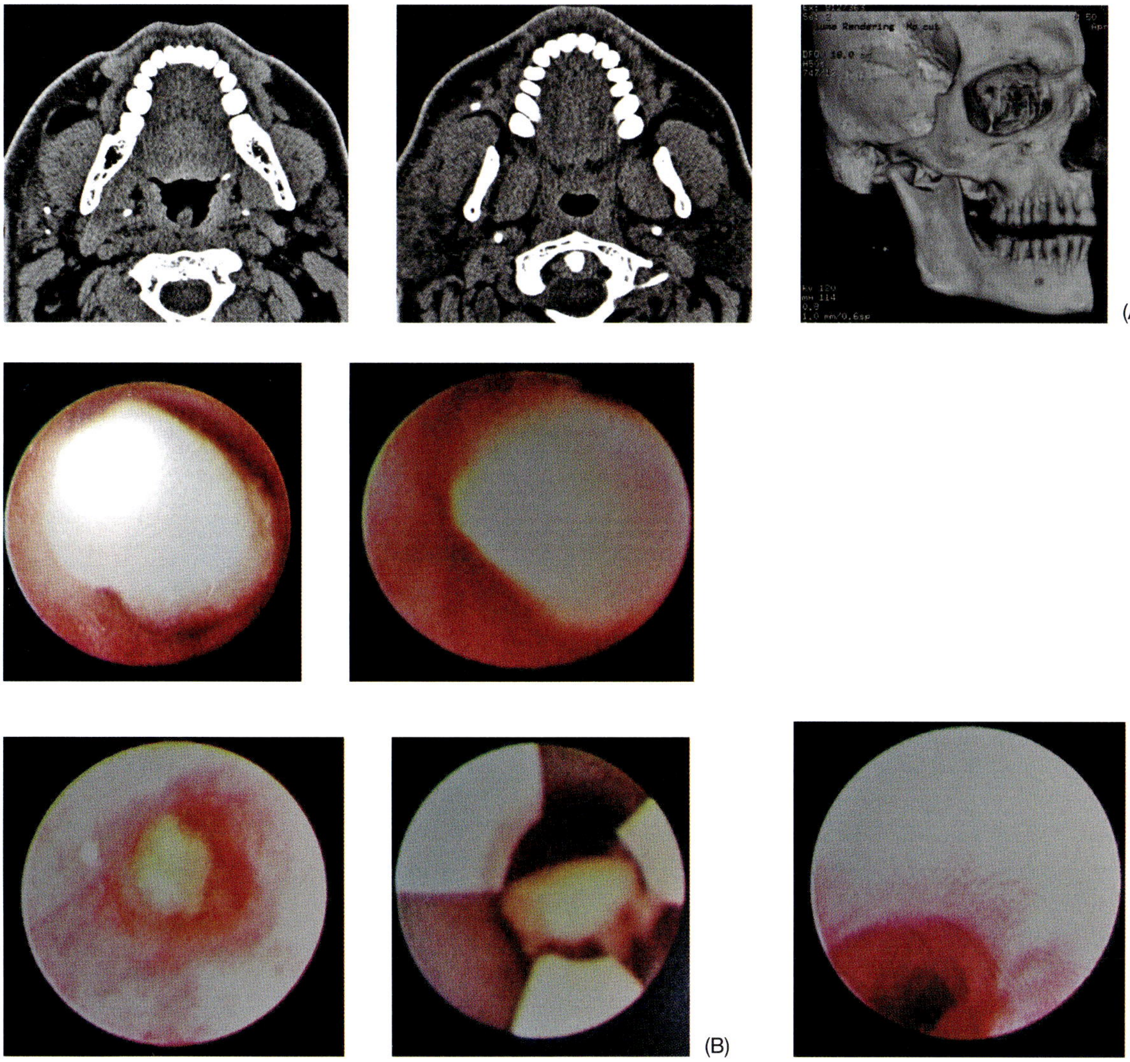

(A)

(B)

(C)

FIGURE 3 • **A,** Axial and 3D CT scan of 53-year-old male patient with four stones in the right parotid gland and duct (*arrows*). **B,** Endoscopic view of the stones. Two stones were removed one with the basket and one was fragmented with the lithotripter. **C,** Patent duct after stone removal.

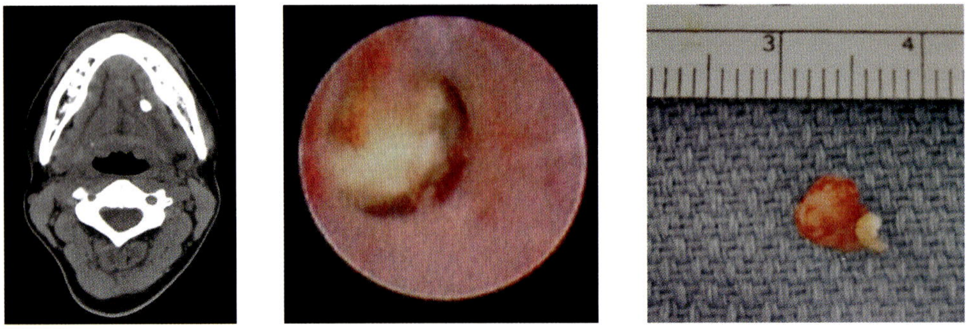

(A) (B) (C)

FIGURE 4 • **A**, CT scan of 43-year-old female patient that demonstrates a 7 mm stone in the left submandibular gland duct. **B**, Endoscopic view of the wedged in the duct wall stone. **C**, Both stone engagement into the basket and fragmentation failed. Stone removal was achieved with the endoscopic-assisted technique, "cut down" of stone with endoscopic localization.

underwent interventional sialoendoscopy had symptomatic relief (89.01%). Mean follow-up was 1 year (range, 6 months to 4 years).[15] To date between the 2 institutions more than 250 patients have been treated using sialoendoscopy.

Complications

Damage, tears, or perforation of the duct, during repeated manipulation of the endoscope and surgical instruments, is a known complication.[7,10,11] This can result in a false passage, which may compromise the success of the procedure. Furthermore, subsequent lingual nerve damage, saliva extravasation, ranula formation, and infection may occur.[11] In the retrospective study reported here, 2 of 91 patients experienced transient lingual nerve paresthesia, 1 patient developed TMJ arthralgia, and 1 remained intubated overnight for airway management due to extravasation of irrigation fluid.

DISCUSSION

Duct Access and Dilatation

Ideally, dilatation of the duct is accomplished through the natural orifice with lacrimal or salivary dilators of increasing diameter. A small papillotomy incision with a number 11 scalpel blade or a carbon dioxide laser may be necessary to facilitate access to the duct in cases of scarring or stricture of the duct orifice. In our case series, a papillotomy was performed in 5 of 91 patients. Nahlieli (see Chapter 8, "Interventional Sialoendoscopy: also suggests longitudinal incision of the duct to facilitate access for the endoscopy or placement of the endoscope into the proximal duct through the incision used to remove an anterior stone.[18]

The ability to reliably dilate the duct and to access it without causing perforation or other damage is essential for successful sialoendoscopy. At Massachusetts General Hospital (MGH), a polyethylene tube or an angiocatheter (14 or 16 gauge) is inserted into the duct and secured with a suture to protect it and maintain access during sialoendoscopy.

To assess the feasibility of using dilating cardiovascular stents to dilate and to maintain access to the parotid and submandibular ducts during sialoendoscopy, a study on 10 Yorkshire pig cadaver heads was performed. In the experimental group (10 parotid and 10 submandibular ducts), a Cypher coronary stent (Cordis Corporation, Miami Lakes, FL) was inserted, expanded to 3.5 mm, and left in place as a permanent scaffolding for the widened duct. The Nahlieli Storz Endoscope (Karl Storz, Tuttlingen, Germany) was inserted into the stented duct and navigated to the hilum of the gland (Figure 5). Navigation of the duct system was repeated 5 times. In the control parotid ($n = 10$) and submandibular ducts ($n = 10$), the endoscope was inserted and navigated to the hilum immediately after dilatation (and repeated five times), without the placement of a coronary stent. Insertion and handling of the

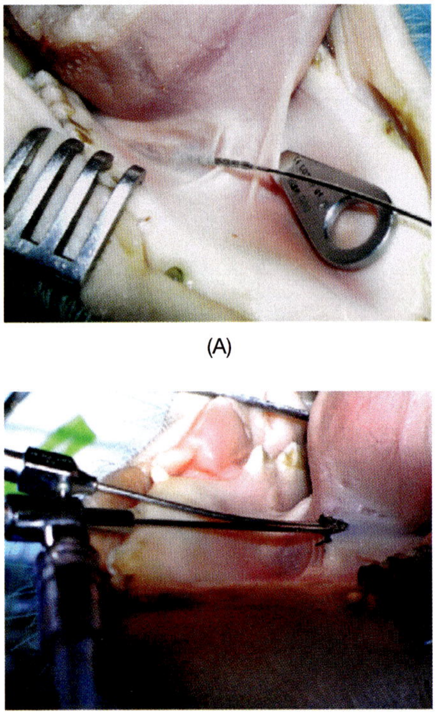

(A)

(B)

FIGURE 5 • **A**, Insertion of the stent with its guiding wire into the right Wharton's duct. A dilator has been introduced in the left submandibular gland duct. **B**, Insertion of the scope (with the sheath) through the stented duct (Nahlieli Storz Endoscope, Karl Storz, Tuttlingen, Germany)

endoscope (2.3 mm sheath) through the stented salivary duct was fast (<20 minutes) and easily achieved in the experimental group (10 parotid and 10 submandibular gland ducts). Insertion of the endoscope was successful in all control parotid and submandibular ducts but with greater difficulty. Use of stents made dilatation and access considerably easier and allowed better manipulation of the scope and instruments. The need for repeated dilatation was eliminated, therefore decreasing the risk of a false passage. Stents should remain in the duct for 2 to 4 weeks postoperatively to allow salivary flow and to prevent stenosis. Expandable stents have potential therapeutic value, by maintaining duct diameter in areas of strictures and straighten kinks.

Stone Removal versus Fragmentation

Sialoendoscopy is a technically demanding procedure. It is performed with very delicate and fragile instruments through an access site of 1 to 4 mm in diameter. Stone retrieval, especially for large stones (greater than 4 mm in diameter), is the most challenging part of the procedure. Nahlieli (see Chapter 8, "Interventional Sialoendoscopy") recommended stone removal in one piece with the basket when the stone is <5 mm in size and can be bypassed by the inactivated basket device.[18] The basket is then activated and slowly withdrawn, thereby engaging the stone. Marchal recommends that 3 mm to 4 mm stone diameter is the borderline size to choose between stone retrieval in one piece or stone fragmentation.[1]

Fragmentation of large stones can be achieved by a variety of methods including external (extracorporeal) or internal lithotripsy, shockwave or laser (CO_2, holmium, dye laser).[7,19–25] In 1990, Iro and colleagues first described extracorporeal shockwave lithotripsy (ESWL) for sialolithiasis and reported success rates of 75% for the parotid and 40% for the submandibular gland.[23,24] Escudier and colleagues reported 33% complete success and 35% symptom free of extracorporeal lithotripsy in 122 patients with sialolithiasis.[22,25]

Intracorporeal shockwave lithotripsy and lasers are used in combination with sialoendoscopy. There are no adequate studies to evaluate the efficacy of these modalities. Escudier reports a success rate of 30% for both external and internal shockwave lithotripsy.[22] Data from urological and gastrointestinal literature indicate that, endoscopic techniques (ureterendoscopy and cholangioscopy), in combination with intracorporeal laser lithotripsy tend to replace ESWL due to higher success rates.[26–34] ESWL is reserved for endoparenchymal stones in these specialties.

It has been demonstrated that laser lithotripsy has higher success rates for stone fragmentation when compared to shockwave lithotripsy. The time required for stone disintegration and the possibility of salivary duct wall injury are limiting factors of shockwave lithotripsy. Holmium laser is widely used for fragmentation of ureteral stones.[30] Some case reports have been published on lithotripsy of salivary stones with holmium laser.[31,32] Our preliminary experience with this contact laser system shows that it is effective and safe (in some stones). Stone fragmentation with holmium laser or shockwave lithotripsy has been successful. The Erbium:yttrium aluminum garnet (Er:YAG) laser is now considered the most efficient for hard tissue ablation[33,34] and may be promising in the field of laser-tripsy of salivary stones.

Endoscopic-Assisted Stone Removal

Large, impacted, and densely mineralized stones are difficult to remove or fragment. In these cases endoscopic-assisted removal under the guidance of the endoscope light and the monitor is helpful. This is also referred to as the "slit" or "slitting" or McGurk procedure (see Chapter 8, "Interventional Sialoendoscopy," Algorithm #1). Five patients in our original series underwent this a modified "McGurk" procedure. In all these patients the stones were wedged in the duct wall and located in the proximal half of Wharton's duct.

CONCLUSIONS

The field of sialoendoscopy is in its infancy and has gained popularity as patients demand and surgeons turn to less invasive treatments. To date, clinical investigators have clearly demonstrated the feasibility and high success rates of endoscopy in diagnosis and treatment of obstructive sialadenitis. To date, we have treated an additional 175 patients using these techniques. Courses are being offered in Boston and in Europe to familiarize surgeons with sialoendoscopy. Interventional sialoendoscopy may eventually be the treatment of choice for obstructive salivary gland disease as long as the clinician and the patient accept the risks and limitations.

References

1. Marchal F, Dulguerov P, Becker M, et al. Specificity of parotid sialendoscopy. Laryngoscope 2001;111:264–71.
2. Teymoortash A, Ramaswamy A, Werner JA. Is there evidence of a sphincter system in Wharton's duct? Etiological factors related to sialolith formation. J Oral Science 2003;45:233–5.
3. Marchal F, Dulguerov P. Sialolithiasis management: the state of the art. Arch Otolaryngol Head Neck Surg 2003;129:951–6.
4. Escudier MP, McGurk M. Symptomatic sialadenitis and sialolithiasis in the English population, an estimate of the cost of hospital treatment. Br Dent J 1999:8;186:463–6.
5. Iro H, Zenk J, Escudier MP, Nahlieli O, et al. Outcome of minimally invasive management of salivary calculi in 4,691 patients. Laryngoscope 2009;119:263–8.
6. Gundlach P, Hopf J, Linnarz M. Introduction of a new diagnostic procedure: salivary duct endoscopy (sialendoscopy) clinical evaluation of sialendoscopy, sialography, and X-ray imaging. Endosc Surg Allied Technol 1994;2:294–6.
7. Nahlieli O, Baruchin AM. Long-term experience with endoscopic diagnosis and treatment of salivary gland inflammatory diseases. Laryngoscope 2000;110:988–93.
8. Koch M, Zenk J, Bozzato A, Bumm K, Iro H. Sialoscopy in cases of unclear swelling of the major salivary glands. Otolaryngol Head Neck Surg 2005;133:863–8.
9. O'Brien CJ. Current management of benign parotid tumors—the role of limited superficial parotidectomy. Head Neck 2003;25:946–52.
10. Marchal F, Dulguerov P, Becker M, Barki G, Disant F, Lehmann W. Submandibular diagnostic and interventional sialendoscopy: new procedure for ductal disorders. Ann Otol Rhinol Laryngol 2002;111:27–35.
11. Katz P. New techniques for the treatment of salivary lithiasis: sialoendoscopy and extracorporal lithotripsy: 1773 cases. Ann Otolaryngol Chir Cervicofac 2004;121:123–32.

12. Nahlieli O, Shacham R, Bar T, Eliav E. Endoscopic mechanical retrieval of sialoliths. Oral Surg Oral Med Oral Pathol Oral Radiol Endod 2003;95:396–402.
13. Nahlieli O, Shacham R, Shlesinger M, Eliav E. Juvenile recurrent parotitis: a new method of diagnosis and treatment. Pediatrics 2004;114:9–12.
14. Konigsberger R, Feyh J, Goetz A, et al. Endoscopic controlled laser lithotripsy in the treatment of sialolithiasis. Laryngorhinootologie 1990;69:322–3.
15. Katz PH. Endoscopy of the salivary glands. Ann Radiol 1991;34: 110–3.
16. Nahlieli O, Neder A, Baruchin AM. Salivary gland endoscopy: a new technique for diagnosis and treatment of sialolithiasis. J Oral Maxillofac Surg 1994;52:1240–2.
17. Papadaki ME, McCain JP, Kim K, et al. Interventional sialoendoscopy: early clinical results. J Oral Maxillofac Surg 2008; 66:954–62.
18. Nahlieli O. Salivary gland inflammatory disorders in children. In: Kaban LB, Troulis MJ, editors. Pediatric oral and maxillofacial surgery. Philadelphia: Elsevier Science 2004; 190–3.
19. Marchal F, Becker M, Dulguerov P, Lehmann W. Interventional sialendoscopy. Laryngoscope 2000;110:318–20.
20. McGurk M, Escudier MP, Brown E. Modern management of obstructive salivary gland disease. Ann R Australas Coll Dent Surg 2004;17:45–50.
21. McGurk M, Prince MJ, Jiang ZX, King TA. Laser lithotripsy: a preliminary study on its application for sialolithiasis. Br J Oral Maxillofac Surg 1994;32:218–21.
22. Escudier MP. The current status and possible future for lithotripsy of salivary calculi. Atlas Oral Maxillofac Surg Clin North Am 1998;6:117–32.
23. Iro H, Schneider T, Nitsche N, Waitz G, Ell C. Extracorporeal piezoelectric lithotripsy of salivary calculi. Initial clinical experiences. HNO 1990;38:251–5.
24. Iro H, Waitz G, Nitsche N, et al. Extracorporeal piezoelectric shock-wave lithotripsy of salivary gland stones. Laryngoscope 1992;102:492–4.
25. Escudier MP, Brown JE, Drage NA, McGurk M. Extracorporeal shockwave lithotripsy in the management of salivary calculi. Br J Surg 2003;90:482–5.
26. Anagnostou T, Tolley D. Management of ureteric stones. Eur Urol 2004;45:714–21.
27. Neuhaus H, Zillinger C, Born P, et al. Randomized study of intracorporeal laser lithotripsy versus extracorporeal shock-wave lithotripsy for difficult bile duct stones. Gastrointest Endosc 1998; 47:327–34.
28. Preminger GM. Management of lower pole renal calculi: shock wave lithotripsy versus percutaneous nephrolithotomy versus flexible ureteroscopy. Urol Res 2006;34:108–11.
29. Strohmaier WL, Schubert G, Rosenkranz T, Weigl A. Comparison of extracorporeal shock wave lithotripsy and ureteroscopy in the treatment of ureteral calculi: a prospective study. Eur Urol 1999;36:376–9.
30. Farkas A, Peteri L, Lorincz L, et al. Holmium: YAG laser treatment of ureteral calculi: a 5-year experience. Lasers Med Sci 2006;21:170–4.
31. Katz P, Fritch MH. Salivary stones: innovative techniques in diagnosis and treatment. Curr Opin Otolaryngol Head Neck Surg 2003;11:173–8.
32. Kerr PD, Krahn H, Brodovsky D. Endoscopic laser lithotripsy of a proximal parotid duct calculus. J Otolaryngol 2001;30: 129–30.
33. Gonzalez C, van de Merwe WP, Smith M, Reinisch L. Comparison of the erbium-yttrium aluminum garnet and carbon dioxide lasers for in vitro bone and cartilage ablation. Laryngoscope 1990;100:14–7.
34. Raif J, Vardi M, Nahlieli O, Gannot I. An Er:YAG laser endoscopic fiber delivery system for lithotripsy of salivary stones. Lasers Surg Med 2006;38:580–7.

10

Biology of Distraction Osteogenesis

Maria Papadaki, DDS, MD / Maria J. Troulis, DDS, MSc /
Leonard B. Kaban, DMD, MD

OUTLINE

GENERAL CONCEPTS

Distraction osteogenesis (DO) is a unique method of generating new bone, using the body's healing capacity to avoid the need for a bone graft. A corticotomy or osteotomy is made and a distraction device is fixed to the bone across the cut. Activation of the device results in tension and stretching of the callus between the bone segments. New bone is formed across the gap as the cut edges are separated. Conceptually, distraction osteogenesis can be considered a form of in vivo tissue engineering.

Codivilla, in 1905, was the first to describe limb lengthening by DO.[1] Ilizarov popularized the technique in the 1950s by reporting its use in large numbers of patients and by publishing the results of his experimental studies in large animals. Ilizarov is considered the "father" of DO.[2,3] Snyder, in 1972, was the first to use DO to expand the mandible in a canine model.[4] Subsequently, McCarthy et al (1992) used DO for mandibular lengthening in four syndromic patients. The work of McCarthy's group stimulated tremendous interest in DO and changed the way many craniomaxillofacial deformaties are correctly managed. Perrott et al, in 1993, reported mandibular widening by DO (Figure 1).[5,6]

CLINICAL APPLICATIONS

Distraction osteogenesis (DO) is now a commonly used technique for skeletal expansion of the lower jaw, including symmetric and asymmetric mandibular lengthening in the vertical, sagittal, and transverse planes, and correction of mandibular discontinuity defects (transport DO) (see Chapter 12, "Applications of Distraction Osteogenesis for the Mandible: An Overview" through Chapter 16, "Alveolar Distraction Osteogenesis").[7] Expansion of the midface by DO has also been frequently reported at the LeFort I and III levels (see Chapter 17, "Distraction Osteogenesis for Correction of Maxillary Deformities" and Chapter 18, "Midface Distraction Osteogenesis").[8,9] DO has become one of the standard techniques for the correction of syndromic and nonsyndromic congenital deformities (eg, hemifacial microsomia, Treacher Collins syndrome, Nager syndrome, midface hypoplasia secondary to cleft lip/palate and syndromic craniosynostosis) and acquired midface and mandibular deficiencies that require advancement of more than 10 mm.

Advantages and Disadvantages

In the past, skeletal expansion was achieved by the creation of osteotomies, acute bone repositioning and autogenous bone grafts to fill the gaps. Complications such as infection, donor site morbidity, skeletal relapse, and bone graft loss or nonunion, were, not infrequently, reported in these cases.[10,11] In contrast, DO offers unique clinical benefits that include the possibility of large magnitudes of bone advancement, simultaneous soft-tissue histogenesis, and decreased risk of damaging adjacent nerves and tooth buds. Furthermore, DO is less invasive, requires shorter operating time, and eliminates donor site morbidity. Disadvantages of DO include lengthy treatment protocols, difficulty in accurately controlling the vectors of movement, and dependence on patient and parent cooperation. Ongoing studies are addressing these limitations.[11]

The biologic process of new bone formation as a result of distraction osteogenesis has attracted great interest. Ilizarov himself referred to a "new discovered biological regularity" that achieves bone and soft tissue genesis and growth.[2] An understanding of

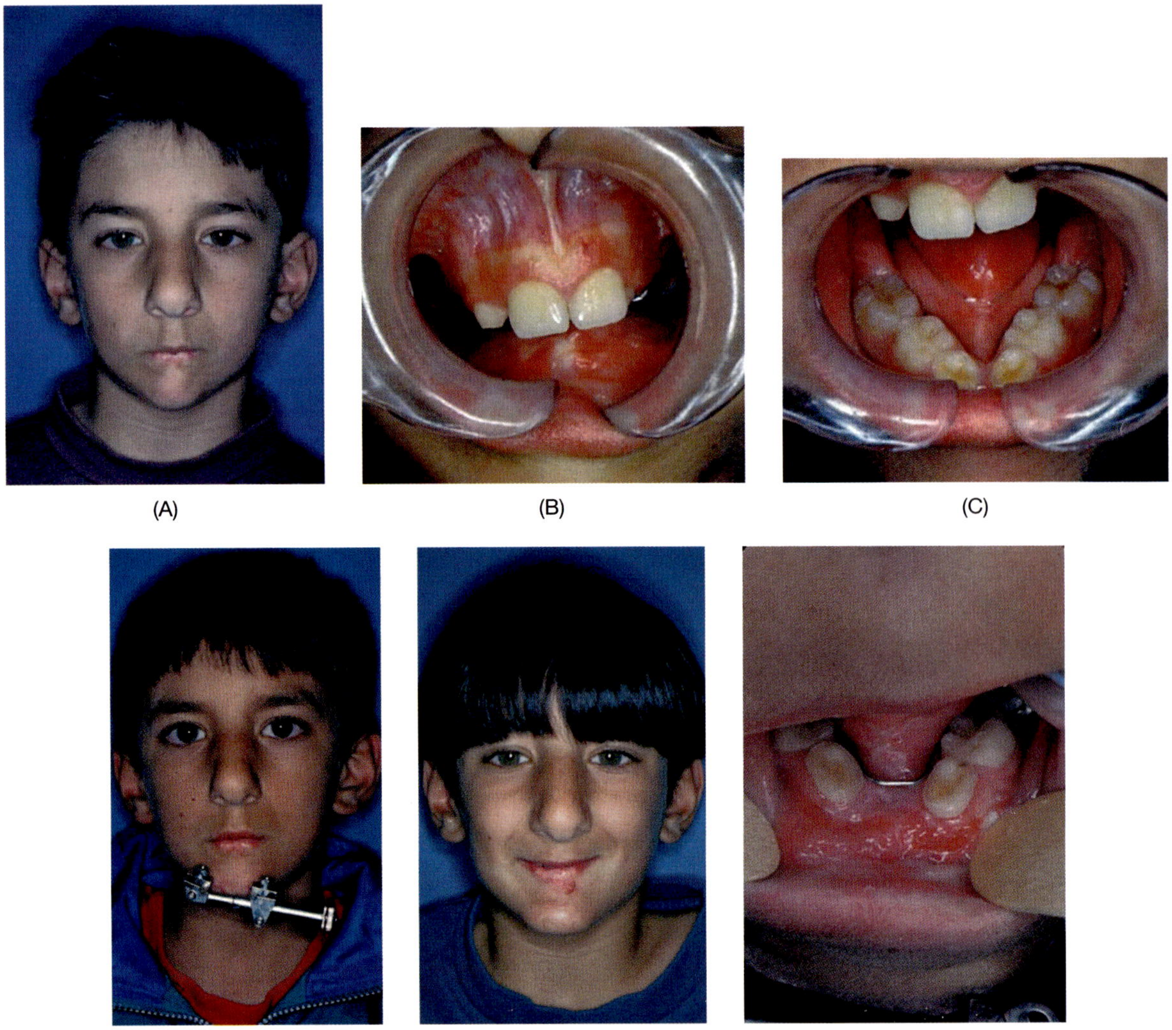

(A) (B) (C) (D) (E) (F)

FIGURE 1 • First case of symphysis widening by DO in a child with Aglossia-Adactylia Syndrome reported by Perrott, Kaban and Vargervik in 1993. *A*, *B*, and *C*, Frontal and intraoral photographs of the patient showing the narrow mandible and dental arch. *D*, Extraoral distraction device in place during treatment. *E*, Frontal picture of the patient post treatment. *F*, intraoral photograph shows widening of the dental arch with new bone and soft tissue formation between the two premolars. Reproduced from Perrott et al.[6] with permission from Elsevier.

the biology of the facial DO wound may allow surgeons to manipulate it to increase the rate, quality, and quantity of bone formation while decreasing the treatment period. Thus, new, shorter distraction protocols could be developed. Since 1997, biology of DO has been studied in a minipig animal model developed at the Skeletal Biology Research Center of the Massachusetts General Hospital, Department of Oral and Maxillofacial Surgery (Figure 2). In this model, the distraction protocol consists of zero-day latency, 1 mm/day rate, rhythm of 0.5 mm twice per day and of fixation two times the number of mm (or days) of distraction.[11]

BIOLOGIC PARAMETERS

Biologic parameters that influence bone formation and its biomechanical properties following DO include the nature of the corticotomy, duration of latency, rate of distraction, rhythm, and duration of fixation. To understand the biology of DO requires documentation of the clinical healing process, the histology of the DO wound, the cellular populations responsible for bone formation, and the growth factors involved in the phases of DO, that is, latency, active distraction, and consolidation or fixation. The basic biology of distraction osteogenesis was discovered by Ilizarov, in long bone animal models, during the 1950s. His papers described the importance of preservation of the periosteum and intramedullary blood flow by doing a "low energy" corticotomy. He documented the importance of the rate and rhythm and the rigidity and duration of fixation. Both the long bone and craniomaxillofacial distraction protocols in use today are all based on the work of Ilizarov.[2]

Corticotomy/Osteotomy

Ilizarov demonstrated that successful bone formation by distraction osteogenesis depends on the preservation of the periosteum during corticotomy.[3] It is conceivable that the osteogenic potential of the periosteum outweighs the osteogenic potential of the

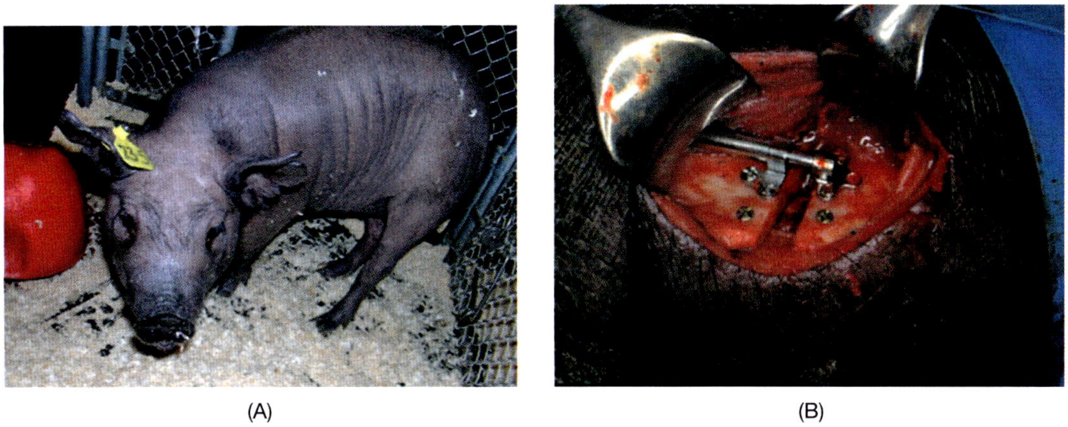

FIGURE 2 • *A*, Yucatan minipigs have been used since 1997, at the Skeletal Biology Research Center of the Massachusetts General Hospital, Department of Oral and Maxillofacial Surgery, to study the biology of DO. *B*, Osteotomy on the minipig mandible and fixation of the distraction device. The device has been activated intraoperatively to confirm the completeness of the osteotomy. Reproduced from Troulis et al. with permission from Elsevier.[18]

endosteum. This idea has led to the construction of supra-periosteal distraction devices and to the development of minimally invasive surgical techniques. In distraction of rat and rabbit long bones and sheep mandibles, new bone formation was predominantly seen in the subperiosteal region. However, there are also reports documenting simultaneous endosteal and periosteal bone formation in these species.[12] Other studies (rat, mouse, and dog) describe periosteal bone formation occurring in the initial treatment phase, whereas endosteal bone formation is more prominent after about 2 weeks of treatment.[12,13] Aronson et al reported that with increasing age there is less endosteal bone formation, whereas periosteal bone formation remains unaltered.[14] Studies in our laboratory have demonstrated a relatively low percentage of endosteal bone fill and a significant role of periosteal bone formation in the healing of the mandibular distraction wound.[12,15]

Latency

Latency period is defined as the time from the corticotomy/osteotomy to the start of activation of the distraction device. Hematoma formation between and around the bone segments occurs in this period as a result of blood vessel disruption. Actually, latency period does not differ from the first stage of fracture healing. Within 3 days, hematoma proceeds to the formation of granulation tissue that consists of inflammatory cells, fibroblasts, collagen, and capillaries. By the fifth day, soft callus is formed that is characterized by increased vascularity, mesenchymal cells and fibrous tissue.[17]

A latency period of 7 to 21 days has been reported in long bone DO, and 4 to 10 days has been the standard for craniomaxillofacial DO.[2,3,5,18,19] The guidelines are based on Ilizarov's original experiments in the long bones. The rational for latency period is to permit soft callus formation that will be stretched in the next phase of distraction osteogenesis.

However, the anatomy of the craniomaxillofacial region, especially its rich blood supply, is significantly different from that of long bones. Therefore, it may be possible that a shorter or even zero-day latency would be feasible in the facial skeleton. Tavakoli et al[20] performed mandibular DO in sheep and found that latency periods of zero, 4, and 7 days resulted in equivalent bone formation on histologic evaluation. The biomechanical properties and bone density of the regenerated bone were equal in all groups. Aronson and Shen also found that bone regeneration and consolidation were best after zero-day latency in metaphyseal and diaphyseal long bone lengthening.[21]

Similarly, experiments in our laboratory demonstrated that mandibular DO could be successfully accomplished with no latency period.[19] A porcine model (Yucatan minipigs) was developed and used for the study of mandibular DO (Figures 2 and 3). For analysis of latency, mandibular osteotomies were distracted after zero ($n = 3$) or 4 ($n = 2$) days at a rate of 1 mm/day (7 days) with 24 days fixation. Clinical and radiographic evaluation of groups showed that latency of zero or 4 days had equivalent healing (Figure 3). These results were confirmed by histologic and histomorphometric evaluation of the bone.[15]

Furthermore, maxillary DO in the porcine model ($n = 3$ minipigs) with zero-day latency (1 mm/day, 12 days) showed excellent clinical stability, high radiographic bone fill scores and tissue continuity with many bone islands histologically, at the end of a 24-day fixation period (Figure 4).[16]

Thus, soft callus formation may not be necessary to start active distraction. Early application of tension forces at the stage of hematoma formation results in new bone formation.

Rate

Rate refers to the total number of millimeters of lengthening per day. The typical rate in both long bones and the maxillofacial skeleton is 1 mm per day divided into two doses (rhythm = 2). Multiple investigators, beginning with Ilizarov, have reported that a 1 mm/day rate produces superior bone formation when compared to 2 mm and 4 mm/day.[3,7,8,11,15,19]

For analysis of rate, 12 minipigs underwent mandibular DO in our laboratory. Osteotomies were distracted at 1 mm ($n = 4$), 2 mm ($n = 4$), or 4 mm ($n = 4$) per day to produce a 12-mm gap with 24 days fixation. DO wounds were assessed in vivo with bimanual palpation to detect mobility across the gap at the end

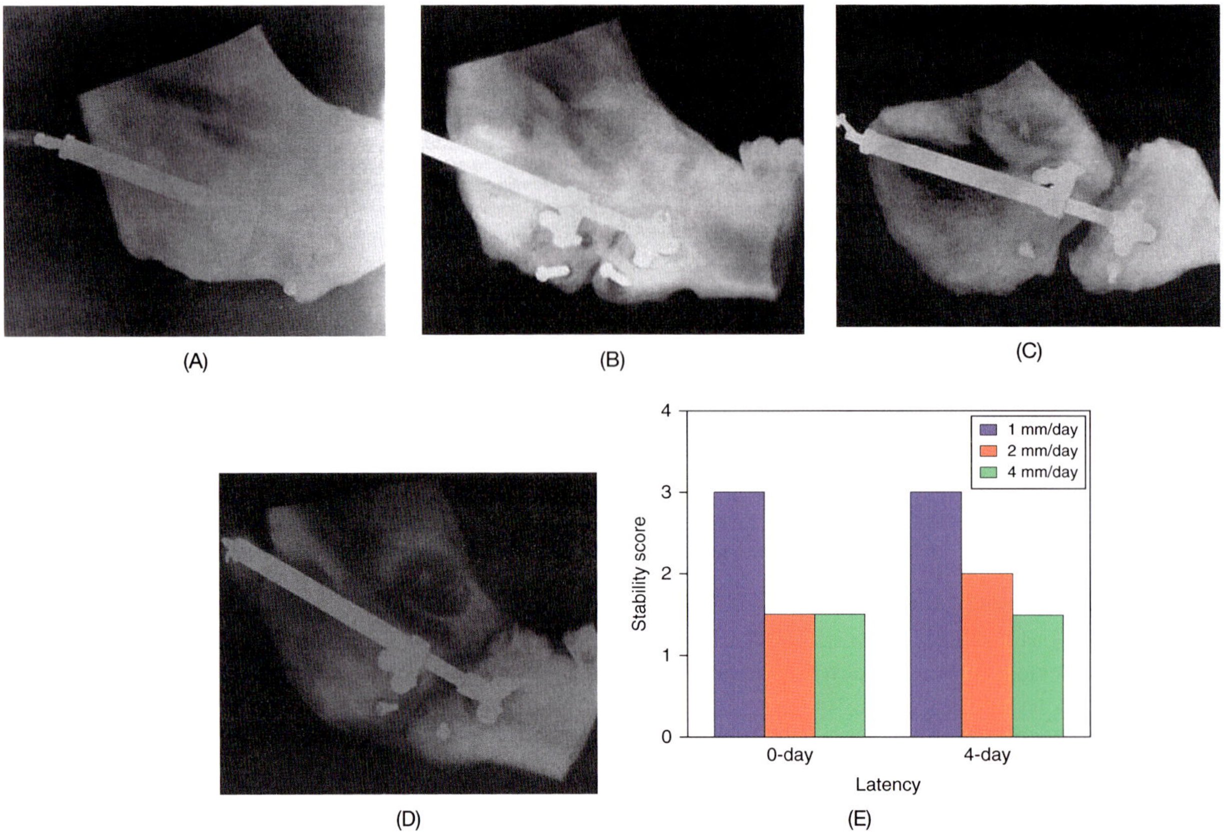

FIGURE 3 • Seventeen minipigs underwent unilateral distraction osteogenesis of the mandible to define the effects of latency and rate on bone formation. For analysis of Latency (L), mandibles were distracted after 0 (n = 3) or 4 (n = 2) days at a rate of 1 mm/day with 14 days fixation. For analysis of rate, osteotomies were distracted after 0 or 4 days L at 1 mm/day (n = 2 0L, n = 2 4L), 2 mm/day (n = 2 0L, n = 2 4L) or 4 mm/day. Clinical and radiographic evaluation of groups showed that latency of 0 or 4 days had equivalent healing. ***A***, Ex vivo radiographic assessment at the end of fixation period shows uniform opacification across the gap at the 0 latency with 1mm/day distraction rate group. ***B***, Ex vivo radiograph after 4L, 1 mm/day rate shows uniform opacification across the gap except a small area inferiorly. ***C***, minimal opacification is demonstrated after 0L, 2 mm/day rate at the ex vivo radiograph. ***D***, Ex vivo radiographic assessment at the end of fixation period after 0 L, 4 mm/day rate shows incomplete and irregular bone formation. ***E***, Stability was greatest at the 1 mm/day group either with 0 L or 4 L. Reproduced from Troulis et al. with permission from Elsevier.[19]

of fixation period. Harvested specimens were evaluated by gross appearance and bimanual palpation. Standardized radiographic techniques were used to estimate bone density. Stability was greatest at the 1 mm/day group compared to 2 mm/day and 4 mm/day rate (Figure 3).[19]

Histologic examination of the regenerate stained with Giemsa-Eosin showed no difference in the type of bone (woven versus lamellar) in relation to the rate of distraction, although the percent area of new bone and the presence of osteoblasts were inversely related to rate. In all specimens with bone formation, the bone was of the woven type. In addition, at zero, 8, and 24 days of fixation, the rate of 1 mm per day was associated with a greater quantity of osteoblasts within the distraction wound than a rate of 2 or 4 mm/day.

Histomorphometric evaluation of the bone in the three groups showed higher bone fill in the 1 mm/day group.[15] On the 24th day of fixation, bone fill percentage ranged from 25% to 67% in the group of 4 mm/day rate, 5% to 68% in the group of 2 mm/day, and 85% to 100% in the 1 mm/day group. Interestingly, in the group of 4 mm/day, bone fill percentage increased as the fixation period increased up to 76.7% on the 29th fixation day. Therefore, it may be possible to achieve osteogenesis at a faster rate than the standard 1 mm/day. However, this would require prolonging the fixation period by a considerable amount of time, which would not be desirable.

Rhythm

The rhythm is the number of times per day the distractor is activated. It may vary from 1 to 4 activations per day in standard distraction protocols.

Ilizarov was the first to refer to high distraction rhythm with a motorized distraction device that continuously elongated the bone 1 mm over a 24-hour period.[2] He presented evidence to show that the new device produced more rapid osteogenesis.

Continuous DO requires a device that is programmed to activate independently and constantly at a predetermined rate without patient activation. Hypothetically, a continuous rhythm

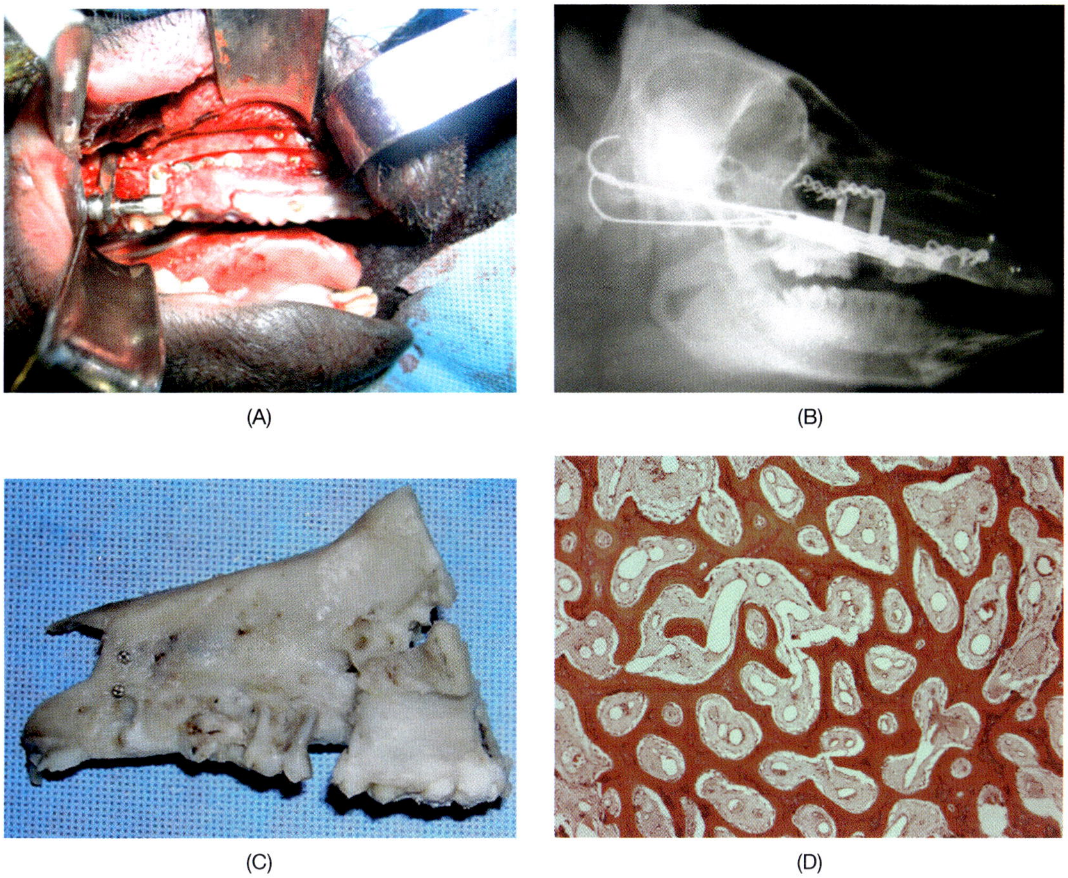

FIGURE 4 • Three Yucatan minipigs underwent bilateral distraction osteogenesis of the maxilla to study the biology of DO. The distraction protocol used was 0 days latency, 1 mm/day rate, 12 days distraction period and 24 days fixation. ***A***, Intraoperative photograph shows the osteotomy at the Le Fort I level and the fixated device on the right side. The osteotomy ended between the first and second molar. ***B***, Lateral cephalogram at the end of fixation period. ***C***, Photograph of the ex vivo specimen shows complete ossification along the zone of advancement. ***D***, Histological examination of the distraction gap at the sagittal level shows numerous trabeculae (H&E, X40).[16]

could allow for a DO rate greater than 1 mm/day. The advantages of continuous DO with an automated device would be a shorter treatment time and elimination of patient performed activation.

Animal studies have been performed to evaluate continuous distraction using motor-powered devices. In 1996, Schmelzeisen et al demonstrated a miniature electric-powered distractor fixed to the mandible and connected to a subcutaneous battery pack. The gearing fixed to the motor provided a force of 10 N (Newtons) and a timer module provided impulses of 10 ms (milliseconds) every 6 minutes. The authors encountered some technical difficulties with gear breakage, mandibular fracture, and infections postoperatively.[22] However, this study confirmed that continuous distraction with a totally buried, motor-driven device is possible.

Ploder et al, in 1999,[23] reported on implanted motor driven devices for mandibular distraction. Investigators achieved mandibular lengthening of 13.6 mm by continuous distraction osteogenesis in three sheep over a period of 14 days. The device was a newly designed, totally buried electromechanical distractor. The micromotor and gearing were built in a titanium case and connected via a flexible coiled lead, embedded in a silicone elastomer, to a 3.6 V lithium cell battery. The device was magnetically activated (a reed-contact from which a buzzer sent acoustic checks).

More recently, Kessler et al (2000) designed a microhydraulic distraction cylinder. It was assembled from a cylinder and a piston (both fabricated from medical steel) and united by a fixation plate. The investigators used the device in four animals but performed continuous distraction in two pigs and intermittent distraction in the other two pigs. The pressure peaks were measured during distraction. In the intermittent distraction group, pressure peaks of up to 25×10^5 Pa (Pascals) were reached but fell after 4 to 6 hours to 8 to 9×10^5 Pa. In the continuous distraction group, a mean pressure level of 12 to 15×10^5 Pa was measured.[24]

Finally, the same group compared the pressure needed for continuous versus intermittent distraction using the microhydraulic distractor in minipigs.[25] They found that the pressure needed for continuous distraction was considerably lower (1200 to 1300 Kilopascal-kPa) than that for intermittent lengthening (up to 2500 kPa). There was some evidence that bone healing was accelerated with continuous compared to intermittent distraction as shown by ultrasonography and scanning electron microscopy. The hypothesis that continuous distraction osteogenesis may decrease the treatment period is accepted by many investigators. Advances in micromotor technology will eventually result in practical continuous DO devices for clinical use. Preliminary data from our laboratory is encouraging that an automated, continuous, curvilinear distractor is possible. Furthermore, these devices may be more comfortable and allow for an overall shorter treatment period. (Figure 5).

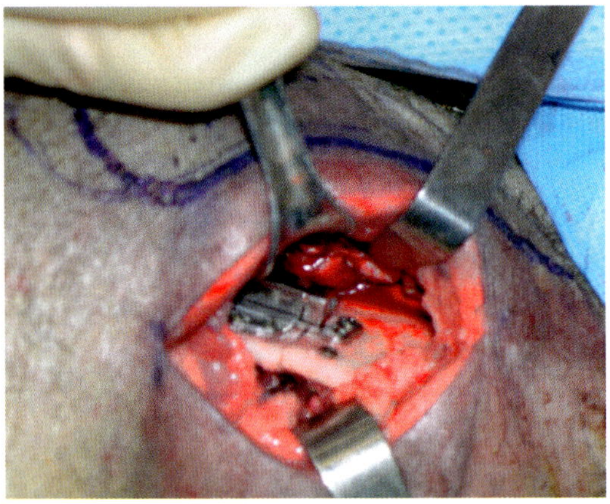

(A)

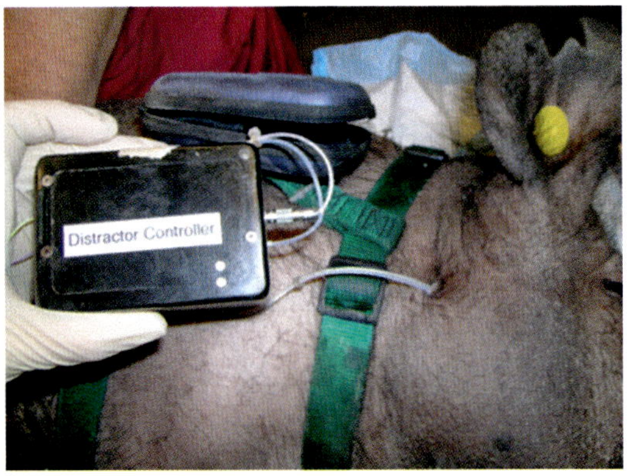

(B)

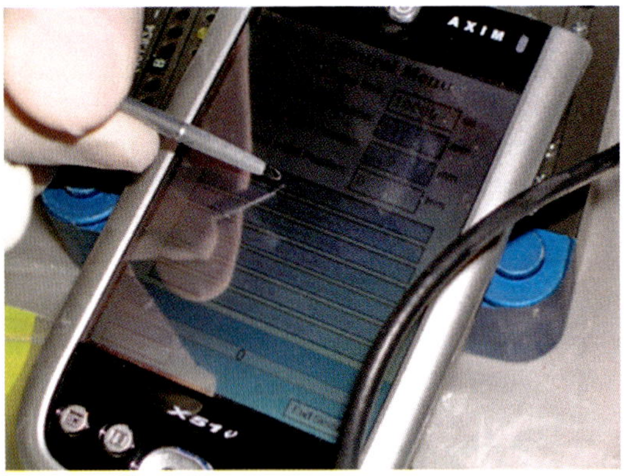

(C)

FIGURE 5 • Six minipigs underwent successful continuous distraction osteogenesis of the mandible. *A*, Intraoperative photograph shows the osteotomy, the fixated device and the marker screws. *B*, The device is connected to the distractor contol box. This is maintained in a pocket on the back of the minipig. *C*, The control box is connected with the PDA. A software is used to control the rate and measure the gap. (NIH-SBIR 4R44DE019047)[36]

Fixation Period

The fixation or consolidation period extends from the end of active distraction to the removal of the device when adequate bone fills the gap. As described by Ilizarov and Aronson in long bones, the duration of fixation required is at least twice the number of millimeters of distraction (at a rate of 1 mm/day) in days. Empirically, most clinicians maintain fixation for approximately 6 to 12 weeks. Mineralization of the regenerate occurs during fixation.[7,8,11,15,19]

A series of studies were performed in the Skeletal Biology Research Center at MGH to document the progression and pattern of bone formation in the porcine model in relation to duration of fixation (zero, 8, 16, and 24 days). The distraction wounds were evaluated by histology, plain radiographs, ultrasound, three-dimensional computerized tomography, and biomechanical testing.[26–28] Radiographs and sagittal histologic sections (from the center of mandible) were evaluated by computer morphometrics and by a semi-quantitative bone formation score. Three-dimensional data from computed tomographic scans for the percent of bone fill were also obtained (Figure 6).

Multiple sagittal CT sections were obtained to delineate volume of bone in the gap, expressed as % bone fill.[26] In each section, the distraction gap was identified and the bone volume was calculated. It was demonstrated that the mean percent bone volume and radiographic bone fill scores increased with fixation time from 16.7% at zero days, to 64% at 24 days fixation (Figure 6). The lowest value (13.8%) was found in a mandible that was distracted 4 mm/day. The highest percent volume of bone (64%) occurred at a rate of 1 mm/day and the end of fixation. This bone volume remained below control values (81.3%). However, clinical stability and radiographic bone fill scores were high. Histologic examination also showed that the percent area of new bone within the distraction wounds increased with increasing duration of fixation but was considerably less than control non-operated mandibles.[12,26]

CT bone volume measurements correlated well with the values of bone fill estimated from plain radiographs. The CT measurements of percent bone volume also were consistent with biomechanical stiffness data in the same experimental model previously reported. There was, however, no significant correlation between the percent volume of bone on CT scans and the ultrasound data on the healing distraction wound reported by Thurmuller et al.[27] Ultrasound is sensitive enough to identify changes in the DO wound (eg, osteoid formation) as early as the first or second week of distraction (second or third week after the osteotomy). Cortex formation can be detected 8 weeks after the osteotomy. Ultrasonography, however, documents the progress of new bone formation only until bony bridging has occurred. After that, the ultrasound beam cannot penetrate the wound to help determine volume or maturity of bone.

BIOLOGIC OUTCOMES

Histologic and Cellular Profile

Despite differing distraction protocols (latency, rate, gap size, fixation period), histologic studies on distraction osteogenesis have consistently shown that, independent of species (mouse,

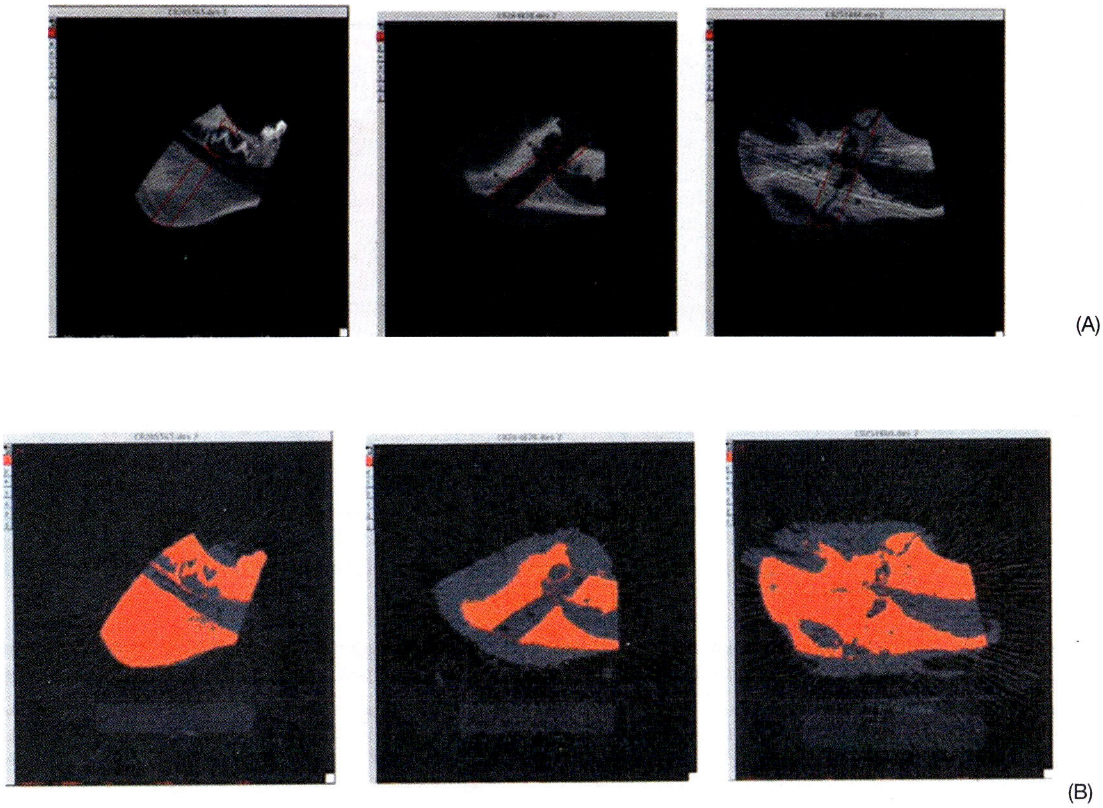

FIGURE 6 • *A*, Sagittal sections of CT scans with the distraction gap identified and marked with rectangles. *B*, CT scans of the same specimens. Areas of bone, defined by a standard threshold, are shown in red. The mean percent area of new bone (pooled sample) increased with fixation: 16.8% at 0 days to 64% at 24 days. At all time points, CT values (as defined by standard threshold and seen in red) and bone fill scores were higher at 1 mm/day rate. Reproduced from Zimmermann et al. with permission from Elsevier.[26]

rat, rabbit, dog, sheep, goat, minipig, human), the deposition of matrix, osteoid, and new woven and/or lamellar bone occurs within a scaffold of fibers that are oriented parallel to the direction of distraction. This is true for long tubular bones as well as for bones of desmal origin, for example, the mandible.[12] The architecture of three distinct histologic zones, which has been described for long bone distraction and for mandibular lengthening in dogs and sheep, was also demonstrated in our model of porcine mandibular distraction.

Histologic examination by Yates et al[29] revealed hematoma or fibrin clot on the fourth day of distraction. A few blood vessels were present in wound tissues in two animals. On the seventh day of distraction, the DO wounds contained granulation tissue and hematoma. Fibroblasts and many blood vessels were present at this time point. On the fourth day of fixation, granulation tissue and blood vessels were seen in the regenerate. Islands of woven bone were observed in two animals. Alkaline phosphatase staining confirmed the presence of cuboidal osteoblasts lining trabeculae and spicules of bone. No chondrocytes were observed.

Histologic evaluation by Zimmerman et al showed that bone was forming from the margins of the osteotomy toward the center of the mandibular distraction wound.[12] The DO wound demonstrated three zones: the zone of the original bone, the zone of newly formed woven bone starting from the osteotomy margins, and the fibrous interzone (Figure 7). Of note was that bone formation was particularly pronounced near the tooth bud of the unerupted molar and surrounding the inferior alveolar nerve. The expression of cytokines (eg, TGFb, IGF, bFGF, NGF), which are known to be osteoinductive and/or osteoconductive in bone formation during fracture healing and/or distraction osteogenesis may be particularly high in these regions. In addition, in the vicinity of the large diameter inferior alveolar artery the necessary blood supply, known to be crucial for new bone formation in DO, may be readily available. This finding may be unique to the mandible as it is the only bone that contains a large, central neurovascular canal.

Ongoing research in our laboratory involves identification of cell populations in the DO wound at different time points during distraction using histology and immunohistochemistry techniques.

Nineteen minipigs underwent unilateral mandibular distraction osteogenesis. Specimens were harvested at four time points: mid-DO, end-DO, mid-fixation, and end-fixation. The specimens were divided through the exact midpoint of regenerate zone making two tissue blocks for each animal. Two histologic slides from each tissue block were stained with standard Hematoxylin-Eosin (total of 68 H&E slides). One histologic slide from each tissue block was stained with Hematoxylin/Alcian Blue/Sirius Red in order to differentiate cartilage from bone. Slides have been evaluated under light microscopy.[30]

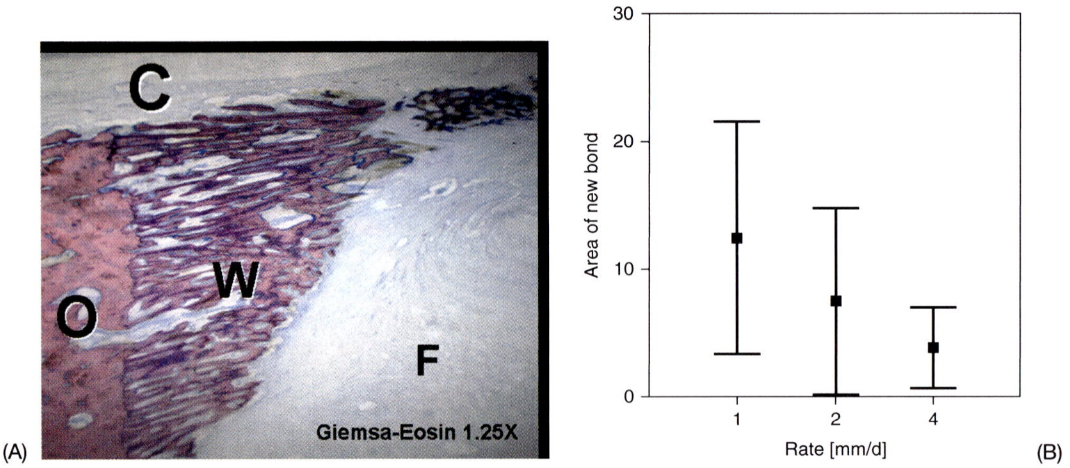

FIGURE 7 • ***A***, Bone deposition from margins toward center. Bone occurred first at IAN and tooth bud regions. Only intramembraneous ossification observed (Giemsa-Eosin 1.25X, C: Inferior alveolar canal, W: woven bone, F: fibrous tissue, O: original bone). ***B***, Mean percent area of new bone (all time points): 12.4% at 1 mm/day, 7.5% at 2 mm/day, 3.8% at 4 mm/day. At all time points, bone was highest at 1 mm/day rate. Reproduced from Zimmermann CE, et al. with permission from Elsevier.[12]

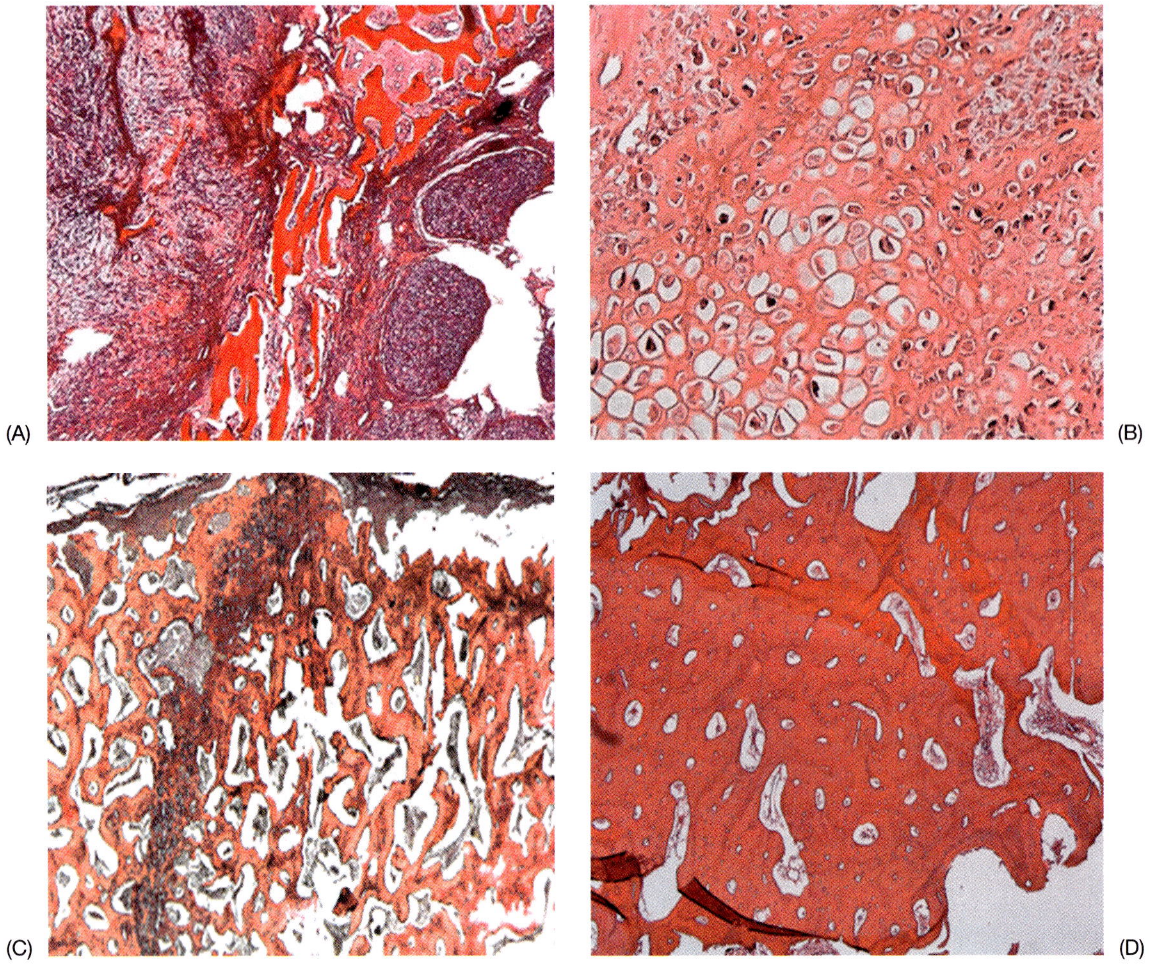

FIGURE 8 • **A**, In the mid-DO group there was the least amount of bone and the largest number of inflammatory and mesenchymal cells with islands of hematoma and fibrous tissue. Numerous ossifying centers were apparent on the periosteal and endosteal surfaces. This photo shows areas of hematoma and fibrous tissue with a small area of primary bone (H&E, X100). ***B***, In the end-DO group there were small islands of chondrocyte-like cells scattered throughout the wound. The ossifying centers were joined together, making a trabecular bone-like network with numerous areas of fibrous tissue in between (H&E, X200). ***C***, Mid-fixation group: Islands of chondrocyte-like cells were more numerous than in the end-DO group. There were fewer areas of predominantly fibrous tissue and more areas of dense bone formation. The darker greenish area is cartilage (Hematoxylin, Alcian Blue, and Sirius Red, x100). ***D***, In the end-fixation group, the islands of chondrocyte-like cells and areas of fibrous tissue were almost completely gone and replaced by dense bone.

In the mid-DO group, as expected, there was the least amount of bone and the largest number of inflammatory and mesenchymal cells with islands of hematoma and fibrous tissue. In the end-DO group, there were small islands of chondrocyte-like cells scattered throughout the wound. The ossifying centers were joined together, making a trabecular bone-like network with numerous areas of fibrous tissue in between. In the mid-fixation group, islands of chondrocyte-like cells were more numerous than in end DO. There were fewer areas of predominantly fibrous tissue and more areas of dense bone formation. In the end-fixation group, the islands of chondrocyte-like cells and areas of fibrous tissue had almost completely disappeared and replaced by dense bone (Figure 8).

In summary, the sequence consisted of hematoma formation and fibrous tissue that were replaced by cartilage and bone. Then cartilage and the remaining fibrous tissue progressively disappeared and were replaced by bone. This appeared to be a combination of both endochondral and membranous bone formation. Ongoing work is in progress in order to confirm these findings.

Ossification Process

Numerous authors hypothesize that osteogenesis occurs via intramembranous bone formation. However, others counter that the mechanism is endochondral, or perhaps a combination of endochondral and intramembranous formation. Finally it may be that neither intramembranous nor endochondral bone formation occur but rather a distinct and unique process. Li et al (1999) have used histology, immunohistochemistry, and nonradioactive in situ mRNA (in a rabbit model) to document the presence of an overlapping cartilage-bone phenotype in cells of the cartilage-bone transitional areas in the DO gap. In this study chondrocytes appeared to be directly transformed into cells that form bone matrix.[31]

Endochondral ossification has been observed together with intramembranous bone formation in limb and mandible lengthening. Some authors consider the occurrence of cartilage or fibrocartilage to be species-specific, strain-related, a result of low blood supply, and/or instability during the fixation period.[12]

In our zero-day latency porcine model of mandibular lengthening, histologic examination of regenerate ($n = 24$) stained with Giemsa-Eosin showed formation of new woven bone in 70% of the specimens by intramembranous ossification alone. Cartilage or fibrocartilage was not found in any of the specimens.

Our more recent studies have shown islands of chondrocyte-like cells present in the DO wound at the end of distraction, in the middle and at the end of fixation period, in both the mandible and maxilla. Staining of the specimens was performed with Hematoxylin-Eosin and the cartilage specific stain Hematoxylin/Alcian Blue/Sirius Red. In the mandible, chondrocyte-like cells progressively disappear from mid fixation to end fixation (Figure 8). In contrast, after DO at the Le Fort I level, chondrocyte-like cells were present in the maxilla at the end of fixation but were absent at the end of DO and the middle of fixation (Figure 9).

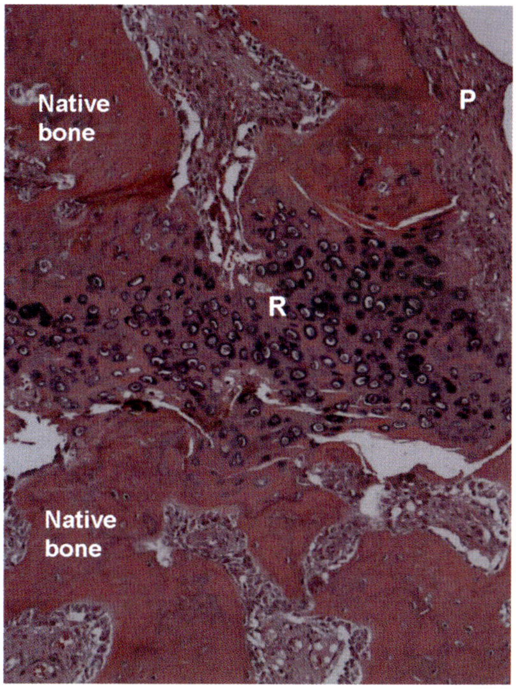

FIGURE 9 • Chondrocyte-like cells were identified in the distraction regenerate at the end of fixation period after DO of the minipig maxilla (**P**: periosteum, **R**: regenerate, H&E x100).

Growth Factors

To date, the molecular mechanisms that govern bone formation during DO are poorly understood. Yates et al using the Yucatan porcine model at the Skeletal Biology Laboratory of MGH found that transforming growth factor β (TGF-b) and bone morphogenetic protein 4 (BMP-4) were at the highest levels during the earliest time periods (ie, latency and active distraction) of distraction and decreased throughout the treatment period, reaching the lowest concentrations during fixation.[29] Mandibles were harvested on the 4th–7th days of distraction and on the 4th day of fixation in minipigs who underwent hemimandibular distraction at a rate of 1 mm/day with no latency. RNA was isolated from DO wounds and a control mandible (nonoperated). Genes for insulin-like growth factor (IGF-I), TGF-b, and BMP-4 were expressed with PCR. Qualitative and quantitative analysis was performed. IGF-I and TGF-b mRNAs were detected in all tissue samples from all animals at all time points. BMP-4 mRNA was detected in all wound samples on the fourth and seventh day of distraction but only in one of the three animals on the fourth day of fixation. On the fourth day of distraction TGF-b in the DO wound was 450% of the control and BMP-4 was 360% of the control. On the seventh day of distraction, however, TGF-b in the DO wound was 82% and BMP-4 was 60% of the control. Finally, on the fourth day of fixation TGF-b was 60%, whereas BMP-4 mRNA was detected in traces in one animal and not detected in two animals.

Comparisons between small animal models and the above large animal data, however, are confounded by differences between the experimental systems, such as species, anatomic site, and distraction protocol. The potential effects of these variables on gene expression results are evident in two recent studies. In a rat model of mandibular distraction TGF-b mRNA was found elevated during latency, distraction, and fixation periods.[32] Specifically, TGF-b gene expression was elevated 1.7-fold (compared to nonoperated, control bone) after a 3-day latency period. Woven bone was not observed in the distraction wound at that time point. Expression of TGF-b mRNA increased further (2.5-fold) during distraction, when osteogenesis was evident. In comparison, BMP-4 mRNA levels were elevated approximately three-fold in a rat model of femoral distraction after 4 days of latency.[33] In situ hybridization showed that BMP-4 was expressed primarily in chondrogenic cells of the soft callus; new bone was not observed at that time point. BMP-4 mRNA expression returned to control bone levels after 7 days of latency, but was greatly increased (>60-fold) after distraction was initiated. Both chondrogenic and osteogenic cells expressed BMP-4 during the second, larger increase in BMP-4 expression. Different animal models, latency periods, rates of distraction, and tissue heterogenicity may involve different patterns of gene expression.

SOFT TISSUE EFFECTS

Skeletal advancement by DO is potentially more stable than traditional techniques because concurrent soft-tissue expansion may improve long-term skeletal stability. Castano et al documented, that increased muscle length in a minipig distraction model was due to increase in masseter muscle size by myocyte proliferation (hypertrophy).[34] The animal model for the study composed of sixteen Yucatan minipigs who underwent an osteotomy at the junction of the body and ramus unilaterally and fixation of the device perpendicular to the osteotomy. Distraction was carried out according to the following protocols: group 1 (12 animals): zero- or 4-day latency and 1, 2 or 4 mm/day ($n = 2$ animals in each group) for an elongation of 12 mm and 24 days fixation and group 2 (4 animals): zero- or 4-day latency and 1 mm distraction per day ($n = 2$ animals in each group) for 7 days and 14 days of fixation. At the end of fixation period, tissues were harvested from masseter muscle overlying the osteotomy and the nonoperated side (Figure 10). Proliferation of myocytes was estimated using immunohistochemical localization with antibodies against proliferating cell nuclear antigen (PCNA). Numerous PCNA-positive myocytes were observed in the muscle samples on the distracted side (16.8% of myocyte nuclei were PCNA positive, with higher results in the group with 12-mm elongation) compared to the contralateral control side (2.8%). Hypertrophy of the soft tissue envelope has clinical significance in long-term stability for patients undergoing distraction procedures.

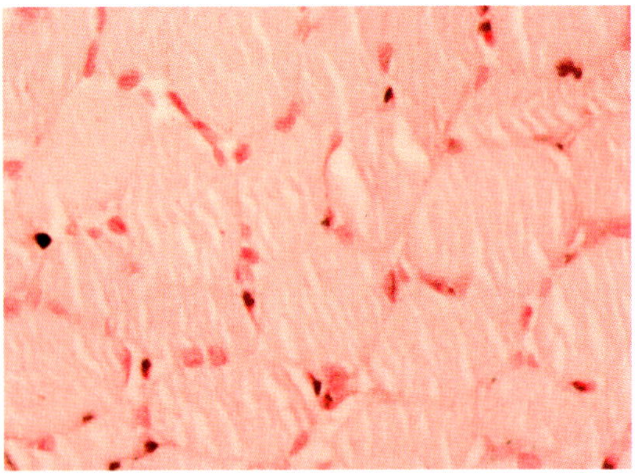

(A)

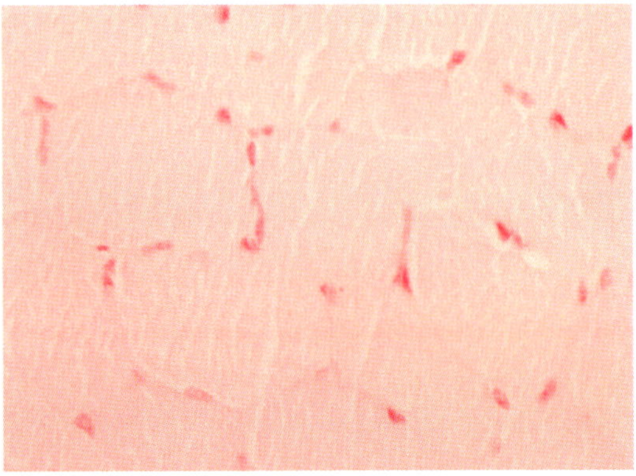

(B)

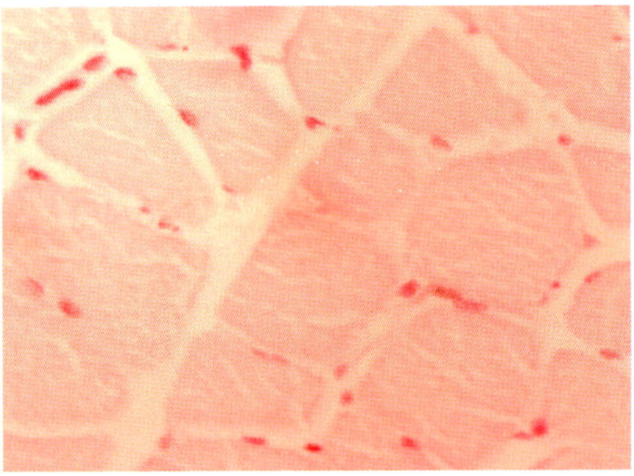

(C)

FIGURE 10 • Muscle overlying distracted mandible **(A)** showed a 6-fold increase of PCNA-positive myocytes (16.8% +/− 11.3%) as compared to the control side **(B)** or as compared to the sham animals **(C)**. ***A***, Regenerate. 4 days Latency, 2 mm/day rate. ***B***, Control unoperated. ***C***, Control osteotomy without DO. Reproduced from Castano et al. with permission from Elsevier.[34]

BIOMECHANICAL PROPERTIES OF DO WOUND

Biomechanical properties and stability of the distraction wound are related to the quality and quantity of bone formed. Experimental data from our laboratory and others demonstrated that stiffness of the DO wound is proportionate to bone volume demonstrated on CT. Kaban et al studied the biomechanical stiffness of an experimental mandibular distraction wound and correlated it with radiographic and ultrasound data.[28] Stiffness describes the force needed to achieve deformation of a structure and stiffness of a healing bone wound correlates with strength. Twenty-four Yucatan minipigs underwent unilateral mandibular distraction (Figure 11). The protocol used was zero-day latency, rates of 1, 2 and 4 mm per day (12 mm distraction gap) and 24-day fixation period. Radiographic and ultrasound bone fill scores were obtained at zero, 8, 16, and 24 days of the neutral fixation. After sacrifice, the harvested mandible was immobilized in a metal jig and inserted into an Instron material testing machine.

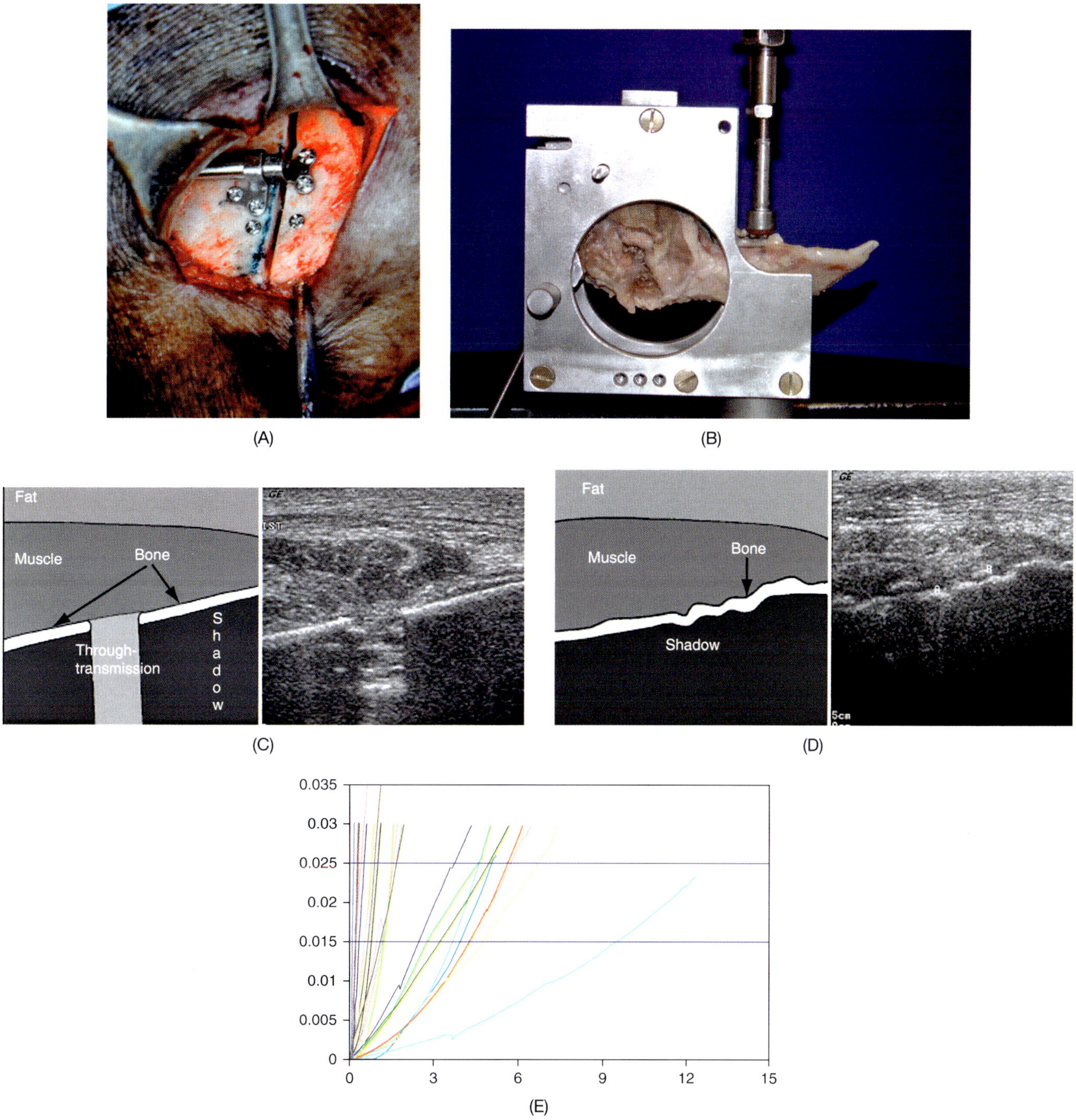

FIGURE 11 • Biomechanical testing of the minipig mandibular DO wound. ***A***, Twenty-four Yucatan minipigs underwent unilateral mandibular distraction. ***B***, Testing Apparatus. After sacrifice, the harvested mandible was immobilized in a metal jig and inserted into an Instron material testing machine. ***C***, Ultrasound (US) was performed by an experienced ultrasonographer blinded to all study parameters. The apparatus used was a GE 700 high-speed linear B scan with a 10 MHz transducer. The US beam was oriented perpendicular to the bone and scanned top to bottom giving axial slices. The diagram and the US illustrate a bone fill score of 0. The gap margins are clear, and there is total through transmission of the beam. ***D***, The diagram and the US illustrate a score of 3, ie, complete bone bridging across the gap and no visible gap margins. Greater bone fill by US was correlated with increased biomechanical strength. ***E***, This graph shows the force (Y axis) vs displacement (X axis) for the control and experimental hemimandibles. Note that the maximum displacement for the control mandibles was 1 mm, whereas for the experimental mandibles it was 2.5 mm–6.0 mm. For the most part, specimens at 0-day fixation had no resistance to displacement. Specimens at 24 days of fixation showed stiffness closer to that of the control side. Reproduced from Kaban et al. with permission from Elsevier.[28]

The occlusal plane was oriented parallel to the floor and a downward force, to approximate occlusal function, was applied at a rate of 2 mm/minute at the premolar occlusal surface.[35] Force (kilo-Newton) and displacement (mm) were recorded at a rate of 10,000 points/second. Two control mandibles were tested to near breaking point, which occurred at 2 KN the maximum. The maximum displacement for the control mandibles was 1 mm, whereas for the experimental mandibles it was 2.5 mm to 6.0 mm. For the most part, specimens at zero-day fixation had no resistance to displacement. Specimens at 24 days of fixation showed stiffness closer to that of the control side.

The results of this study indicated that stiffness of the DO wound increased with duration of neutral fixation regardless of the rate of distraction. The stiffness values of the wounds at all time points were greater in the group of animals distracted at 1 mm/day versus those distracted at 2 mm and 4 mm/day. In addition, greater bone fill by plain radiographs and ultrasonography was correlated with increased biomechanical strength. Bone penetration depth on ultrasound was inversely related to stiffness, as expected. However, at the end of 24 days, neutral fixation, even in the most clinically stable group, stiffness across the distraction wound was only 25.5% of biomechanical strength of control bone.

Biomechanical strength of the wound is, of course, the critical factor in being able to remove a distraction device without risk of acute relapse or deformation at the distracted site. In this study model, it was demonstrated that clinical stability and evidence of bone fill by plain radiographs and ultrasound occur at a time when the distracted mandible has not nearly reached the stiffness of the unoperated side. This raises questions regarding the management of patients at the time of removal of distraction devices. Future studies to determine the progression of mechanical strength, documenting duration to 100% normal, would be important. This would allow surgeons to develop evidence-based management protocols for diet and physical activity for these patients. Such protocols would be especially important for those patients undergoing large movements by DO.

CONCLUSION

Distraction osteogenesis offers a distinct advantage for clinical applications including elimination of donor site morbidity (Chapter 12, "Applications of Distraction Osteogenesis for the Mandible: An Overview," Chapter 13, "Clinical Application of Curvilinear Distraction Osteogenesis for Correction of Mandibular Deformities," and Chapter 15 "Transport Distraction Osteogenesis"), expansion of large magnitudes (Chapter 12, "Applications of Distraction Osteogenesis for the Mandible: An Overview," Chapter 13, "Clinical Application of Curvilinear Distraction Osteogenesis for Correction of Mandibular Deformities," and Chapter 15 "Transport Distraction Osteogenesis," Chapter 15, "Transport DO," and Chapter 16 "Alveolar DO"), decrease risk to vital structures (Chapter 17, "Distraction Osteogenesis for Correction of Maxillary Deformities"); and soft tissue accommodate (Chapter 12, "Applications of Distraction Osteogenesis for the Mandible: An Overview," Chapter 13, "Clinical Application of Curvilinear Distraction Osteogenesis for Correction of Mandibular Deformities," Chapter 17, "Distraction Osteogenesis for Correction of Maxillary Deformities," Chapter 18 "Midface DO," Chapter 15, "Transport DO," and Chapter 16 "Alveolar DO").

One significant limitation of DO is the lengthy treatment protocols: latency + gap size/rate + fixation = treatment time.

Once the biology is better understood, the biologic system might potentially be manipulated (growth factors, rate manipulation) to perform successful DO in a shorter treatment time while increasing the quality and quantity of bone formation.

Bibliography

1. Codivilla A. On the means of lengthening, in the lower limbs, the muscles and tissues which are shortened through deformity. Am J Orth Surg 1905;2:353–7.
2. Ilizarov GA. The principles of the Ilizarov method. Bull Hosp Jt Dis Orthop Inst 1988;48:1–8.
3. Ilizarov GA. The tension-stress effect on the genesis and growth of tissue: part I. The influence of stability of fixation and soft tissue preservation. Clin Orthop 1989;238:249–56.
4. Snyder CC, Levine G A, Swanson HM, Browne EZ. Mandibular lengthening by gradual distraction: preliminary report. Plast Reconstr Surg 1973;5:506–9.
5. McCarthy JG, Schreiber JS, Karp N, et al. Lengthening of the human mandible by gradual distraction. Plast Reconstr Surg 1992;89:1–8.
6. Perrott DH, Berger R, Vargervik K, Kaban LB. Use of a skeletal distraction device to widen the mandible: a case report. J Oral Maxillofac Surg 1993;51:435–9.
7. Constantino PD, Friedman CD. Distraction osteogenesis. Application for mandibular regrowth. Otolaryng Clin N Am 1991;24: 1433–45.
8. Mofid MM, Manson PN, Robertson BC, et al. Craniofacial distraction osteogenesis: a review of 3278 cases. Plast Reconstr Surg 2001;108:1103–14.
9. McCarthy JG, Stelnicki EJ, Mehrara BJ, Longaker MT. Distraction osteogenesis of the craniofacial skeleton. Plast Reconstr Surg 2001;107:1812–7.
10. Younger EM, Chapman MW. Morbidity at bone graft donor sites. J Orthop Trauma 1989;3:192–5.
11. Troulis MJ, Padwa BP, Kaban LB. Distraction Osteogenesis: Past, present, future. Facial Plast Surg 1998:14:205–15.
12. Zimmermann CE, Thurmuller P, Troulis MJ, et al. Histology of the porcine mandibular distraction wound. Int J Oral Maxillofac Surg 2005;34:411–9.
13. Aronson J, Shen XC, Gao GG, et al. Sustained proliferation accompanies distraction osteogenesis in the rat. J Orthop Res 1997;15:563–9.
14. Aronson J, Gao GG, Shen XC, et al. The effect of aging on distraction osteogenesis in the rat. J Orthop Res 2001;19:421–7.
15. Glowacki J, Shusterman EM, Troulis M, et al. Distraction osteogenesis of the porcine mandible: histomorphometric evaluation of bone. Plast Reconstr Surg 2004;113:566–73.
16. Papadaki ME, Troulis MS, Glowacki J, Kaban LB. A mini pig model of maxillary distraction osteogenesis, 2010;68:2783–2791.
17. Samchukov ML, Cherkashin AM, Cope JB. Distraction Osteogenesis: History and Biologic basis of new bone formation. In: Lynch SE, Genco RJ, Marx RE, editors. Tissue engineering: Applications in Maxillofacial surgery and periodontics. Chicago: Quintessence;1999. p131–146
18. Rachmiel A, Levy M, Laufer D, Lengthening of the mandible by distraction osteogenesis: Report of cases. J Oral Maxillofac Surg 1995;53:838–46.

19. Troulis MJ, Glowacki J, Perrott DH, Kaban LB. Effects of latency and rate on bone formation in a porcine mandibular distraction model. J Oral MaxillofacSurg 2000;58(5):507–513.

20. Tavakoli K, Walsh WR, Smart R, et al. The role of latency in mandibular osteodistraction. J Craniomaxillofac Surg 1998; 26:209–19.

21. Aronson J, Shen X. Experimental healing of distraction osteogenesis comparing metaphyseal to diaphyseal sites. Clin Orthop 1993;301:25–30.

22. Schmelzeisen R, Neumann G, von der Fecht R. Distraction osteogenesis in the mandible with a motor-driven plate: a preliminary animal study. Br J Oral Maxillofac Surg 1996;34(5):375–8.

23. Ploder O, Mayr W, Schnetz G, Unger E, Ewers R, Plenk H Jr. Mandibular lengthening with an implanted motor-driven device: preliminary study in sheep. Br J Oral Maxillofac Surg 1999; 37:273–6.

24. Kessler P, Wiltfang J, Neukam FW. A new distraction device to compare continuous and discontinuous bone distraction in mini-pigs: a preliminary report. J Craniomaxillofac Surg 2000;28:5–11.

25. Wiltfang J, Kessler P, Merten HA, Neukam FW. Continuous and intermittent bone distraction using a microhydraulic cylinder: an experimental study in minipigs. Br J Oral Maxillofac Surg 2001;39:2–7.

26. Zimmermann CE, Harris G, Thurmuller P, Troulis MJ, Perrott DH, Rahn B, Kaban LB. Assessment of bone formation in a porcine mandibular distraction wound by computed tomography. Int J Oral Maxillofac Surg 2004;33(6):569–74.

27. Thurmuller P, Troulis M, O'Neill MJ, Kaban LB. Use of ultrasound to assess healing of a mandibular distraction wound. J Oral Maxillofac Surg 2002;60(9):1038–44.

28. Kaban LB, Thurmuller P, Troulis MJ, Glowacki J, Wahl D, Linke B, Rahn B, Perrott DH. Correlation of biomechanical stiffness with plain radiographic and ultrasound data in an experimental mandibular distraction wound. Int J Oral Maxillofac Surg 2003; 32:296–304.

29. Yates KE, Troulis MJ, Kaban LB, Glowacki J. IGF-I, TGF-beta, and BMP-4 are expressed during distraction osteogenesis of the pig mandible. Int J Oral Maxillofac Surg 2002;31:173–8.

30. Tayebaty FT, Williams WB, Baumann A, Abukawa H, Troulis MJ, Kaban LB. Histologic and Histomorphometric Analysis of the Porcine Mandibular Distraction Wound. J Oral Maxillofac Surg 2006; 64:43–4.

31. Li G, Simpson AH, Triffitt JT. The role of chondrocytes in intramembranous and endochondral ossification during distraction osteogenesis in the rabbit. Calcif Tissue Int 1999;4:310–17.

32. Mehrara BJ, Rowe NM, Steinbreck DS, Dudziak ME, Saadeh PB, McCarthy JG, Gittes GK, Longaker MT. Rat mandibular distraction osteogenesis: Molecular analysis of transforming growth factor beta-1 and osteocalcin gene expression. Plast Reconstr Surg 1999: 103:536–47.

33. Sato M, Ochi T, Nakase T, Hirota S, Kitamura Y, Nomura S, Yasui N. Mechanical tension-stress induces expression of bone morphogenetic protein (BMP)-2 and BMP-4 but not BMP-6, BMP-7 and GDF-5 Mrna, during distraction osteogenesis. J Bone Min Res 1999:14:1084–95.

34. Castano FJ, Troulis MJ, Glowacki J, Kaban LB, Yates KE. Proliferation of masseter myocytes after distraction osteogenesis of the porcine mandible. J Oral Maxillofac Surg 2001;59(3):302–7.

35. Perrott DH, Rahn B, Wahl D, Linke B, Thuruller P, Troulis M, Glowacki J, Kaban LB. Development of a mechanical testing system for a mandibular distraction wound. Int J Oral Maxillofac Surg 2003;32:523–7.

36. Magill, J.C., Byl, M.F., Goldwaser, B., Papadaki, M., Kromann, R., Yates, B., Morency, J.R., Kaban, L.B., and Troulis, M.J., "Automating Skeletal Expansion: An Implant for Distraction Osteogenesis of the Mandible," J Med. Devices, 3, 014502 (2009)

Three-Dimensional Treatment Planning for Distraction Osteogenesis

11

Krishna C. Yeshwant, MD / Edward B. Seldin, MD, DMD

OUTLINE

INTRODUCTION

The success of skeletal reconstruction using distraction osteogenesis (DO) depends on careful preoperative planning, accurate intraoperative implementation, and detailed postoperative analysis of the achieved movements. Craniomaxillofacial surgeons have traditionally used tracings based on superimposed two-dimensional (2D) cephalograms to support each of these tasks. This approach has been the standard because the lateral cephalometric tracing is simple, accurate, and reproducible and there is a large body of literature and longitudinal experience associated with this technique.[1] However, projection of the facial anatomy onto the mid-sagittal plane does not permit adequate evaluation and planning for patients with asymmetric and complex deformities involving multiple planes. A three-dimensional (3D) analysis is required for this group of patients.[2,3] Improved surgical techniques such as minimally invasive endoscopic access and the use of buried distraction devices further increase the need for more sophisticated planning tools. In addition, treatment planning methods must take into account the inability to make mid-treatment adjustments with the use of buried, miniature distraction devices.[4,5]

Advances in computer graphics hardware and software have enabled surgeons to more quickly and accurately visualize a patient's anatomy using computed tomographic (CT) and magnetic resonance imaging (MRI) data. However, these rich data sets are generally presented as a series of 2D slices, which surgeons scroll through to develop a mental 3D model of the craniomaxillofacial anatomy or of a pathologic process (eg, tumor). This "slice-at-a-time" approach to viewing data is invaluable in many areas of surgery and medicine, but it does not allow craniomaxillofacial surgeons to visualize the facial skeleton in its 3D entirety. Thus, the technique of digitally reconstructing a 3D CT model from a series of 2D CT slices was developed and has become an increasingly valuable tool in OMFS[6] (Figure 1).

Because 3D CT images allow a surgeon to readily appreciate a patient's skeletal deformity as it presents clinically, rather than as a series of 2D CT slices, it was a natural transition to use this tool for planning DO and other reconstructive procedures requiring multidimensional movements. There are currently two general approaches to 3D planning for DO: (1) CT data is used to create an acrylic model for planning, and (2) CT data is used to create a virtual model on the computer that can be manipulated for planning.

In the first approach, a 3D CT is downloaded to a stereolithographic printer that fabricates a custom 3D acrylic model representing the patient's anatomy. The surgeon can visualize the patient's skeletal anatomy and simulate the operation on the model. Unfortunately, these models are often expensive; they can only be used one time; there is no way to overlay and compare multiple possible plans and only a rudimentary quantitative analysis of the planned movements is possible. Furthermore, acrylic models only allow visualization of the bony tissue and there is no way to integrate movement, growth, and soft tissue data.

In response to these shortcomings, several groups have developed the second approach, that is, software systems to create and manipulate 3D CTs virtually for surgical planning.[3,7,8] These systems enable surgeons to: assess craniomaxillofacial (CMF) anomalies using 3D imaging data; quantitatively plan required multiplanar corrections; compare multiple treatment plans; and to accurately prescribe the characteristics of surgical navigation

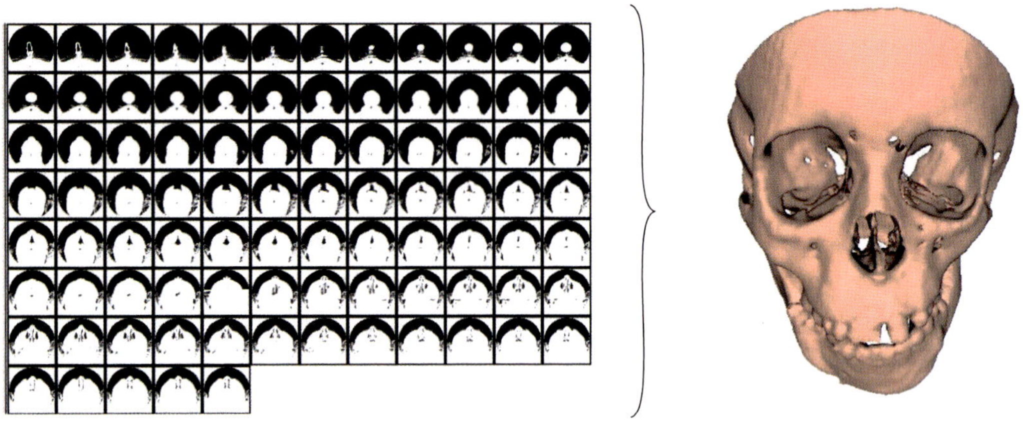

Two-Dimensional CT Scan Slices

Three-Dimensional CT Reconstruction (3D CT)

FIGURE 1 • A series of 2D CT scan slices (left) were used to reconstruct the 3D CT demonstrated on the right.

stents, occlusal splints and specific distraction devices (ie, distraction vector).[9] Some planning software can also drive intraoperative guidance systems that enable a surgeon to directly interact with the preoperative plan in the operating room.

In this chapter, we discuss the use of software-driven systems to plan, execute, and analyze the outcomes of skeletal expansion by distraction osteogenesis. Benefits, and shortcomings of these systems, as well as potential future improvements in this important technology are reviewed.

THREE-DIMENSIONAL TREATMENT PLANNING FOR DISTRACTION OSTEOGENESIS

The process of developing a 3D treatment plan and ultimately executing that plan in the operating room consists of four general steps: (1) data acquisition, (2) model generation, (3) development of a specific treatment plan, and (4) transfer of the plan to the operating room. Although most of these steps are fundamentally the same among the available 3D treatment planning systems, each software package differs in the algorithms and features offered in the treatment planning step.

Data Acquisition

An axial head CT scan (1 mm cuts) is obtained to assess skeletal anatomy and morphology. In order to enable optimal segmentation (described below), the CT scan is ideally performed with the patient wearing a bite splint to separate the upper and lower teeth.

Metal dental restorations (amalgam fillings, gold crowns, etc), removable prostheses, and stainless steel orthodontic appliances cause severe distortion of the CT scan, making the segmentation process difficult and less precise. Some software packages have algorithms that mitigate this concern by automatically removing interference caused by dental work.[9] Nevertheless, even with these algorithms, better image quality is obtained in patients with no metal dental devices in the mouth.

Finally, in growing children, the CT scan should be acquired as close to the time of the operation as possible, to ensure accuracy of the model and the surgical plan. No existing 3D treatment planning systems currently account for skeletal changes that occur with patient growth.

Model Generation

Once the CT data has been acquired, the next step is to extract the relevant information from the scan. The patient's CT data is transferred to the clinician's computer via network, CD, or other media and loaded into the selected treatment planning software. The process of producing a 3D CT from the selected CT scan involves three steps: (1) window and level adjustment, (2) segmentation, and (3) reconstruction.

Window and Level Adjustment

Window and level adjustment is the process of filtering the scan data so that the borders of the facial skeleton are emphasized without obscuring the scan images (Figure 2). This is often a quick step, but can be complicated by the presence of artifact from dental devices.

Segmentation

Segmentation is the process of identifying the relevant anatomic structures from the CT scan. During segmentation, each anatomic element of a CT slice is associated with a label corresponding to a tissue type or anatomic structure. For instance, the mandibular portions of the CT scan may be labeled as "mandible" using a blue label while the skull base is labeled as "skull base" using a tan label (Figure 3). If necessary, other objects visible on CT, such as soft tissue, the globe, and devices can be identified using different labels during this step. Segmentation of the mandible is often problematic when the maxillary and mandibular teeth overlap or are in contact. Properly identifying the condylar head and the temporal fossa can also be difficult when the head is resting in the fossa. Additionally, the interferences caused by dental appliances can dramatically impede the segmentation process, forcing the user to guess where the patient's anatomy is located. These inaccuracies are important to note because quantitative decisions

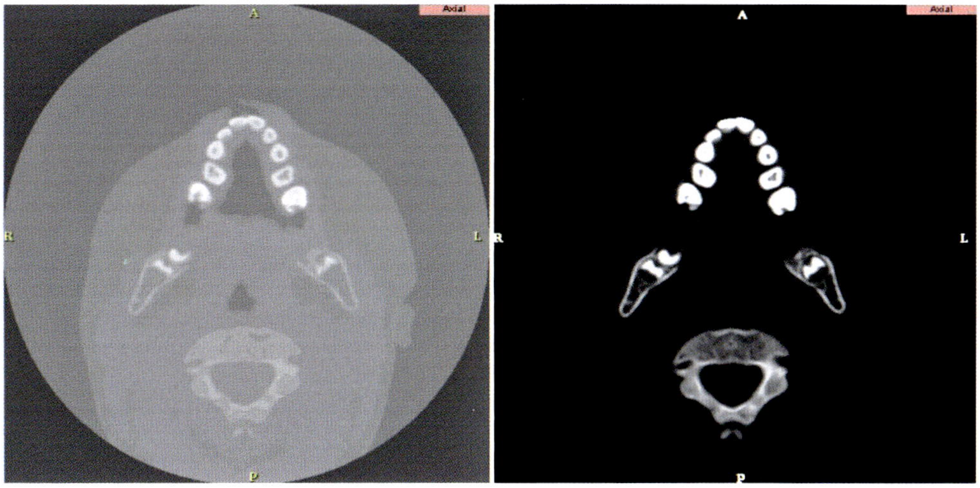

FIGURE 2 • A single slice from an axial CT image is shown on the left. Note how this original image includes soft tissue densities along with high density areas. Having both high and low density images superimposed makes it more difficult for the user and for the computer to identify the high density areas of interest to craniofacial surgeons. Thus, the original image is filtered using the window and level adjustment process to produce the image on the right that emphasizes the high density entities in the scan (bone and tooth here).

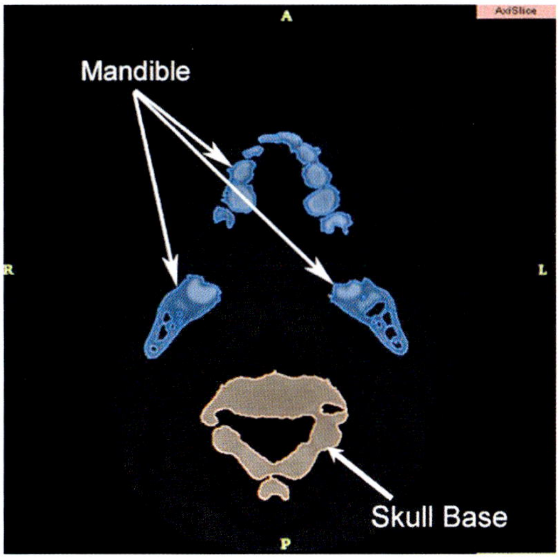

FIGURE 3 • This axial CT slice is segmented using a label maps in which a unique label is assigned to each tissue type. If a bite splint is not used, then the user must manually identify anatomic structures by outlining them in the appropriate label map color.

regarding device placement, osteotomy orientation, anatomic movements, and intraoperative navigation are made based on the quality of the 3D CT model. The accuracy of the surgical plan and execution ultimately depend on the quality of this segmentation step. It is therefore critical that objects are segmented accurately.

Segmentation can be performed manually, semi-automatically, or automatically. In the manual approach, a technician traces the outline of anatomic structures on each slice by hand using a mouse. This approach can take hours and may not produce an accurate reconstruction because it is difficult to trace the exact outlines of an irregular set of borders consistently from one slice to the next.

The semi-automated approach is a two-stage process that enables a more accurate representation of the patient's anatomy. In the first stage, an algorithm that identifies anatomic structures of interest by intensity, position, or some other heuristic is applied to each slice in the CT scan, automatically labeling the anatomic structures according to its analysis. In the second stage, the user reviews each slice and corrects the labeling as needed using simple drawing tools such as fill, add, and subtract. This semi-automated approach is powerful because the computer does much of the repetitive work of identifying all structures of a particular density (like bone and tooth for instance), allowing the technician to then focus his/her time on delineating intricate anatomy. If the patient used a bite guard during the scan and did not have any intraoral restorations or devices that cause CT interference, this second stage can often be completed within a few minutes.

There is a great deal of interest in developing fully automated segmentation algorithms because both the manual and the semi-automated approaches become increasingly time-consuming when applied to CT series with particularly large amounts of data. However, fully automated segmentation of pathologic anatomic structures is arguably the most difficult problem in 3D medical image processing and the general problem remains unsolved.[10] There is a large body of literature describing increasingly sophisticated algorithms to automatically segmenting medical images with varying degrees of success.[11–15] Although no segmentation algorithm exists that can automatically label a head CT of a patient with significant pathology, fully automated segmentation algorithms have been described for normal anatomy; however, these algorithms have not been tested on a large number of patients.[16]

Reconstruction

After segmentation is completed, the label data is fed into an algorithm that reconstructs virtual 3D models of the patient's mandible and skull base (Figure 4). Other anatomic features,

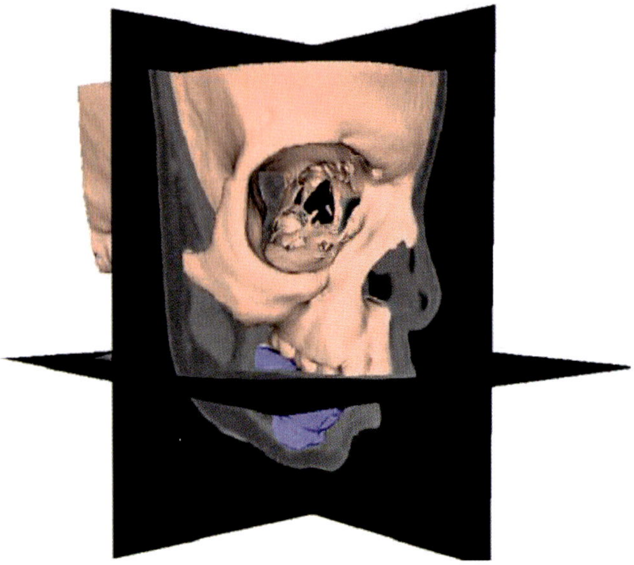

FIGURE 4 • The reconstructed 3D image is generated by using an algorithm to combine the 2D slices from a CT scan. In this figure the 3D CT is viewed with planes along the axial, sagittal, and coronal planes to illustrate that the model is directly based on the source CT scan.

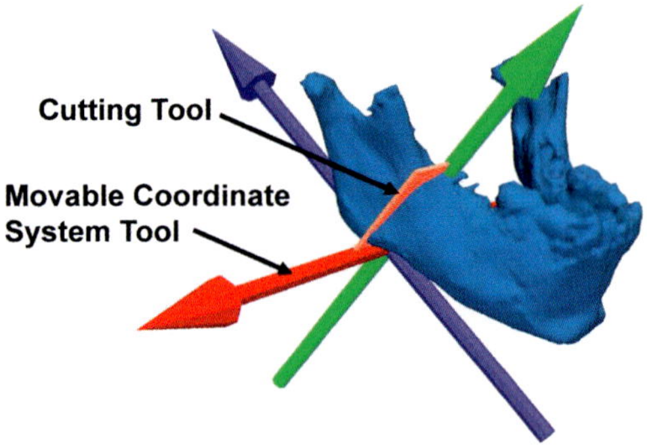

FIGURE 5 • The cutting tool (red rectangle) is positioned with respect to the patient's 3D CT using the movable coordinate system tool (large arrows). Each of the movable coordinate system's colored axes can be used to translate or rotate the connected object in a different direction.

devices, and objects that were segmented are also reconstructed into 3D models. These reconstructed virtual models provide a detailed depiction of the relationships between craniofacial structures. They can be printed or displayed on the computer monitor and the surgeon can use the treatment planning software to view the anatomy from any vantage point.

Treatment Planning

With all of the necessary virtual anatomic models created, it is now possible to use a 3D treatment planning software package to manipulate these models to create a treatment plan. Treatment planning involves four major steps: (1) alignment of models, (2) assessment of the deformity, (3) plan formulation and (4) distractor prescription and simulated placement.

Because patients are usually not placed in a standardized position when the head CT is acquired, it is important to begin the treatment planning process by aligning the reconstructed models into an anatomically appropriate orientation. This can be done by identifying landmarks along the mid-sagittal plane on the reconstructed models (eg, menton, nasion, and the anterior nasal spine) along with landmarks in the coronal plane of the models (eg, right and left orbitale).[18] The treatment planning software can then use these landmarks to align the associated skeletal models to the true vertical and horizontal planes.

Landmarks can be used to quantitatively assess the patient's deformity by calculating relative facial proportions. Because no well-established 3D cephalometric database exists, the surgeon cannot compare a patient's anatomy to a normative standard, as is usually done in 2D treatment planning. A few groups have proposed a set of 3D craniofacial landmarks that could be used to establish the needed normative 3D cephalometric data. However this landmark data has not been collected on a large enough population to allow for treatment planning at this time.[17,18] The absence of standard 3D cephalometric data is a fundamental issue in 3D treatment planning today. Without normative data, in some cases (eg, asymmetries) it may not be possible to distinguish normal from abnormal anatomy. However, when the surgeon is confident regarding the type of correction required for a particular patient, assessment using relative facial proportions is sufficient.

Once the 3D CT is reviewed and the patient's deformity is quantified using landmark measurements, the surgeon can formulate any number of surgical treatment plans. In the 3D Slicer software package (developed in collaboration with the Harvard Surgical Planning Laboratory and used at Massachusetts General Hospital), an osteotomy is simulated with a cutting tool that is positioned with respect to the 3D CT using a movable coordinate system (Figure 5). Any arbitrary osteotomy may be simulated using these systems including mandibular, maxillary, orbital, or alveolar osteotomies. In mandibular DO, a cutting tool is positioned with respect to the ramus using the coordinate system tool. Once the desired osteotomy position is located, a cut operation is applied and two topologically closed models are created. In bilateral mandibular cases this operation is repeated on the contralateral side.

Once the osteotomies are simulated, the various treatment planning systems diverge in their approaches. In one approach, the surgeon selects the distraction device and then uses the software system to simulate the placement of that device on the virtual patient model.[7] The software system can then illustrate the distraction path (vector) based on the nature of the distraction device and its placement determined by the surgeon (see Chapter 13, "Clinical Application of Curvilinear Distraction Osteogenesis for Correction of Mandibular Deformities," Figure 8). The movement of the bony fragments that are being distracted is displayed on the computer screen resulting in a simulated image of the final distraction position. The surgeon can manipulate the plan by altering the distractor location and orientation or by changing the distraction distance. Although this approach is intuitive, in that the simulated movement of the bone fragments is directly related to the simulated placement and movements of the distraction device, it does not give the surgeon complete freedom in choosing optimal fragment placement.

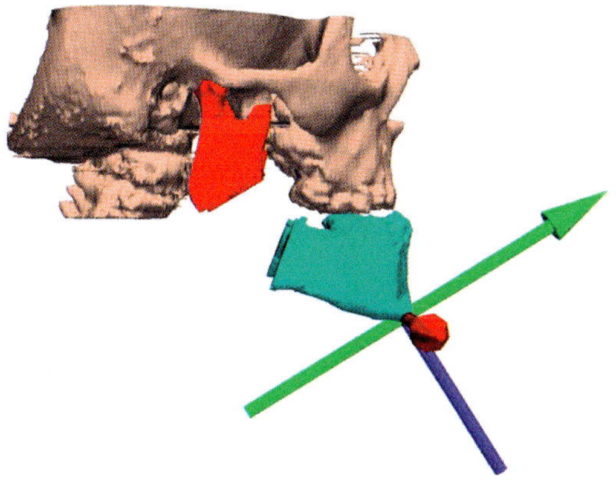

FIGURE 6 • The cut models can be repositioned into their final positions using the movable coordinate system by attaching the models and moving the coordinate system using the mouse.

In the alternative approach, which is used in the 3D slicer system, the bony fragments are first repositioned using a movable coordinate system (Figure 6). The software checks for bony interference ("collision") as the models are moved. Once the bony fragments are located in their ideal positions, the surgeon identifies areas of the anatomy where it would be possible and desirable to attach a distraction device. Appropriate devices, along with the necessary movement parameters, are then automatically determined and reported by the planning software. The benefit to this approach is that the surgeon can concentrate on determining ideal bone placement rather than focusing on device placement, thus making the treatment planning process simpler and more robust.

The "cut-and-move" interface used in each of these planning systems is a virtual replication of the stereolithographic model approach described earlier. However, because all of the elements of the treatment planning are virtual, there are several advantages now available to the surgeon. First, multiple treatment plans can be rapidly developed for each patient with no additional cost for each new plan. This allows surgeons to investigate different treatment scenarios before choosing one. Virtual planning also enables surgeons to view several treatment plans superimposed for comparison (Figure 7A). This tool is useful not only for comparing different treatment plans, but also for comparing multiple preoperative scans and for postoperative outcomes assessment, that is, comparison of the simulated versus the actual result (Figure 7B). Software-based treatment planning also allows for more accurate measurements and for more sophisticated approaches to device prescription, as is necessary in curvilinear distraction.[19-20] Finally, some groups have begun to integrate jaw movement data into the 3D CT, thus allowing for simulation of the functional effects of various treatment plans.[16]

Transfer to the Operating Room

The primary concern in DO procedures, from a treatment planning perspective, is the ability to place the distraction device in the desired anatomic location. There is considerable flexibility in locating the osteotomy (ie, corticotomy) site because the distraction path, not the osteotomy, determines the final outcome. On the other hand, placing the distraction device must be done precisely. This is best achieved by preoperatively calculating the location of the drill holes where the distraction device will be fixed to the bone. In most cases it is appropriate to simply make the osteotomy between the proximal and distal drill holes.

There are three major techniques that can be used to identify the planned locations of the drill holes in the operating room: (1) surgeon's judgment based on intraoperative visualization (manual), (2) drill guide (see Chapter 13, "Clinical Application of Curvilinear Distraction Osteogenesis for Correction of Mandibular Deformities," Figure 11), and (3) navigation system. As the surgeon progresses from the use of intraoperative judgment to drill guide to navigation, more accuracy is achieved, but at a higher cost.

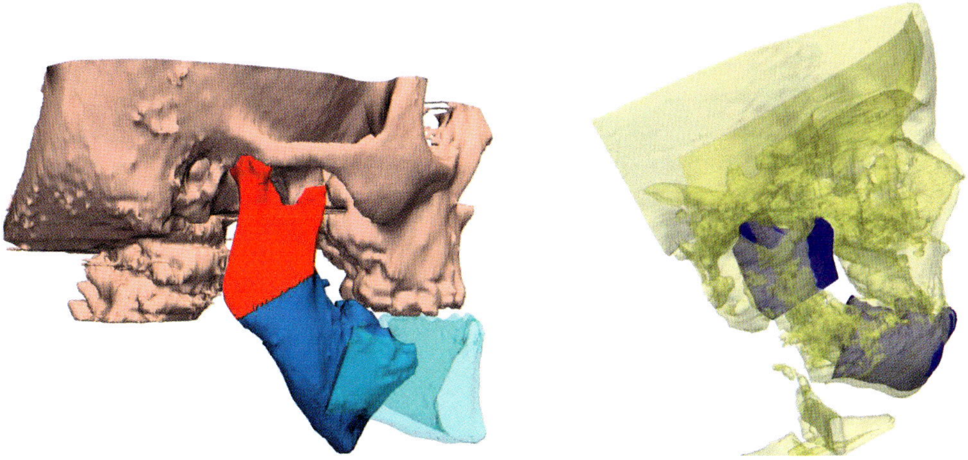

(A) (B)

FIGURE 7 • *A*, Multiple models can be superimposed using the treatment planning system. This allows for comparison of multiple treatment plans, assessment of postoperative device placement, or comparison of predicted movement to the patient's original anatomy. *B*, The actual post-distraction outcome is represented in transparent yellow and the predicted post-distraction outcome overlaid in solid blue.

Although the manual approach is the cheapest, easiest to implement, and can result in a successful outcome in the hands of an experienced surgeon, it is also the most prone to error. Difficulty in accurately identifying landmarks intraoperatively commonly prevents the manual approach from achieving predictable outcomes. Because even minor errors in distractor placement can produce a dramatic impact on the vector of movement and the ultimate position of the bone, it is critical to use a more precise approach to ensure accuracy.

Custom-made drill guides can be used to more precisely identify distractor drill hole locations. These custom-made stents can be fabricated based on models produced by the 3D treatment planning software.[7] Alternatively, the 3D CT can be used to generate a stereolithographic model that can in turn be used to fabricate stents using traditional techniques. In either case, a drill guide may require an extraoral incision and dissection because the guide must be placed over a large contour of the bone surface to ensure correct identification of the drill hole locations.

The most advanced approach to locating drill holes is an intraoperative 3D navigation system. There are many commercial navigation systems currently available, and some of the treatment planning systems have navigation technology built-in.[21, 22] Most systems can be classified as optical, radiofrequency-based, or mechanical. Optical and radiofrequency-based systems both require a tracking tower in the operating room, a wand or tracking device on the surgical instrument being tracked (a drill in this case), and a tracking device on the patient's anatomy (Figure 8). The tower has two sensors that are used to triangulate the position of the wand and the patient tracking device. An important step in the use of these navigation systems is registration of anatomic landmarks with the treatment plan. The desired landmarks are identified on the patient intraoperatively and they are correlated with virtual landmarks on the 3D treatment planning CT. When both sets of landmarks are identified, the navigation system registers the patient's coordinate system with the 3D CT based treatment plan. The navigation system is able to compensate for intraoperative patient movement by using data from the patient tracking device. The tracking system is used to identify the exact locations of the required drill holes on the patient's anatomy based on the treatment plan. This approach is ideal because it allows screws to be placed without direct visualization of the anatomy, thus enabling minimally invasive approaches. At the Massachusetts General Hospital, with our partners at Physical Sciences Inc (Andover, MA) we are working on a Navigation system called OsteoMark™. This system can be used to transfer treatment plans from 3D-surgical planning CTs (Figure 9). NIH-SBIR 2R44 DE019322 (PI, Maria Troulis and John MaGill).[21]

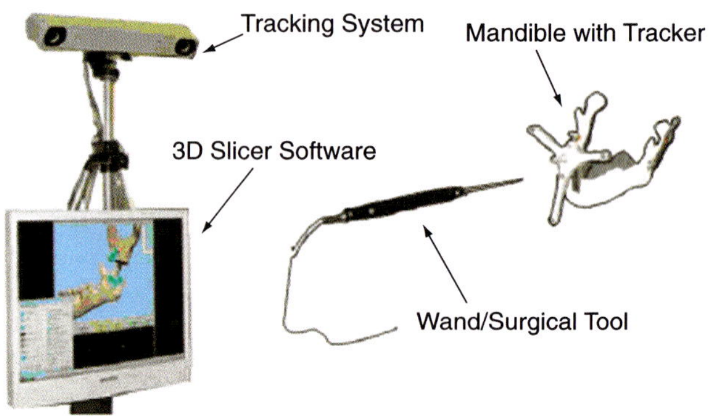

FIGURE 8 • This optical surgical navigation system consists of a tower (labeled tracking system here), a wand, and a tracker. As the surgeon moves the wand, the tracking system locates the virtual position of the surgeon's tool in the treatment planning system. This approach allows for drill holes to be placed within 0.5 mm of their planned location. (Image courtesy of R. Ellis, PhD).

Mechanical navigation systems involve a robotic arm that is registered to the patient's anatomy in a similar fashion to that described above.[22] The true value of these systems, however, is more evident in orthognathic surgery when the robot can be used to relocate a bone fragment to a desired location. Video-assisted navigation systems that have achieved good surgical outcomes have also been described.[23]

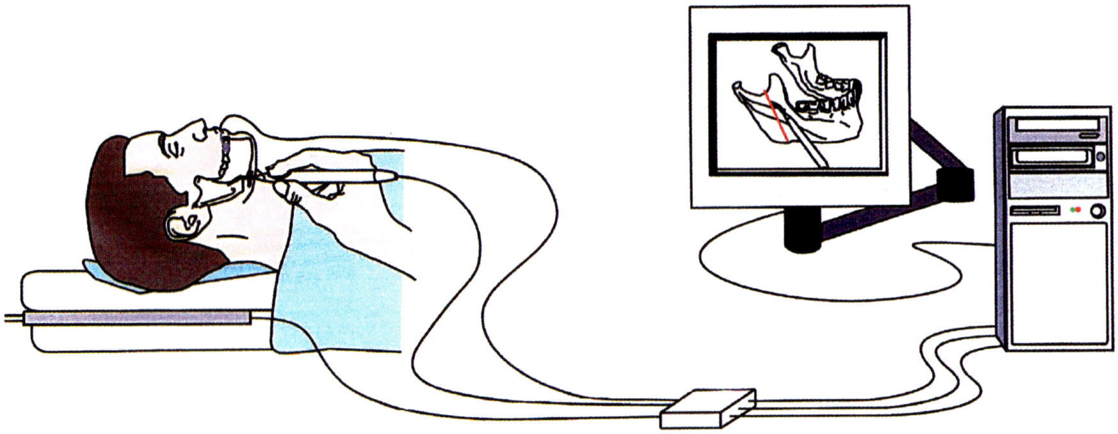

FIGURE 9 • The hand tool (pencil) is used to mark the bone. Its position is tracked with an electromagnetic sensor. A second sensor is attached to a tooth or dental splint during the surgery, providing a registration of the patient to the planning images.[21]

CONCLUSIONS

In this chapter, the state of 3D treatment planning for distraction osteogenesis procedures has been presented. Although traditional 2D treatment planning approaches have worked well for many patients, patients that present with complex asymmetric deformities require a 3D approach. Much work has been done in this area, but there are still crucial components that require further development.

The development of a normative set of 3D cephalometric landmarks is critical if 3D treatment planning is to be used on large patient populations. Although some groups have proposed landmarks that could be used to build such data sets, there are no databases currently available. Surgeons must instead use facial proportion measurements to assess deformities.

The time and computer expertise currently required to produce a treatment plan is a significant barrier for most surgeons. The inability of systems to account for growth over time and for soft-tissue deformation further limits the applicability of existing systems. Finally, navigation systems must become cheaper and less cumbersome if they are to be used on a broader scale.

Efforts to address these issues are continually underway. In particular, improved user interfaces and segmentation algorithms will continue to make treatment planning systems more accessible to surgeons and less time consuming.[8] Normative 3D cephalometric data will offer users a standard by which to compare patient cases with and will offer a means of assessing growth over time. Several groups are actively working on improving algorithms and hardware to accurately simulate soft-tissue movement to improve preoperative prediction of surgical outcomes. Finally, a small group of companies have taken the challenge to improve existing intraoperative navigation systems.

Although many issues remain, 3D treatment planning undoubtedly offers significant advantages over traditional 2D methods. Craniomaxillofacial surgeons have traditionally been at the forefront of developing these new techniques and technologies, and there is little question that this will continue to be the case in the future. As the limitations discussed in this chapter are addressed and as hardware prices continue to trend down and performance trends up, access to these tools will continue to improve and the innovations of today will soon become the standards of practice tomorrow.

References

1. Farkas LG. Anthropometry of the head and face. 2nd ed. New York: Raven Press; 1994.
2. Troulis MJ, Everett P, Seldin EB, et al. Development of a three-dimensional treatment planning system based on computed tomographic data. Int J Oral Maxillofac Surg 2002;31:349–57.
3. Everett P, Troulis MJ, Seldin EB, et al. A 3-D System for Planning and Simulating Minimally-Invasive Distraction Osteogenesis of the Facial Skeleton. Proceedings of 3rd International Conference On Medical Robotics, Imaging and Computer Assisted Surgery; 2000 Oct 11–14; Pittsburgh, PA, USA: Springer; 2000.
4. Troulis MJ, Perrott DH, Kaban LB. Endoscopic mandibular osteotomy, and placement and activation of a semiburied distractor. J Oral Maxillofac Surg 1999;57:1110–3.
5. Seldin EB, Troulis MJ, Kaban LB. Evaluation of a semiburied, fixed-trajectory, curvilinear, distraction device in an animal model. J Oral Maxillofac Surg 1999;57:1442–6.
6. Lorensen W, Cline H. Marching cubes: a high resolution 3D surface construction algorithm. ACM Computer Graphics 1987; 21:38–44.
7. Gateno J, Teichgraeber JF, Aguilar E, et al. Computer planning for distraction osteogenesis. Plast Reconstr Surg 2000;105:873–82.
8. Keeve, E, Jansen T, Krol Z, et al. JULIUS—An Extendable Software Framework for Surgical Planning and Image-Guided Navigation. Fourth International Conference on Medical Image Computing and Computer-Assisted Intervention; 2001 Oct 14–17, Utrecht, The Netherlands: Springer; 2001.
9. Gateno J, Xia J, Teichgraeber JF, et al. The precision of computer-generated surgical splints. J Oral Maxillofac Surg 2003;61:814–7.
10. Tylor RH, Lavallee S, Burdea GC, et al., eds. Data Acquisition and Segmentation. In: Computer-Integrated Surgery: Technology and Clinical Applications. Cambridge, MA: MIT Press; 1996:4.
11. Haralick RM, Shapiro SG. Survey: Image segmentation techniques. Comput Vision Graphics Image Proc 1985;29:100–32.
12. Monga O. An optimal region growing algorithm for image segmentation. Int J Patter Recog Artif Intel 1987;1:351–75.
13. Ayache, N. Boissonnat JD, Cohen L, et al. Steps toward the automatic interpretation of 3d images. NATO ASI Series on 3D Imaging in Medicine 1990;F 60:107–20.
14. Cohen I, Cohen D, Ayache N. Using deformable surfaces to segment 3-D images and infer differential structures. Comput Vision Graphics Image Proc 1992;56:242–63.
15. Rueda S, Gil JA, Pichery R, et al. Automatic segmentation of jaw tissues in CT using active appearance models and semi-automatic landmarking. Proceedings of the Ninth International Conference on Medical Image Computing and Computer-Assister Intervention; 2006 Oct 1–6, Copenhagen, Denmark. Springer; 2006.
16. Enciso R, Memon A, Fidaleo DA, et al. The virtual craniofacial patient: 3D jaw modeling and animation. Study Health Technol Inform 2003;94:65–71.
17. Cutting, C, Taylor RH, Bookstein F, et al. Comprehensive three-dimensional cephalometric system for the planning. Proceedings of the Fourth Biannual Meeting of the International Society of Cranio-Maxillofacial Surgery, 1991 June 13–16, Santiago de Campostella, Spain; 1991.
18. White, CL (Troulis, MJ., supervisor). Development and standardization of landmarks for three-dimensional treatment planning [dissertation]. Boston, MA: Harvard University; 2002.
19. Yeshwant K, Seldin EB, Gateno J, et al. Analysis of skeletal movements in mandibular distraction osteogenesis. J Oral Maxillofac Surg 2005;63:335–40.
20. Ritter L, Yeshwant K, Seldin EB, et al. Range of curvilinear distraction devices required for treatment of mandibular deformities. J Oral Maxillofac Surg 2006;64:259–64.
21. Bouchard C, MaGill J, Nikonovskiy V, Byl M, Murphy L, Kaban LB and Troulis MJ: Osteomark: A surgical nangation system for oral and maxillofacial Surgeny. Int J Oral Maxillofac Surg 2012;41:265–70.
22. Taylor RH, Cutting CB, Kim Y, et al. A model-based optimal planning and execution system with active sensing and passive manipulation for augmentation of human precision in computer-integrated surgery. Proceedings of the 1991 International Symposium on Experimental Robotics, 1991, June 25–27, Toulouse, France: Springer-Verlag; 1991.
23. Altobelli DE, Kikinis R, Mulliken JB, et al. Computer-assisted three-dimensional planning in craniofacial surgery. Plast Reconst Surg 1993;92:576–85.

Applications of Distraction Osteogenesis for the Mandible: An Overview

12

Carl Bouchard, DMD / Maria J. Troulis, DDS, MSc / Leonard B. Kaban, DMD, MD

OUTLINE

INTRODUCTION

As the sophistication of devices, surgeons' familiarity and comfort with the concept, understanding of the biology and surgical techniques have improved, distraction osteogenesis (DO) has become an alternative treatment for skeletal deformities that require bone expansion. It is a minimally invasive alternative to osteotomy and bone grafting procedures and has become a frequently used technique for a variety of mandibular movements.

Because bone expansion in any direction can be achieved by DO, its use in patients with congenital and acquired deformities is increasing. For the purpose of this chapter, we classify DO based on the direction of bone lengthening: 1) sagittal: mandibular lengthening parallel to the body of the mandible; 2) vertical: mandibular lengthening parallel to the ramus of mandible; 3) transverse: mandibular widening; and 4) multiplanar or multidirectional DO.

The distraction protocol for mandibular DO used at Massachusetts General Hospital includes a latency period of 48 hours, distraction rhythm of two to four times per day and a rate of 1 mm per day. The consolidation period is at least twice the number of days of distraction or twice the number of millimeters of lengthening in days.

MANDIBULAR LENGTHENING IN THE SAGITTAL PLANE

The most common deformity requiring mandibular lengthening primarily in the sagittal plane is non-syndromic mandibular retrognathism. These are the typical orthognathic surgery patients who require 6 mm–10 mm of advancement to correct Class II malocclusion or patients with severe retrognathism requiring more than 15 mm of mandibular lengthening. The advantages of DO versus acute advancement of the mandible are: 1) an operation of lesser magnitude than standard mandibular advancement by sagittal split osteotomy or by osteotomy and bone graft, 2) the ability to accomplish advancements of larger magnitude than the standard operations, 3) decreased chance of permanent paresthesia of the inferior alveolar nerve, and 4) less resistance of the soft tissue envelope to gradual versus acute lengthening and therefore potentially less relapse. There is evidence that the muscles may respond to gradual lengthening by undergoing hyperplasia and hypertrophy[1,2] rather than the stretching that occurs with acute advancement.

Retrognathia

Nonsyndromic Mandibular Retrognathia

Patients who exhibit non-syndromic mandibular retrognathism may have a normal or deficient mandibular growth rate. The group with a normal growth rate is born with a small mandible or a mandible that is positioned posteriorly relative to the maxilla because of the location of the glenoid fossa. In these patients, the mandible grows proportionately with the midface but not enough to compensate for the initial mandibular deficiency. The deformity remains stable and does not worsen with time. Because of the normal growth rate of the mandible in these children, mandibular lengthening can be accomplished before the completion of growth and the correction will be maintained[3–6] (Figure 1).

Other patients have mandibular deficiency secondary to radiation therapy or traumatic injury.[6] The mandible grows insufficiently over time and the deformity becomes progressively worse. There are usually bone and soft tissue abnormalities and the resistance of the soft tissue to mandibular advancement may adversely affect long-term stability. Distraction osteogenesis simultaneously expands the bone and soft tissue in these cases.

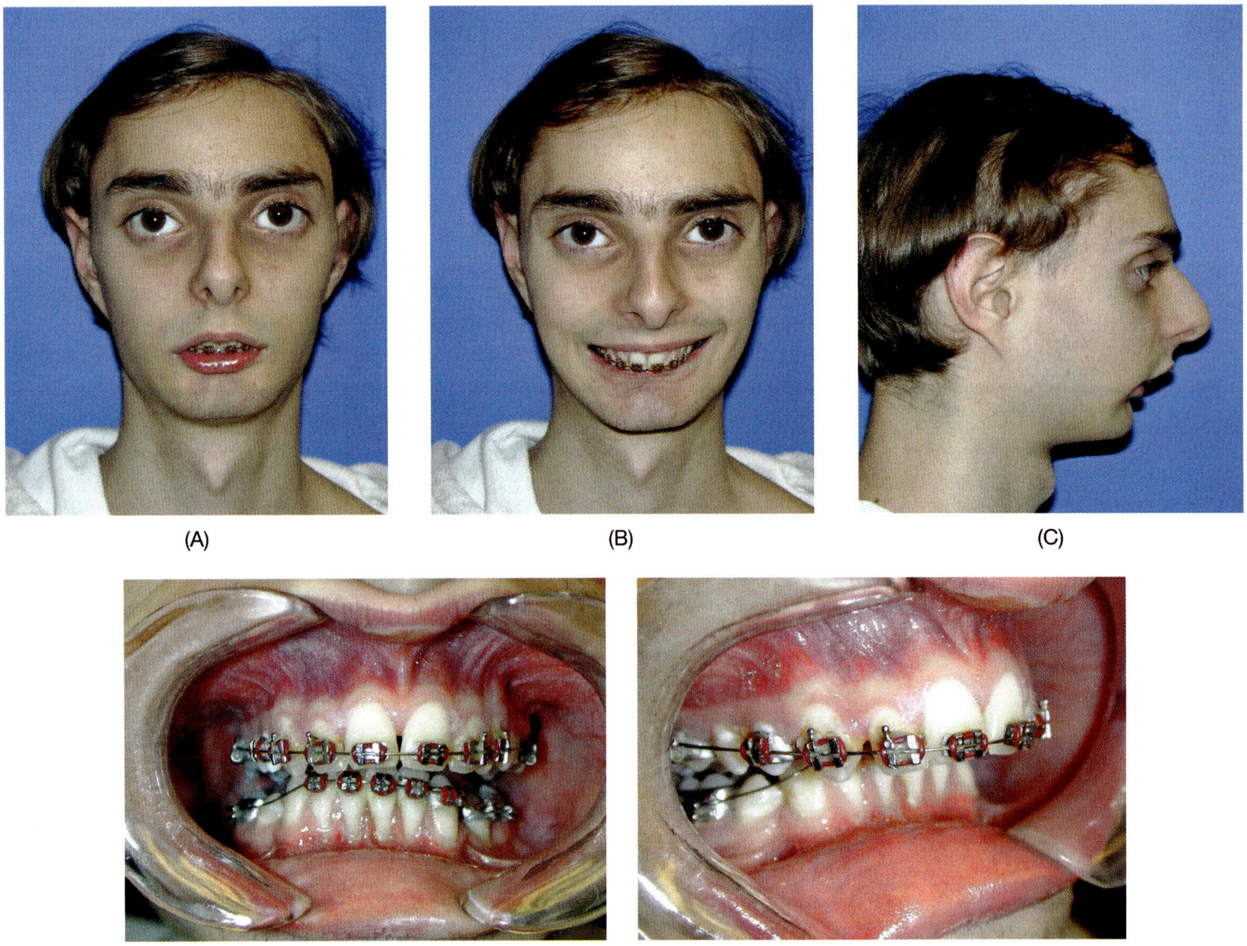

(A) (B) (C)

(D) (E)

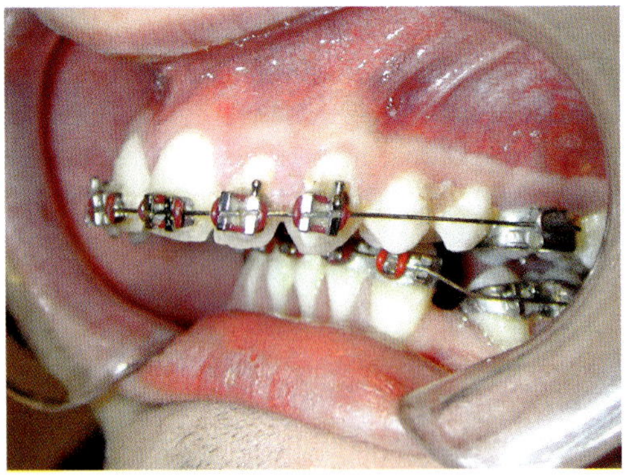

(F)

FIGURE 1 • **Mandibular DO for sagittal hypoplasia.** 16-year-old patient with a non-syndromic sagittal mandibular hypoplasia. (A) Frontal, (B) smiling, and (C) lateral photographs demonstrating the severe mandibular sagittal hypoplasia, vertical maxillary excess, obtuse nasiolabial angle, convex facial profile, and increased labio-mental fold. (D)(E)(F) **Pre-operative** intraoral photographs during the orthodontic decompensation phase - illustrating the severe overjet and the accentuated curve of spee. (*continued*)

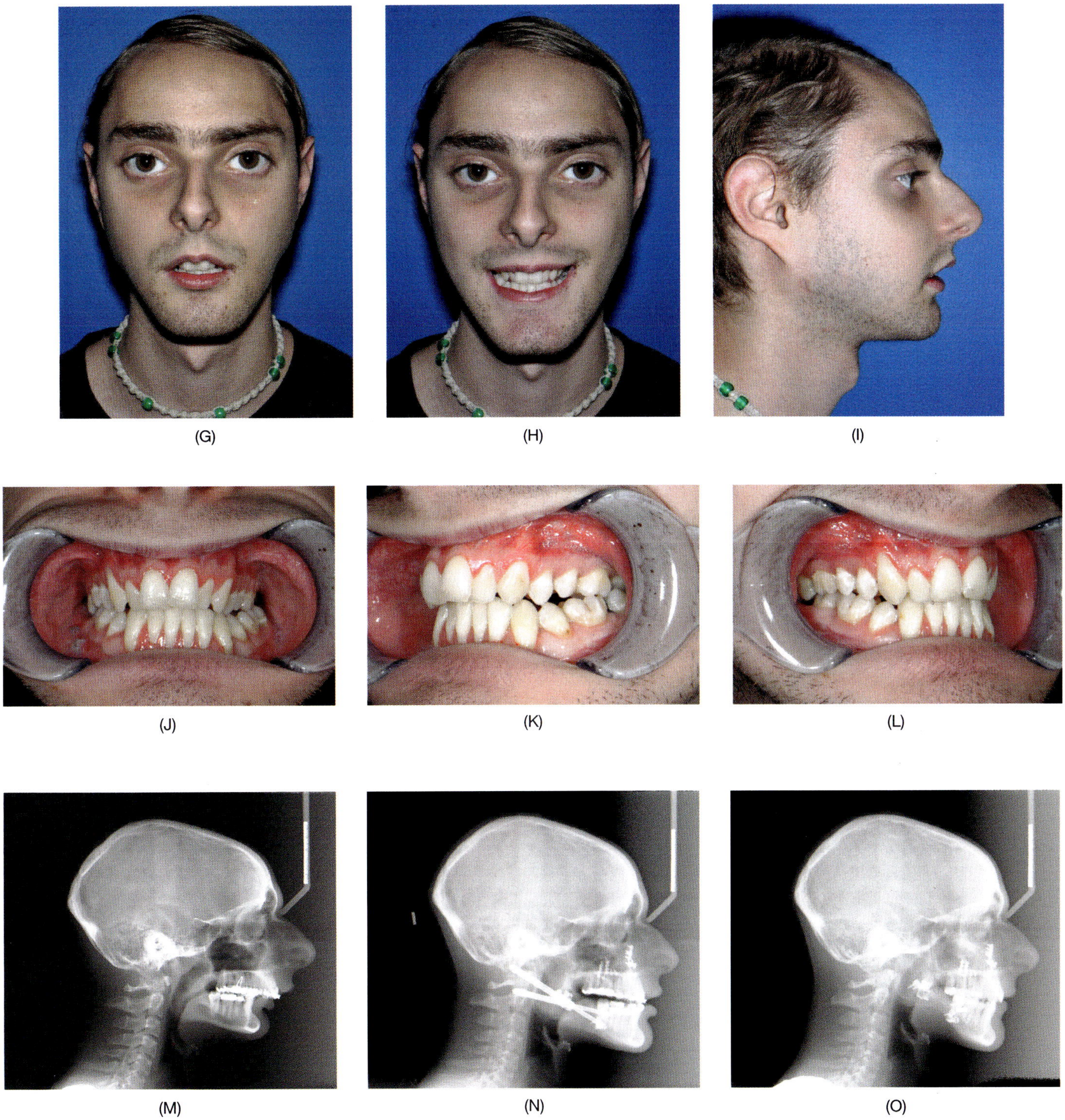

FIGURE 1 • (*continued from previous page*) (G)(H)(I) 3-year **post-operative** pictures of the same patient, with a normalized facial contour and an overall increase in facial proportions. (J)(K)(L) **Post-operative** intraoral photographs. No relapse was observed and the overjet remains normal. Serial cephalometric radiographs at the initial presentation (M), end of distraction (N), and 3 years postoperatively (O) (*continued*)

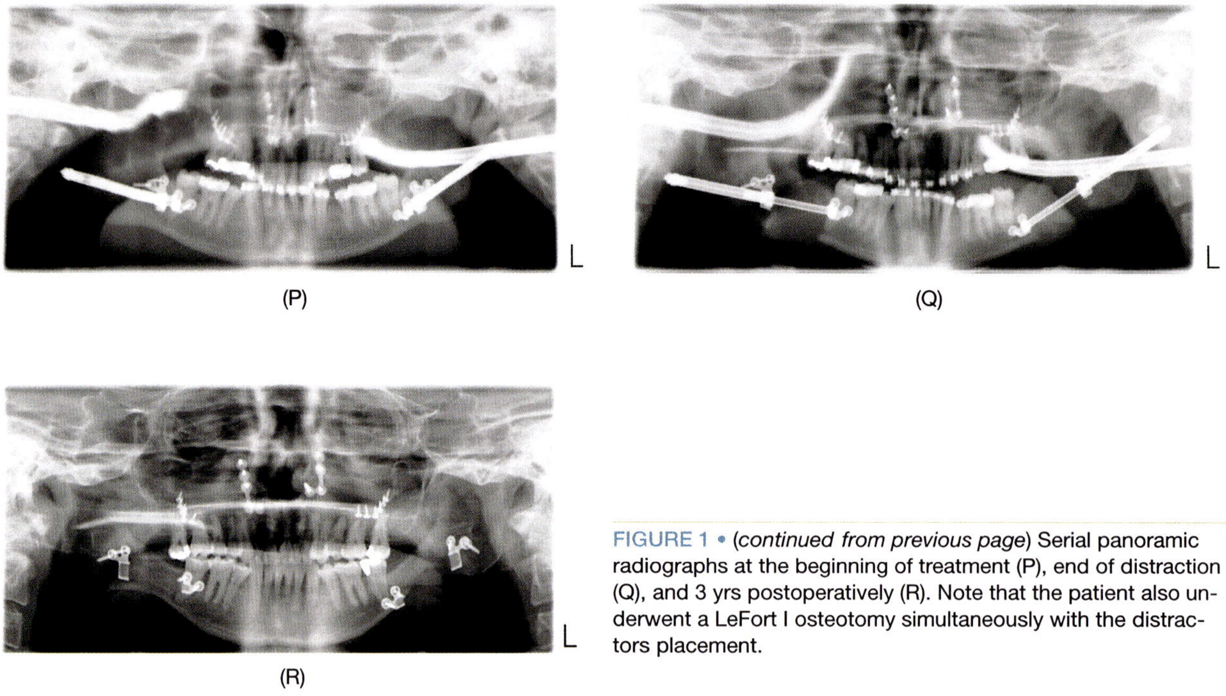

FIGURE 1 • (*continued from previous page*) Serial panoramic radiographs at the beginning of treatment (P), end of distraction (Q), and 3 yrs postoperatively (R). Note that the patient also underwent a LeFort I osteotomy simultaneously with the distractors placement.

Robin Sequence and Syndromic Micrognathia

Robin Sequence is a constellation of physical findings consisting of: micrognathia (small retruded mandible), glossoptosis (tongue retroposition), characteristic "U-shaped" cleft palate, and upper airway obstruction. The patients may also have feeding difficulties. Robin Sequence may be non-syndromic or syndromic. Those patients who do not have an underlying syndrome may exhibit catch-up growth and may be managed by training the parents in home care, positioning, and feeding methods. They often will not require an operation. Robin Sequence is associated with multiple syndromes. The most common are: Stickler syndrome, velocardiofacial syndrome, and Treacher Collins syndrome. Syndromic patients are more likely than non-syndromic patients to require tracheostomy or mandibular advancement to manage airway obstruction.

Mandibular advancement by distraction osteogenesis has been advocated for treatment of micrognathia and resultant obstructive sleep apnea (OSA) and feeding difficulties in patients with syndromic Robin Sequence[7–9] or other congenital anomalies (Figure 2). Burstein and colleagues[9] reported a series of 15 infants and 5 children with Robin Sequence that were treated with distraction osteogenesis. Satisfactory results were achieved in 19 patients and the status of one patient was not reported. Monasterio and associates[10] demonstrated that distraction osteogenesis improved swallowing disorders and sleep apnea in 18 patients with syndromic Robin Sequence.

Although DO to prevent tracheostomy or to facilitate decannulation may produce satisfactory results in these patients, the ideal timing of treatment remains controversial. Children less than 2 years of age are not optimal candidates for treatment by distraction. Technical issues include deficient quantity of bone stock to accept pins or screws and difficulty in identifying developing tooth follicles. Family adherence to the treatment protocol can also prove to be a difficult challenge in this age group.

Severe Micrognathia and Airway Obstruction in Children

Distraction osteogenesis to advance the maxilla and/or mandible has been used for more than a decade in the treatment of OSA[7,8,11–13] (Figure 3). Steinbacher and associates[11] reported a series of 5 tracheostomy-dependent children with severe micrognathia who underwent mandibular advancement by distraction osteogenesis. The results of this study indicate that mandibular advancement by DO is a potentially viable treatment option for tracheostomy-dependent children with upper airway obstruction secondary to micrognathia. Cohen and colleagues[14] performed mandibular distraction in 16 patients with an average of 25 mm advancement. They found that DO produced clinical improvement in signs and symptoms of sleep apnea in 7 out of 8 children previously unresponsive to medical management. Sidman and associates[7] reported a series of 11 children with severe airway obstruction secondary to tongue-base obstruction who underwent DO of

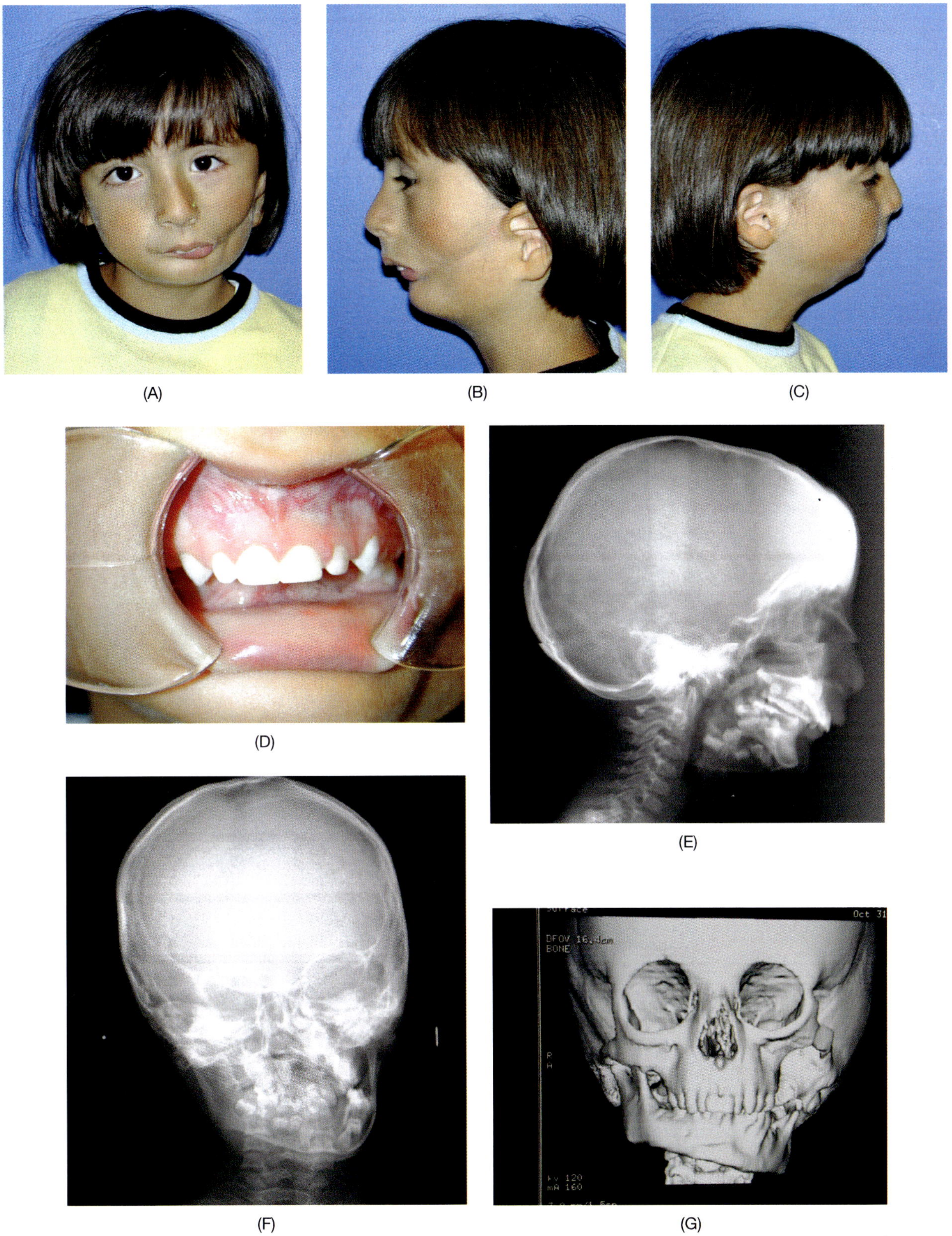

FIGURE 2 • **Mandibular DO in patients with congenital anomalies.** 7-year-old male patient with a non-specific first and second branchial arch anomaly and an incomplete left facial cleft Tessier # 7 that was repaired in another center. (A) Pre-operative frontal, (B) left, and (C) right facial photographs demonstrating the convex facial profile and the increased labio-mental fold. The lower lip is everted under the upper central incisors. The repaired left facial cleft is apparent on the left lateral photograph. (D) Preoperative intraoral photograph demonstrates increased overjet, (E) Pre-operative lateral cephalogram, (F) postero-anterior cephalogram, and (G) frontal view of 3D CT scan. The left inferior border of the mandible is considerably overgrown relative to the right side. (*continued*)

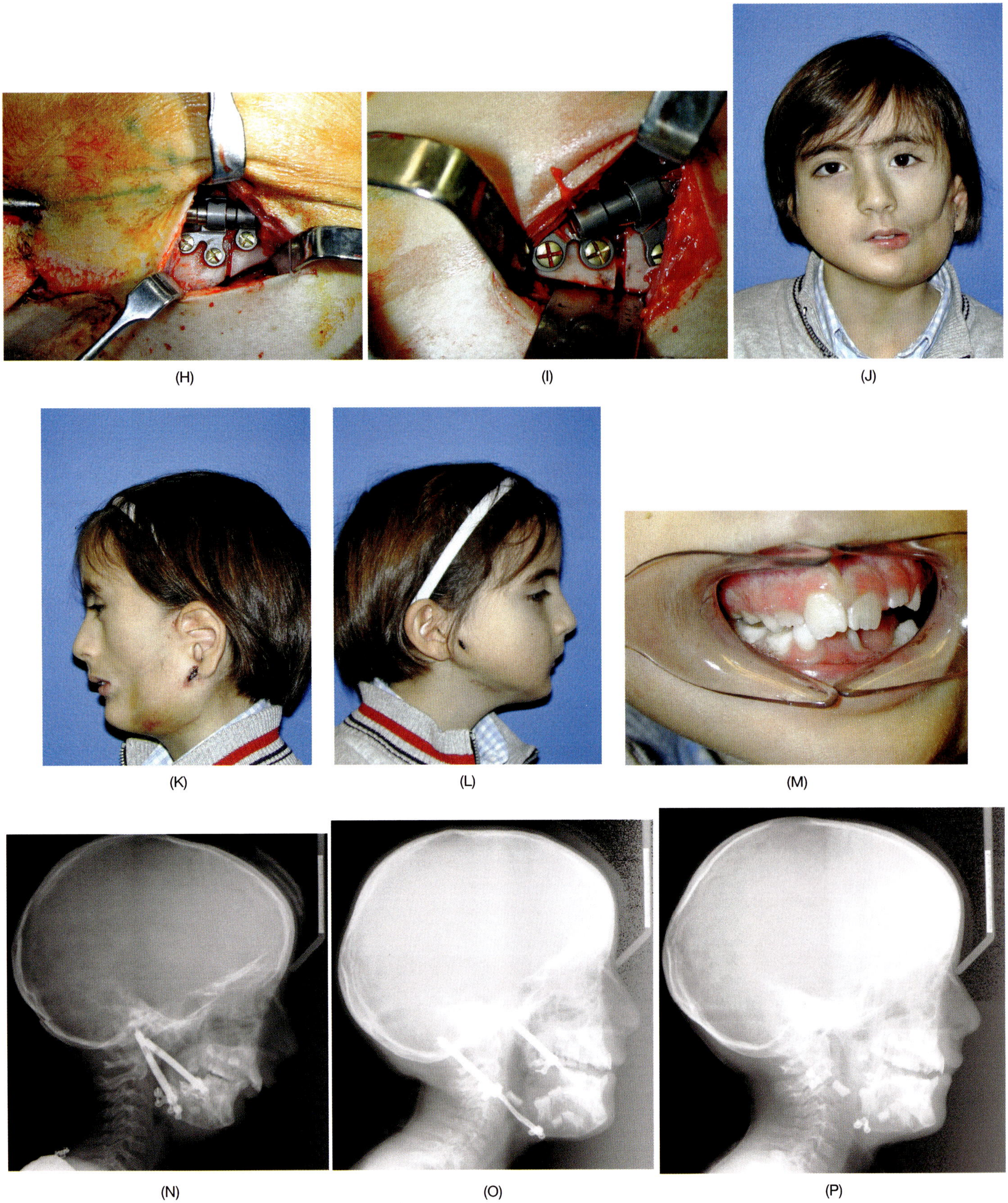

FIGURE 2 • (*continued from previous page*) (H) Right and (I) left Intra operative views of osteotomies and distractors in place. (J) Frontal (K) left and (L) right postoperative pictures after distractor removal. The facial contour and the labio-mental fold are now normalized. The chin point is coincident with the facial midline. The inferior mandibular border asymmetry persists and is scheduled to be corrected after eruption of the permanent molars. (M) Dental occlusion after distraction. (N) Serial cephalometric radiographs at the beginning of distraction, (O) the end of distraction, and (P) after distractor removal. (Also see Chapter 13, Figures 3 and 4.)

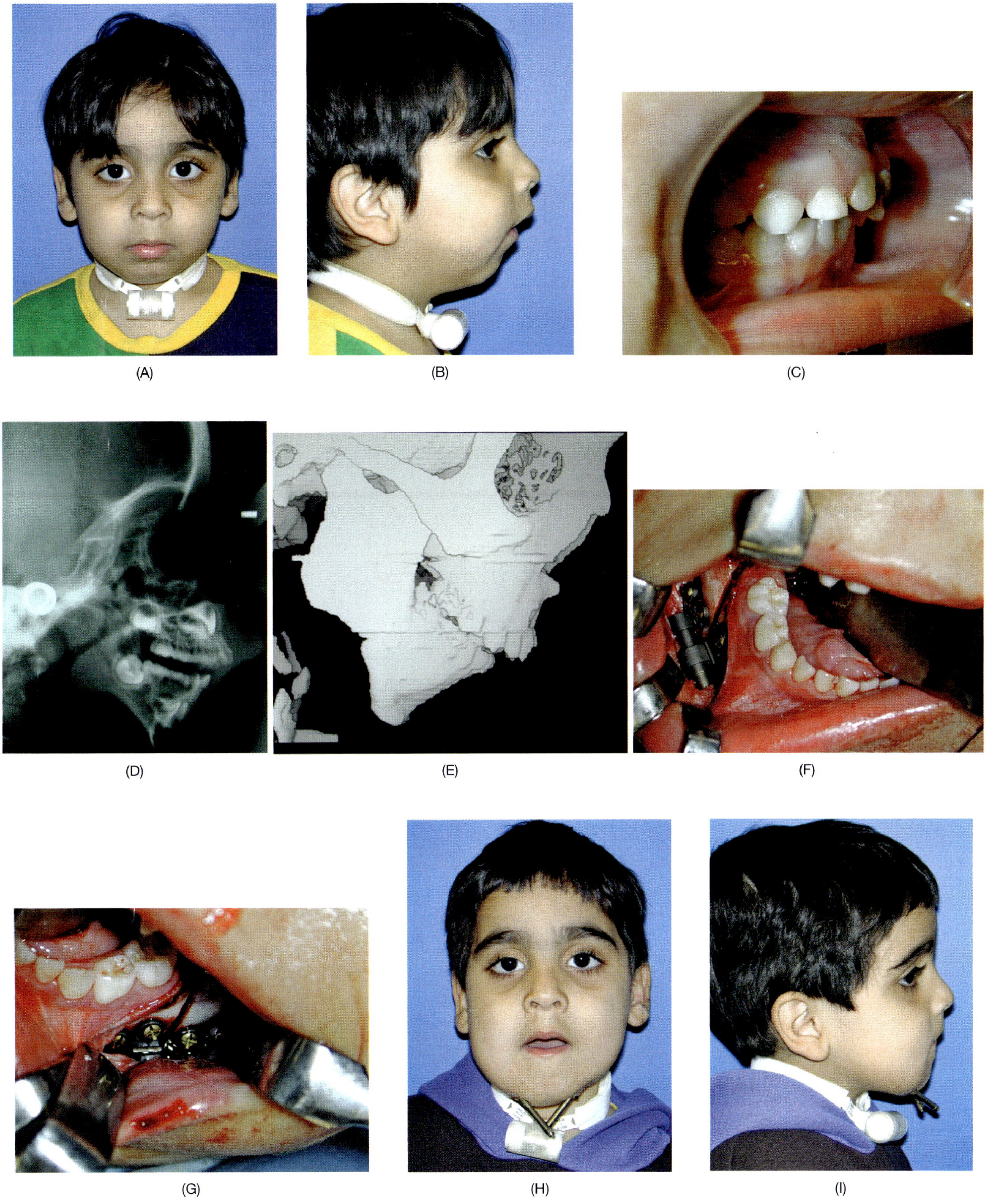

FIGURE 3 • Severe micrognathia and airway obstruction. 3-year-old male with severe micrognathia and airway obstruction. (A) Frontal, (B) lateral, and (C) intraoral pre-operative photographs demonstrating a convex facial profile and a class II dental occlusion. This patient was tracheostomy-dependent. (D) Cephalometric radiograph and (E) 3D CT scan illustrating the mandibular sagittal hypoplasia and the short ramus height. Intraoperative pictures of (F) the right and (G) left osteotomy and distractor. (H) End of distraction frontal, (I) lateral, (*continued*)

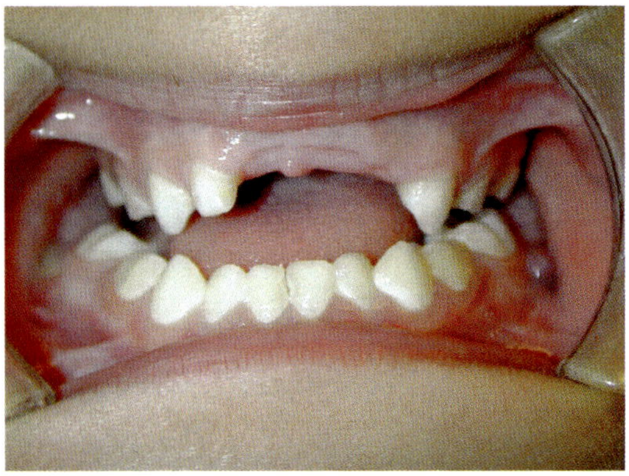

(J)

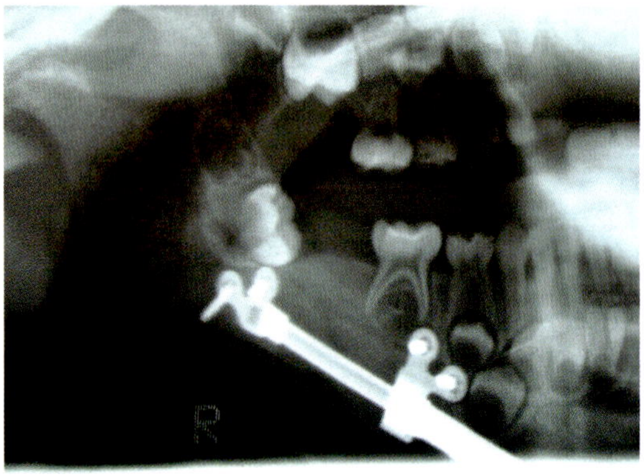

(K)

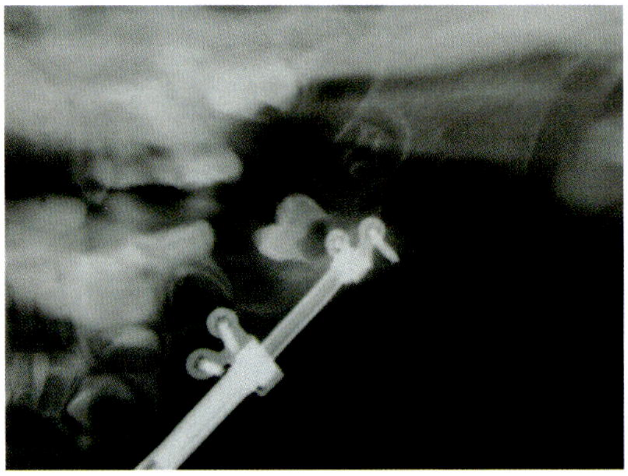

(L)

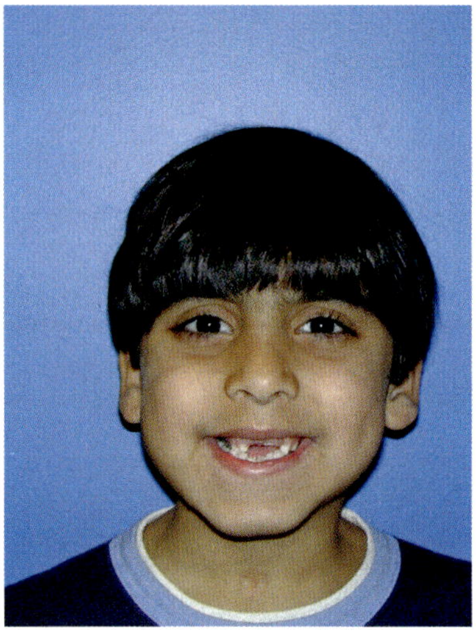

(M)

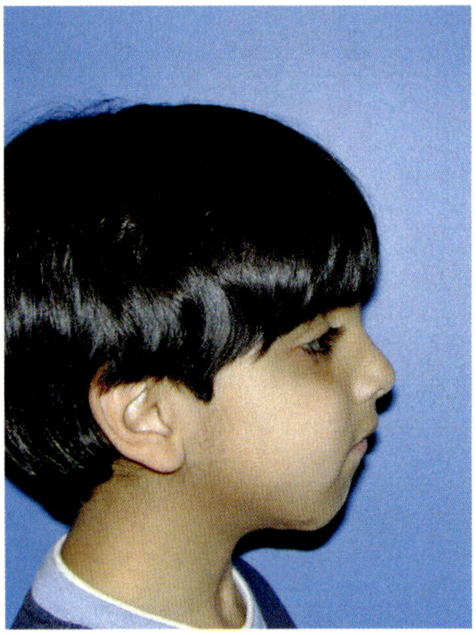

(N)

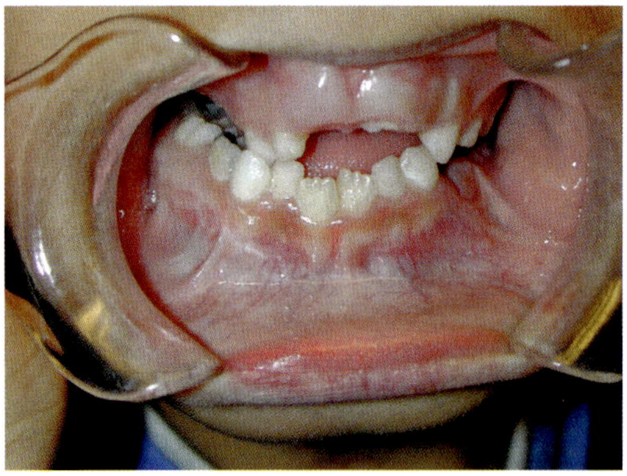

(O)

FIGURE 3 • (*continued from previous page*) and (J) intraoral photograph. The defect was overcorrected to compensate for relapse and provide enough movement to allow decannulation. End of distraction radiographs. The right (K) and (L) left distractors are still in place and the regenerate is calcified. (M) Post-operative frontal, (N) lateral and (O) intraoral photographs. The results are stable and no relapse was noted. He was permanently decannulated shortly after the end of distraction. Reproduced in part from Steinbacher et al.[11] with permission from Elsevier.

the mandible. The results of this study showed that DO can be used either to eliminate the need for tracheostomy in micrognathic children or to achieve decannulation in these patients. Wang and colleagues[15] reported a series of 28 patients with obstructive sleep apnea (OSA) syndrome who underwent mandibular distraction osteogenesis. Nine patients with severe, six with moderate, and eight with mild obstructive sleep apnea were cured by mandibular lengthening. The other five patients with either severe or moderate OSA exhibited improvement to mild OSA.

RECONSTRUCTION OF CONTINUITY DEFECTS (TRANSPORT)

Mandibular reconstruction after bone loss is a great challenge. Grafts, flaps, and alloplastic materials have been used to reconstruct the mandible. Recently, DO has been shown to be a successful technique not only in the surgical treatment of congenital and acquired mandibular hypoplasia, but also in the reconstruction of segmental mandibular defects.[16–18] (See Chapter 15, "Transport Distraction Osteogenesis.)

Costantino and colleagues[19] demonstrated that bifocal distraction osteogenesis was effective in the reconstruction of a segmental mandibular defect in an animal model. Annino and associates[20] demonstrated the feasibility of curved trifocal distraction osteogenesis for reconstruction of mandibular symphyseal defects in an animal model. Block and colleagues[18] reported 4 patients treated by bifocal distraction osteogenesis with satisfactory results. Sawaki and associates[21] reported the successful trifocal distraction osteogenesis for reconstruction of mandibular defect.

Distraction osteogenesis has been advocated for replacement of hard and soft tissue loss due to facial injuries.[22–24] Shvyrkov and colleagues[22] reported on 33 males with bone and soft tissue defects of the mandible that were initially managed by debridement and collapsing the proximal segments. They then used transport disk distraction to generate bone and soft tissue across the gap. Of these patients, 28 presented within a few hours of injury, and the remaining 5 had old injuries. The only complication occurred in the group with old injuries, where 4 patients developed abscesses that required drainage, but these did not interfere with the process of osteogenesis. All 33 patients had good functional and aesthetic results within 3 to 4.5 months.[22]

Distraction osteogenesis has also been used for reconstruction of mandibular defects following tumor ablation.[16,17,25] Kuriakose and colleagues[16] demonstrated that mandibular reconstruction with distraction osteogenesis is a potentially useful technique in selected patients with segmental mandibular continuity defects after ablative cancer surgery (Figure 4).

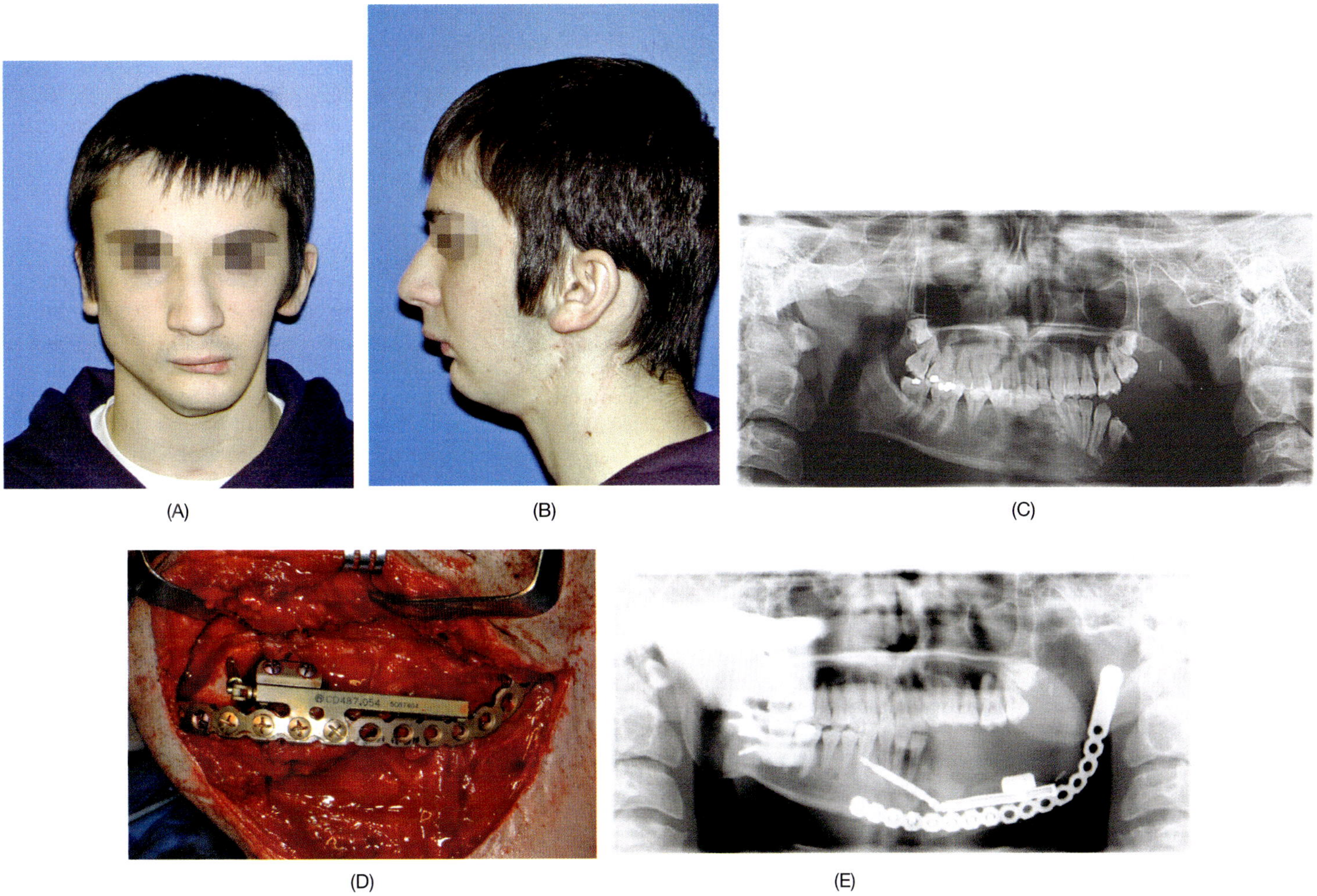

FIGURE 4 • **Reconstruction of continuity defect.** 14-year-old male patient who underwent an en bloc resection of a Ewing's sarcoma on his left mandible at the age of 8. He was left with a continuity defect after failure of a free vascularized fibular flap. (A) Facial and (B) lateral pre-operative photographs showing a retrognathic mandible deviated to the left. (C) Panoramic x-ray prior to distraction. (D) Distractor in place with the transport disc secured to the upper portion of the reconstruction plate. (E) Panoramic x-ray during the consolidation phase, (*continued*)

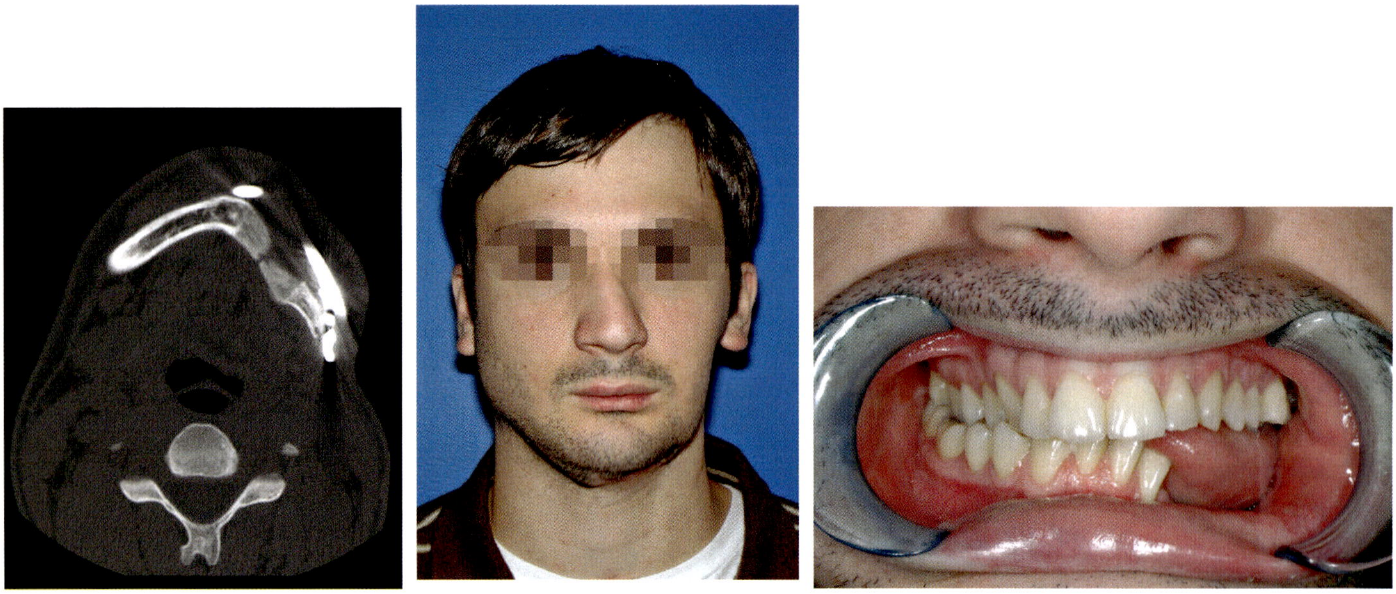

(F) (G) (H)

FIGURE 4 • (*continued from previous page*) and (F) CT scan demonstrating the regenerate partially calcified. (G) Normalization of the facial contour. (H) Stable dental occlusion. Patient will eventually be reconstructed with an implant-supported prosthesis.

VERTICAL LENGTHENING

Ramus Lengthening

Patients with hemifacial microsomia and bilateral craniofacial microsomia have skeletal deformities that include hypoplastic TMJs, short mandibular rami, and decreased posterior face height.[26,27] DO is a useful technique for bone lengthening in these patients[28–31] (Figure 5). The technique is applicable only in patients who have enough bone stock in the mandibular ramus condyle unit (RCU) to allow for the corticotomy and fixation of the distraction device (Type I and IIA mandible). DO is particularly useful in patients with hemifacial microsomia. It is the only technique that allows RCU lengthening and counterclockwise rotation of the mandible of the magnitude required in these patients. Shetye and colleagues[32] documented that while distraction effectively produces the desired mandibular lengthening, it does not improve subsequent growth of the deficient mandible in children. Therefore, the mandible must be overcorrected and an open bite must be created on the affected side. This is maintained by an appliance that is gradually modified by the orthodontist to allow for vertical lengthening of the maxilla by gradual eruption of the teeth and dentoalveolar and midface growth. Skeletal stability has been satisfactory, and prevention or elimination of tracheostomy has been documented.[11–13]

Bimaxillary DO

The indications for bimaxillary distraction are: 1) end-stage type I and IIA HFM where the correction is going to be carried out purely by lengthening the short side, 2) secondary bimaxillary DO in adults or growing children with HFM who had inadequate correction by other methods, and 3) adults with end-stage Treacher Collins or Nager syndrome or other patients with a short posterior face height (Figure 6). As a result of distraction, both the maxillary and mandibular occlusal planes become horizontal, and facial asymmetry is corrected.[29,33] Scolozzi and colleagues[29] demonstrated that simultaneous maxillo-mandibular distraction osteogenesis resulted in a high rate of success in correcting facial asymmetry as well as occlusal plane cant.

Alveolar DO

Distraction osteogenesis can be used to restore vertical alveolar bone deficiency secondary to bone resections, atrophy after tooth loss, or trauma[34–35] (Figure 7). Alveolar distraction osteogenesis offers some advantages over traditional bone grafting techniques: 1) no donor site morbidity, 2) bone quality is ideal for implant placement,[36] and 3) the result is more predictable.[36] Chiapasco and colleagues[36] reported a series of 37 patients who underwent vertical distraction osteogenesis for the correction of vertically deficient alveolar ridges in 4 different centers. The results of this study demonstrate that distraction osteogenesis is a reliable technique for the correction of vertically deficient edentulous ridges. The regenerated bone appeared to withstand the functional demands of implant loading. The survival and success rates of the implants placed in the distracted areas were consistent with those reported in the literature regarding implants placed in native bone in this patient population. (See Chapter 16, "Alveolar Distraction Osteogenesis.")

TRANSVERSE MANDIBULAR EXPANSION OR WIDENING OF THE SYMPHYSIS OF THE MANDIBLE

Mandibular Widening

Transverse mandibular skeletal deficiency is a common clinical problem resulting from narrow basal and dentoalveolar bone. Crowding and bilateral posterior buccal crossbites are the

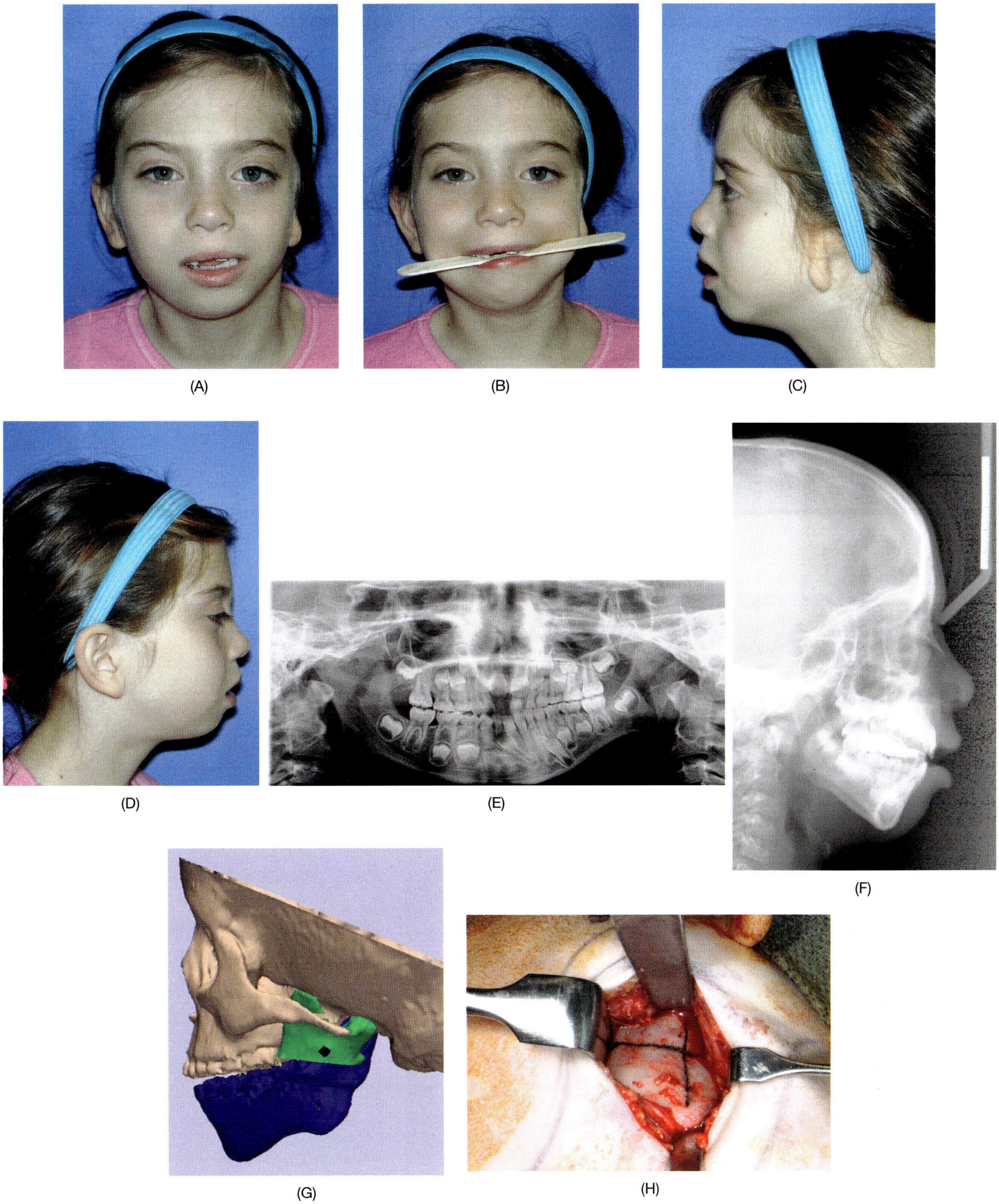

FIGURE 5 • **Ramus lengthening.** 7-year-old female patient with left hemifacial microsomia (Omens 7, Kaban II-A classification). (A) Frontal picture showing deviation of the chin point to the left and hypoplastic soft tissue. (B) Marked occlusal cant upward to the left. (C) Microtia grade III on the left and (D) normal right ear. (E) Panoramic x-ray demonstrating a short ramus-condyle unit on the left. (F) The lateral cephalogram shows the double shadow of the mandible indicative of the asymmetry. (G) 3D surgical planning. Note that the coronoid process had to be removed to prevent interference with the skull base. (H) Intraoperative view of the proposed osteotomy. *(continued)*

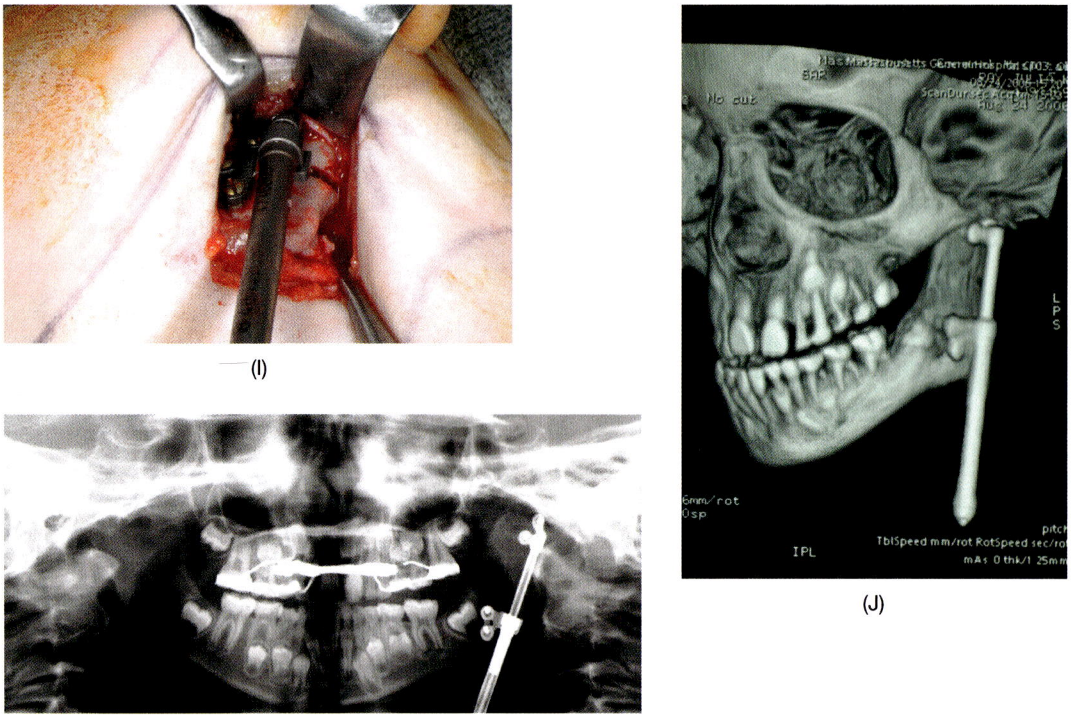

(I) (J) (K)

FIGURE 5 • (*continued from previous page*) (I) Distractor in place. (J) 3D CT images after the end of distraction. The regenerate is partially calcified. (K) Panoramic x-ray. An occlusal appliance maintains the newly created open bite. This will allow extrusion of the maxilla and correction of the occlusal cant.

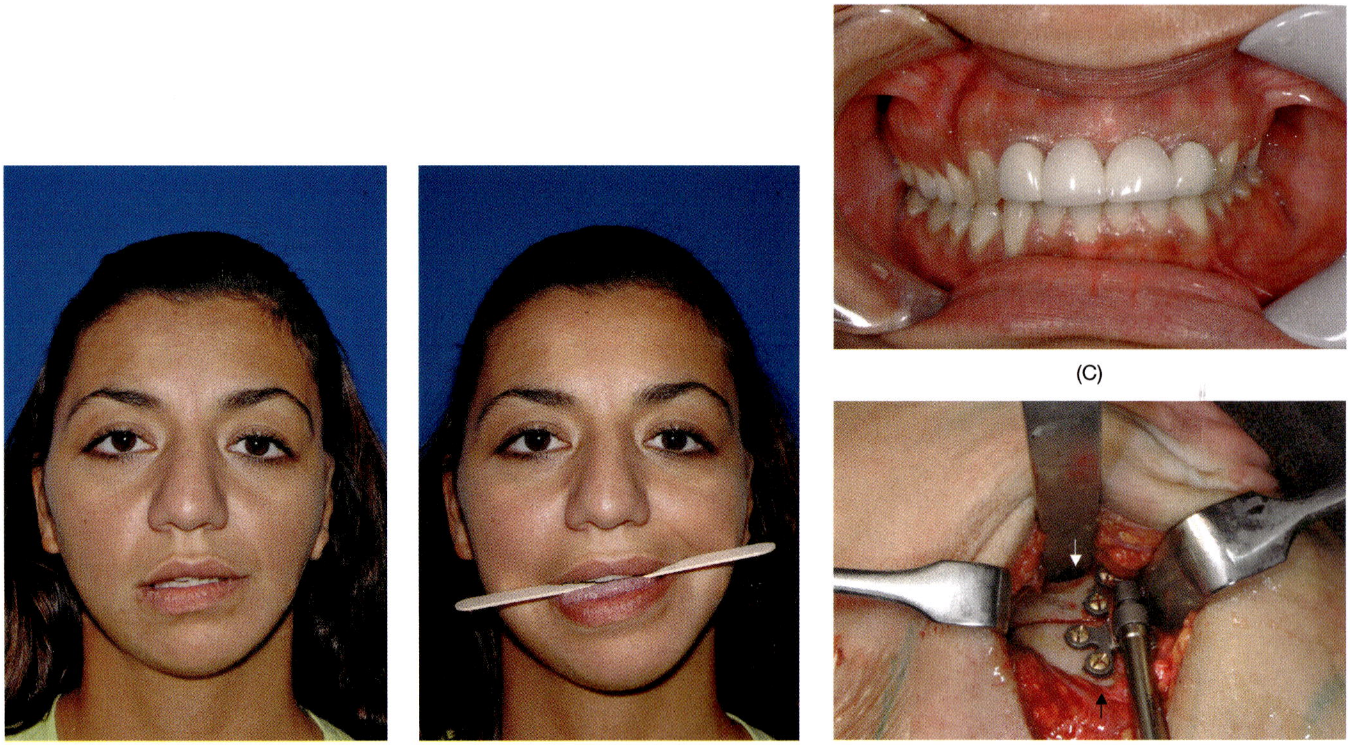

(A) (B) (C) (D)

FIGURE 6 • **Bimaxillary distraction.** 29-year-old female with end-stage left hemifacial microsomia (Omens 6, Kaban II-A classification). (A) Her frontal photograph demonstrates with her left commissure canted up. Her chin point is deviated to the left of the facial midline with a soft tissue deficit overlying the posterior left mandible. (B) The patient has 5 degrees of occlusal cant. (C) The patient underwent a bimaxillary distraction to correct the asymmetry of the mandible and the maxilla simultaneously. Lefort I osteotomy cuts were performed without downfracturing the maxilla. (D) The osteotomy of the left ramus was performed and the unidirectional distractor (Synthes CMF, West Chester, PA) secured on both sides of the cut (white arrow = sigmoid notch, black arrow = mandibular angle). Patient was placed in maxillo-mandibular fixation during the distraction. (*continued*)

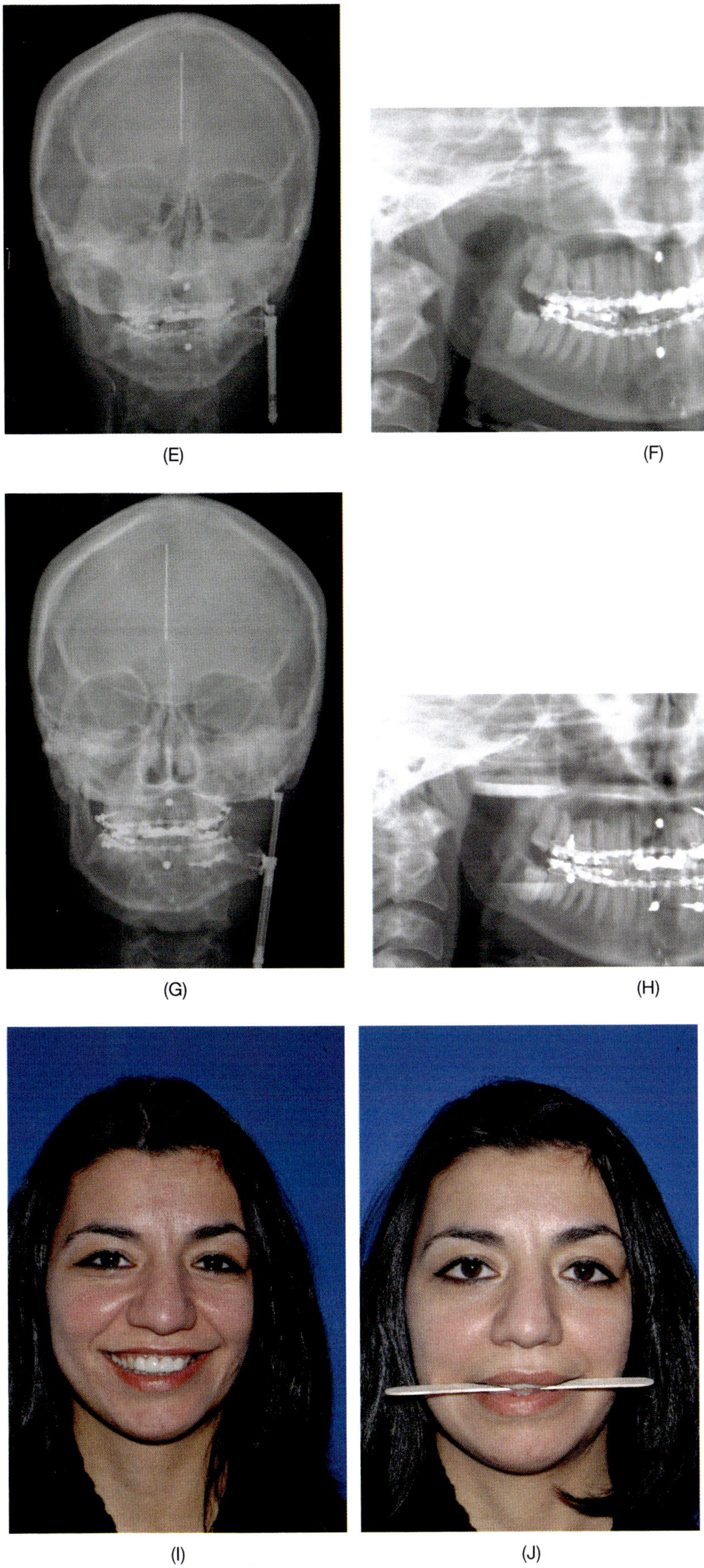

FIGURE 6 • (*continued from previous page*) Immediate post-operative (E) cephalometric and (F) panoramic radiographs. (G) End of distraction cephalometric and (H) panoramic radiographs. (I + J) Facial photographs 1 year after end distraction. The commissures are leveled and the chin point is centralized. The occlusal cant was also corrected. (Also see Chapter 17.)

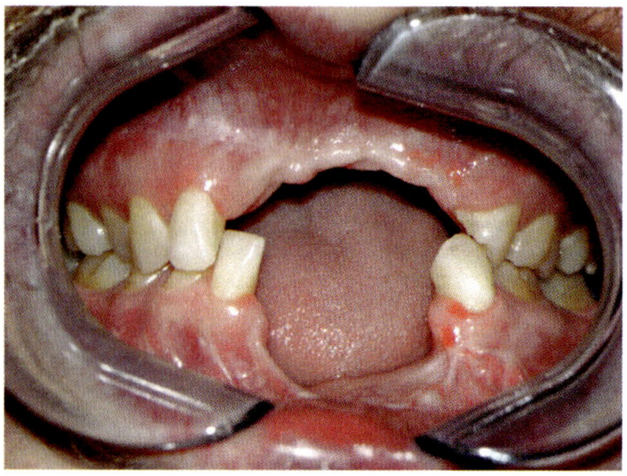
(A)

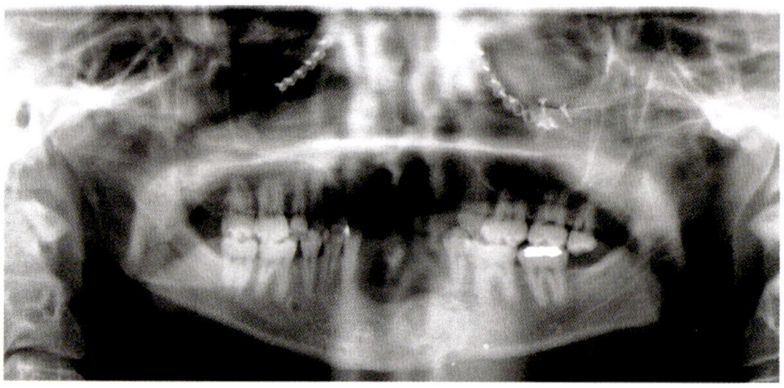
(B)

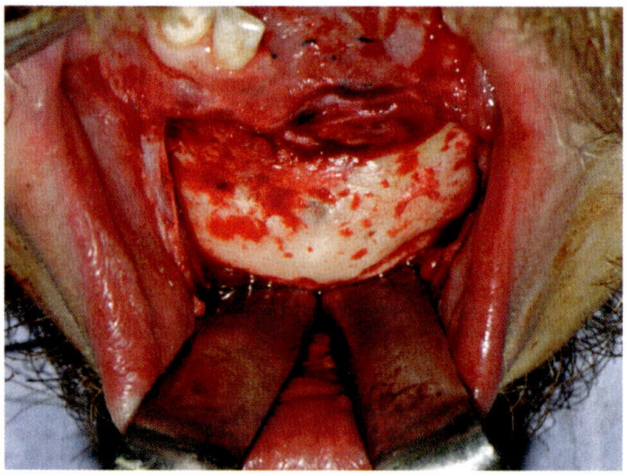
(C)

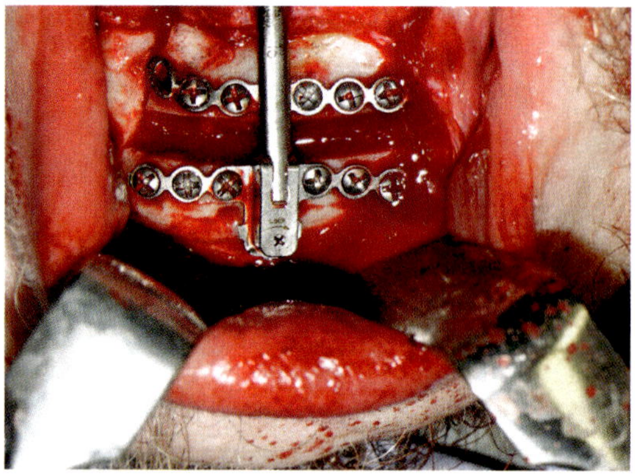
(D)

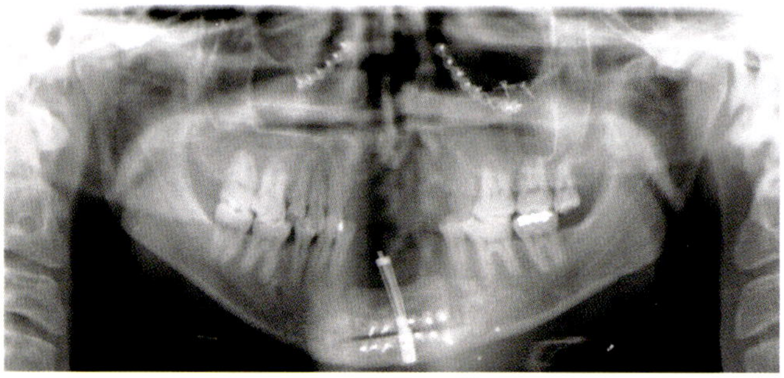
(E)

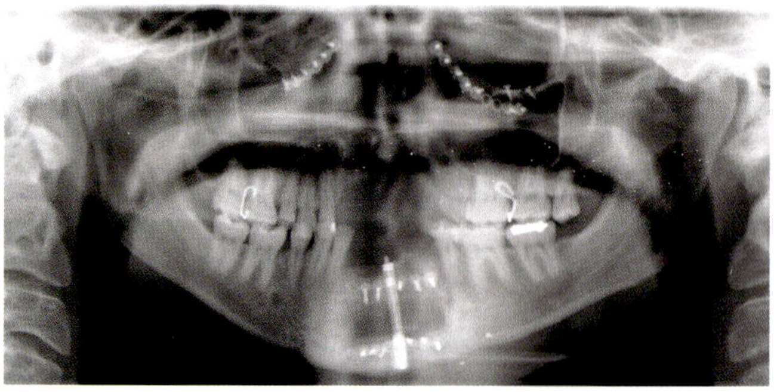
(F)

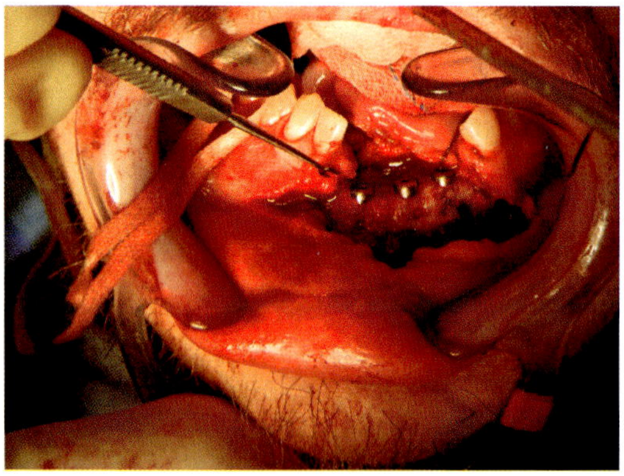
(G)

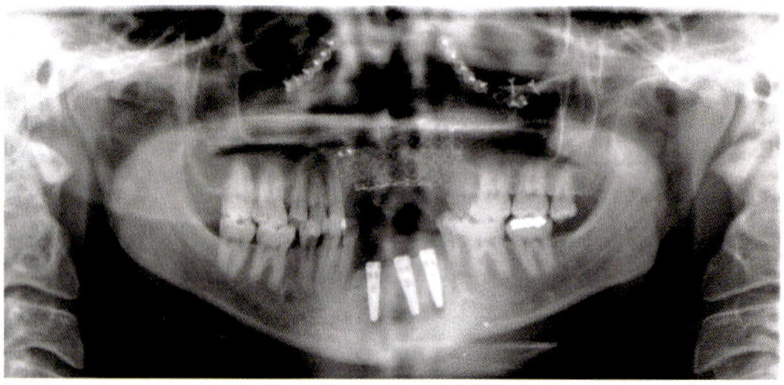
(H)

FIGURE 7 • **Alveolar distraction.** 24-year-old male who sustained multiple facial fractures in a motor vehicle accident. He lost part of his anterior alveolus processes both on the maxilla and the mandible. (A) Pre-operative intraoral pictures showing the bony defect and the missing teeth in the anterior region. (B) Panoramic x-ray. (C) The anterior region of the mandible was exposed through an intraoral approach. (D) Distractor in place. (E)(F) Pre- and post-distraction x-rays. (G)(H) Reconstruction with dental implants. (Also see Chapter 16.)

clinical manifestations of transverse mandibular deficiency (Figure 8). Traditionally, this problem has been treated by extraction of teeth, dental arch compensation, or orthognathic surgery. In 1993, Perrott and associates[37] introduced mandibular widening with distraction osteogenesis to solve transverse deficiencies. The goal of distraction osteogenesis of symphysis is to create basal and alveolar bone, keratinized gingival and mucosa. King and colleagues[38] and Tae and associates[39] reported that mandibular widening with DO can successfully treat unilateral crossbite.

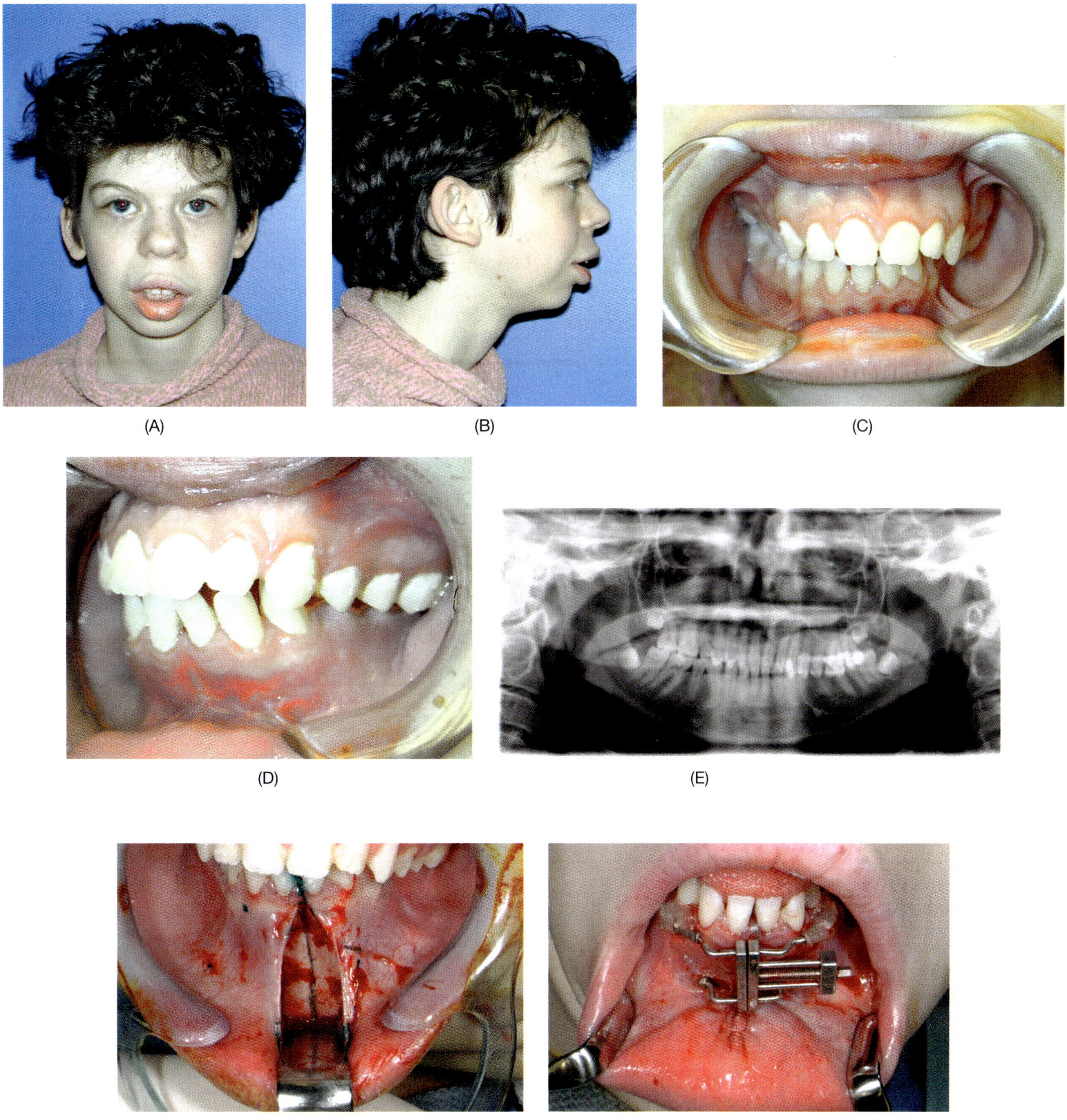

(A) (B) (C) (D) (E) (F) (G)

FIGURE 8 • **Mandibular widening**. 16-year-old female patient with Williams syndrome and presenting with a severe maxillomandibular width discrepancy. (A) Frontal and (B) lateral pre-operative pictures showing a retruded mandible and lip incompetence. (C)(D) She had protrusive maxilla and an increased overjet. (E) Pre-operative panoramic x-ray. (F) Planned osteotomy. (G) Distractor (Dynaform™, Leibinger, Dallas, TX) in place and soft tissue closure. (*continued*)

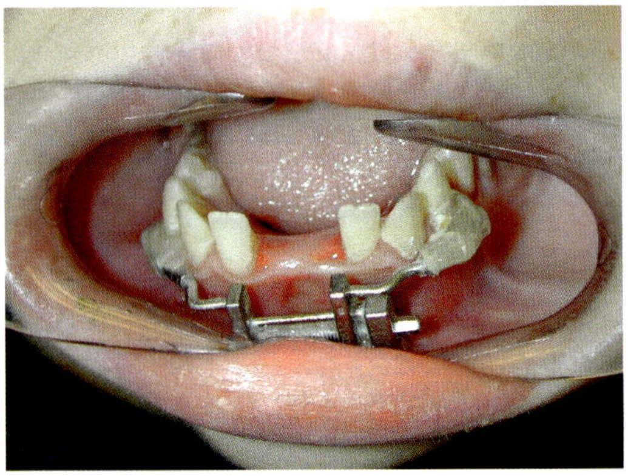

(H)

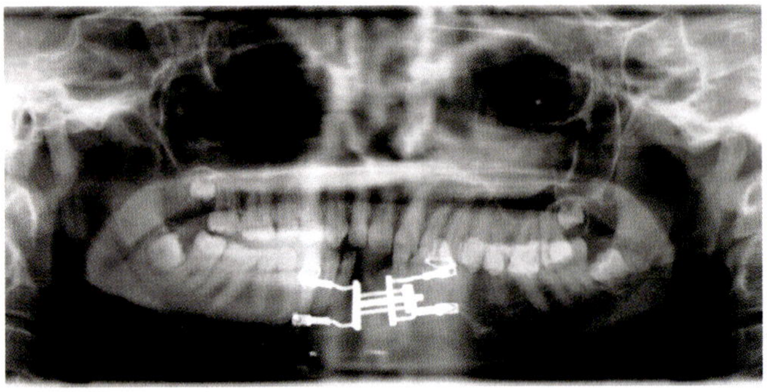

(I)

FIGURE 8 • (*continued from previous page*) (H) Intraoral picture and (I) panoramic radiograph during the consolidation phase. Note the amount of distraction and gain in soft tissue (mucosa).

MULTIVECTOR DISTRACTION OSTEOGENESIS

It is generally accepted that most mandibular corrections require multidirectional movements. Currently, there are three methods of achieving multiplanar distraction: 1) use of external, pin-retained devices, with multiple joints to facilitate lengthening in multiple directions, 2) use of a univector device to distract the jaw, removal of the device prior to consolidation, and then molding the regenerate into the desired position and use of plates or maxillomandibular fixation to allow for bone formation, and 3) use of a curvilinear distraction device with a specific radius of curvature, pitch, and elongation track (Figure 9). (See Chapter 13, "Clinical Application of Curvilinear Distraction Osteogenesis for Correction of Mandibular Deformities".) A curvilinear device with a specific radius of curvature can be constructed to correct any mandibular deformity. The use of a semi-buried, fixed-trajectory curvilinear distraction device, propelled by a rack and worm gear, in a mini-pig model has been demonstrated.[40] However, because a buried device does not permit mid-course correction, accurate treatment planning is required to ensure achievement of the desired multidirectional movement. (See Chapter 13, "Clinical Application of Curvilinear Distraction Osteogenesis for Correction of Mandibular Deformities".)

DISCUSSION

Distraction osteogenesis has been shown to be an excellent technique to achieve the skeletal expansion required for a variety of mandibular deformities. Distraction osteogenesis can be substituted for acute mandibular lengthening in situations requiring large movements not achievable by standard osteotomies, acute lengthening and bone grafting. Distraction osteogenesis decreases the need for bone grafting for large (>10 mm) mandibular advancements; one can achieve 20 mm or more of advancement without a bone graft and the associated donor site morbidity, scarring, and potential for infection. Soft tissue also has the ability to grow linearly along lines of tension. This is referred to as distraction histogenesis. Skin, muscle, nerves, and vascular tissue are generated, not stretched. The advantage is obvious, especially for severe retrognathia, in which the stretched soft tissue envelope can contribute to relapse when a traditional mandibular osteotomy is performed for a large (>10 mm) advancement. Due to the process of distraction histogenesis, resistance to advancement by the soft tissue envelope is decreased. Schreuder and colleagues[41] noted that bilateral sagittal split osteotomy (BSSO) and distraction osteogenesis are the most common techniques currently applied to surgically correct mandibular retrognathia. These investigators reviewed the literature on BSSO and mandibular distraction osteogenesis with emphasis on the influence of age and post-surgical growth, damage to the inferior alveolar nerve, and post-surgical stability and relapse. Although randomized clinical trials are lacking, some support was found in the literature for distraction osteogenesis having advantages over BSSO in the surgical treatment of low and normal mandibular plane angle patients needing greater advancement (greater than 7 mm). (See Chapter 14, "Distraction Osteogenesis Versus Bilateral Sagittal Split Osteotomy for Mandibular Advancement.") In all other mandibular retrognathia patients, the treatment outcomes of distraction osteogenesis and BSSO seemed to be comparable. There is a need for randomized clinical trials comparing the two techniques in all types of mandibular retrognathia, in order to provide evidence-based guidelines for selecting which retrognathia cases are preferably treated by BSSO or distraction osteogenesis.

Early results indicate that the incidence of permanent paresthesia of the inferior alveolar nerve (IAN) is less with DO when compared to standard operations and the relapse rate is less. Whitesides and associates[42] report that stable mandibular advancements of 10 mm and greater can be successfully accomplished by distraction osteogenesis without producing significant damage to the IAN.

Distraction osteogenesis is not a simple technique and there is a considerable learning curve. It requires considerable planning to ensure accurate vectors of movement and the patients require close supervision during the period of active distraction. With experience, the overall complication rate is low. Complications include: fibrous non-union or premature union of bone, infection that may hinder osteogenesis, noncompliant patient with treatment failure, scarring of the skin with external devices,

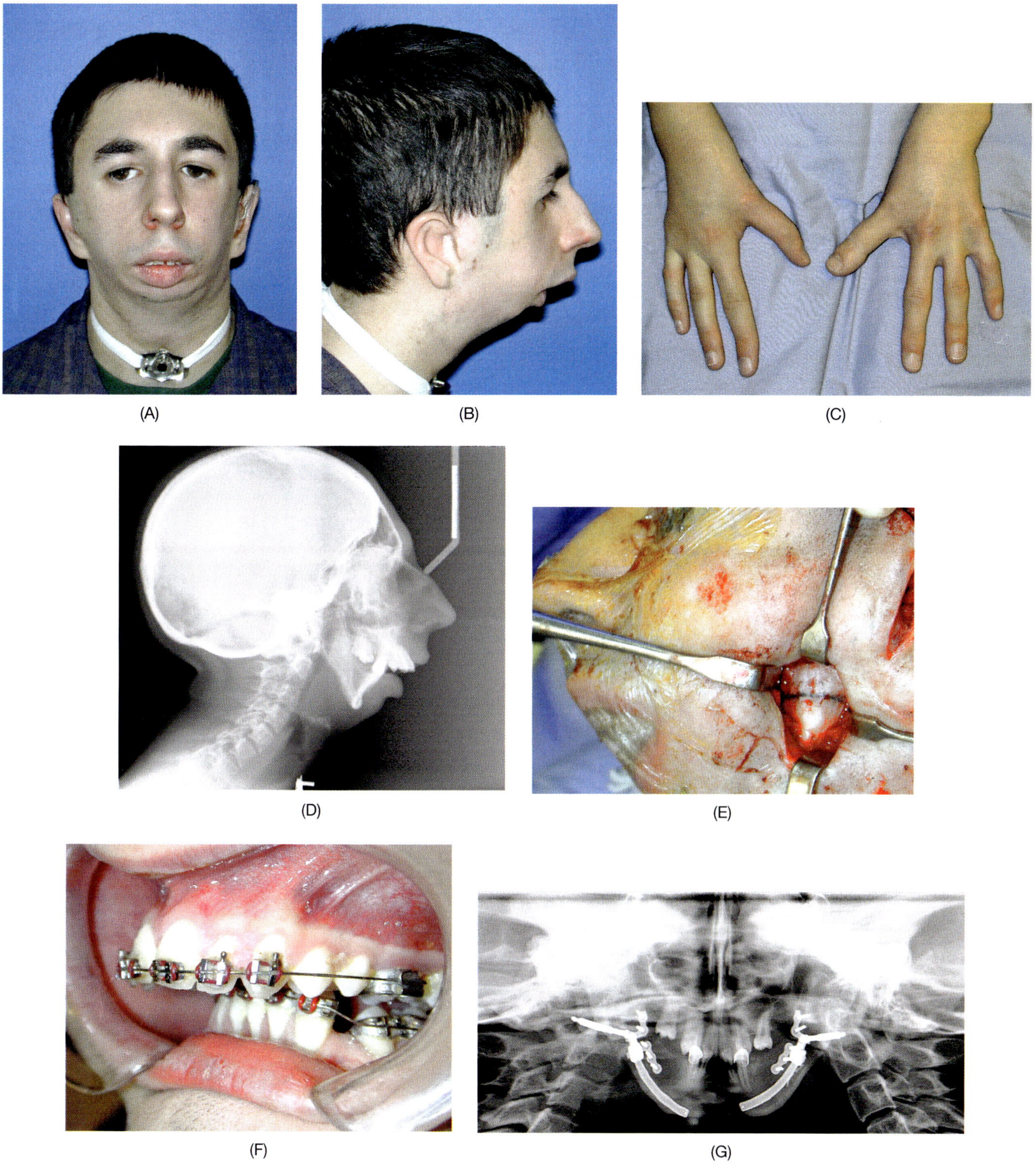

FIGURE 9 • **Curvilinear distractor**. 20-year-old male with Nager syndrome presenting with severe mandibular and temporomandibular joint hypoplasia. He was tracheostomy-dependent due to the severe mandibular retrognathia. (A) Pre-operative frontal and (B) lateral photographs demonstrating facial characteristics of patients with this syndrome: down-slanting palpebral fissures, lack of lower eyelashes, flat malar eminences, convex facial profile secondary to mandibular hypoplasia. (C) Index fingers are missing. (D) Pre-operative cephalometric x-ray. (E) Intraoperative view of the proposed osteotomy. (F) Buried curvilinear distractor in place. (G) Immediate post-operative panoramic and (*continued*)

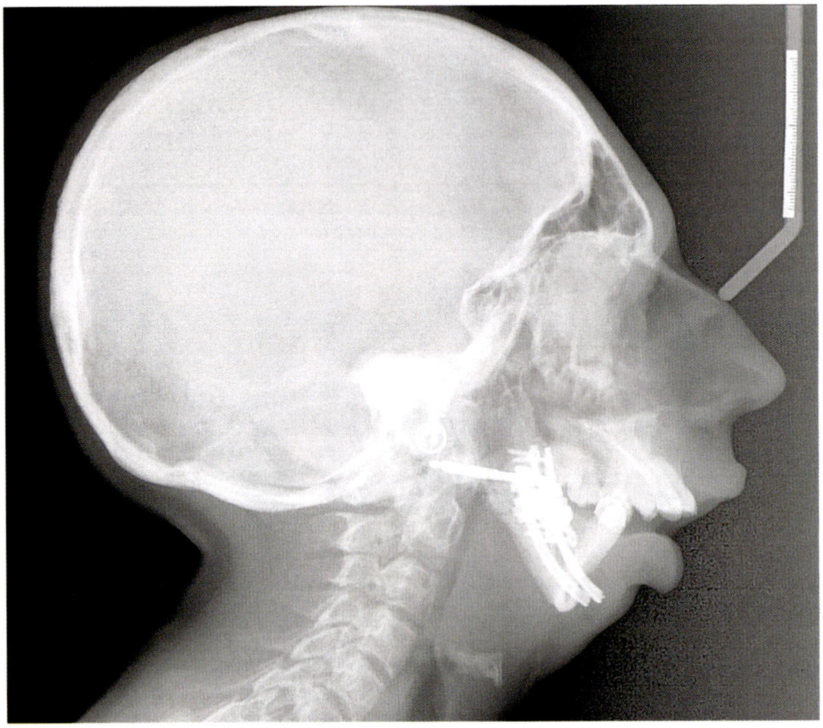

(H)

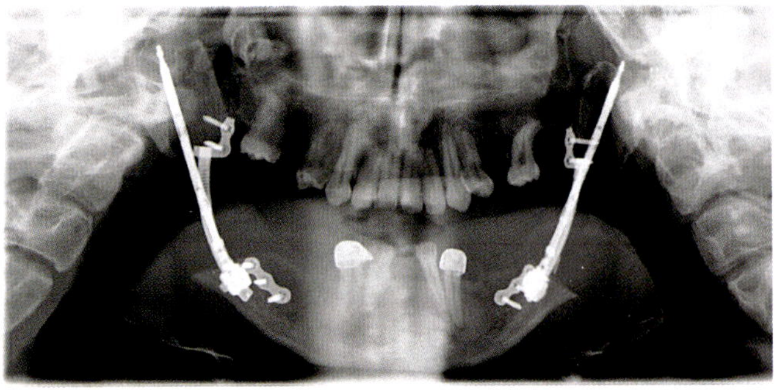

(I)

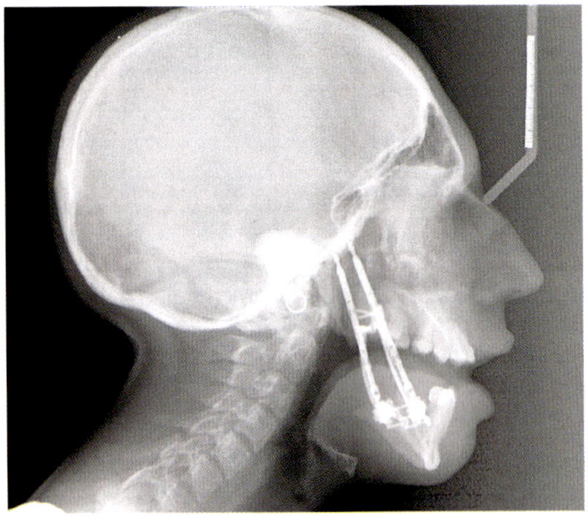

(J)

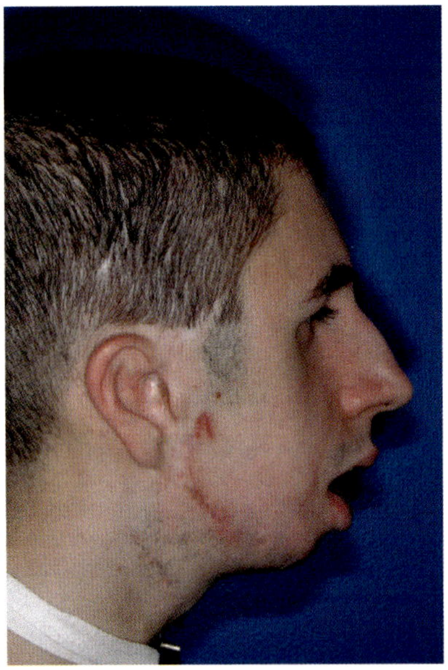

(K)

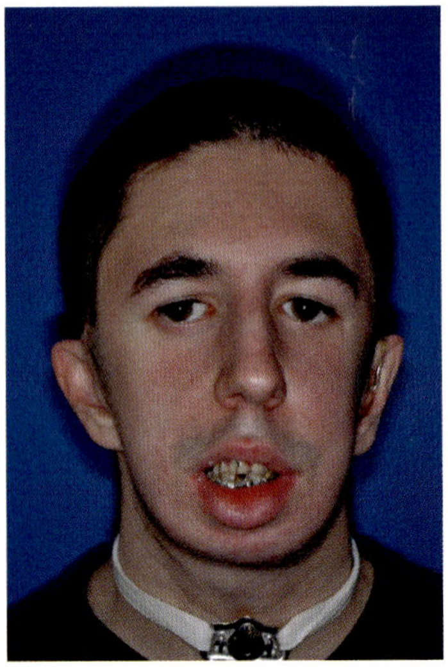

(L)

FIGURE 9 • *(continued from previous page)* (H) cephalometric x-rays. (I) Panoramic and (J) cephalometric radiographs during the consolidation phase. (K) Facial and (L) lateral photographs showing the final result. (Also see Chapter 13.)

hardware failure, and malocclusion or worsening facial deformity secondary to inaccurate vectors of movement, and psychosocial issues related to the recovery (length of treatment time and patient's physical appearance).

With technologic advancements, distraction devices have become smaller and more sophisticated than early versions. Distraction osteogenesis may even be teamed with endoscopic techniques to allow the placement of these devices with minimal surgery. Future improvements in device design (for example, automatic continuous distraction devices), three-dimensional treatment-planning software and navigation technology will improve outcomes. Preliminary studies of rabbits have shown that distraction performed in the presence of recombinant human bone morphogenetic protein placed into the distraction site accelerates bone formation and the potential for the use of growth factors to enhance the rate and quality of bone formation may be of considerable benefit in the future.

The authors would like to acknowledge Fardad Tayebaty, former AO/Synthes student fellow, for his assistance in manuscript preparation.

References

1. Sato M, Maruoka Y, Kunimori K, et al. Morphological and immunohistochemical changes in muscle tissue in association with mandibular distraction osteogenesis. J Oral Maxillofac Surg 2007;65:1517–25.
2. Castaño FJ, Troulis MJ, Glowacki J, Kaban LB, Yates KE. Proliferation of masseter myocytes after distraction osteogenesis of the porcine mandible. J Oral Maxillofac Surg 2001;59:302–7.
3. Snow MD, Turvey TA, Walker D, Proffit WR. Surgical mandibular advancement in adolescents: postsurgical growth related to stability. Int J Adult Orthodon Orthognath Surg 1991;6:143–51.
4. Wolford LM, Schendel SA, Epker BN. Surgical-orthodontic correction of mandibular deficiency in growing children (long term treatment results). J Maxillofac Surg 1979;7:61–72.
5. Freihofer HP Jr. Results of osteotomies of the facial skeleton in adolescence. J Maxillofac Surg 1977;5:267–97.
6. Wolford LM, Karras SC, Mehra P. Consideration for orthognathic surgery during growth. I. Mandibular deformities. Am J Orthod Dentofacial Orthop 2001;119:95–101.
7. Denny AD, Talisman R, Hanson PR, Recinos RF. Mandibular distraction osteogenesis in very young patients to correct airway obstruction. Plast Reconstr Surg 2001;108:302–11.
8. Sidman JD, Sampson D, Templeton B. Distraction osteogenesis of the mandible for airway obstruction in children. Laryngoscope 2001;111:1137–46.
9. Burstein FD, Williams JK. Mandibular distraction osteogenesis in Pierre Robin sequence: application of a new internal single-stage resorbable device. Plast Reconstr Surg 2005;115:61–7.
10. Monasterio FO, Molina F, Berlanga F, et al. Swallowing disorders in Pierre Robin Sequence: its correction by distraction. J Cranio fac Surg 2004;15:934–41.
11. Steinbacher DM, Kaban LB, Troulis MJ. Treatment of tracheostomy-dependent children with severe micrognathia. J Oral Maxillofac Surg 2005;63:1072–79.
12. Moore MH, Guzman-Stein G, Prodman TW, et al. Mandibular lengthening by distraction for airway obstruction in Treacher Collins syndrome. J Craniofac Surg 1994;5:22–5.
13. Cohen SR. Early decannulation with bilateral mandibular distraction for tracheostomy-dependent patients (Discussion). Plast Reconstr Surg 1999;103:58–59.
14. Cohen SR, Simms C, Burstein FD. Mandibular distraction osteogenesis in the treatment of upper airway obstruction in children with craniofacial deformities. Plast Recostr Surg 1998;101:312–8.
15. Wang X, Wang XX, Liang C, et al. Distraction osteogenesis in correction of micrognathia accompanying obstructive sleep apnea syndrome. Plast Reconstr Surg 2003;112:1549–57.
16. Kuriakose MA, Shnayder Y, DeLacure MD. Recostruction of segmental mandibular defects by distraction osteogenesis for mandibular reconstruction. Head Neck 2003;25:816–24.
17. Rubio-Bueno P, Naval L, Rodriguez-Campo F, Gil-Díez JL, Díaz-González FJ: Internal Distraction Osteogenesis with a Unidirectional Device for Reconstruction of Mandibular Segmental Defects, J Oral Maxillofac Surg 63:598–608, 2005.
18. Block MS, Otten J, McLaurin D, et al. Bifocal distraction osteogenesis for mandibular defect healing: Case reports. J Oral Maxillofac Surg 1996;54:1365–70.
19. Costantino PD, Shybut G, Friedman CD, et al. Segmental mandibular regeneration by distraction osteogenesis. Arch Otolaryngol Head Neck Surg 1990;116:535–45.
20. Annino DJ Jr, Goguen LA, Karmody CS. Distraction osteogenesis for reconstruction of mandibular symphyseal defects. Arch Otolaryngol Head Neck Surg 1994;120:911–6.
21. Sawaki Y, Hagino H, Yamamoto H, Ueda M. Trifocal distraction osteogenesis for segmental mandibular defect: a technical innovation. J Cranio-Maxillofac Surg 1997;25:310–5.
22. Shvyrkov MB, Shamsudinov AH, Sumarokov DD, et al. Non-free osteoplasty of the mandible in maxillofacial gunshot wounds: Mandibular reconstruction by compression osteodistraction. Br J Oral Maxillofac Surg 1999;37:261–7.
23. Shuker S. Management of severe facial injuries by local tissue traction. J Oral Maxillofac Surg 1994;52:817–20.
24. Labbe D, Nicolas J, Kaluzinski E, et al. Gunshot wounds: reconstruction of the lower face by osteogenic distraction. Plast Reconstr Surg 2005;116:1596–603.
25. Kunkel M, Wahlmann U, Reichert TE, Wegener J, Wagner W. Reconstruction of mandibular defects following tumor ablation by vertical distraction osteogenesis using intraosseous distraction devices. Clin Oral Implants Res 2005;16:89–97.
26. Kaban LB, Mulliken JB, Murray JE. Cranifacial Abnormalities. In: Welch KJ, et al, editors: *Pediatric Surgery.* 4th ed. Chicago: Year Book Medical Publishers; 1986, p. 423–38.
27. Kaban LB. Congenital and acquired growth abnormalities of the temporomandibular joint. In: Keith DA, editor. *Surgery of the temporomandibular joint.* Boston: Blackwell Scientific Publishers; 1988. p. 55–90.
28. Polley JW, Figueroa AA. Distraction Osteogenesis: its application in severe mandibular deformities in hemifacial microsomia. J Craniofac Surg 1997;8:422–30.
29. Scolozzi P, Herzog G, Jaques B. Simultaneous maxillo-mandibular distraction osteogenesis in hemifacial microsomia: a new technique using two distractors. Plast Reconstr Surg 2006;117: 1530–41.
30. Wiens JL, Forte RA, Wiens JP. The use of distraction osteogenesis to treat hemifacial microsomia: A clinical report. J Prosthet Dent 2003;89:11–4.
31. Tharanon W, Sinn DP. Mandibular distraction osteogenesis with multidirectional extraoral distraction device in hemifacial

microsomia patients: three-dimensional treatment planning, prediction tracings, and case outcomes. J Craniofac Surg 1999;10:202–13.

32. Shetye PR, Grayson BH, Mackool RJ, McCarthy JG. Long-term stability and growth following unilateral mandibular distraction in growing children with craniofacial microsomia. Plast Recostr Surg 2006;118:985–95.
33. Padwa BL, Kearns GJ, Todd R, Troulis M, Mulliken JB, Kaban LB. Simultaneous maxillary and mandibular distraction osteogenesis with a semiburied device. Int J Oral Maxillofac Surg 1999;28:2–8.
34. Hwang SJ, Jung JG, Jung JU, Kyung SH. Vertical alveolar bone distraction at molar region using lag screw principle. J Oral Maxillofac Surg 2004;62:787–94.
35. Block MS, Chang A, Crawford C. Mandibular alveolar ridge augmentation in the dog using distraction osteogenesis. J Oral Maxillofac Surg 1996;54:309–14.
36. Chiapasco M, Consolo U, Bianchi A, Ronchi P. Alveolar distraction osteogenesis for the correction vertically deficient edentulous ridges: a multicenter prospective study on humans. Int J Oral Maxillofac Implants 2004;19:399–407.
37. Perrott DH, Berger R, Vargervik K, Kaban LB. Use of a skeletal distraction device to widen the mandible: a case report. J Oral Maxillofac Surg 1993;51:435–9.
38. King JW, Wallace JC. Unilateral Brodie bite treated with distraction osteogenesis. Am J Orthod Dentofacial Orthop 2004;125: 500–9.
39. Tae KC, Kang KH, Kim SC. Unilateral mandibular widening with distraction osteogenesis. Angle Orthod. 2005;75:1053–60.
40. Seldin EB, Troulis MJ, Kaban LB. Evaluation of a semiburied fixed trajectory curvilinear distraction device in an animal model. J Oral Maxillofac Surg 1999;57:1442–6.
41. Schreuder WH, Jansma J, Bierman MW, Vissink A. Distraction osteogenesis versus bilateral sagittal split osteotomy for advancement of the retrognathic mandible: a review of the literature. Int J Oral Maxillofac Surg 2007;36:103–10.
42. Whitesides LM, Meye RA. Effect of Distraction Osteogenesis on the Severely Hypoplastic Mandible and Inferior Alveolar Nerve Function. J Oral Maxillofac Surg 2004;62:292–7.

Clinical Application of Curvilinear Distraction Osteogenesis for Correction of Mandibular Deformities

13

Leonard B. Kaban, DMD, MD / Edward B. Seldin, DMD, MD / Bonnie Padwa, DMD, MD / Maria J. Troulis, DDS, MSc

OUTLINE

INTRODUCTION

This chapter is adapted from the paper with the same title published in the *Journal of Oral & Maxillofacial Surgery* in May, 2009.[1]

Early results with distraction osteogenesis to correct maxillofacial deformities have been compromised by inadequate planning, inaccurate skeletal movements, and poor occlusal and esthetic outcomes.[2] In addition, translational and rotational bony movements have been considered separately, resulting in cumbersome and complex external devices with multiple joints.[3,4] Activation of these devices requires sophisticated patient participation to achieve the desired movements.

Trauner[5] was the first to apply Euclidean principles to the planning of orthognathic surgical corrections. For the purposes of planning movements for distraction osteogenesis, Seldin and associates[6] observed that combinations of translational and rotational movements could be resolved into a simple rotation about a unique axis determined by algebraic or geometric means. The geometric axes for a proposed surgical correction define the simplest path of the skeletal movement. Once the axes are known, the radius of curvature to define the shape and orientation of a proposed distraction device, in relation to an osteotomy, can be calculated. Thus, a simple curvilinear device with a specific radius of curvature can be constructed to correct any mandibular deformity. Understanding this principle allows the development of novel devices that are not limited to linear movements and that do not necessarily depend on screw-type propulsion mechanisms. We have demonstrated the use of a semi-buried, fixed-trajectory curvilinear distraction device, propelled by a rack and worm gear, in a mini-pig model[7] (Figures 1, 2, 3). Curvilinear devices to produce similar movements have subsequently been described by others.[7] Because a buried device does not permit mid-course correction, accurate treatment planning is critical to ensure achievement of the desired final skeletal position. To help with proper positioning of the osteotomy and device placement and orientation, we have developed a computed tomography (CT)-based three-dimensional (3D) treatment planning system.[8–12]

In this chapter, we describe the clinical application of semi-buried, fixed-trajectory, curvilinear distraction devices used in coordination with a CT-based 3D treatment planning system to correct mandibular deformities (Figure 4).[1] The details of the treatment planning system are described in Chapter 11.

FIGURE 1 • The classic Euclidian construction for finding the center of a circle. Perpendicular bisectors of cords of a circle intersect at the center. (From Seldin et al., *JOMS* 1999;57:1443; used with permission from Elsevier.)

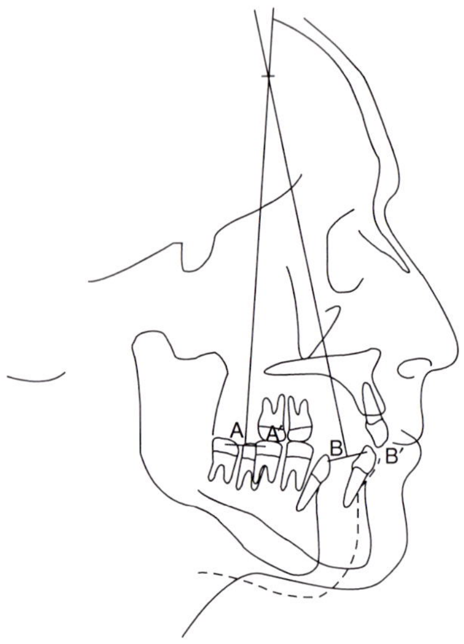

FIGURE 2 • Superimposed pre-operative (solid line) and post-operative (dotted line) cephalometric tracings with the axis of rotation for the correction in the sagittal plane determined by Euclid's method. (From Seldin et al., *JOMS* 1999;57:1443; used with permission from Elsevier.)

THREE-DIMENSIONAL TREATMENT PLANNING AND CONSTRUCTION OF A NAVIGATION DEVICE

A 3D virtual model of the skull and mandible is reconstructed from standard 1 mm axial and coronal CT cuts. The 3D data are imported into the treatment planning system (Osteoplan™) developed in collaboration with the Harvard Surgical Planning Laboratory (SPL).[8,9,10,11]

Reproducible landmarks and regions of the osteotomies on each side of the mandible are identified on the 3D scans using Osteoplan™.[8,9,11,12,13] With the cutting tool, osteotomies are performed and the mandible is moved to the proposed post-operative position. The system warns the operator of any potential bony collisions (interferences) along the path of movement. These bony interferences can then be modified. For example, in patients with skeletal type II hemifacial microsomia, the elongated coronoid process may collide with the skull base before the shorter condyle reaches its destination, thereby preventing the desired mandibular ramus elongation. A coronoidectomy is performed to eliminate this problem (Figures 5, 6, 7). The axis of the proposed movement on each side, the radius of curvature of the distraction devices, and their lengths and locations are all calculated by the software program and illustrated on the images (Figures 8, 9, 10).

Correct placement and orientation of the osteotomies and the distraction devices is critical. A 3D acrylic model is created from the CT data and an acrylic stent is made to fit onto the ramus of the mandible. Pre-operatively, the osteotomy is drawn on the stent and then holes are drilled through the stent into the mandibular model. This localizes the osteotomy and the distraction device. The procedure can be carried out on the model to ensure the correct movement. Intra-operatively, the stent, with the drilled holes, is placed on the mandibular ramus, duplicating the position on the model. It serves as an intra-operative template for placing the osteotomy and drilling the holes that accurately localize the distraction device (Figures 11, 12).

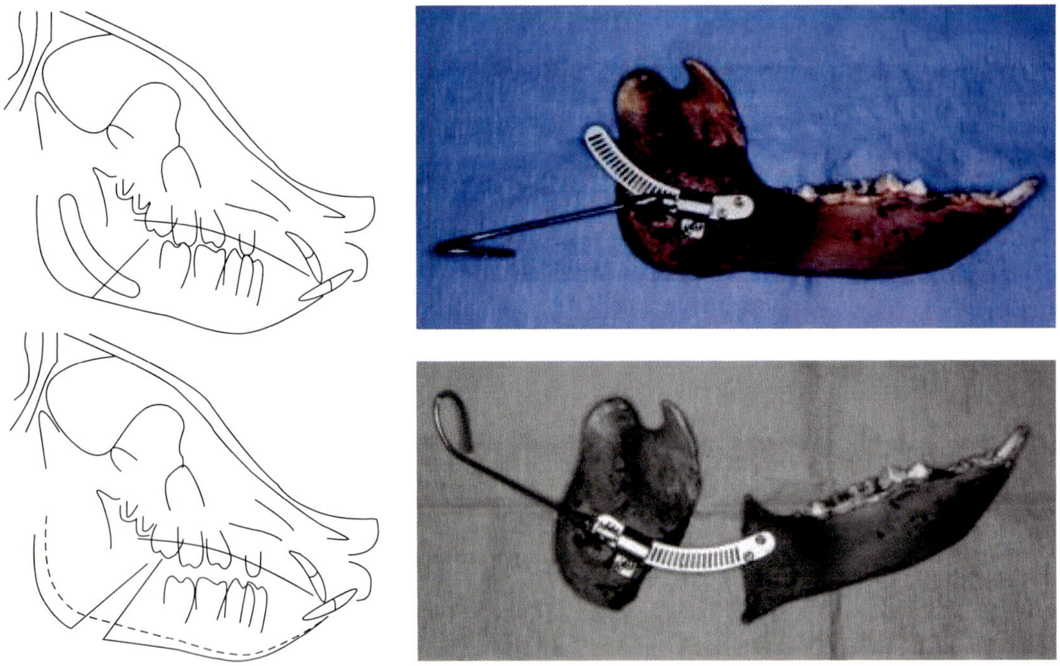

FIGURE 3 • (left top) Pre-operative and (left bottom) end distraction lateral cephalometric tracings of pig who underwent curvilinear distraction. (right top) Prototype curvilinear device on pig mandible. (right bottom) Prototype curvilinear device fully activated to demonstrate trapezoidal distraction gap. (From Seldin et al.; JOMS 1999; 57:1446; used with permission from Elsevier.)

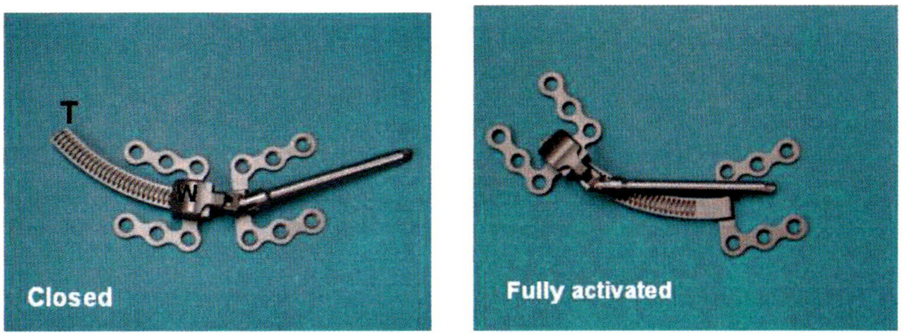

FIGURE 4 • Current curvilinear distraction device for left side: Radius of curvature 5 cm, elongation parameter 25 mm. (W, worm gear housing; T, curved track indicating path along which distractor and distal mandible move.) (From Kaban, et al.; Curvilinear Distraction Osteogenesis. JOMS 2009; 67:997; used with permission from Elsevier.)

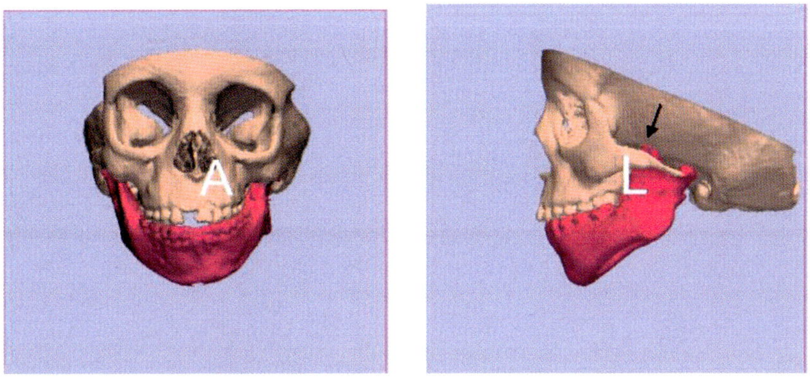

FIGURE 5 • (Left) Frontal and (right) left lateral view of patient with type I hemifacial microsomia prior to left ramus lengthening and creation of an open bite. (Arrow) Long coronoid process relative to condyle. (From Kaban, et al.; Curvilinear Distraction Osteogenesis. JOMS 2009; 67:997; used with permission from Elsevier.)

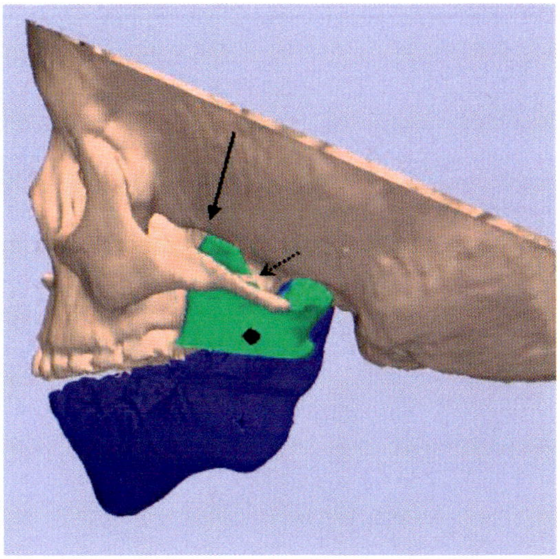

FIGURE 6 • Left coronoid collides with skull base (arrow) prior to completion of desired ramus lengthening. (Dotted arrow) Coronoid osteotomy and then removal of coronoid allows room for movement of the condyle to the skull base. (From Kaban et al.; Curvilinear Distraction Osteogenesis. JOMS 2009; 67:998; used with permission from Elsevier.)

SURGICAL PROCEDURE

Under nasoendotracheal anesthesia, the ramus of the mandible is exposed via a submandibular incision. The stent is fit over the mandible, the bone cut drawn with a sterile pencil and the holes for the device drilled. Using a reciprocating saw, the bone cut is made and the distraction device is fixed to the mandible with 2 mm diameter screws. The osteotomy is completed with an osteotome and the device is turned until a wedge-shaped gap is demonstrated. Then the distractor is turned back to the neutral position (Figures 8–12). The wounds are irrigated and closed in layers.

Distraction is begun after a latency period of 48–96 hours and carried out at a rate of 1 full turn per day (approximately 1 mm per day). The neutral fixation period is at least two times the number of millimeters of distraction in days or until the defect is filled in with bone by CT scan (Figures 13, 14).

RESULTS

In a preliminary study, this promising technique was used in 11 patients who had mandibular osteotomies and placement of custom-made bilateral curvilinear distraction devices to correct severe mandibular retrognathism with or without open bite Figure 4–7). There were 6 females and 5 males with a mean age of 11.5 years (range = 15 months−39 years) at the time of distraction.

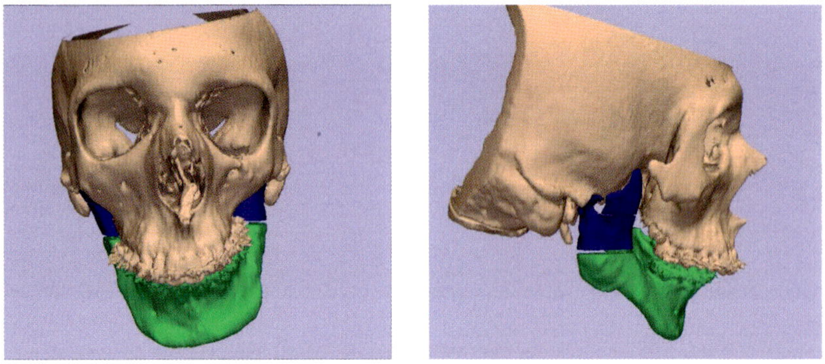

FIGURE 7 • 3D treatment planning.

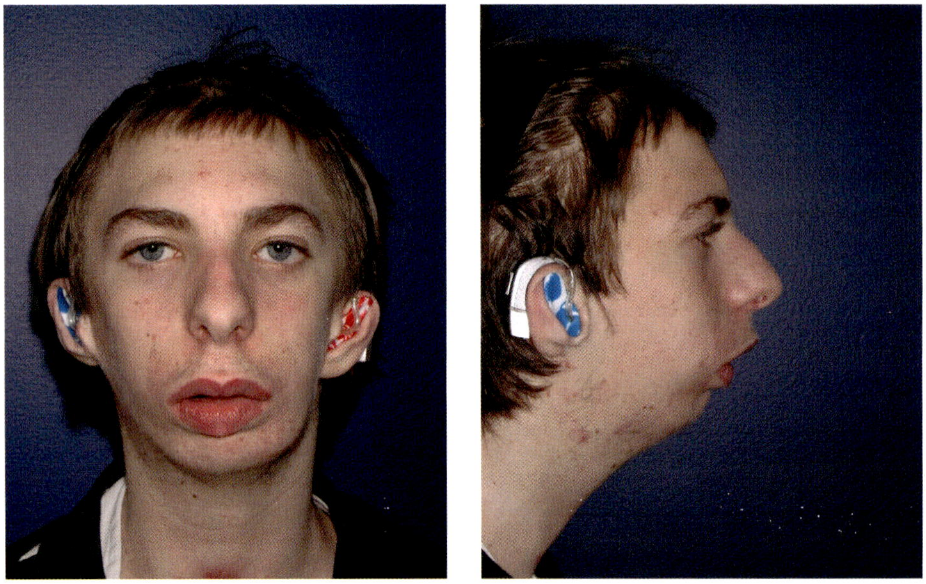

FIGURE 8 • (left) Frontal and (right) lateral views of patient with severe micrognathia and obstructive sleep apnea. The goals were to lengthen posterior face height; advance and rotate the mandible in a counterclockwise direction. At the time of device removal, an LFI osteotomy would be performed.

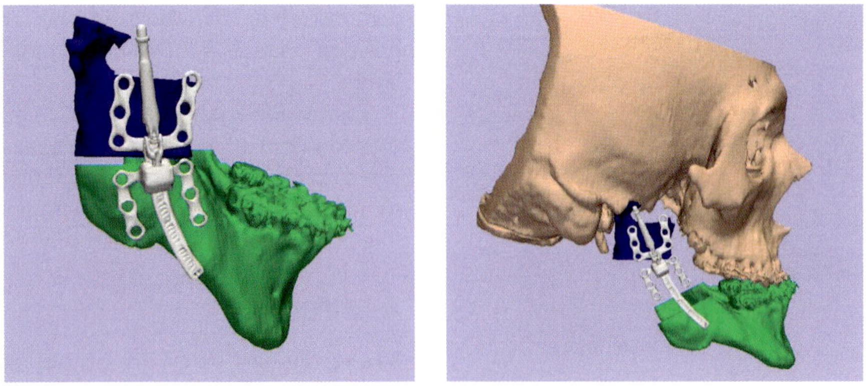

FIGURE 9 • Using Osteoplan™, the osteotomy and desired movements were accomplished on the virtual model. The distractor parameters were calculated and the device manufactured. Any combination of footplates on the proximal and distal limbs of the device are fixed to the bone. The others are cut off.

- Left side
  - —Actual measures: 75 mm radius of curvature (ROC), elongation 37 mm
  - —Distractor chosen: 70 mm ROC, 40 mm length (choose this one for best symmetry)
- Right side
  - —Actual measures: 55 mm ROC (exactly), 28 mm elongation
  - —Distractor chosen: 70 mm ROC, 30 mm elongation

FIGURE 10 • Distractor prescription for the desired movement, calculated by the software program.

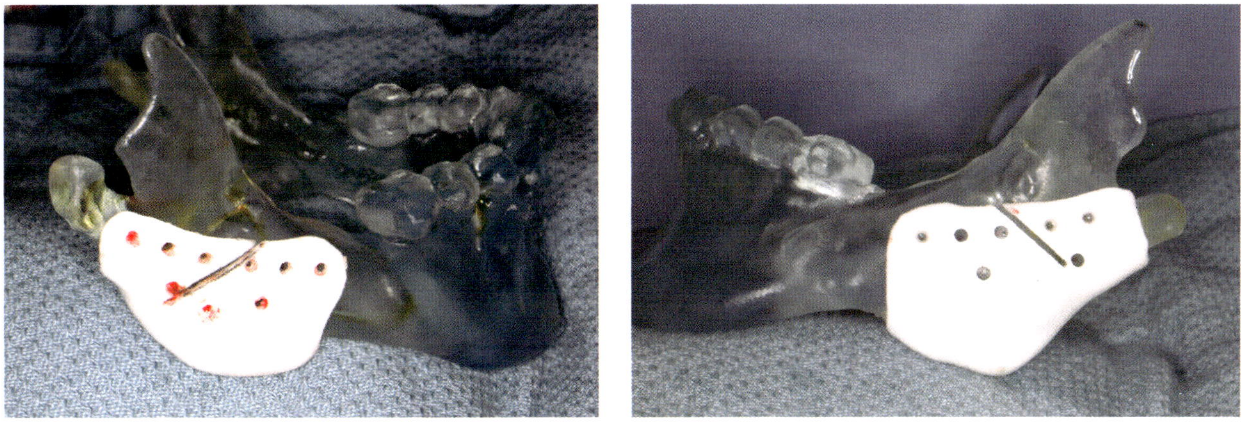

FIGURE 11 • A mold of the ramus on the 3D model. The osteotomies are drawn and the distractor oriented on the model. Holes are drilled to localize the footplates, thereby orienting the osteotomy and the device properly.

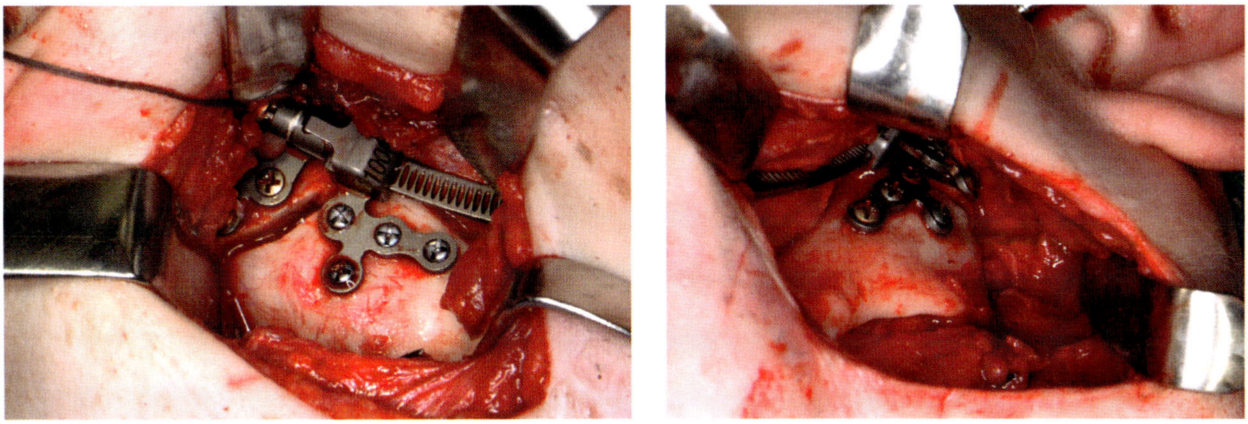

FIGURE 12 • Right and left distractors in place across the osteotomies.

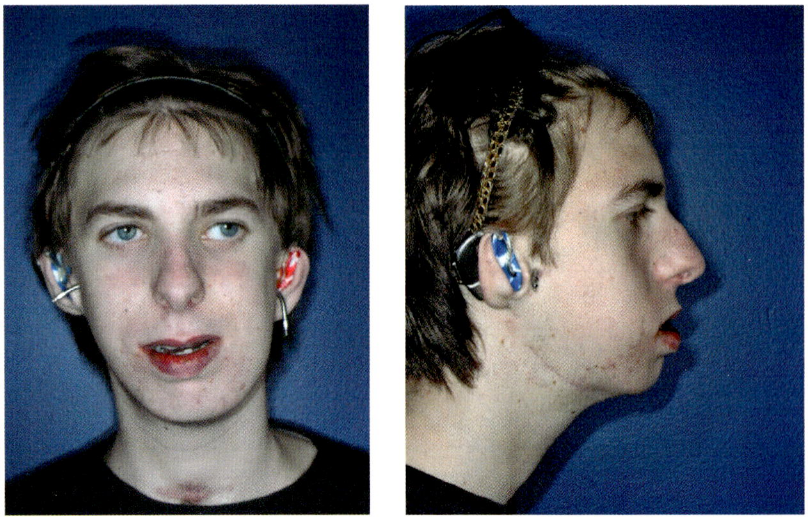

FIGURE 13 • (left) Frontal and (right) lateral views at end of distraction.

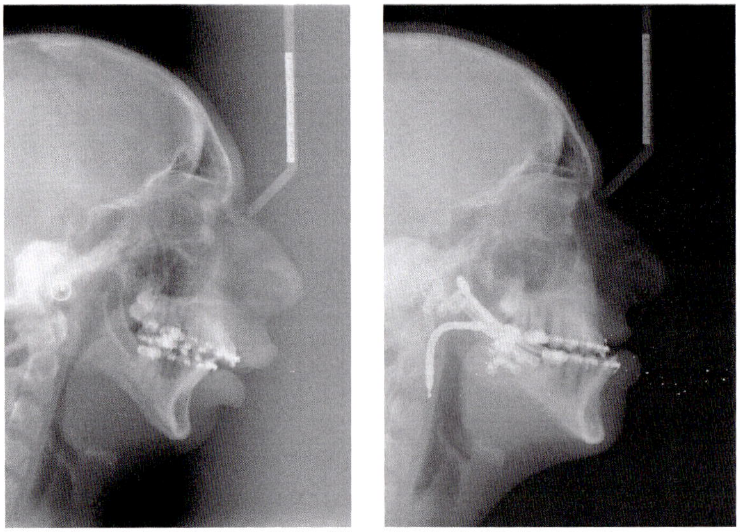

FIGURE 14 • (left) Pre- and (right) end-distraction lateral cephalograms. Posterior airway space increased from 5.03 mm to 18.96 mm and distance from hyoid to mandible decreased from 19.3 mm to 12.9 mm. (Normal values 11 mm and <15 mm, respectively.)

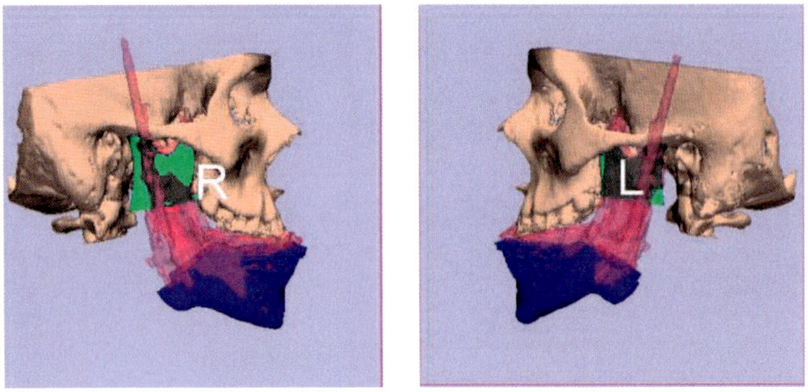

FIGURE 15 • Overlay of predicted (blue) and end-fixation (pink) position of mandible demonstrating that the desired movement was achieved.

Of the 11 patients, 9 were 16 years or younger and 6 were less than 10 years of age. The diagnoses included Treacher Collins (n = 2) and Nager (n = 3) syndromes, hemifacial microsomia or bilateral craniofacial microsomia (n = 2), retrognathia and open bite secondary to mandibular fractures and ankylosis (n = 1), and micrognathia (n = 3, 2 syndromic, 1 non-syndromic). All patients underwent bilateral distraction.

Distractor placement was documented on anterior-posterior (AP), lateral cephalometric, and panoramic radiographs. The patients were evaluated by clinical examination every other day and by serial radiographs twice per week or more frequently if necessary. At the end of the fixation period, the final position of the jaw was documented on AP, lateral cephalometric, and panoramic radiographs. A repeat 3D CT scan was obtained and the treatment planning system was used to superimpose the pre-operative, predicted, and actual outcomes (Figure 15). Distractor placement, the arc of movement of the mandible, and the final mandibular position occurred along the correct axis and provided the desired clinical result in 9 out of 11 patients (Figures 16–20). In the other 2, distractor placement resulted in less than ideal jaw position. The devices were removed approximately halfway through the fixation period, prior to complete bone formation. The desired jaw position was achieved by "molding" the regenerate and immobilizing the jaw until healing occurred. The use of an acrylic template, made on a 3D model, and fitted to the mandible with pre-drilled holes to determine distractor placement has eliminated this problem.

DISCUSSION

In this chapter, we illustrated the use of a fixed trajectory curvilinear distraction device propelled by a rack and gear mechanism, to accomplish 3D mandibular movements. The device is simple, easy to use, and small enough to be implanted with only the activation rod being visible externally. Standard distraction devices are propelled by screw-type mechanisms that move in only one plane of space. Hence, multi-dimensional movements are accomplished by constructing complex devices with multiple joints to achieve movement in each plane separately. The distractors are large and cumbersome and, of necessity, must be placed extra-orally. Multivector, extra-oral distraction devices may be difficult for the patient and surgeon to manage and errors often occur during active distraction.

The surgical plan for the patients reported here was to advance and rotate the mandible in a counter-clockwise direction, to vertically lengthen the ramus and to maximally advance the chin (pogonion). This movement also has a profound effect on posterior airway space. The patient illustrated in Figures 8–15 had obstructive sleep apnea. His pre-operative posterior airway space was 5.03 mm; post-operatively it increased to 18.96 mm. He had no further oxygen desaturations or apneic episodes. The child illustrated in Figures 16–20 had significant improvement in posterior airway space as demonstrated in the lateral cephalogram.

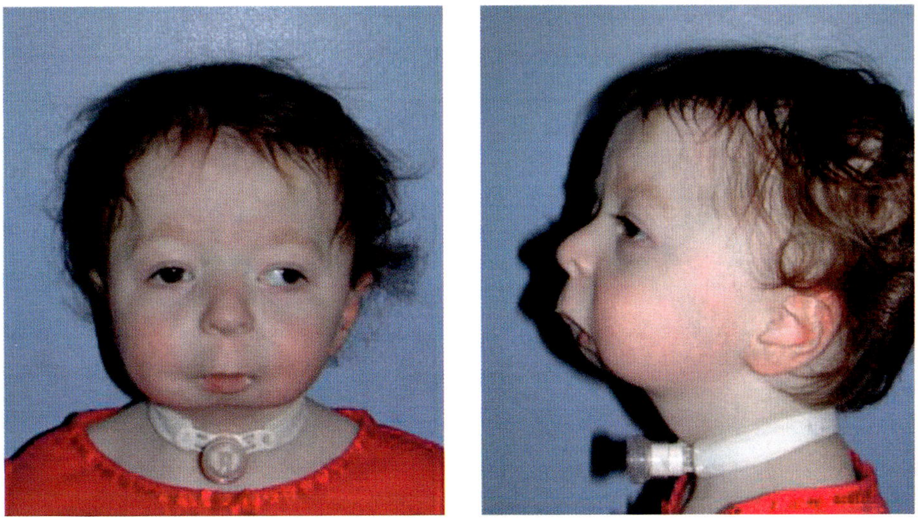

FIGURE 16 • (left) Frontal and (right) lateral views of a 2-year-old child with severe micrognathia and obstructive sleep apnea. She is tracheostomy-dependent. (From Kaban et al.; Curvilinear Distraction Osteogenesis. JOMS 2009; 67:1006; used with permission from Elsevier.)

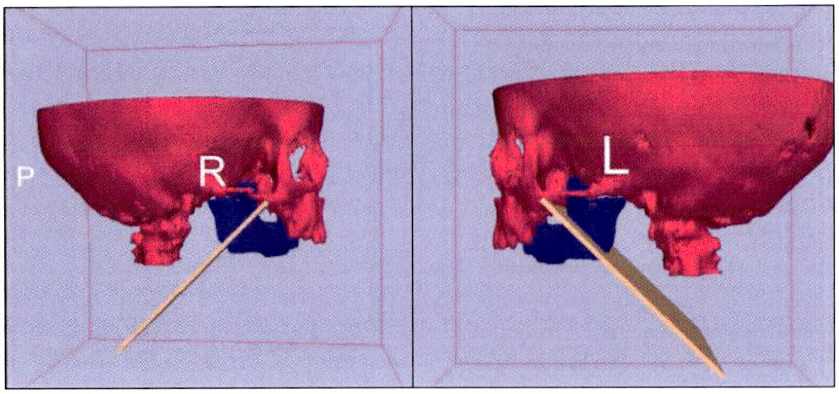

FIGURE 17 • Three-dimensional model with cutting tool demonstrated. The osteotomy is made and the mandible segmented.

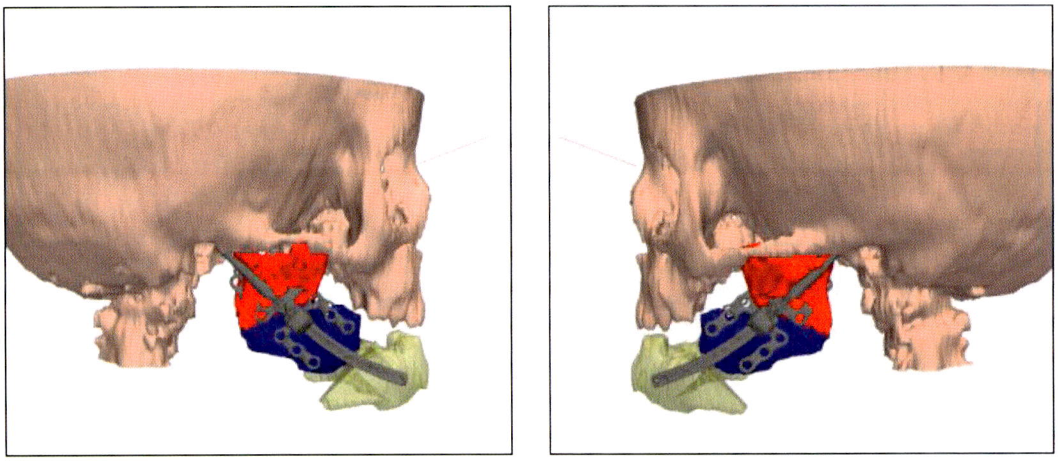

FIGURE 18 • Movement and device prescription calculated by software:
Movement: 7.0 cm radius of curvature; 2.3 cm elongation
Distractors: 7.0 cm radius of curvature; 3.0 cm elongation
(From Kaban et al.; Curvilinear Distraction Osteogenesis. 2009; 67:1006; used with permission from Elsevier.)

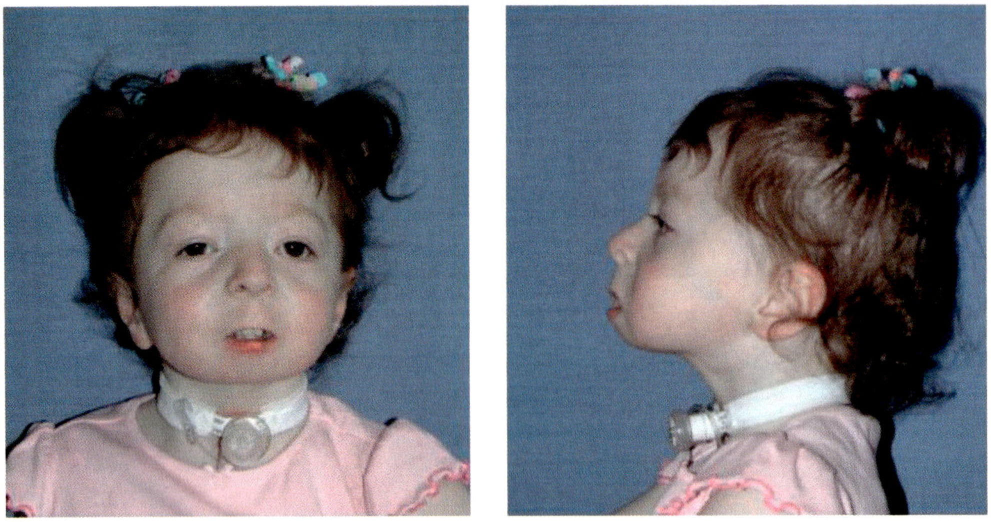

FIGURE 19 • After distractors have been removed. Patient will have a repeat sleep study and eventually be decannulated. (From Kaban et al.; Curvilinear Distraction Osteogenesis. JOMS 2009; 67:1007; used with permission from Elsevier.)

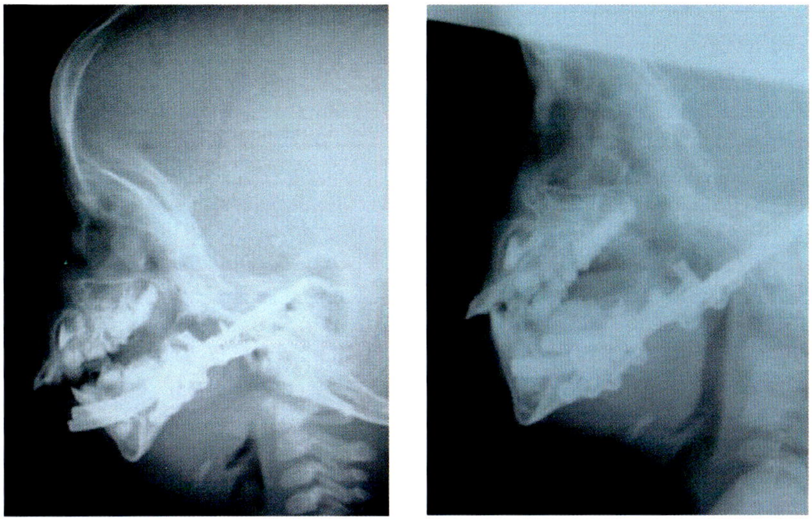

FIGURE 20 • (left) Mid-distraction and (right) end distraction lateral cephalograms. Note the marked increase in posterior airway space. (From Kaban et al.; Curvilinear Distraction Osteogenesis. JOMS 2009; 67:1007; used with permission from Elsevier.)

The treatment planning system developed along with this distraction device (Osteoplan™)[8–11] accurately described the arcs of rotation and radii of curvature of the distractors, the location of the distractors, and the orientation of the osteotomies to achieve the proposed mandibular movements. The post-operative radiographs and CT scans demonstrated the accuracy of this method (Figure 15).

Although the distractors used in this series of patients did not carry out movements in the horizontal plane (because there was no need for horizontal movement), this is possible using the curvilinear distractor design described in this chapter. Figure 21 illustrates such a curvilinear distractor that also moves in the horizontal plane.

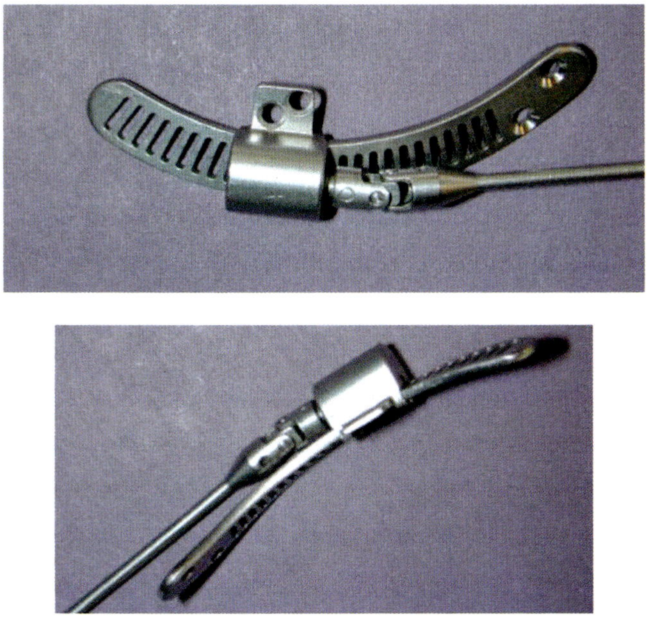

FIGURE 21 • (Top and bottom) Prototype curvilinear device capable of horizontal movement. (From Kaban et al.; Curvilinear Distraction Osteogenesis. JOMS 2009; 67:1008; used with permission from Elsevier.)

The successful use of a semi-buried, fixed-trajectory curvilinear distraction device requires accurate treatment planning that is accomplished by the CT-based treatment planning software package (Osteoplan™) described in this chapter. Using a computer model, we have demonstrated that there is a finite range of devices required to correct all mandibular deformities: radii of curvature 3, 4, 5, 7 mm and straight line.[12] Therefore, a custom device for each patient is not required.

In the future, it will be possible to drive these miniature distraction devices by micromotors. They may also be totally buried, powered by batteries and remotely activated. Our group is currently developing such a device, funded by a National Institutes of Health SBIR, phase II grant.[13] Long-term outcome studies to document the efficacy of distraction osteogenesis as a technique and to document the efficacy of new distraction devices should be carried out.

References

1. Kaban LB, Seldin EB, Padwa B, Yeshwant K, Troulis MJ. Clinical application of curvilinear distraction osteogenesis for correction of mandibular deformities. J Oral Maxillofac Surg 2009; 67: 996–1008.
2. Troulis MJ, Padwa B, Kaban LB. Distraction osteogenesis: Past, present and future. Facial Plast Surg 1998;14:205–15.
3. Gateno J, Teichgraber JF, Aguilar E. Distraction osteogenesis: a new surgical technique for use with the multiplanar mandibular distractor. Plast Reconstr Surg 2000;105:883–8.
4. Gateno J. Accuracy of custom stereolithographic templates for the installation of an external multiplanar distraction device. J Oral Maxillofac Surg 1999;57:96.
5. Trauner R. Kiefer-und Gesichtschirurgie. Urban & Schwarzenberg. Munchen-Berlin-Wien, 1973.
6. Seldin EB, Troulis MJ, Kaban LB. Evaluation of a semiburied fixed-trajectory curvilinear distraction device in an animal model. J Oral Maxillofac Surg 1999;57:1442–6.

7. Schendel SA, Linck DW. Mandibular distraction osteogenesis by sagittal split osteotomy and intraoral curvilinear distraction. J Craniofac Surg 2004;15:631–5.
8. Everett PC, Seldin EB, Troulis M, Kaban LB, Kikinis R. A 3-D system for planning and simulating minimally-invasive distraction osteogenesis of the facial skeleton. In: Delp Sl, DiGioia AM, Jaramaz B, editors. MICCAI 2000. Proceedings of the 3rd International Conference on Medical Image Computing and Computer-Assisted Intervention. 2000 Oct; Pittsburgh, PA. New York: Springer-Verlag; 2000. pp. 1029–39.
9. Yeshwant K, Seldin EB, Kikinis R, Kaban LB. A computer-assisted approach to planning multidimensional distraction osteogenesis. Atlas Oral Maxillofac Surg Clin N Am 2005:13;1–12.
10. Troulis MJ, Everett P, Seldin EB, et al. Three-dimensional treatment planning system based on computed tomographic data. Int J Oral Maxillofac Surg 2002;31:349–57.
11. Yeshwant K, Seldin EB, Gateno J, et al. Analysis of skeletal movements in mandibular distraction osteogenesis. J Oral Maxillofac Surg 2005;63:335–40.
12. Ritter L, Yeshwant K, Seldin EB, et al. Range of curvilinear distraction devices required for treatment of mandibular deformities. J Oral Maxillofac Surg 2006;64:259–64.
13. Kaban LB, Papadaki M, Goldwaser B, et al. Development of a buried, automated, continuous, curvilinear distraction device. NIH Grant 2 R 44 DE 014803-02, and NIH Grant 4 R 44 DE 019047.

Distraction Osteogenesis versus Bilateral Sagittal Split Osteotomy for Mandibular Advancement

14

Carl Bouchard, DMD / Zachary Abramson, DMD / Gabriel Bendahan, DMD / Leonard B. Kaban, DMD, MD / Maria J. Troulis DDS, MSc

OUTLINE

INTRODUCTION

More than 50% of North American patients undergoing orthodontic treatment have a class II malocclusion.[1] When the malocclusion is a result of mandibular hypoplasia (retrognathia) and the severity is beyond the scope of orthodontic tooth movement alone, the treatment of choice is to advance (lengthen) the mandible.

The bilateral sagittal split osteotomy (BSSO) was described by Obwegeser and Trauner in the English literature in 1957.[2] Although numerous modifications of the original operation have been described, BSSO is one of the most commonly performed and versatile orthognathic surgical procedures in use today (Figure 1). The osteotomy results in broad bony overlap of the segments, providing a large surface area for bone healing. Rigid fixation can be used, which improves patient comfort and reduces post-operative morbidity. In addition, the BSSO can be used to reposition the tooth-bearing segment in multiple directions.[3,4] Despite its popularity and extensive use over the past 50 years, there are some disadvantages of the sagittal split osteotomy. The inferior alveolar nerve is frequently injured and the incidence of paresthesia reported in the literature varies from 0% to 70%.[5] This is particularly true in patients over 40 years of age.[6–10] There is also the risk of an unfavourable split, tooth damage, and skeletal relapse due to the resistance of the soft tissue envelope to skeletal advancement.

Distraction osteogenesis (DO) has become a frequently used technique to expand the mandible in patients with craniofacial anomalies such as hemifacial microsomia and Treacher Collins syndrome.[11–14] In many of these cases, the magnitude of advancement is large and it might not be feasible by standard osteotomy and acute bone movement. In addition, relapse potential is unacceptably high. With experience gained in these major deformities, DO has become a minimally invasive alternative to BSSO for correction of non-syndromic mandibular retrognathia (Figure 2). Because distraction devices have become smaller, less cumbersome, better tolerated by the patient and intraoral placement has become feasible, DO is now included in the armamentarium of surgeons performing orthognathic surgery.[15,16]

In this chapter, we compare the outcomes of BSSO and DO for the treatment of non-syndromic mandibular retrognathism, emphasizing pre- and post-operative care, morbidity, duration of orthodontic treatment, inferior alveolar nerve injuries, and relapse.

PRE- AND POSTOPERATIVE CARE AND ORTHODONTIC TREATMENT

The goals of pre-operative orthodontic treatment are the same regardless of the surgical procedure selected. Teeth have to be aligned, arches coordinated, spaces closed, and midlines properly positioned. Breuning and Van Strijen compared the duration of combined orthodontic and surgical (BSSO versus DO) treatment for mandibular retrognathia.[17] They found the mean overall treatment time to be shorter for patients treated by DO

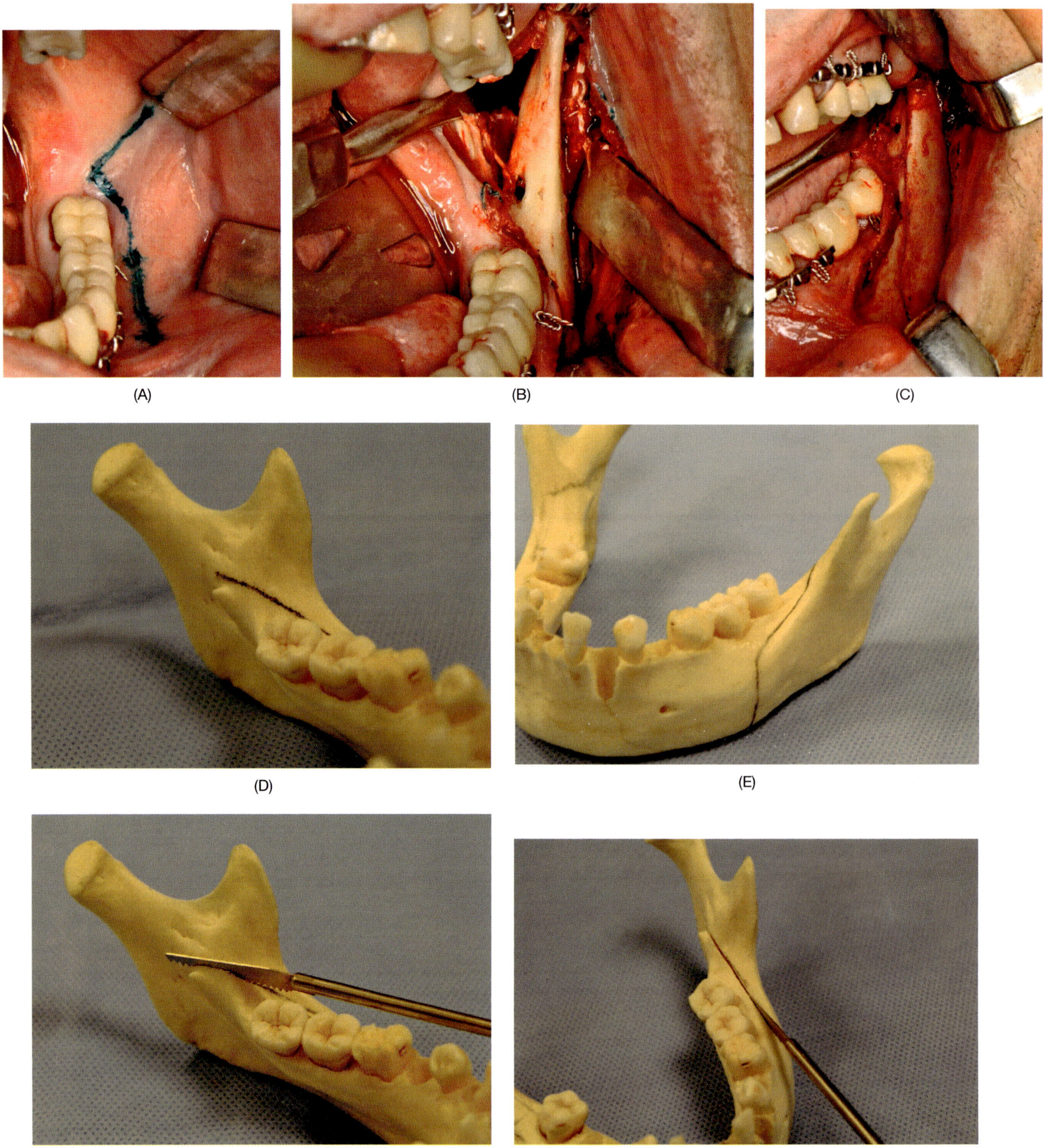

(A) (B) (C) (D) (E) (F) (G)

FIGURE 1 • Bilateral sagittal split osteotomy technique: (A) A dart-shaped incision is made along the left external oblique ridge to provide better exposure and facilitate closure. (B) A subperiosteal dissection of the lateral and the medial aspect of the ramus is performed, (C) intraoperative view of completed cuts are shown. On the model, the execution of each step is outlined: (D) Medial and, (E) Lateral, The osteotomy is marked. The horizontal cut is just above the lingula and 2 mm beyond it. The corticotomy is carried in the sagittal plane to meet the vertical cut in the area of the first molar. (F), (G), (H) Corticotomies are performed with a reciprocating saw. *(continued)*

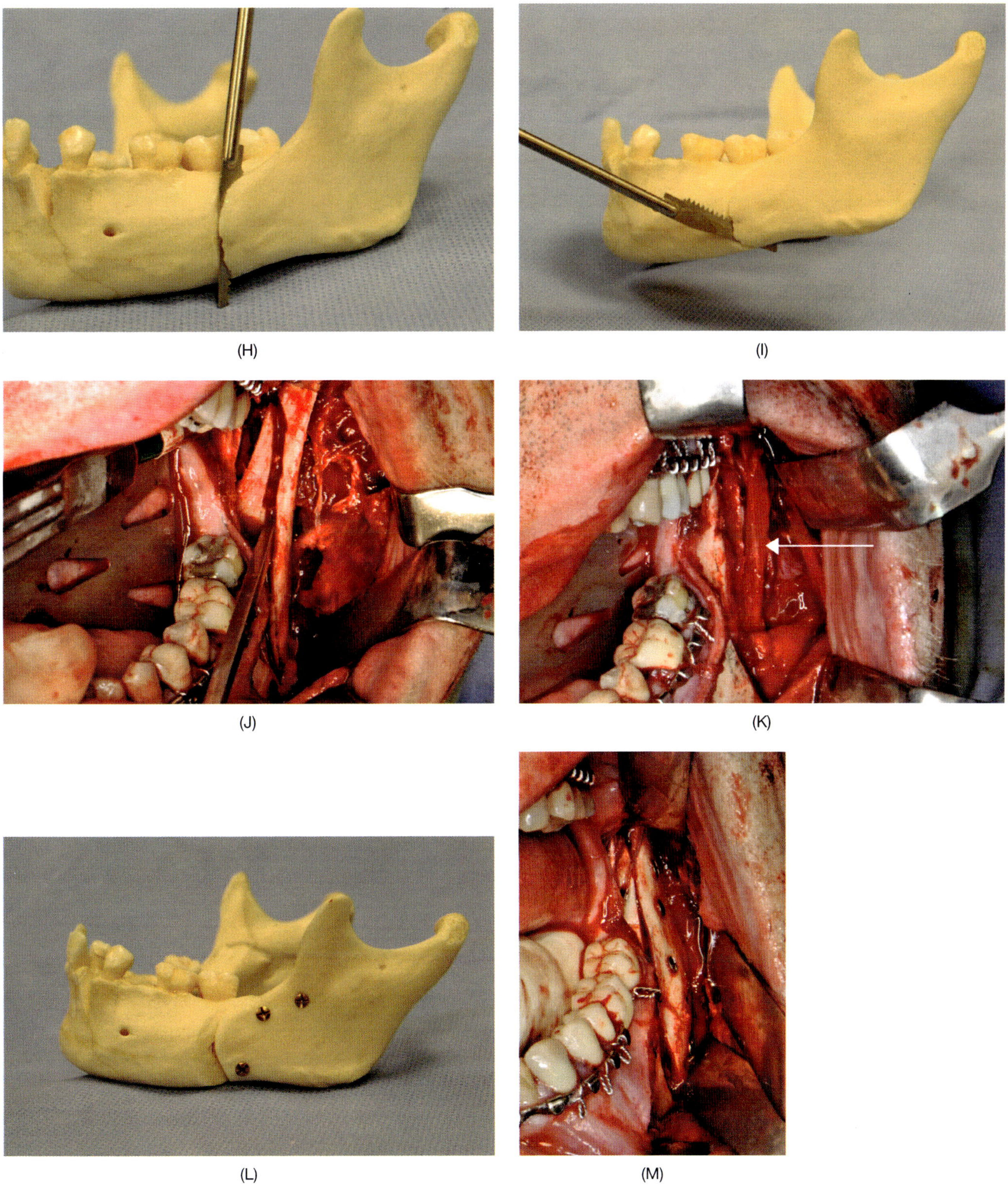

FIGURE 1 • (*continued from previous page*) (I) The inferior cortex is osteotomized to minimize the risk of an unfavorable split. (J) Intraoprative view: The distal and proximal segments are being separated with osteotomes. (K) Once the split completed, the inferior alveolar nerve is visualized (arrow). The 2 segments are rigidly fixated with plates or with bicortical screws as shown here (L) on the model and (M) intraoperatively.

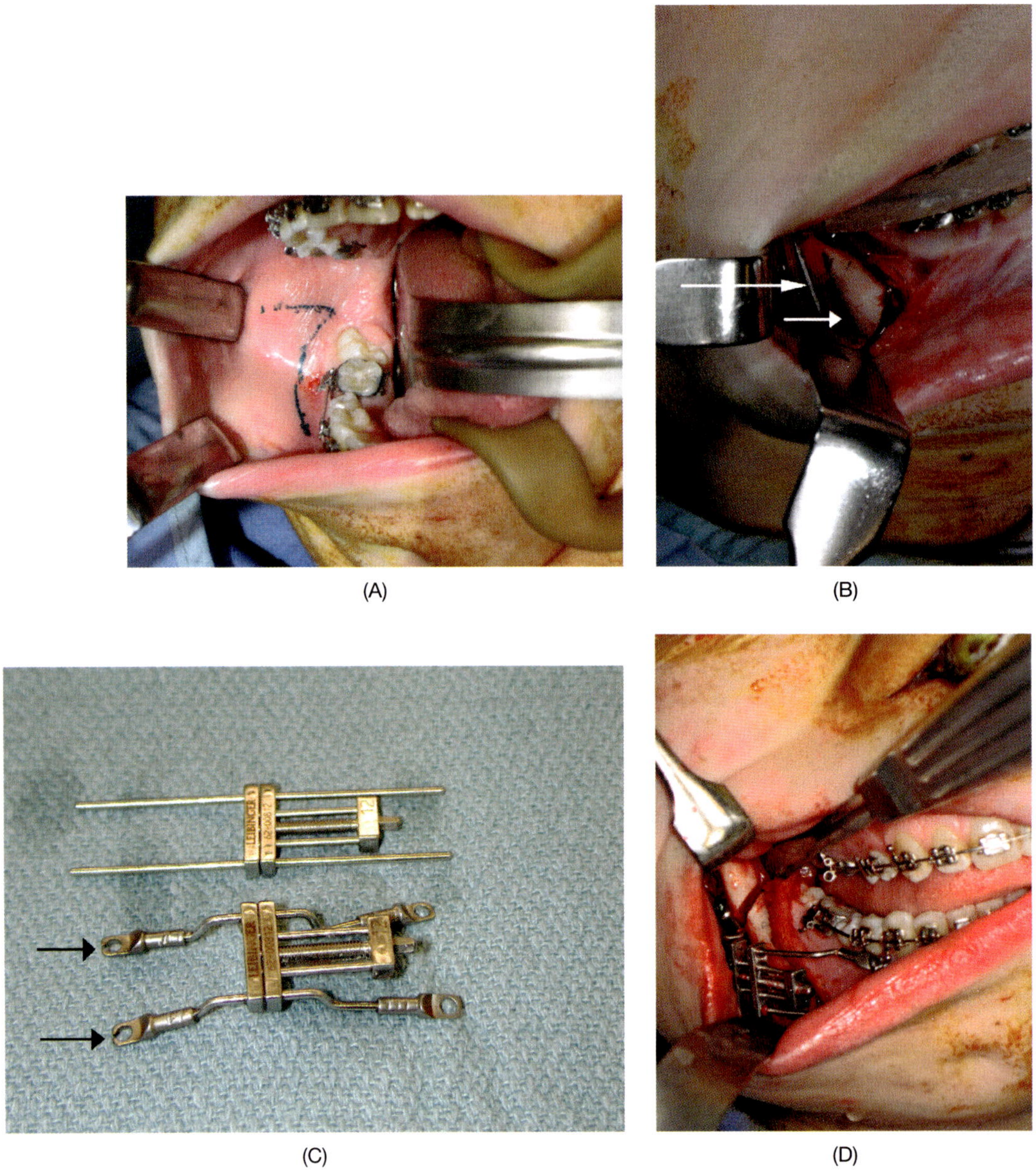

FIGURE 2 • Mandibular distraction technique. (A) The incision and dissection are similar to those used for BSSO, except for the medial dissection, which is only tunneled in the area of the lingual corticotomy. In addition, the dart is horizontal and does not extend along the anterior border of the ramus. (B) Once the dissection is completed, a surgical navigation device (long arrow) on the teeth and with a metal extension can be used to mark the proposed osteotomy (short arrow). The device is an acrylic splint indexed to the teeth with posterior metallic extensions at the desired angulation. The correct vector is determined pre-operatively with mounted models and cephalometric radiographs. The osteotomy is located through the alveolar socket of the third molar or just behind the last molar. The mandible is separated with osteotomes after the corticotomies have been completed. (C) For a single vector of distraction in the sagittal plane, the Dynaform™ device is used (Leibinger LP, Dallas, TX). The device (top) is bent to conform to the patient's mandible (bottom). Note that the proximal footplates are notched (arrows), rendering them detachable. (D) The appliance is secured across the osteotomy with two screws on the proximal segment and one screw on the lower aspect of the distal segment. The remaining footplate is attached to the adjacent tooth with a circumdental wire and acrylated. Wounds are closed and the activating mechanism is left exposed to allow activation of the device.

(28.6 months) than for those treated by BSSO (34.7 months). Both groups had comparable malocclusions and were treated by the same orthodontist. Therefore, the difference in treatment time was attributed to the variation in orthodontic requirements for the two surgical techniques used. The number of post-operative visits for patients undergoing DO is usually greater than that for BSSO because of the necessity to closely monitor the progression of mandibular advancement during the phase of activation and the stability during maturation of the bone regenerate. In a recent cohort study at the Massachusetts General Hospital (MGH), 10 patients each were treated by BSSO or DO to advance the mandible for retrognathism and class II malocclusion. The number of post-operative visits was significantly greater, by an average of 4.8 visits (21.1 for the DO group and 16.3 for the BSSO group, $p = 0.009$), for the DO group.[18]

It has also been reported that DO requires more patient compliance when compared to conventional operations for mandibular advancement.[19,20] Patients must be chosen carefully to make sure that they are compliant with the protocol and that they are able to activate the appliance as directed.

INFERIOR ALVEOLAR NERVE INJURY

A significant disadvantage of the sagittal split osteotomy is the incidence of temporary or permanent altered sensation in the inferior alveolar nerve distribution. This is secondary to the anatomic relationship of the inferior alveolar nerve (IAN) to the osteotomy cuts. Damage to the IAN may occur during: 1) retraction of the soft tissues to gain access to the medial surface of the ramus and the lingula; 2) the medial osteotomy; 3) the anterior vertical osteotomy (DalPont modification); 4) splitting of the mandible if the nerve is not completely free, if it remains in the proximal fragment, or if it is damaged by instruments; and 5) placement of rigid internal fixation to stabilize the two segments.[7,21]

Differences in the reported methods of assessment of nerve function, and in the descriptions of the type and magnitude of nerve deficit after BSSO, make it difficult to establish the exact incidence of this complication. Almost all patients have some temporary altered sensation that will usually resolve in 2 to 6 months after the operation. However, according to Cunningham,[22] 20% to 25% of patients have some long-term altered sensation. In a systematic review, Bays[5] reported that neurosensory loss after BSSO with internal rigid fixation varies from 0% to 75% with a mean of 35% for subjective reporting and 33% for objective testing at a mean follow-up of 21 months. MacIntosh reported 9% of patients with paresthesia at 1 year in a series of 236 patients evaluated by pinprick.[8] He also reported that the incidence of sensory deficit increased linearly with age. Similar findings have been reported by August and colleagues,[6] Upton,[10] and Nishioka and colleagues.[9] Increasing thickness of the mandible and decrease in quantity of medullary bone with age may account for a greater incidence of nerve injury in older patients.[6]

Damage to the IAN during distraction may result from direct intraoperative injury during the osteotomy and/or placement of fixation pins or screws or indirect injury from excessive stretching of the nerve during large advancements, compression by post-operative edema or hematoma and/or a narrowing of the canal due to bone ingrowth during consolidation.[23,24]

Michieli and Miotti found no observable histologic alterations in IAN fibres after 5 mm and 15 mm of mandibular lengthening by distraction in a dog model.[25] In 2002, Wang et al performed a series of experimental studies of DO in monkeys and examined nerve function and structure immediately post distraction and at 2, 4, 6, and 12 weeks post DO.[26] Nerve conduction studies revealed an increased latency and decreased amplitude immediately after surgery which returned to normal by 12 weeks. Histologic analysis demonstrated edema, segmental demyelinization, and axon loss following DO with a return to normal histologic anatomy by 12 weeks.

Van Strijen performed a retrospective clinical study of 70 patients (140 sides) who underwent mandibular lengthening by distraction osteogenesis. Twenty-three patients (33 sides, 23.6%) had altered sensation in the distribution of the IAN when tested with sharp and blunt stimuli.[27] However, only 2 patients (3 sides, 2.1%) had altered sensation 12 months after the removal of the distraction devices. In a series of 5 patients (10 sides) who underwent large successful mandibular advancement (10 to 14 mm) using DO, Whitesides and Meyer tested neurosensory function by measuring 2-point discrimination, response to painful stimulus, and brush stroke. There was no altered sensory function in this group.[28] Based on these reports, it appears that DO results in a low rate of permanent neurosensory deficit for mild and even large advancements. This has been confirmed by extensive animal studies on nerve function after DO.

In the study performed at MGH, neurosensory function was assessed among the two treatment groups (BSSO and DO).[18] The distraction group was composed of 4 females and 6 males with a mean age of 25 years (range, 12–53.3). This was comparable to the sagittal split group, which had 6 females and 4 males with a mean age of 25 years and 11 months (range, 15.3–50.6). However, the DO group appeared to have two subgroups; all patients were either younger than 20 ($n = 7$) or older than 40 ($n = 3$) years of age. In the early post-operative period, 26% of patients in the BSSO group and 14% in the DO group reported altered sensation in the distribution of the IAN. These subjective findings were confirmed by the surgeon's objective clinical evaluation. At 1 year post-operatively, 10% of BSSO and 4% of DO patients had subjective sensory deficits confirmed by nerve testing.[18]

RELAPSE

Relapse is defined as post-operative movement of the jaw toward the pre-operative position in any of the three planes of space. Proffit ranked mandibular advancement alone as being the second most stable procedure in orthognathic surgery after maxillary impaction.[29] Relapse has been associated with increasing patient age, high mandibular plane angle, magnitude of the advancement, post-operative increase in the posterior lower face height, and displacement of the condyle from the fossa.[30,31] One other limitation might be the inability of the muscles to be acutely stretched without the significant risk of relapse.[31-33]

Schendel and Epker found that displacement of the condyle from the fossa was a significant risk factor for relapse.[34] Lake and associates confirmed that observation.[35] They also demonstrated a relationship between the amount of advancement and the amount of relapse.[35] Another factor affecting relapse in mandibular advancement is the type of fixation used. Several authors have demonstrated greater stability when rigid fixation was used when compared to wire fixation.[33,36,37]

Relapse is rarely reported with mandibular lengthening using DO, although there is much less data than for BSSO. In a sample of 152 patients with congenital micrognathia who underwent distraction osteogenesis to lengthen the mandible, Swennen reported relapse in only 6 patients (4%).[38] In a series of 50 patients who underwent mandibular advancement by DO (mean, 8.2 mm; range, 5 mm–13 mm), Van Strijen and colleagues reported that 22% of patients exhibited some degree of relapse ranging from 2 mm to 6 mm. Of these patients, 72% had a high mandibular plane angle. The authors concluded that high-angle patients are at risk of relapse when DO is used for mandibular advancement.[39] Whitesides and Meyer evaluated occlusal and skeletal outcomes in 5 patients with severe mandibular retrognathia who underwent a mandibular advancement using DO (ranging from 10 mm–14 mm). All patients maintained class I occlusion at 1 year post-operatively and were considered stable.[28]

Relapse reported after mandibular lengthening using DO is most commonly attributed to premature removal of external devices with inadequate time for consolidation of the regenerate.[40,41] Rigidity and stability of the distraction device and adequate consolidation time are important factors required for formation of the bone regenerate.[42] Other authors have reported that relapse after mandibular lengthening of 10 mm to 20 mm with DO is minimal or nonexistent if adequate distraction protocol and bony consolidation periods are used.[13,28,39] Buried intraoral distraction devices offer greater patient acceptance and more adequate time for bony consolidation than external devices. As distraction devices are made less cumbersome, patient acceptance may improve and DO may be applied for the treatment of less severe deformities such as non-syndromic mandibular retrognathism.[39,43]

In the study comparing DO and BSSO at the MGH previously discussed, the severity of the malocclusion was comparable and the mean advancement between T0 (pre-operative) and T1 (immediate post-operative) was similar among both DO and BSSO groups: 6.5 mm for the DO group and 5.3 mm for the BSSO group.[18] For the same magnitude of initial mandibular advancement, there was no statistically significant difference in relapse between T2 (latest post-operative visit) and T1 for both groups for all the sagittal cephalometric values. However, in the vertical direction, there was a significant difference in overbite (OB) at baseline. Patients with larger OB were offered DO rather than BSSO. Likewise, there was a significant difference in the net correction (T0–T2) between the two treatment groups when the overbite was compared. As expected, the bite opened significantly in the DO group when compared to the BSSO group. These effects on the vertical dimension and overbite in patients undergoing distraction of the mandible have been also demonstrated by other authors.[44] Breuning and colleagues concluded that opening of the bite should be expected when distraction is performed to correct mandibular hypoplasia, and thus recommended hypodivergent patients as good candidates for this procedure. The author presented multiple factors as potentially responsible for bite opening. However, distraction vector was the major cause of anterior bite opening. The results of our study also showed no statistical difference detected in sagittal measurements in both groups.[18] However, as noted, an intended increase in the mandibular plane angle divergence was observed in the DO group. This indicates that patients with deep bite and mandibular retrognathia are good candidates for mandibular advancement using DO.

OTHER CLINICAL OUTCOMES

Van Strijen and associates compared duration of operation and length of hospital stay in DO versus BSSO patients. The distraction group had a longer operative time by 37% when compared to BSSO but 1 less day of hospitalization.[17]

In the study performed at the MGH, there was no significant difference in the duration of the operation when comparing DO (mean = 2 hours 38 min) versus BSSO (mean = 2 hours 37 min). The length of hospital stay was 1 day for both DO and BSSO patients.[18]

To our knowledge, no study comparing the frequency and severity of complications during BSSO and DO, for standard advancement, has been performed to date. In the MGH study, both groups had similar incidence of complications.[18] Most complications were minor and did not require a second operation.

RECOMMENDATIONS

Reflecting on the advantages and disadvantages of DO versus BSSO, we recommend the use of DO to lengthen the mandible for: 1) growing children or young adults with non-syndromic mandibular retrognathism and deep bite requiring moderate advancement (<10 mm) (Figure 3); 2) adults over 40 who are not willing accept the risk of inferior alveolar nerve injury deficit after BSSO (Figure 4); 3) infants and children with severe micrognathia to prevent a tracheostomy or to improve the airway to permit decannulation prior to speech development or starting school. (See Chapter 12.)

Teenagers with mandibular retrognathism and deep bite requiring mandibular advancement are also candidates for DO. BSSO in growing children under the age of 17 can be performed safely; however, certain challenges unique to the teenage group are encountered: 1) the lingula is more posteriorly and superiorly positioned, which may require extension of the medial cut to the posterior border to give more surface area of contact; 2) difficulty in splitting the mandible due to the elasticity and thinness of the cortical bone and relatively short posterior vertical body height; 3) rigid internal fixation may not be possible because of the developing tooth buds.[45] For growing children 13 to 15 years of age, mandibular advancement by DO poses the advantage of performing the corticotomy through the third molar tooth bud extraction socket, obviating a separate procedure to extract the teeth. In addition, the tendency of DO

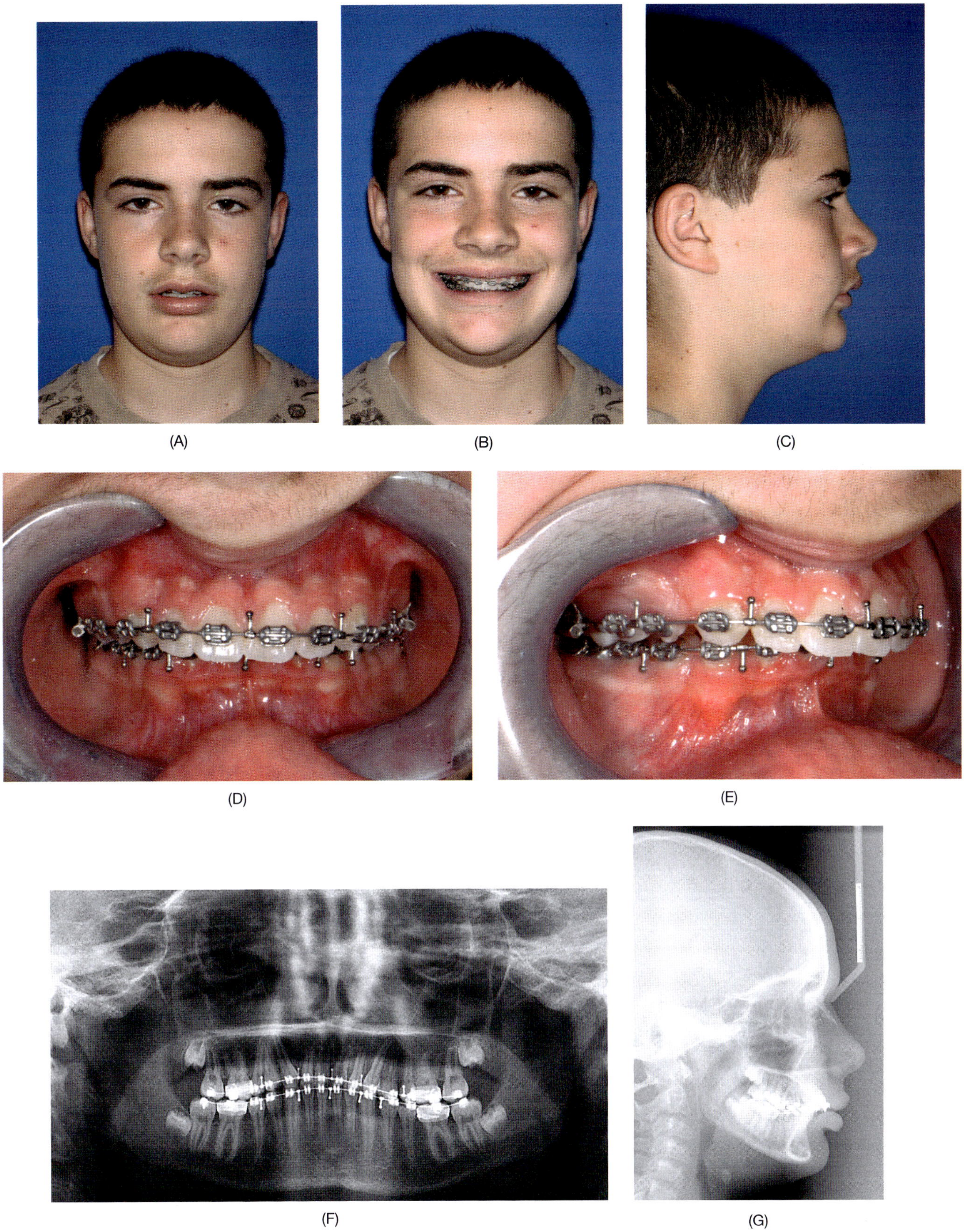

(A) (B) (C) (D) (E) (F) (G)

FIGURE 3 • Mandibular retrognathism in a non-syndromic patient. (A) Pre-operative frontal, (B) smiling, and (C) lateral facial photographs of a 14-year-old male. The patient has a deep labial mental fold and a short lower facial height. (D)(E) Intraoral photographs showing an increased overjet and overbite and a class II malocclusion. (F) Pre-operative panoramic and (G) Lat cephalometric radiographs *(continued)*

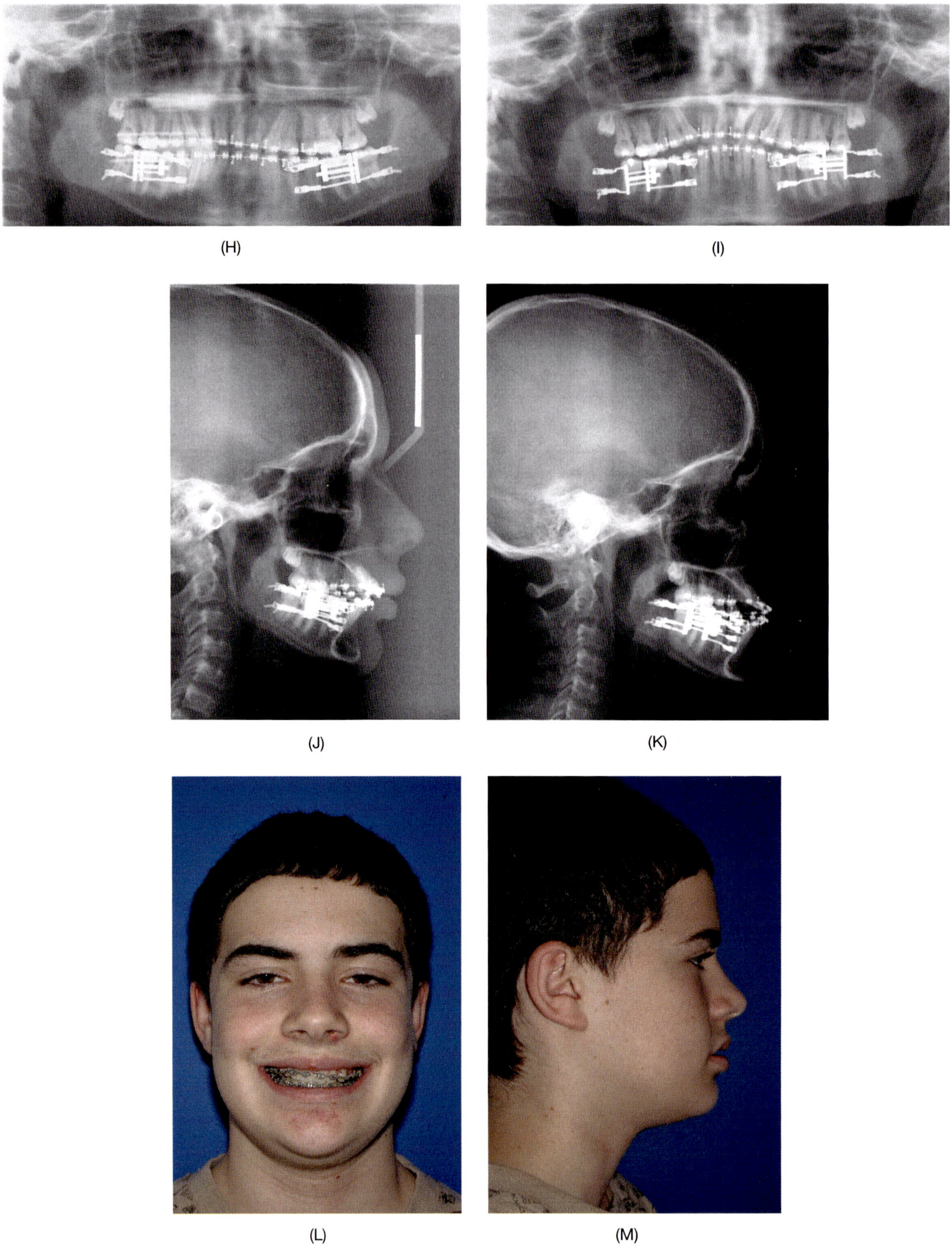

FIGURE 3 • (*continued from previous page*) Note the accentuated curve of spee. (H)(I) Serial panoramic and (J)(K) cephalometric radiographs at the immediate post-operative time and at the end of distraction, respectively. (L) Post-operative frontal and (M) lateral facial photographs. The labial mental fold is normalized and the lower facial height was increased. (*continued*)

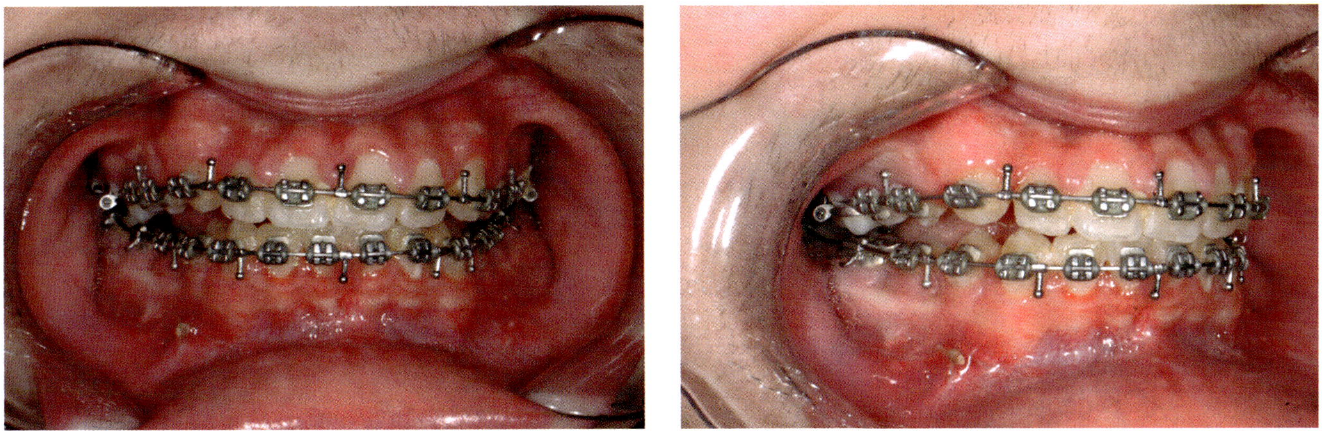

(N) (O)

FIGURE 3 • (*continued from previous page*) (N)(O) Occlusal photographs demonstrating the corrected overjet and overbite. With the increase in lower facial height, the lateral open bites can be corrected by extrusion of the premolars.

to open the bite can facilitate the correction of a class II deep bite.

Patients over 40 who do not accept the risk of IAN injury deficit after BSSO are also candidates for mandibular distraction osteogenesis. Our report confirms those of other investigations that more favorable neurosensory outcomes are associated with mandibular advancement by DO than with BSSO in susceptible patient groups (Age >40, advancement >10 mm).[18,23,28,46]

The third group of patients includes children with severe micrognathia, who would otherwise require conventional osteotomies and bone grafting. (See Chapters 12, 13, 17 and 18.) Many authors have reported the advantages of distraction for large mandibular advancements in syndromic and non-syndromic children with micrognathia and airway obstruction. DO has become the standard for these patients.[11–14]

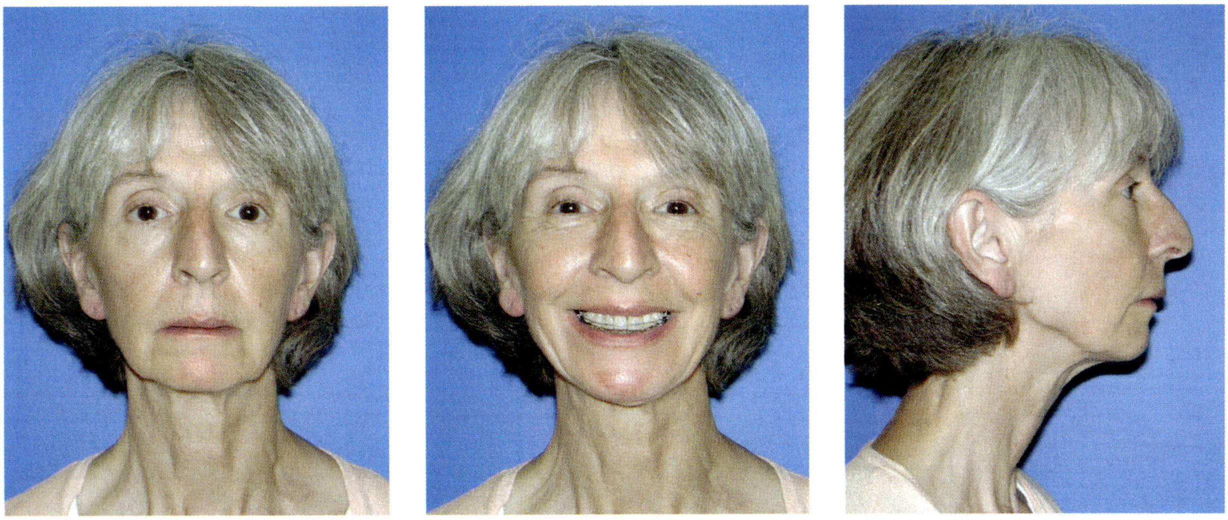

(A) (B) (C)

FIGURE 4 • Mandibular distraction osteogenesis in the adult patient. (A) Pre-operative frontal, (B) smiling, and (C) lateral facial photographs of a 54-year-old female with mandibular retrognathism. Due to her age, it was decided to correct her malocclusion by distraction osteogenesis to minimize the risk of inferior alveolar nerve injury. (*continued*)

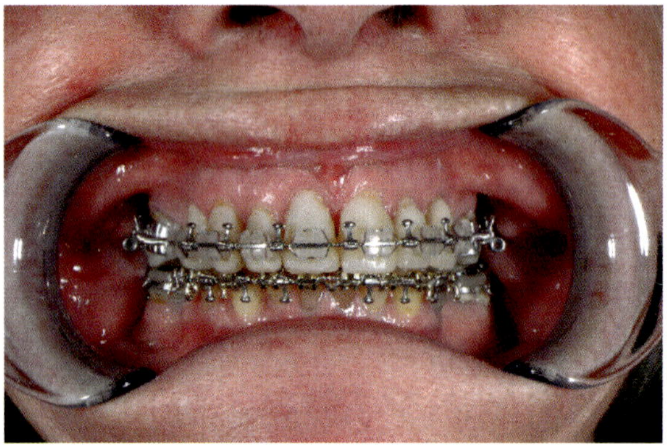

(D)

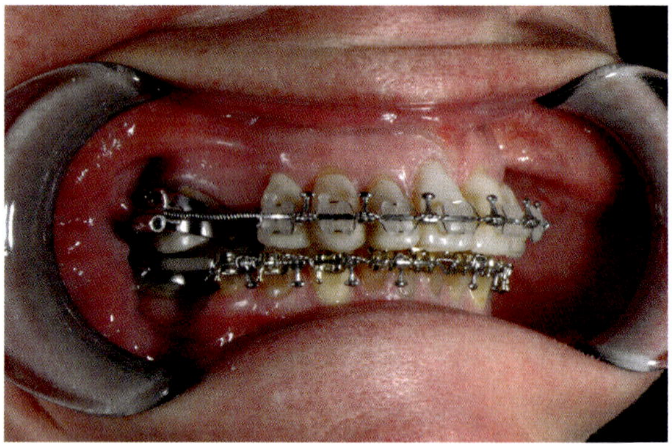

(E)

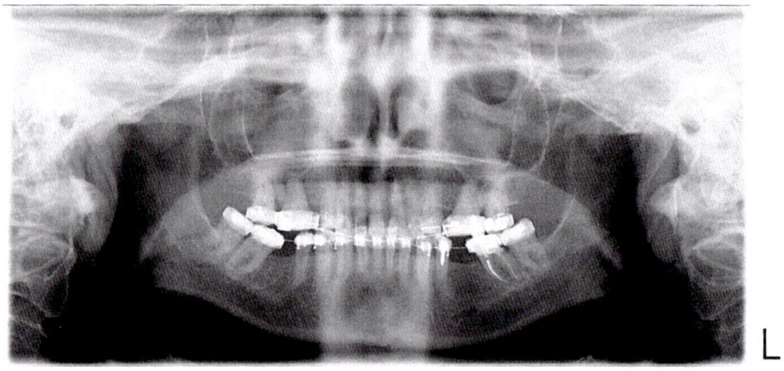

(F)

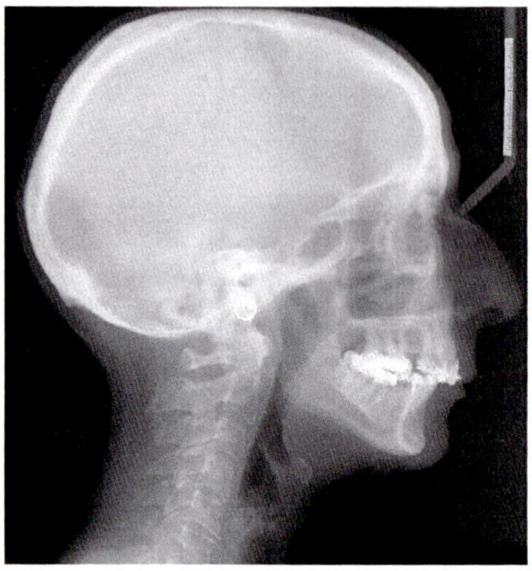

(G)

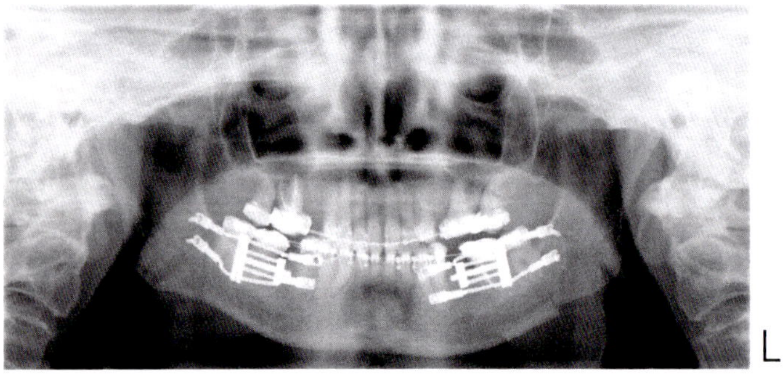

(H)

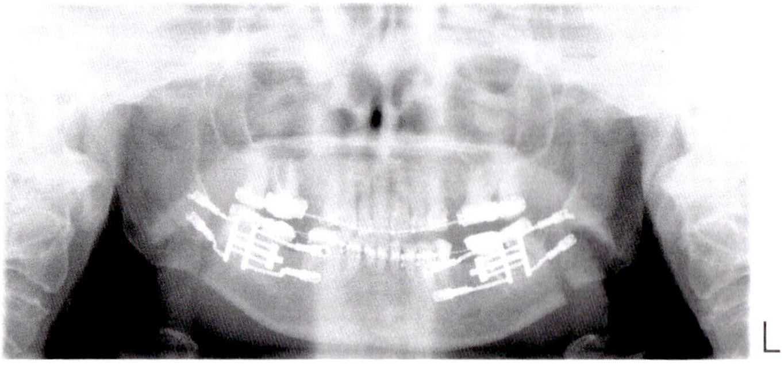

(I)

FIGURE 4 • (*continued from previous page*) (D)(E) Intraoral pre-operative photographs demonstrating the increased overjet and overbite and the class II malocclusion. (F) Pre-operative panoramic and (G) cephalometric radiographs. (H)(I) Serial panoramic immediate postoperatively and at end-DO (*continued*)

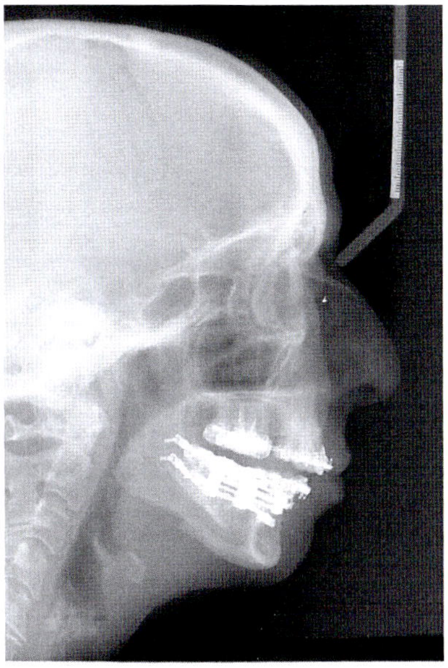

(J)

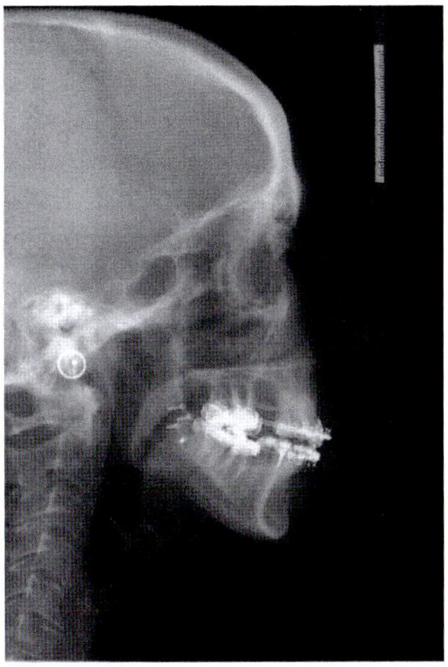

(K)

(L)

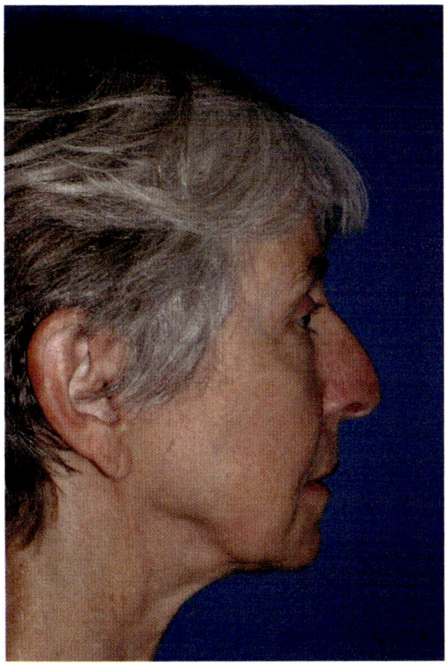

(M)

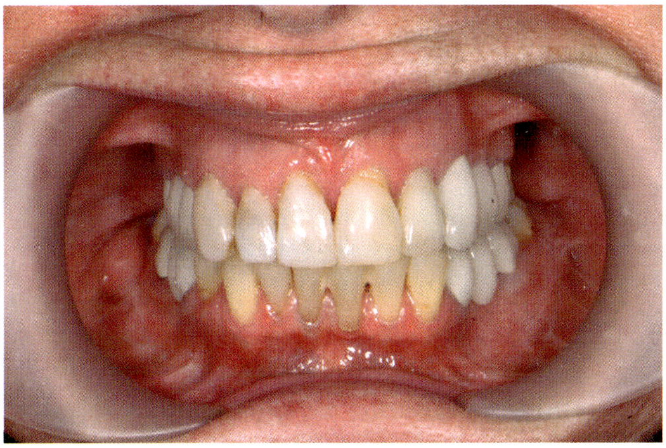

(N)

FIGURE 4 • (*continued from previous page*) (J)(K) cephalometric radiographs at the immediate post-operative time and at the end of distraction (with devices removed), respectively. Note that the mandible was slightly overcorrected to compensate for relapse. (L) 5-year post-operative frontal and (M) lateral facial photographs. (N)(O)(P) 5-year post-operative intraoral photographs demonstrating the corrected overjet and overbite. (*continued*)

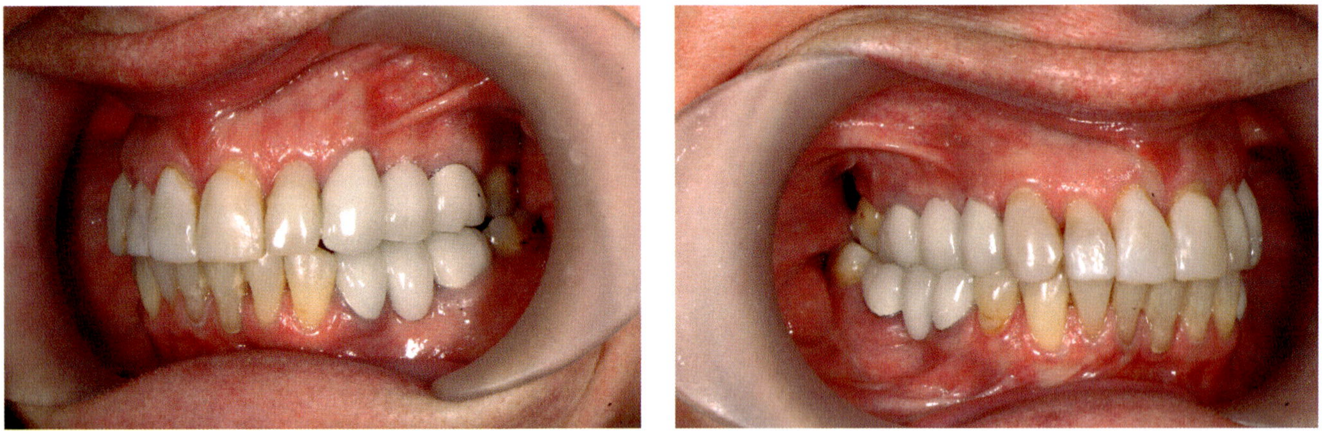

(O) (P)

FIGURE 4 • (*continued from previous page*) Reproduced from: Bouchard C, Lawler M, Kaban LB, Troulis MJ: Minimally invasive orthognathic + condylar surgery. In Cosmesis of the Mouth, Face, and Jaws. S. Guttenberg, Ed. Hoboken, NJ: Wiley-Blackwell, 2012; used with permission.

CONCLUSION

BSSO and DO are both acceptable methods for the treatment of mandibular retrognathia. However, DO may provide specific advantages in patients requiring large advancements (>10 mm–15 mm), patients in whom the bite needs to be opened to correct a short lower face height and in older patients because of a lower incidence of permanent inferior alveolar nerve deficit.

References

1. Bailey LJ, Haltiwanger LH, Blakey GH, Proffit WR. Who seeks surgical-orthodontic treatment: A current review. Int J Adult Orthodon Orthognath Surg 2001;16:280–92.
2. Trauner R, Obwegeser H. The surgical correction of mandibular prognathism and retrognathia with consideration of genioplasty. II. Operating methods for microgenia and distoclusion. Oral Surg Oral Med Oral Pathol 1957;10:787–92.
3. Dalpont G. Retromolar osteotomy for the correction of prognathism. J Oral Surg Anesth Hosp Dent Serv. 1961;19:42–7.
4. Hunsuck E. A method of intraoral open reduction of fractured mandibles. J Oral Surg Anesth Hosp Dent Serv. 1967;25:533–41.
5. Bays R. Complications of orthognathic surgery. Oral Maxillofac Surg Clin North Am 2003;15:229–42.
6. August M, Marchena J, Donady J, Kaban L. Neurosensory deficit and functional impairment after sagittal ramus osteotomy: a long-term follow-up study. J Oral Maxillofac Surg 1998;56:1231–5.
7. Westermark A, Bystedt H, von Konow L. Inferior alveolar nerve function after sagittal split osteotomy of the mandible: correlation with degree of intraoperative nerve encounter and other variables in 496 operations. Br J Oral Maxillofac Surg 1998;36:429–33.
8. MacIntosh RB. Experience with the sagittal osteotomy of the mandibular ramus: a 13-year review. J Maxillofac Surg 1981;9:151–65.
9. Nishioka GJ, Zysset MK, Van Sickels JE. Neurosensory disturbance with rigid fixation of the bilateral sagittal split osteotomy. J Oral Maxillofac Surg 1987;45:20–6.
10. Upton LG, Rajvanakarn M, Hayward JR. Evaluation of the regenerative capacity of the inferior alveolar nerve following surgical trauma. J Oral Maxillofac Surg 1987;45:212–6.
11. Steinbacher DM, Kaban LB, Troulis MJ. Mandibular advancement by distraction osteogenesis for tracheostomy-dependent children with severe micrognathia. J Oral Maxillofac Surg 2005;63:1072–9.
12. Rachmiel A, Aizenbud D, Pillar G, Srouji S, Peled M. Bilateral mandibular distraction for patients with compromised airway analyzed by three-dimensional CT. Int J Oral Maxillofac Surg 2005;34:9–18.
13. Wang X, Wang XX, Liang C, Yi B, Lin Y, Li ZL. Distraction osteogenesis in correction of micrognathia accompanying obstructive sleep apnea syndrome. Plast Reconstr Surg 2003;112:1549–57.
14. Denny AD, Talisman R, Hanson PR, Recinos RF. Mandibular distraction osteogenesis in very young patients to correct airway obstruction. Plast Reconstr Surg 2001;108:302–11.
15. McCarthy JG, Staffenberg DA, Wood RJ, Cutting CB, Grayson BH, Thorne CH. Introduction of an intraoral bone-lengthening device. Plast Reconstr Surg 1995;96:978–81.
16. Diner PA, Kollar EM, Martinez H, Vazquez MP. Intraoral distraction for mandibular lengthening: a technical innovation. J Craniomaxillofac Surg 1996;24:92–5.
17. Van Strijen PJ, Breuning KH, Becking AG, Perdijk FB, Tuinzing DB. Cost, operation and hospitalization times in distraction osteogenesis versus sagittal split osteotomy. J Craniomaxillofac Surg 2003;31:42–5.
18. Bendahan GJ. Distraction osteogenesis versus bilateral sagittal split osteotomy for mandibular advancement [Master of science thesis]. Havard University, Boston; 2005.
19. Molina F, Ortiz Monasterio F. Mandibular elongation and remodeling by distraction: a farewell to major osteotomies. Plast Reconstr Surg 1995;96:825–40.
20. Diner PA, Tomat C, Soupre V, Vazquez MP. [Mandibular advancement: osteotomy or distraction osteogenesis? A case report]. Ann Chir Plast Esthet 2001;46:316–22.
21. White RP Jr., Peters PB, Costich ER, Page HL Jr. Evaluation of sagittal split-ramus osteotomy in 17 patients. J Oral Surg 1969;27:851–5.
22. Cunningham LL, Tiner BD, Clark GM, Bays RA, Keeling SD, Rugh JD. A comparison of questionnaire versus monofilament assessment of neurosensory deficit. J Oral Maxillofac Surg 1996;54:454–9.

23. Makarov MR, Harper RP, Cope JB, Samchukov ML. Evaluation of inferior alveolar nerve function during distraction osteogenesis in the dog. J Oral Maxillofac Surg 56: 1417–23; discussion 24–5, 1998.
24. Troulis MJ, Kaban LB. Complications of mandibular distraction osteogenesis. Oral Maxillofac Surg Clin North Am 2003;15: 251–64.
25. Michieli S, Miotti B. Lengthening of mandibular body by gradual surgical-orthodontic distraction. J Oral Surg 1977;35:187–92.
26. Wang XX, Wang X, Li ZL. Effects of mandibular distraction osteogenesis on the inferior alveolar nerve: an experimental study in monkeys. Plast Reconstr Surg 2002;109:2373–83.
27. Van Strijen PJ, Breuning KH, Becking AG, Perdijk FB, Tuinzing DB. Complications in bilateral mandibular distraction osteogenesis using internal devices. Oral Surg Oral Med Oral Pathol Oral Radiol Endod 2003;96:392–7.
28. Whitesides LM, Meyer RA. Effect of distraction osteogenesis on the severely hypoplastic mandible and inferior alveolar nerve function. J Oral Maxillofac Surg 2004;62:292–7.
29. Proffit WR, Turvey TA, Phillips C. Orthognathic surgery: a hierarchy of stability. Int J Adult Orthodon Orthognath Surg 1996;11: 191–204.
30. Van Sickels JE, Larsen AJ, Thrash WJ. Relapse after rigid fixation of mandibular advancement. J Oral Maxillofac Surg 1986;44: 698–702.
31. Will LA, West RA. Factors influencing the stability of the sagittal split osteotomy for mandibular advancement. J Oral Maxillofac Surg 1989;47:813–8.
32. Wade DB. Surgical-orthodontic stability in retrognathic patients. An implant study. Angle Orthod 1988;58:71–95.
33. Bhatia SN, Yan B, Behbehani I, Harris M. Nature of relapse after surgical mandibular advancement. Br J Orthod 1985; 12:58–69.
34. Schendel SA, Epker BN. Results after mandibular advancement surgery: an analysis of 87 cases. J Oral Surg 1980;38:265–82.
35. Lake SL, McNeill RW, Little RM, West RA. Surgical mandibular advancement: a cephalometric analysis of treatment response. Am J Orthod 1981;80:376–94.
36. Perrott DH, Lu YF, Pogrel MA, Kaban LB. Stability of sagittal split osteotomies. A comparison of three stabilization techniques. Oral Surg Oral Med Oral Pathol 1994;78:696–704.
37. Dolce C, Van Sickels JE, Bays RA, Rugh JD. Skeletal stability after mandibular advancement with rigid versus wire fixation. J Oral Maxillofac Surg 2000;58:1219–27.
38. Swennen G, Schliephake H, Dempf R, Schierle H, Malevez C. Craniofacial distraction osteogenesis: a review of the literature: Part 1: clinical studies. Int J Oral Maxillofac Surg 2001;30:89–103.
39. Van Strijen PJ, Breuning KH, Becking AG, Tuinzing DB. Stability after distraction osteogenesis to lengthen the mandible: results in 50 patients. J Oral Maxillofac Surg 2004; 62:304–7.
40. Ayoub AF, Duncan CM, McLean GR, Moos KF, Chibbaro PD. Response of patients and families to lengthening of the facial bones by extraoral distraction osteogenesis: a review of 14 patients. Br J Oral Maxillofac Surg 2002;40:397–405.
41. Primrose AC, Broadfoot E, Diner PA, Molina F, Moos KF, Ayoub AF. Patients' responses to distraction osteogenesis: a multi-centre study. Int J Oral Maxillofac Surg 2005;34:238–42.
42. Ilizarov GA. The tension-stress effect on the genesis and growth of tissues. Part I. The influence of stability of fixation and soft-tissue preservation. Clin Orthop Relat Res: 1989;238:249–56.
43. Breuning KH, van Strijen PJ, Prahl-Andersen B, Tuinzing DB. Outcome of treatment of Class II malocclusion by intraoral mandibular distraction. Br J Oral Maxillofac Surg 2004;42:520–5.
44. Breuning KH, van Strijen PJ, Prahl-Andersen B, Tuinzing DB. The overbite and intraoral mandibular distraction osteogenesis. J Craniomaxillofac Surg 2004;32:119–25.
45. Kaban LB, Troulis MJ. Orthognathic surgery in the growing child. In: Kaban LB, Troulis MJ., eds. *Pediatric oral and maxillofacial surgery*. Philadelphia, PA: Saunders; 2004:377–99.
46. Hu J, Tang Z, Wang D, Buckley MJ. Changes in the inferior alveolar nerve after mandibular lengthening with different rates of distraction. J Oral Maxillofac Surg 2001;59:1041–5.

15 Transport Distraction Osteogenesis

Suzanne U. Stucki-McCormick, MS, DDS

OUTLINE

INTRODUCTION

Distraction osteogenesis (DO) is a valuable tool to bring bone and soft tissue into an area of tissue deficiency. DO is one of the few techniques that allows for reconstruction of the entire site, utilizing autogenous tissues in their native form, by means of minimal invasive surgical techniques. Classic DO calls for the creation of an osteotomy in an area of bone deficiency and allows for expansion at the site.[1,2] (See Chapters 10 to 14.) Transport DO is unique in that a transport disc is created adjacent to a bone defect and the distraction segment, or transport disc, is advanced through the soft tissue defect to create new regenerate bone and soft tissue.

TRANSPORT DISTRACTION OSTEOGENESIS

Distraction osteogenesis is the slow application of forces across an osteotomy site. As the bone segments are separated, distraction wound healing occurs to create new bone and soft tissue.[3,4] Unique to the distraction process is that the tension stresses stimulate histiogenesis allowing for creation of all structures involved in the site. This is in contrast to bone grafting, whereby the recipient bed is prepared and the donor bone is placed into the recipient bed with closure of the site. The amount of grafting to the recipient bed site is thus limited by the stretch of the local tissues as well as the vascularity of the site. Consequently, in certain instances, both hard and soft tissues are transplanted with a vascular pedicle, ensuring adequate blood supply to the recipient site.

When first described (see Chapter 10), DO was used to expand bone at a specific site, for example the angle of mandible, or ramus for sagittal mandibular advancement.[5–7] (See Chapters 12–14.) Transport DO is unique in that a transport disc is created adjacent to a bone defect and the distraction segment, or transport disc, is advanced through the soft tissue defect to create new regenerate bone and soft tissue (Figure 1).[8,9] The same Ilizarov principles of latency, rate, rhythm, and neutral fixation apply as in classic DO. When treatment planning for a transport DO case, the same Ilizarov algorithm should be followed: 5 to 7 day latency period, 1.0 mm advancement per day, divided into 3 to 4 advancements (0.25mm QID or 0.33mm TID or 0.5mm BID), followed by a period of neutral fixation allowing for the final ossification of the newly created regenerate. The Ilizarov algorithm can be modified to fit the specific clinical situation. For example if the patient is elderly, the latency period may be increased to 7 to 10 days with a distraction rate of 0.5 mm to 0.75 mm per day. Or, if the patient is very young, decreasing the latency to 1 to 3 days with an increased rate of 1.0 mm to 1.5 mm per day may be indicated. If the overlying soft tissue is compromised (thin, poor blood supply, radiation), then the Ilizarov algorithm may be changed to 0.25 mm advancement in the morning, returning the transport disc 0.25 mm to its original AM position midday, then advancing the distraction disc 0.25 mm in the evening for a net advancement of 0.25 mm, thus allowing the soft tissues to recover ("dancing of the regenerate" or "pumping the regenerate"). Conversely, if the site appears to be ossifying rapidly, as often occurs at the end of a long distraction period, the rate may be increased to 1.5 mm to 2.0 mm per day. These examples demonstrate that attention to detail and close patient follow-up that are required to achieve a successful outcome.

Surgical Technique

Transport distraction requires three points of fixation associated with the distraction device: proximal, distal, and in the transport disc. Often transport DO is combined with a reconstruction

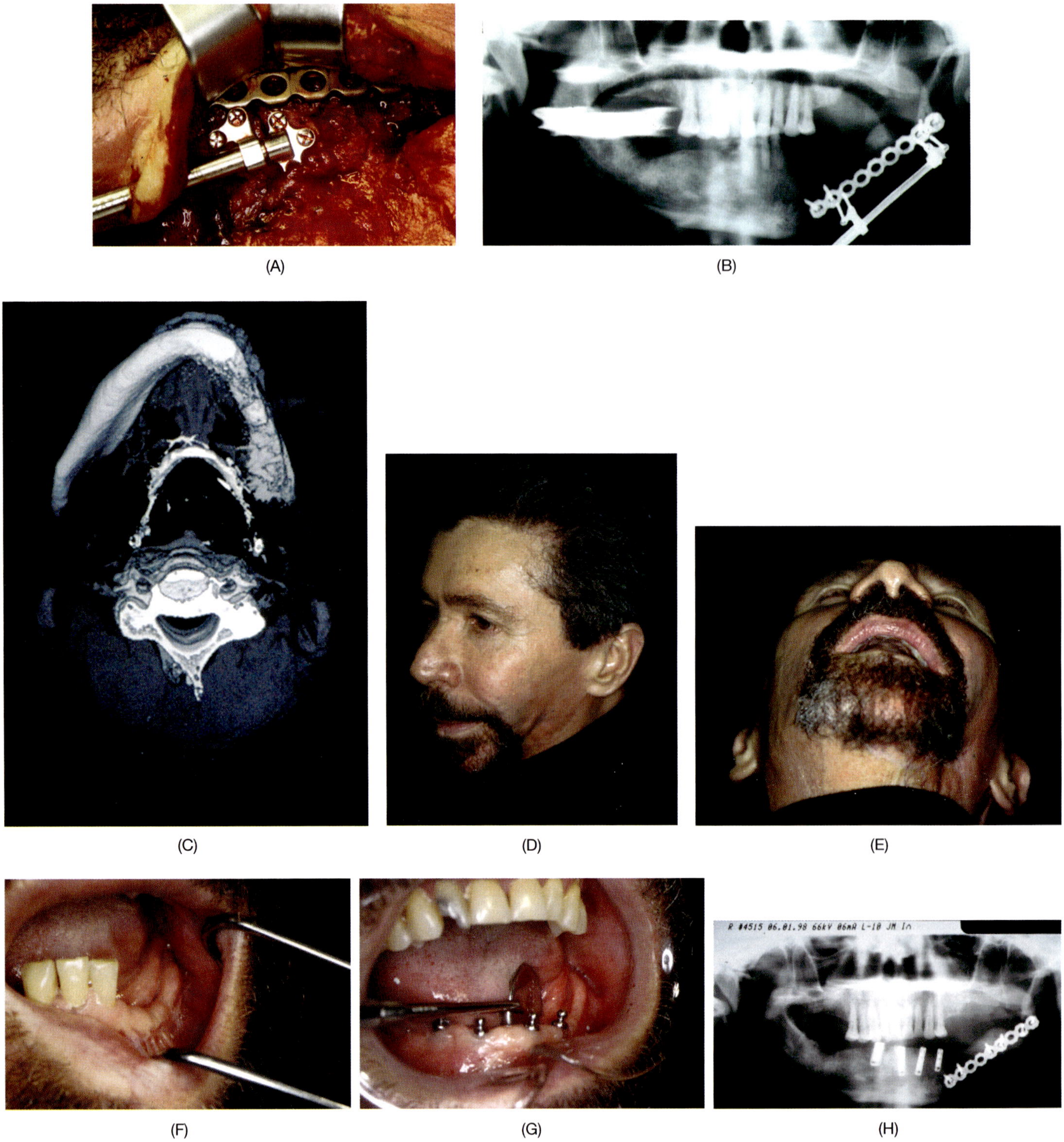

FIGURE 1 • A patient who was status post resection of his mandible for squamous cell carcinoma, underwent transport distraction osteogenesis (A). After placement of the reconstruction plate, the transport disc was created adjacent to the proximal stump of bone. The transport disc is advanced, using an semibuned distraction device, until it reached the distal docking site (B). During the period of neutral fixation, the distraction bone ossifies (C) from the periphery to the central region. The newly created bone assumes the three dimensional shape of the transport disc. The facial contour is maintained, although an semibuned device was utilized (D, E). An alveolus was generated from the transport distraction process, as well as a buccal and lingual sulcus (F). Implants were placed into the bone for final prosthetic reconstruction (G, H).

plate to recreate and maintain the bone gap. For benign tumors, transport DO can be performed at the time of tumor resection, or delayed after primary healing. For malignant tumors, DO reconstruction is often delayed if radiation therapy is considered, or if microvascular free flap techniques are employed or at least to make sure the margins are clear of tumor.

The mandible can be approached from either an intra-oral, or an extra-oral approach. The distraction gap is identified and the proximal and distal stumps are exposed. Care is taken to avoid periosteal stripping of the bone segments. Two of the three points of fixation are created either with the distraction device or by the use of a bridging plate. Some distraction devices incorporate a rigid bridging plate in the transport DO device design (Figure 1).[10,11] Here the bridging plate acts as a direct guide for the DO device, like a railroad car on a train track. Some devices include the use of orthodontic-style nitinol wire to allow for continuous force application, an ideal concept for transport DO.[12,13] Alternatively, a bridging plate can be used with a separate DO device, (Figure 2). Here, care must be taken, when placing the independent DO device, to ensure the proper vector of distraction. It is important that the DO device trajectory remain away from the bone plate, so that the transport disc will not "bump into" the bone bridging plate. Should this occur, the patient may experience difficulty advancing the transport disc, or the leading edge of the transport disc will begin to form a groove on the bone surface adjacent to the rigid fixation plate.

An osteotomy is carefully created 1.5 cm from the edge of the proximal stump, adjacent to the osseous defect in an area of maximal bone height and thickness. Initially, the bone cut is a corticotomy allowing for DO device application. Tunneling and osteotome techniques are utilized to minimize periosteal stripping when periosteum is present and avoid mucosal perforation. The distraction device is applied and the corticotomy is converted to an osteotomy. The device is activated to ensure proper device trajectory. Should this portion of the surgery be less than atraumatic, consideration should be given to increase the latency period from 7–10 days, allowing for adequate callous formation and proper soft tissue healing and vascularization. In some cases, the DO device may be placed transmucosally, with the body of the device positioned in the buccal vestibule. This technique requires acute attention to detail, proper treatment planning and proper follow-up care.

After a latency, the transport disc is distracted, creating new bone and soft tissue in its wake. One advantage of transport DO is that all tissue is re-created including mucosa, a feature unique to DO versus conventional bone grafting techniques. This continues until the transport disc traverses the discontinuity defect and reaches the docking site. Neutral fixation

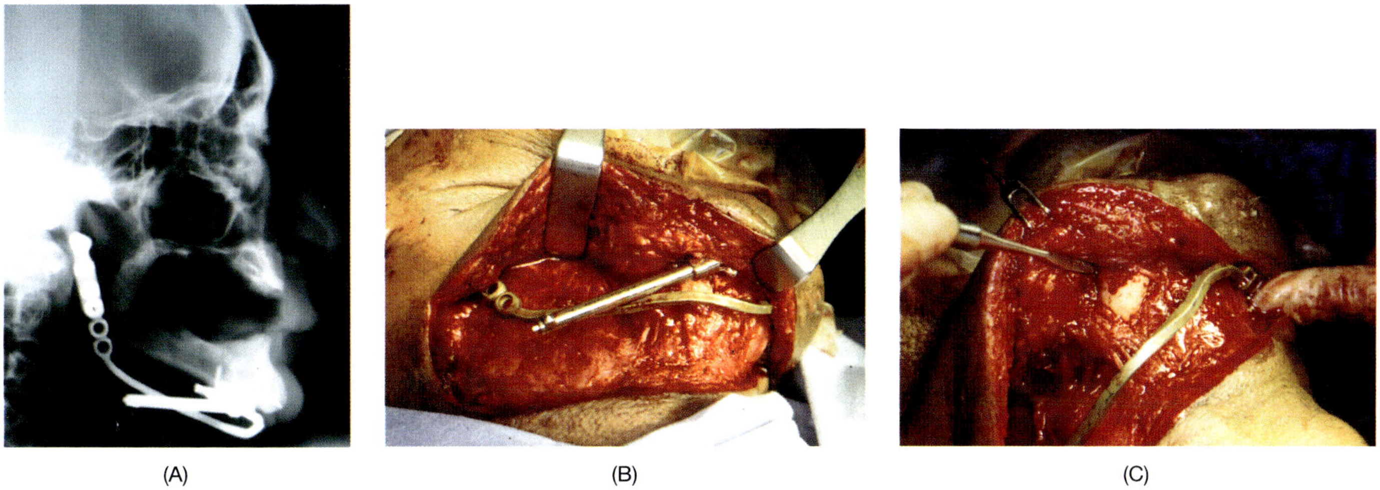

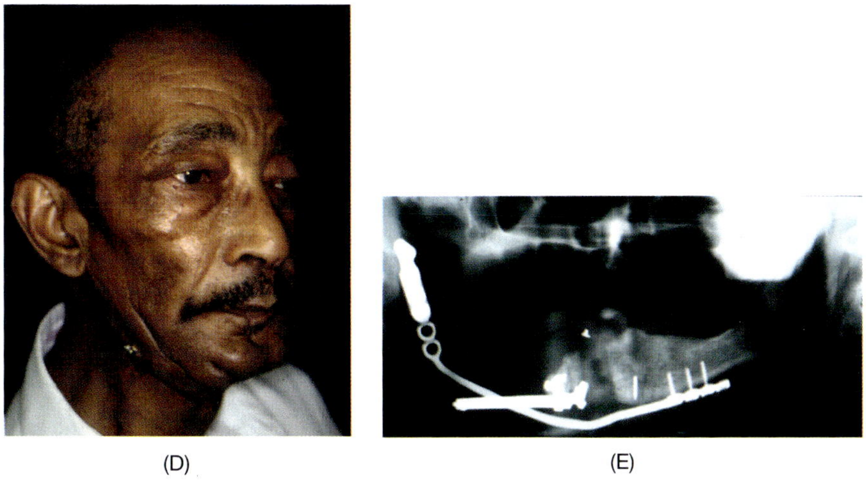

FIGURE 2 • A patient with squamous cell carcinoma underwent hemi-mandibulectomy with placement of a bridging plate (A). The distal end of the proximal bone margin is identified and an osteotomy is created (B). Care is taken to place the osteotomy at the site of optimal bone height as the transport disc acts as the template (C). The submerged distraction device is applied and the vector verified (D). Using a submerged device, facial esthetics can be maintained (E).

maintained until complete ossification occurs. It is important that a new cortical outline be seen radiographically as evidence for osseous consolidation. If a reconstruction plate is utilized, the DO device may be removed at the end of active DO, prior to osseous consolidation. Here the transport disc is rigidly fixated to the reconstruction plate with minimal periosteal stripping. Finally, osseous continuity is established by removing the intervening fibro-osseous cap with consideration of a small localized bone graft if necessary.

Treatment Planning and Case Selection

Treatment planning for transport DO requires a 3D understanding of the bony defect to be reconstructed. (See Chapter 11.) Ideally, the reconstructed site should have the same size and shape as the adjacent, native bone. As the size and shape of the transport disc determines the size and shape of the regenerate bone, care must be given to the location of the osteotomy for the transport disc (Figure 3). The disc must be adequate in width to

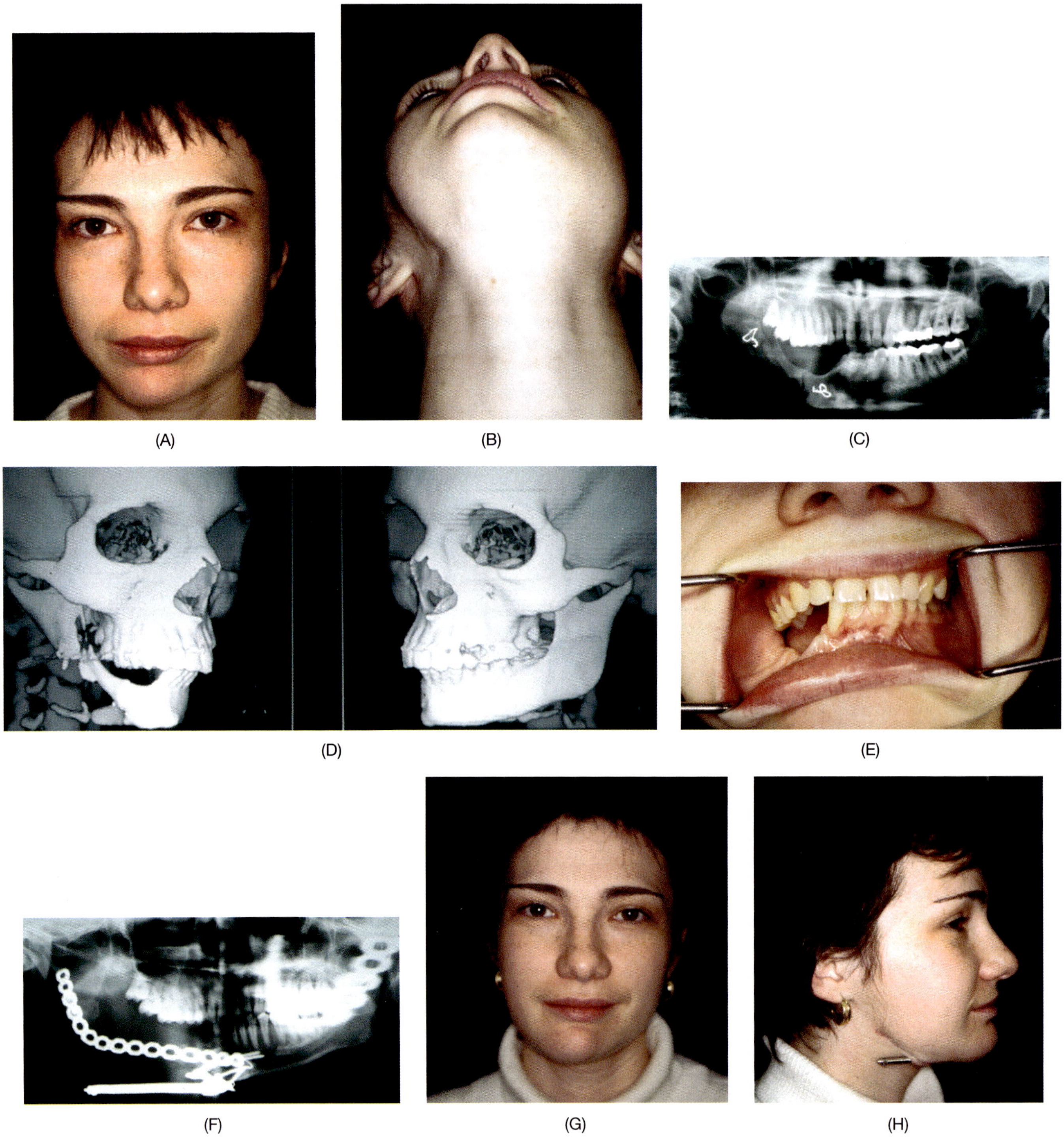

FIGURE 3 • A 24 year old patient presented status post bone graft for a non-union of a prior mandibular fracture (A–E). The submerged device in conjunction with a reconstruction plate (F) allows the patient normal mandibular function during the process of distraction (G–H).

allow for fixation of the distraction device as well as sufficient height to create the required size regenerate. Occasionally a tooth may need to be sacrificed to allow for the proper placement of the transport disc osteotomy. Consideration for coronoidectomy should be entertained to prevent rotation of the proximal segment, when dealing with the ramus.

Transport DO is often used to reconstruct the mandible after traumatic injury or cancer resection. Here the tendency is to consider the mandible as an arc form. In actuality, only the dentoalveolar unit is an arc. When viewed from below, the mandible can be divided into five linear segments, allowing for linear transport DO (Figure 4).[15] This concept is useful, as DO around a curve can be difficult, especially when using a curved device. Recalling geometry, as the outer edge of the transport disc moves around the curve, it is traveling at rate faster than the inner edge of the transport disc (see Chapter 13). If treatment planning calls for 1.0 mm per day DO rate to accommodate new bone synthesis at the outer edge of the transport disc, one runs the risk of advancing the transport disc too slowly on its inner edge, thus allowing for premature ossification (Figure 5). Clinically, patients will note that the device becomes increasingly difficult to turn as time progresses, yet there appears to be no mechanical impedance to the device. Linear DO eliminated this concern by dividing the mandible into segments of ramus to body (right and left), body to parasymphysis (right and left), and chin region (right to left).[16] Using an external device, the regenerate can be easily manipulated to its new vector of orientation by removing the device, reorienting the transport disc, and replacing the device. For external devices with hinges, the varus/valgus movements can be performed. Once in its new position, the transport disc should be allowed to rest for a second, shorter latency period of 1 to 5 days, ensuring adequate vascularization and soft tissue coverage of the site. With an internal device, the transport disc can also be manipulated by adjusting the device, again allowing for a second latency phase. Another option is to section and divide the transport disc into two segments: a larger transport disc is created originally; once it has reached the transition site, a new osteotomy is created, dividing the original transport disc vertically into two. The posterior segment is rigidly fixated to the rigid fixation bar to begin this segment's time of neutral fixation, and the DO device is now

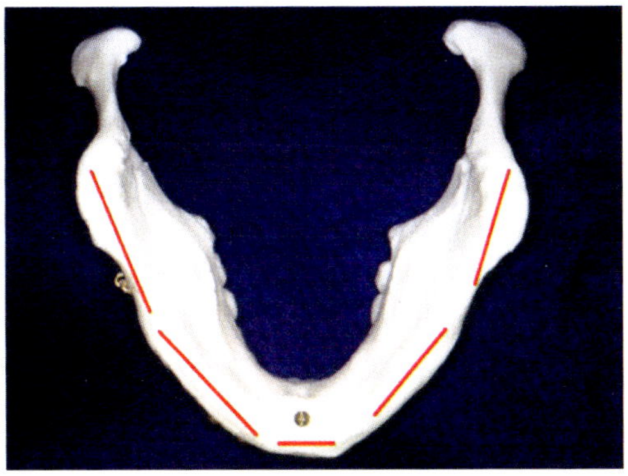

FIGURE 4 • Observing the mandible from below, 5 straight lines (Red) can be identified to produce the curvature of the mandible.

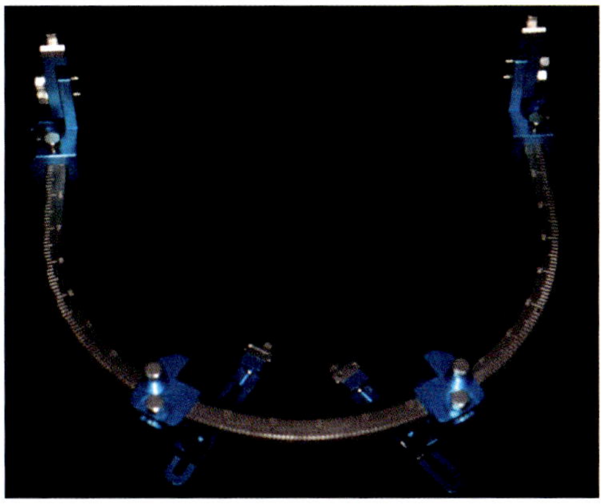

FIGURE 5 • Devices that follow a true arc form are not ideal for distraction osteogenesis. As DO progresses around the arc of the device, the lingual ossifies prematurely and the buccal portion of the transport disc remains patent.

applied to the anterior-most segment. After a 5-day latency, the new transport disc is advanced to continue to regenerate a neomandible.

For large osseous defects, bi-focal distraction may be used, with the creation of two transport discs, one from either side of the discontinuity defect, thus decreasing total treatment time.[11] Careful planning is required for bifocal DO, as the native mandible serves as both the template for bone and for the soft tissues.

During transport DO, the leading edge of the transport disc becomes rounded and covered with a fibrocartilaginous cap. Once the transport disc reaches the docking site, the fibrocartilaginous cap is removed and a rigid fixation bone plate placed, possibly with a small bone graft, to ensure osseous continuity.

Surgical Considerations and Adjunctive Techniques

Transport distraction osteogenesis can be utilized for cases that have undergone post-cancer resection radiation therapy. (These cases usually have an ablated periosteum and have a compromised blood supply.) Prior to DO reconstruction, a course of hyperbaric oxygen (HBO) therapy and a vascularized bone graft was the only reconstructive option. Should HBO therapy not be available, transport DO can still be accomplished (as blood supply is increased at the site). Regardless, soft tissue management becomes the primary issue in the post-radiation patient. As radiation therapy affects both bone and soft tissue, it is the soft tissues that often first show signs of potential wound healing complications. It is best to decrease the rate of DO to 0.5 mm per day and to utilize the technique of advancement, return as described above "dancing of the regenerate" to be incorporated into the overall DO protocol. In this way, the stretch on the soft tissues is minimized, with a planned relaxation period between each DO device activation.

Small-segment transport distraction osteogenesis is the basis for alveolar DO (a subtype of transport DO) for dental implants.[17,18] Here, the dentoalveolar unit can be distracted to increase height and width of the maxilla and mandible in preparation for dental implants. This work is based on the principles of classic transport distraction osteogenesis and these principles can be applied to dentoalveolar DO for small bone segments (Chapter 16).

Transport DO can be applied to resection sites reconstructed with autogenous osseo-cutaneous flaps, especially fibula grafts. Fibula grafts are best used for mandibular resection to reconstruct the inferior border of the mandible, establishing osseous continuity. In place, the fibula graft is shorter in height than the native mandible. Therefore, only the inferior border is reconstructed. The fibula-grafted bone can be secondarily reconstructed with transport/dentoalveolar DO techniques to facilitate dental implant placement.[19–21] This technique has also been utilized for free iliac transfers to reconstruct maxillary defects.[22]

TRANSPORT DISTRACTION OSTEOGENESIS TO CONSTRUCT A NEO-CONDYLE

For transport distraction, an osteotomy is created in the native mandible, adjacent to the area of bone hypoplasia, and the transport disc is advanced to create increased bone volume and new bone stock. As the transport segment is advanced, the leading edge becomes rounded and covered with a fibrocartilaginous cap. This is removed either surgically or mechanically to ensure osseous continuity. Otherwise, a pseudo-joint is created. This concept can be applied to create a neo-condyle from the ramus of the mandible, as has been shown in animal models and clinical settings.[23–33] A reverse L osteotomy is created, extending from the sigmoid notch, posterior to the lingual, with the horizontal limb approximately 1.5 cm above the gonial angle. The regenerate chamber for new bone synthesis is low, near the angle of the mandible (Figure 6). The segment is distracted to create a neo-condyle and elongate the ramus to re-establish posterior vertical height. This transport distraction technique can be applied to various clinical situations whereby creation of a new condyle is indicated, including: ankylosis, degenerative joint disease, arthritis, tumor resection, and idiopathic condylar resorption.

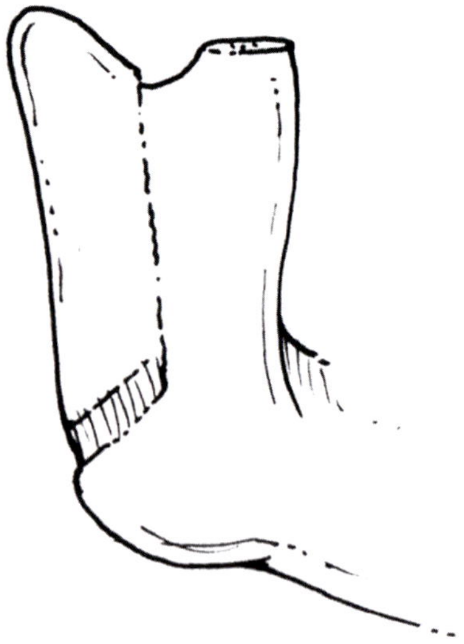

FIGURE 6 • An "Inverted L-Shaped" osteotomy is created to facilitate transport distraction of the posterior ramus. Care is taken to position the vertical cut behind the lingula and the horizontal cut with enough bone stock to allow for application of the DO device.

Surgical Technique

The mandible is approached as usual for orthognathic procedures. The sigmoid notch is visualized and using techniques similar as those for an intra-oral vertical ramus osteotomy, an inverted L osteotomy is created from the sigmoid notch to 1.0–1.5 cm above the angle of the mandible (Figures 6, 7, 8). Initially, the bone cut created is a corticotomy, with the horizontal limb of the bone cut, above the angle of the mandible a complete osteotomy. The width of the segment created by the L osteotomy should approximately that of the anterior-posterior dimension of the neo-condyle, without impinging on the lingua and inferior alveolar nerve.

The distractor is placed vertically, parallel with the posterior border of the mandible, to guide the neo-condyle into the fossa (Figure 8). Once the distraction device is placed, the corticotomy is converted to an osteotomy. The distractor is activated to verify impedance-free movement of the segment.

Distractors with removable foot plates are optimal, allowing for easy device placement and removal. During device placement, the foot plates are secured and the device is removed to allow for ease of conversion of the corticotomy to an osteotomy. As the vector of distraction is primarily vertical, devices with intra-oral activation limbs can be difficult to turn and activate the distractor. Conversely, distraction devices with extra oral activation ports are easy to activate and remove, yet can lead to minimal external scars that are well hidden in the submandibular region (Figure 9).

Surgical Considerations and Distraction Protocol

The distraction protocol follows the basic Ilizarov principles. A 5-day latency is observed to allow for the formation of an adequate callous. The rate of distraction is 1.0 mm per day, divided into two daily advancements of 0.5 mm per activation

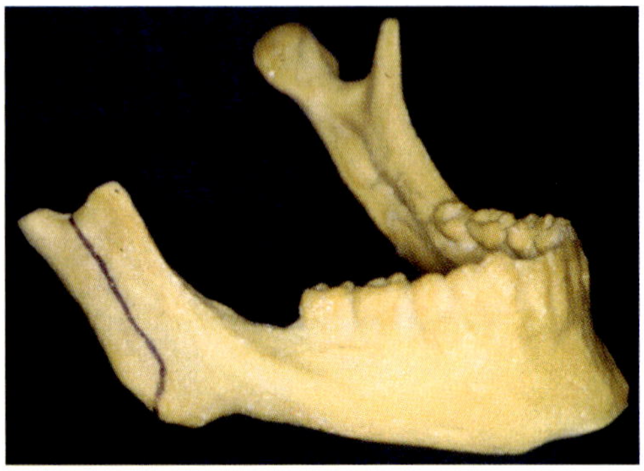

FIGURE 7 • At the time of the osteotomy, a coranoidectomy is performed to alleviate muscle pull from the temporalis.

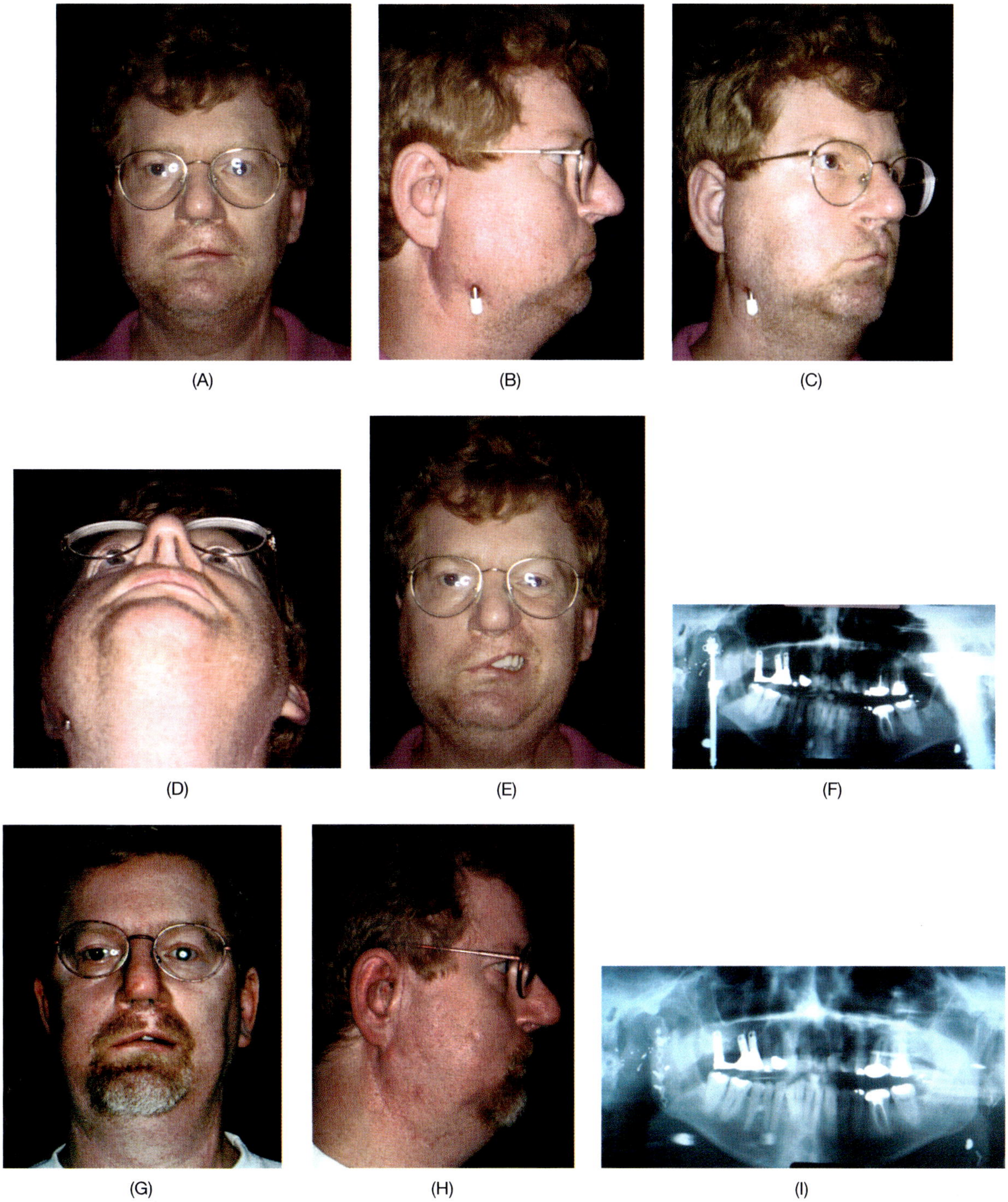

FIGURE 8 • This 31 year old male, sustained two gunshot wounds to the face/neck region resulting in an avulsion of the condylar neck and cranial nerve VII weakness. The distraction device was placed onto the mandibular ramus (A–D). The transport disc was then "closed" by compressing the intervening tissue thus allowing for reorientation of the condyle and directing it into the fossa during the distraction technique (F). The extra oral access site for the submerged device healed without incident (G, H). The condylar stump regenerated new bone from the distraction process (I).

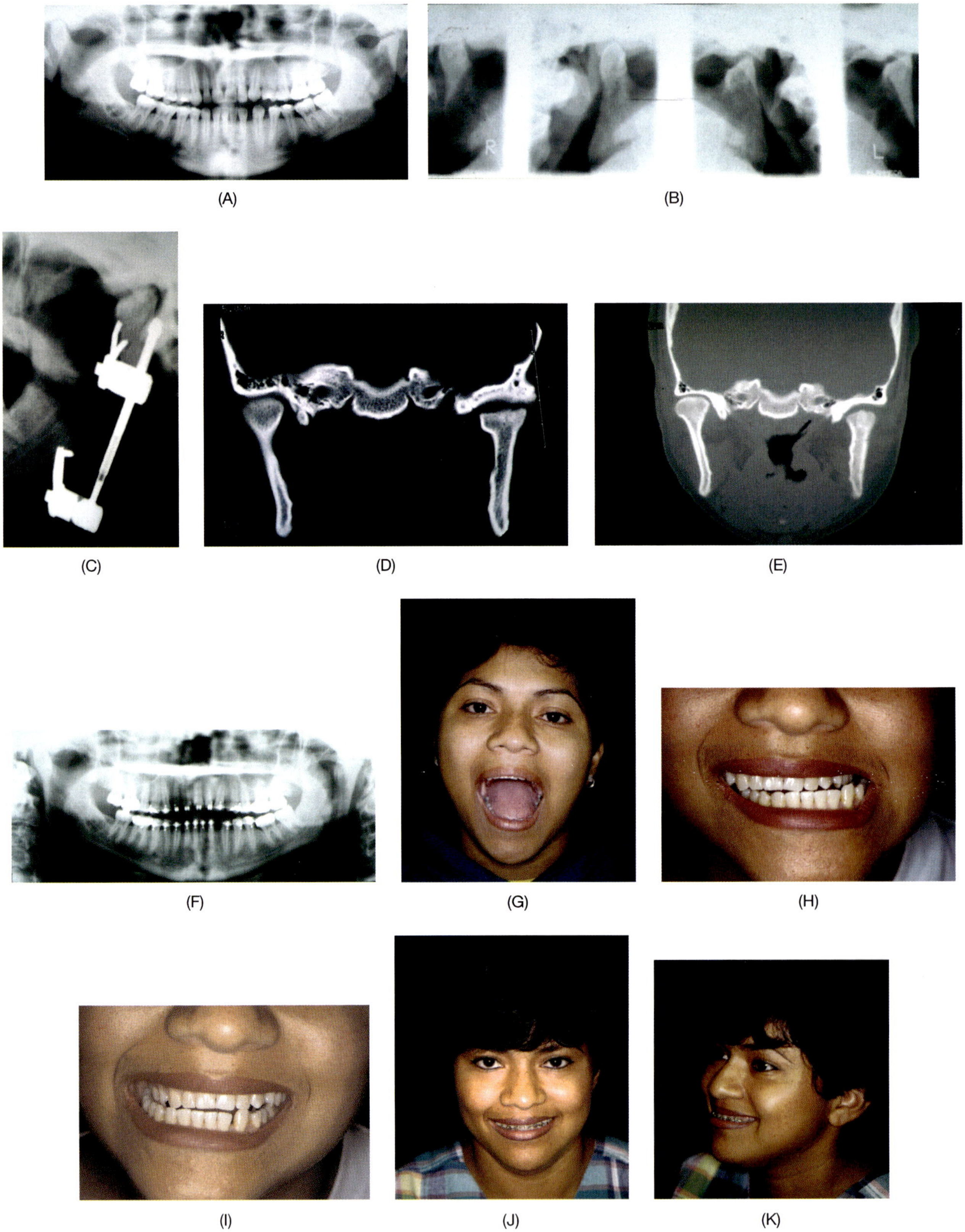

FIGURE 9 • A patient with Langerhans Histiocytosis that affected her left condyle (A, B, C) underwent unilateral extraoral distraction to recreate a neo-condyle (D) Post distraction the condylar head was reestablished, with a cortical outline (D, E, F). Her maximal opening and lateral excursive motions were maintained, with little scarring from the DO external device (G-K).

cycle. Should the patient note pre-auricular pain during the distraction process, the distraction protocol is modified to activate the distractor three or four times a day (0.3 mm TID or 0.25 mm QID) rather than the customary 0.5 mm BID. Neutral fixation is maintained several weeks until a cortical outline can be seen on the posterior border of the mandible. It is important to visualize the cortical outline of the posterior border radiographically (usually panoramic radiograph), regardless of the time required for ossification as predicted from age or amount of distraction regenerate formed. Premature device removal will cause vertical collapse of the under-ossified distraction regenerate secondary to local muscle pull including the masseter and medial pterygoid.

At the end of distraction, during the period of post-distraction fixation, the bony regenerate is at its weakest, for the new bone is incompletely ossified, and the distraction device is maximally activated and easily flexed. The temporalis can cause rotation of the distal segment, such that stripping of the muscle attachment at its insertion onto the coronoid or small coronoidectomy is advocated. For bilateral cases performed simultaneously, this maneuver is particularly important for local muscle-pull results in an open bite (Figure 10). Should this occur, the distraction device can be removed prior to complete ossification of the distraction gap and the "floating bone" concept or "molding the regenerate" utilized to close the open bite.[34] The mandible is rotated counter-clockwise and the proper

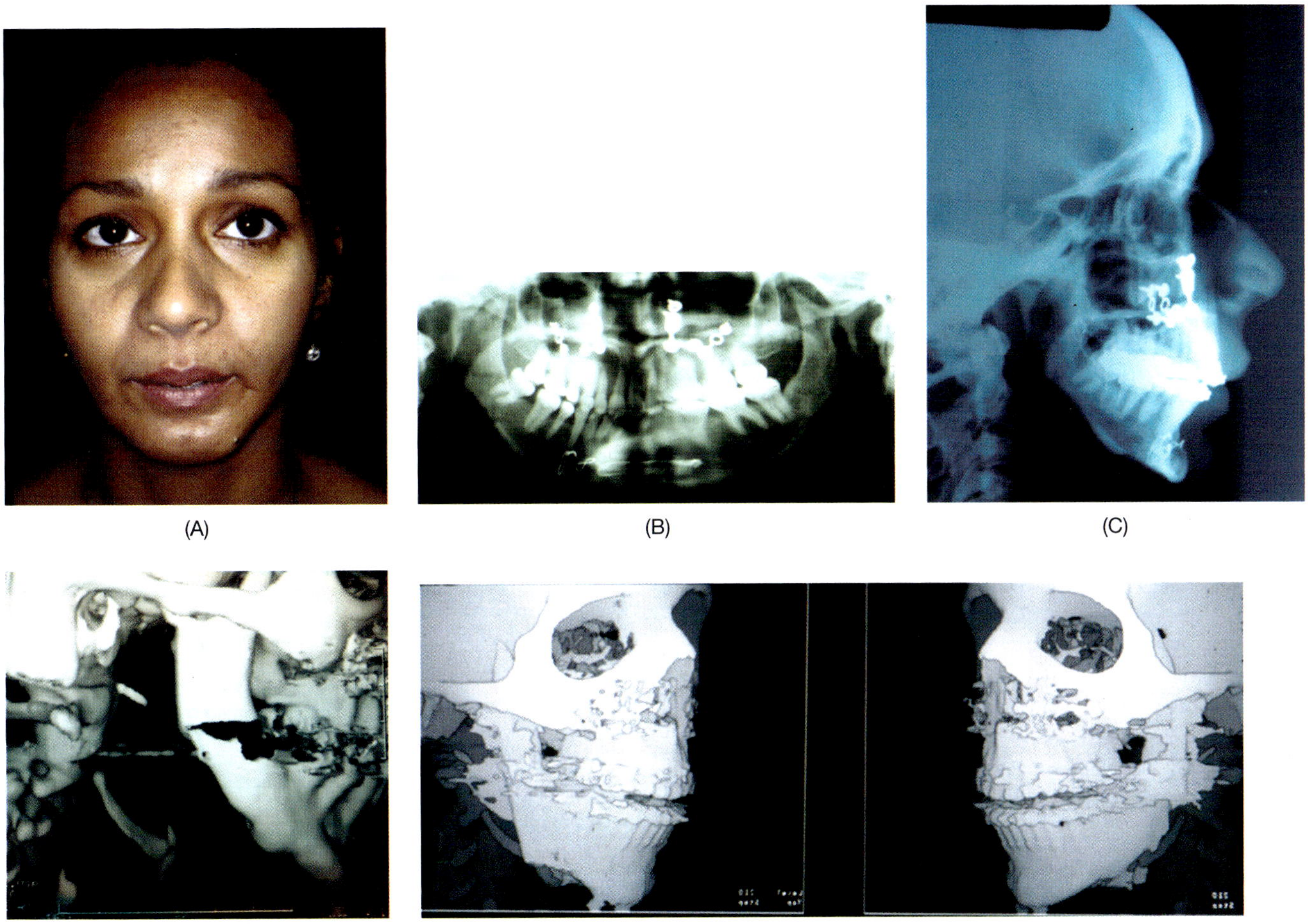

(A) (B) (C) (D) (E)

FIGURE 10 • A 38 year old female with idiopathic condylar resorption after Le Fort osteotomy, bilateral sagittal split osteotomies underwent bilateral transport DO to correct the resultant open bite and foreshortened ramus, and picnotic condyles (A–D). Bilateral inverted L osteotomies were performed and the DO device placed parallel to the posterior ramus (E, F). (*continued*)

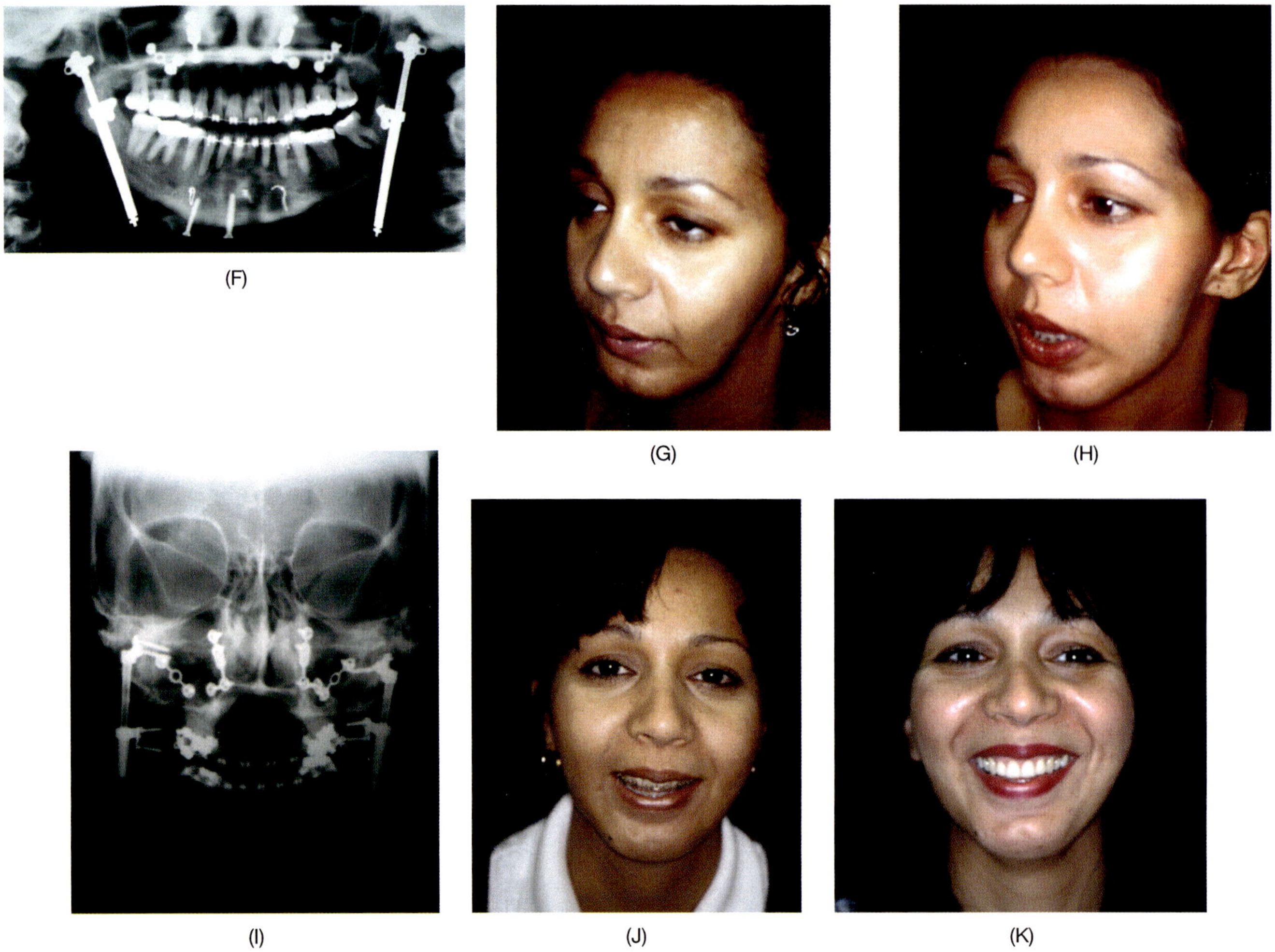

(F) (G) (H) (I) (J) (K)

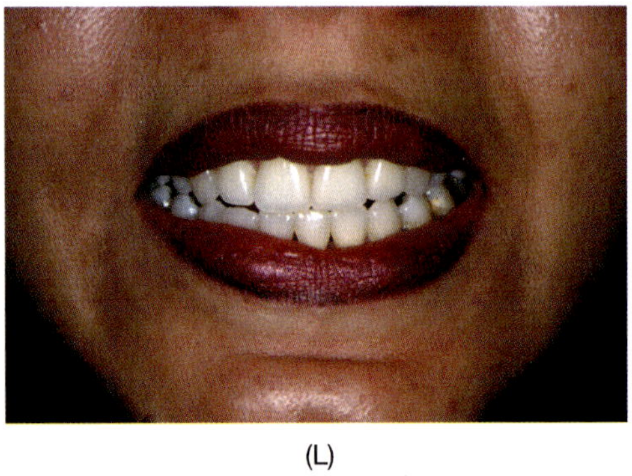

(L)

FIGURE 10 • *(continued from previous page)* During the DO process, the ramus was lengthened and the angle of mandible reestablished as the vertical deformity was corrected (G–J).

occlusion is established. Maxillomandibular Fixation is required, or heavy (6-ounce) elastics.

For bilateral vertical distraction of the mandible to correct the condylar resorption processes (idiopathic condylar resorption, rheumatoid arthritis), the ramus is purposefully over-distracted, creating an edge-to-edge, class III occlusion, and producing extra regenerate matrix that can be molded. The device is held in neutral fixation for 2 to 3 weeks and the device prior to complete ossification of the site. To mold the regenerate, 6-ounce orthodontic elastics are placed in inverted triangles bilaterally (three per side). A maxillary splint is inserted to help guide the mandible into the planned occlusion. Patients are instructed to wear the molding elastics in 4-hour intervals, allowing time for free mandibular movement (that is, 4 hours in elastics, 3 hours off) and monitored closely for complaints of pre-auricular discomfort or limitation of motion.

The use of a maxillary guide splint is inserted at the time of surgery for bilateral cases, as it allows for establishment of the proper occlusion and decreases the load on the neo-condyles. Light class II orthodontic elastics also help to decrease the load on the new articulation.[6] A soft, non-chew diet is recommended until distraction device removal and for 6 weeks thereafter, as the diet is slowly advanced to a soft mechanical and finally to a mechanical chew diet. At the time of device removal, a maxillary flat plane splint is inserted and patients are instructed in its nighttime use.

Regardless of etiology, all patients undergoing condylar transport DO are in active physical therapy during the entire DO process and are also instructed in at-home physical therapy exercises. Success of the transport DO technique to create a neo-condyle depends on mandibular motion.

Case Selection and Treatment Planning

Ankylosis

The concept of reconstruction of a neo-condyle using transport DO requires mandibular movement during DO. Therefore, joint mobility is essential for clinical success. Patients with bony ankylosis are treated in a combined approach: release of the bony ankylosis with concomitant distraction of the mandible. A gap arthroplasty is performed at the time of distraction surgery (Figure 11). The reverse L osteotomy is created in the residual ramus and the distractor applied. As the neo-condyle assumes the form of the fossa, it is important to surgically shape the new fossa well in all three planes of space during the gap arthroplasty portion of the procedure. Inadequate contouring of the fossa results in a neo-condyle that is flat yet functional and a patient that is pain-free masticating a regular diet.

Submerged distractors with external activation ports are ideal for this clinical setting. Submerged devices with intraoral activation ports can be difficult to activate as the

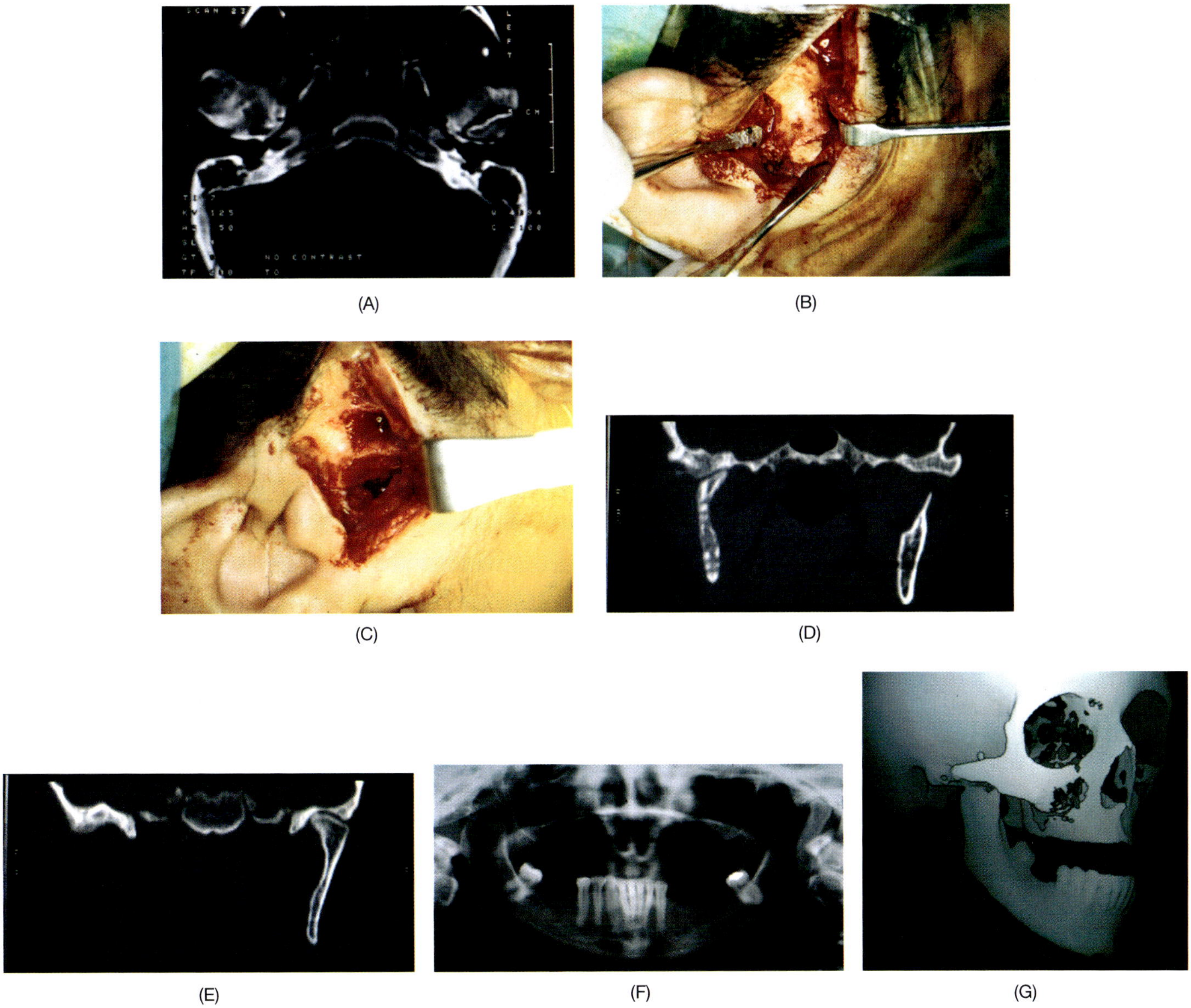

FIGURE 11 • This 43 year old patient underwent resection of her mandibular condyle with immediate chostochondral graft, which ossified outside of the fossa resulting in bony ankylosis (A, B). A gap arthroplasty was performed and unilateral TMJ transport DO was performed (C). Post-distraction a neo-condyle was created (D, E). Although the patient did well, maintaining a normal diet, the five year post-operative radiographs revealed remodeling of the fossa (F, G) presumed to be secondary to the flat fossa created during the gap Arthroplasty (C). Now care is taken to create a round fossa during ankylosis release to alleviate this problem.

activation rod may have as much as a 135-degree bend, to exit in the buccal vestibule. Additionally, the vertical trajectory of the device may be compromised, forcing the clinician to place the device more obliquely to allow for placement of the activation port in the buccal sulcus. Submerged devices with external ports can be placed directly on the bone with a vertical trajectory. Post-surgically, minimal scarring is noted as the device exits low in the submandibular region.

Distraction is initiated after a 5-day latency period and proceeds until the transport disc reaches the glenoid fossa. Patients will remark that they can feel pressure as the articulation of the condyle and fossa is re-established. This is confirmed radiographically. The DO device is held in neutral fixation until a cortical outline can be seen in the regenerate chamber near the angle of mandible, when the device is removed (Figure 12). A flat plane maxillary splint is inserted at the time of device removal and patients are instructed to maintain soft diets for 6 weeks after device removal.

Some have reported a staged approach: ankylosis release then DO secondarily, delayed.[35] This concept is a good treatment alternative, yet active movement and physical therapy is suggested during phase I and phase II surgery. It is not recommended that the distraction process occur first, in an effort to regain the posterior vertical bone height, with a secondary ankylosis release.

Unilateral Degenerative Joint Disease/Arthritis

Patients who have extensive resorptive and arthritic involvement of the mandible with loss of posterior vertical height of the mandible should be considered for vertical distraction of the mandible. Patients are managed in a sequential approach. In phase I, treat the local arthritic involvement of the condyle. In phase II, reconstruct the condyle/mandible using vertical

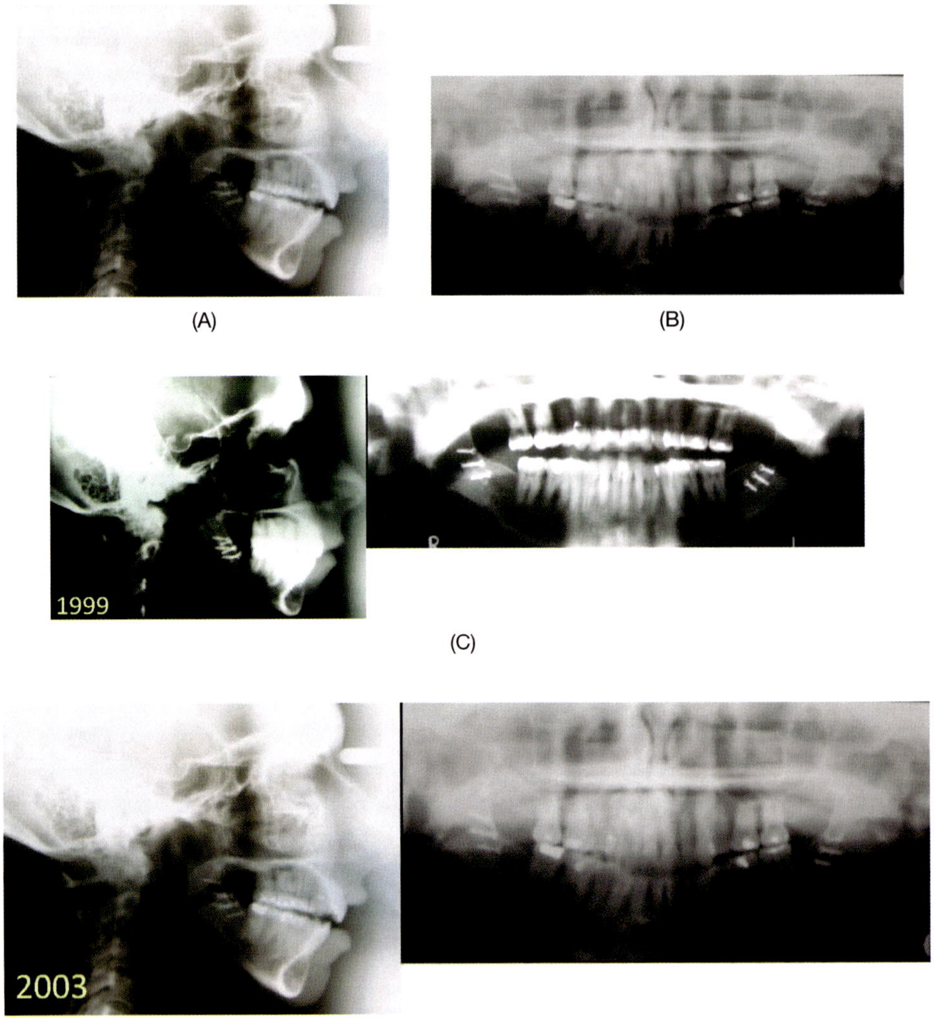

(A) (B) (C) (D)

FIGURE 12 • This 26 year old female with mandibular hypoplasia, prior to orthognathic surgery (A). Note the high mandibular plane angle and posteriorly inclined mandibular condyles (B). In 1999 she underwent bilateral sagittal spit osteotomy with mandibular advancement (C). By 2003 she had "relapse" as the condyles resorbed (D). *(continued)*

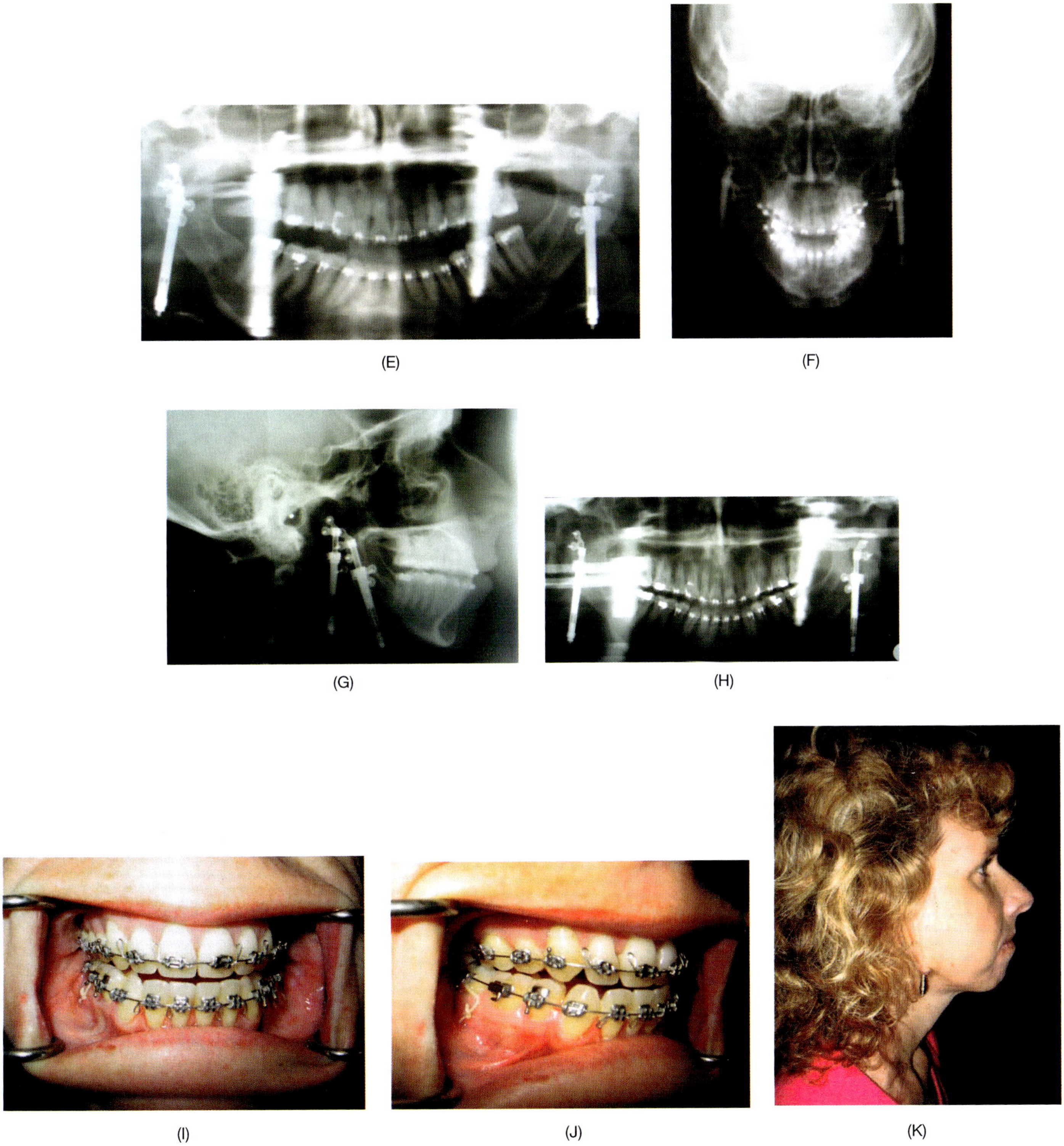

FIGURE 12 • (*continued from previous page*) She presented for treatment of this condition, and bilateral vertical distraction osteogenesis of the condyle was performed (E). The trajectory of the distraction devices are similar in the sagittal orientation (F), yet different in that lateral orientation, to accommodate the difference in the condylar fossa orientation versus the base of skull (G). The posterior ramal height was over corrected creating an edge to edge occlusion (H, I, J, K, L, M). (*continued*)

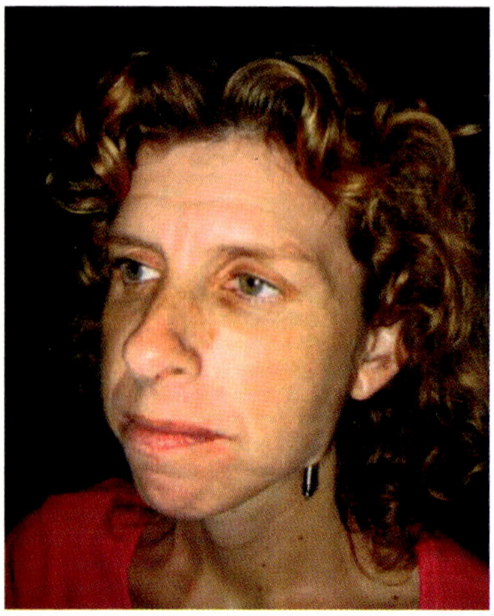

(L)

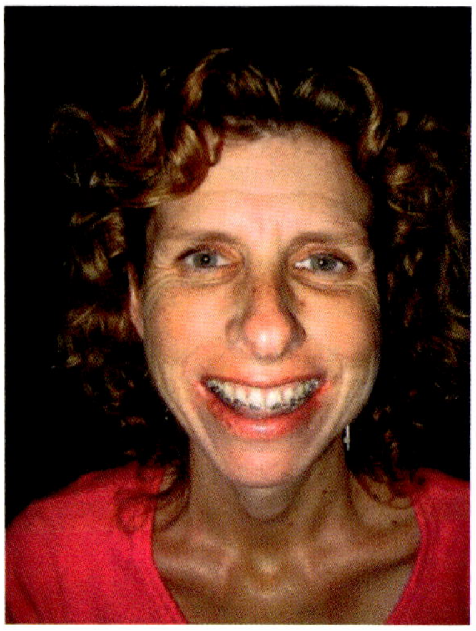

(M)

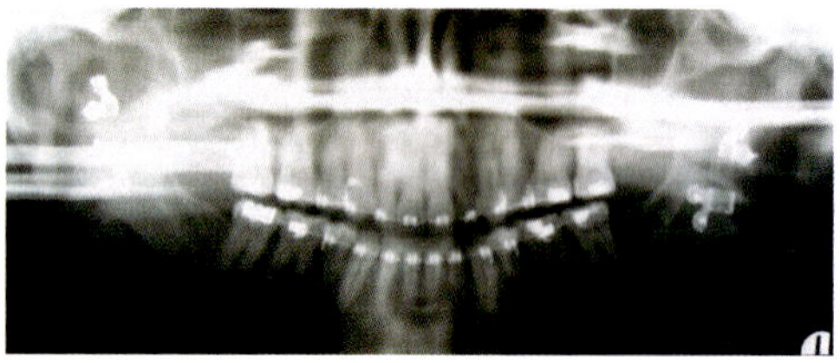

(N)

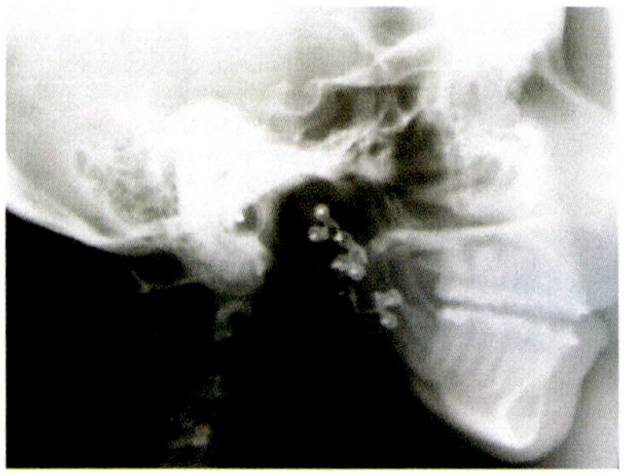

(O)

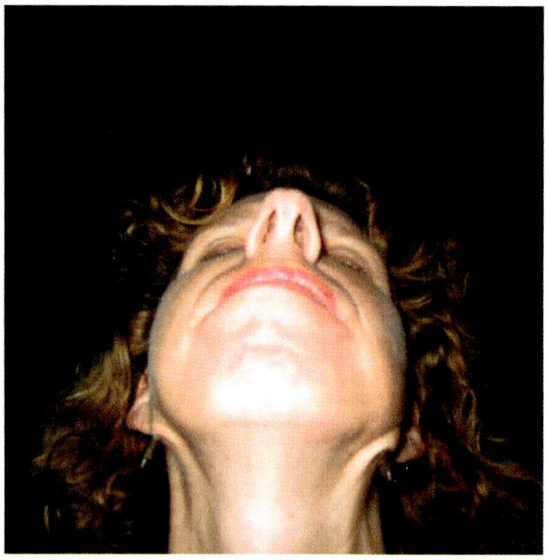

(P)

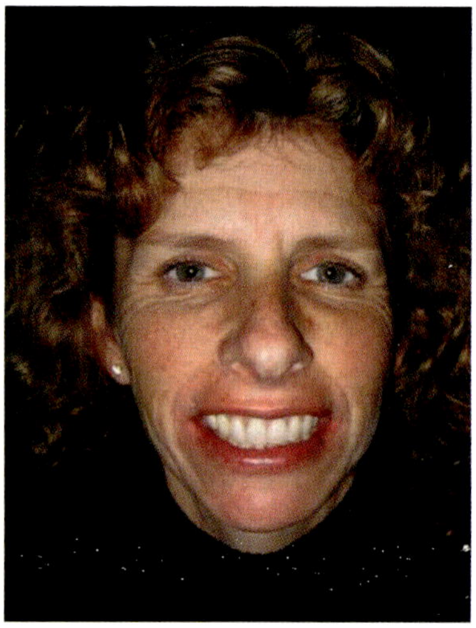

(Q)

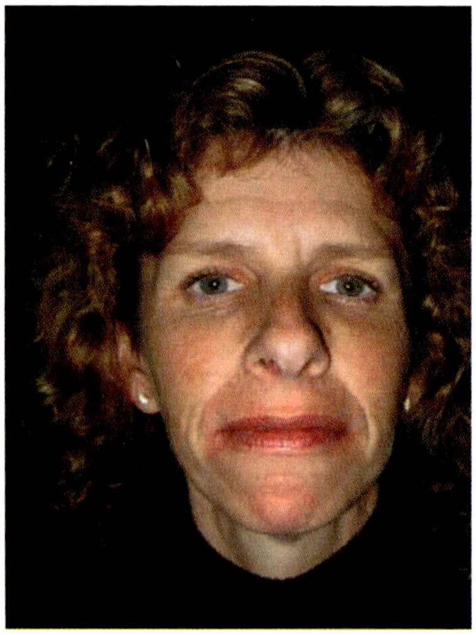

(R)

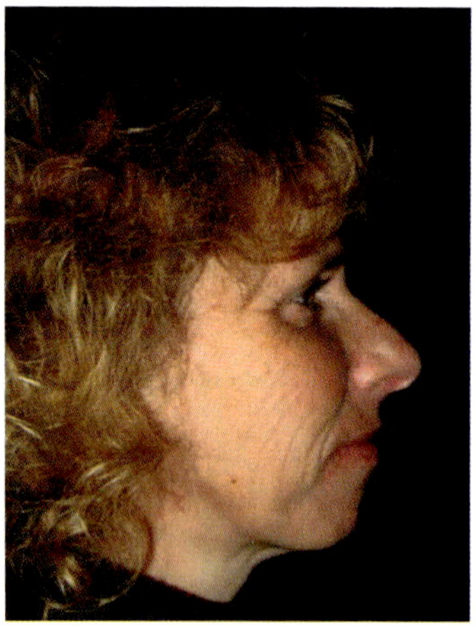

(S)

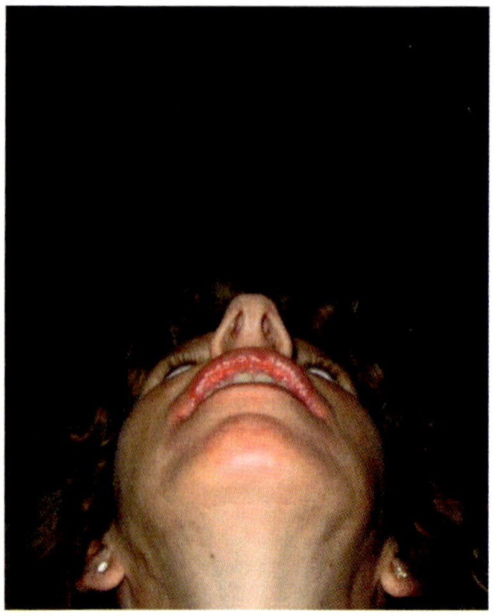

(T)

FIGURE 12 • (*continued from previous page*) Once the distraction devices were removed, elastic traction forces allowed the occlusion to close into Class I. (N–Y *all of the remaining figures*). (*continued*)

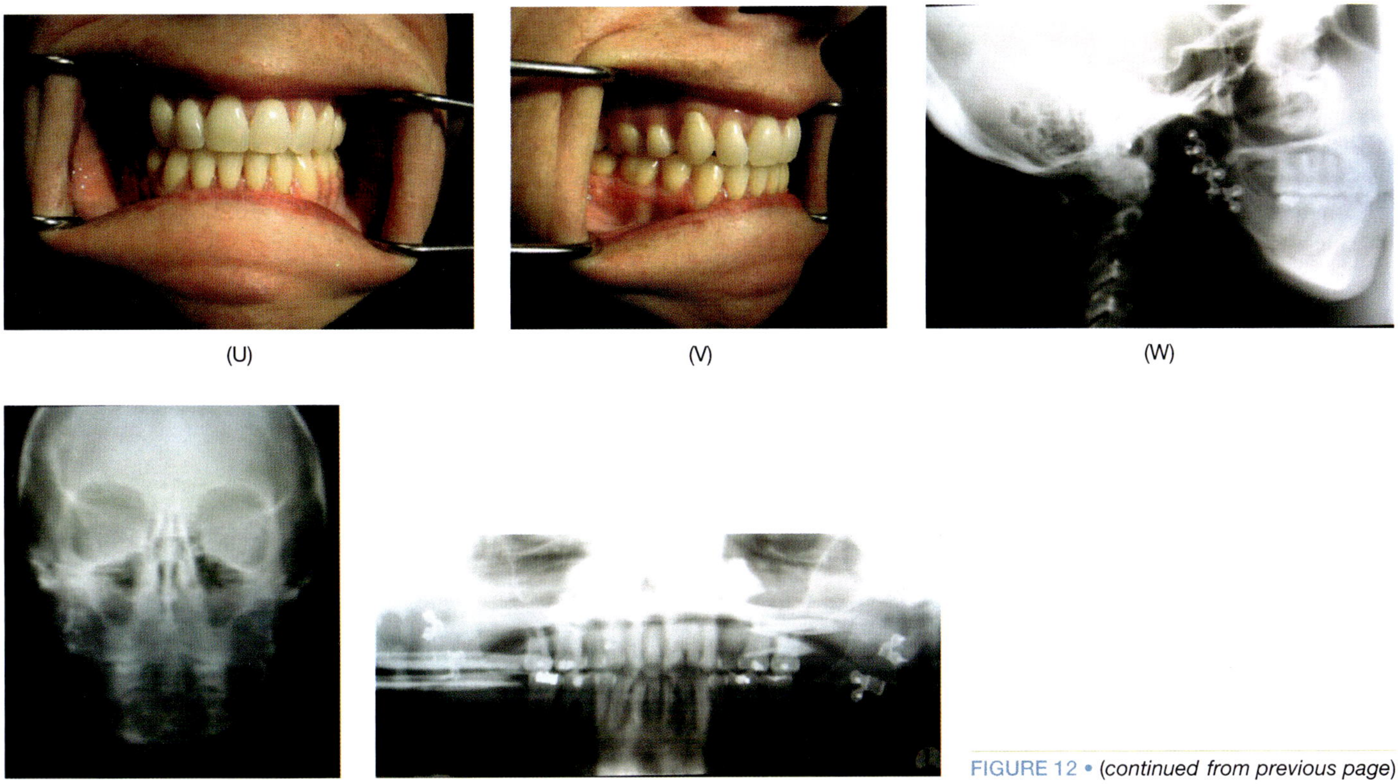

(U) (V) (W)

(X) (Y)

FIGURE 12 • *(continued from previous page)*

distraction of the ramus once the resorptive/arthritic process has stabilized, as evidenced by physical findings: serial radiographs, stabilization of the malocclusion, and Technetium-99 bone scans. Occasionally, if fibrous ankylosis has occurred as a consequence of the disease process, TMJ arthroscopy may be indicated. This can be performed as staged procedures for bilateral arthritic involvement, or simultaneously as the distraction procedure of the non-arthritic side, if the non-arthritic side has undergone compensatory changes as a consequence of mechanical overload from the arthritic side.

The distraction process is utilized to vertically elongate the ramus, and recreate the decreased posterior height of the mandible, including re-establishment of a gonial angle lost as a consequence of the degenerative joint disease process. Vertical elongation of the mandible appears to be stable long term, in contrast to conventional orthognathic surgical elongation of the mandible, which as been found to have relapse.[36,37] Ramal elongation using conventional orthognathic techniques include the use of interpositional bone grafts or other biologic materials to aid in healing and maintenance of the new bone position. These grafting materials require biologic replacement over time, which in this circumstance are under compressive load as a consequence of the stretched masseteric sling, leading to relapse. It appears that the slow application of force over time as afforded by the distraction process induces metaplasia of the muscle sling, in contrast to orthognathic surgery, which immediately repositions the bony segments maxillary stretching two major muscles of the muscles of mastication: masseter and medial pterygoid. Additionally, the distraction gap is held in neutral fixation until the distraction gap is filled with regenerate bone improving long-term stability.

Patients with degenerative joint disease or arthritis often have compensatory changes at the level of the maxilla; these can be treated at the same surgical setting. For patients requiring Le Fort I repositioning, the maxilla is approached first. An interim splint is utilized and the maxilla held in position using rigid fixation. The mandible is then approached. The reverse L osteotomy is created and the distractor applied, either unilaterally or bilaterally. Class II orthodontic elastics are placed during active distraction. During the consolidation phase, the elastic pattern is modified to box elastics.

For patients with unilateral condylar involvement and compensatory maxillary occlusal cants, Le Fort I distraction can be performed at the time of unilateral mandibular distraction. The maxilla is approached as conventional Le Fort I surgery. Corticotomies are performed with complete osteotomies of the pterygoid plate of the affected side (the contra-lateral side acts as a "hinge"), the nasal septum and lateral nasal wall. The maxilla is mobilized but not down-fractured. Box elastics (6-oz.) are placed one on each side at the level of the bicuspid-molars at the time of surgery. The unilateral mandible is treated as described for vertical elongation of the ramus using DO. After a 5-day latency, the distraction is initiated as per protocol. The orthodontic elastics are modified: two Class II elastics on the affected side and one Class II elastic on the contra-lateral side. At night, the elastic force is increased to two elastics on the contralateral side and four elastics on the distracted side. The maxilla is distracted with the mandible. Once distraction is achieved, the elastic pattern is modified to two box elastics per side.

Bilateral Degenerative Joint Disease, Arthritis Condylar Resorption

Patients with condylar resorption (idiopathic, degenerative, rheumatoid arthritis) have loss of posterior vertical height and an anterior open bite as a consequence of the resorptive process of the condyle. Here, the slow application of force over time via DO to generate new bone is ideally suited. Using the same osteotomy, the ramus is over-distracted, creating an edge-to-edge/Class III profile (Figure 13). The distractors are removed early, after 3 weeks of osseous consolidation and the regenerate is molded by using orthodontic elastics (6-oz.) in an alternating triangular pattern to rotate the mandible counterclockwise and close the open bite. Elastic maxillomandibular fixation is not required to mold the regenerate. A maxillary splint is inserted to decrease the load on the temporomandibular joints.

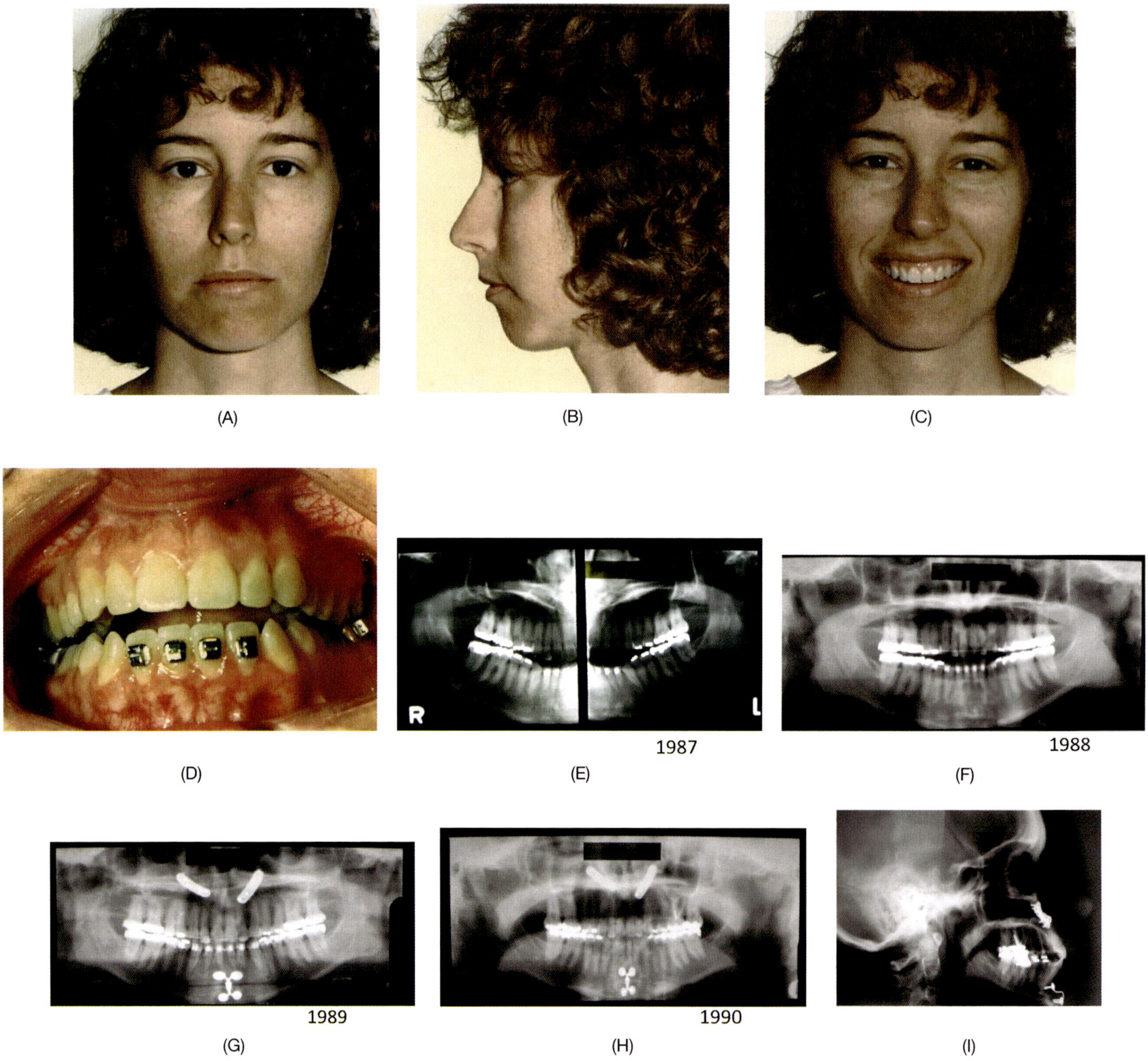

FIGURE 13 • This 21 year old female presented to her orthodontist with an open bite deformity (A, B, C, D). Post-orthodontic therapy she experienced an open bite (F). She underwent LeFort I osteotomy and genioplasty by another surgeon (G). The one year post-operative radiographs reveal the segments to be stable (H, I, J). *(continued)*

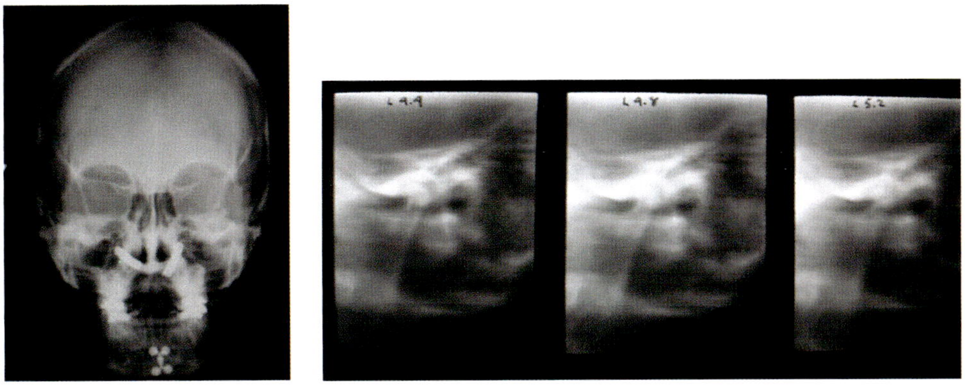

(J) (K)

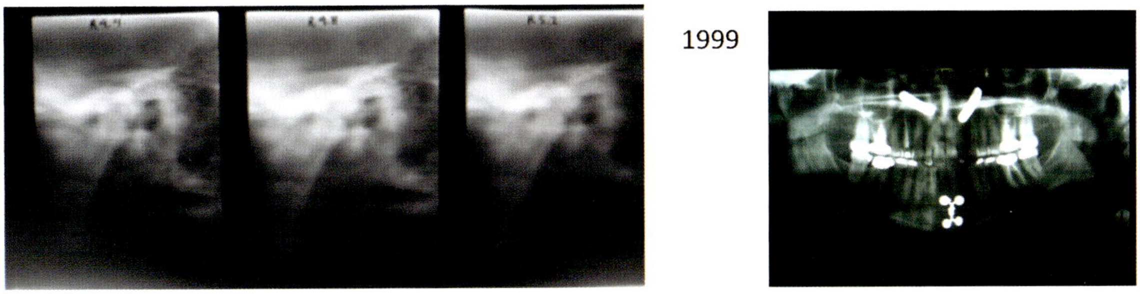

1999

(L) (M)

2000

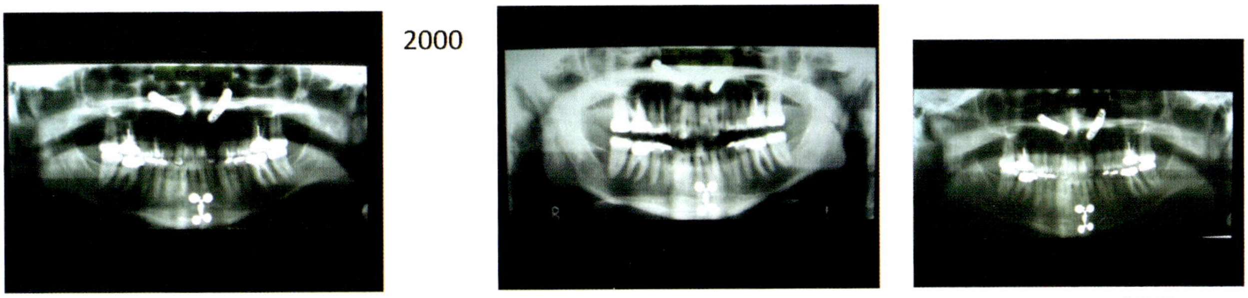

2001 2002

(N) (O) (P)

FIGURE 13 • (*continued from previous page*) Within 10 years, this patient presented with difficulty with mouth opening, yet pain free. The condyles were noted to be picnotic (K, L, M,). For the next 3 years, she was managed non-surgically with physical therapy (N, O, P, Q). (*continued*)

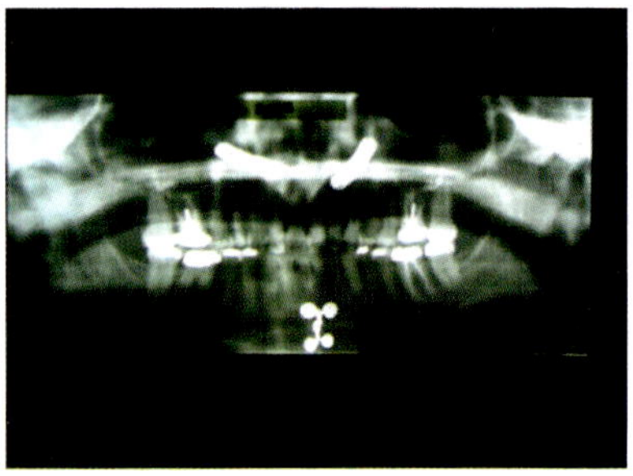

2003

(Q)

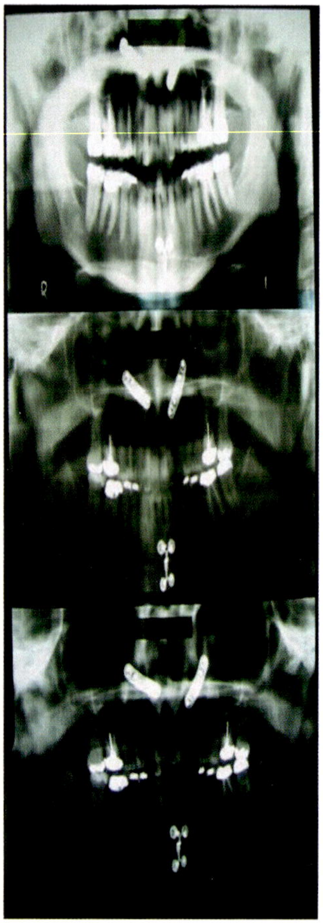

(R)

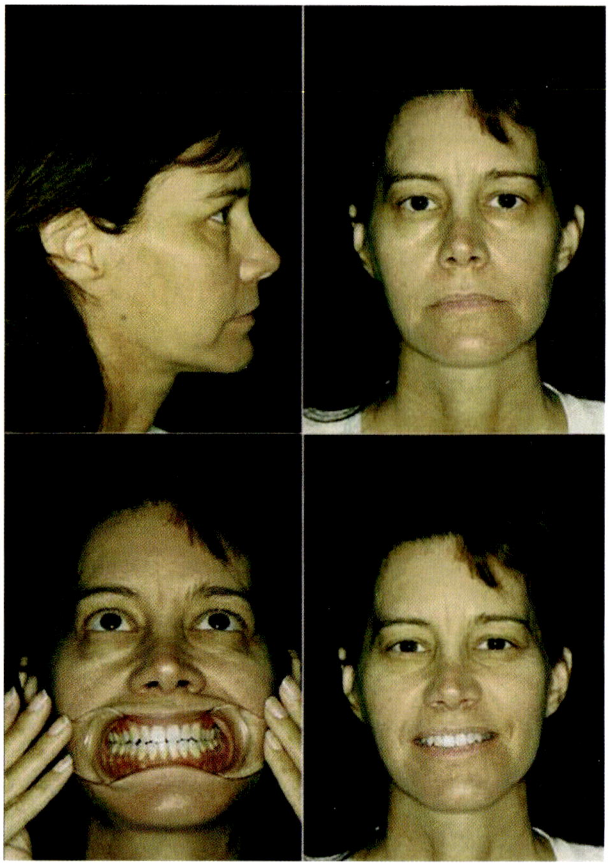

(S)

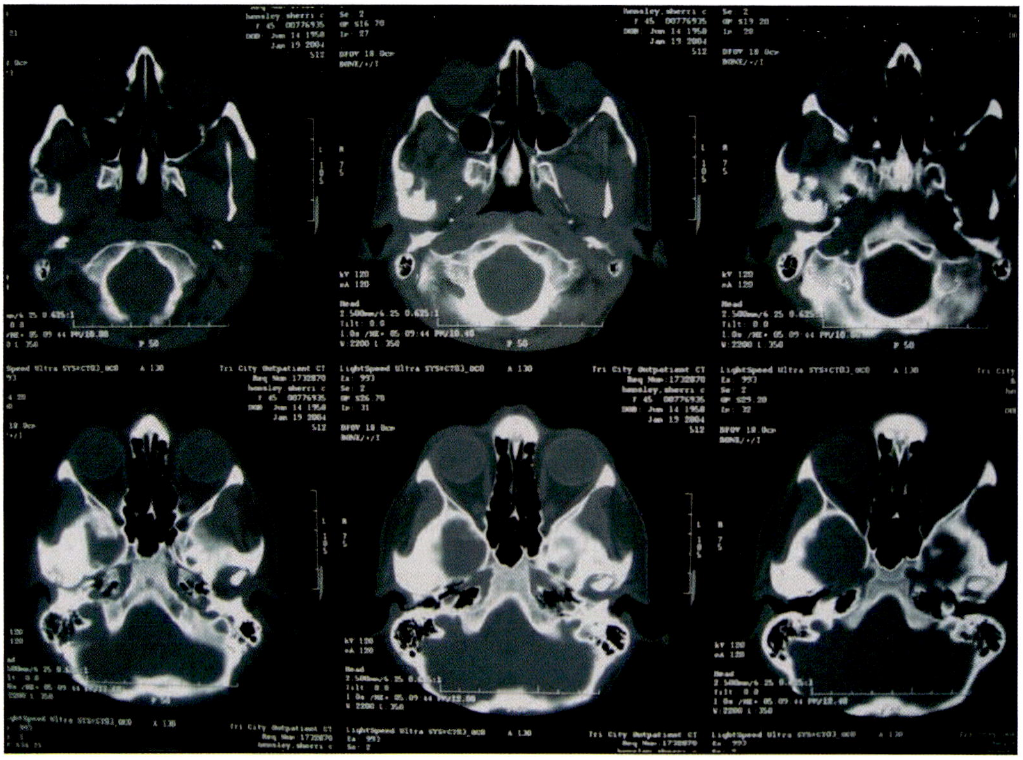

(T)

FIGURE 13 • (*continued from previous page*) By four years post-orthognathic surgery, she was noted to have exuberant bone formation of the right mandibular condyle with severe limited mouth opening (R, S, T, U). (*continued*)

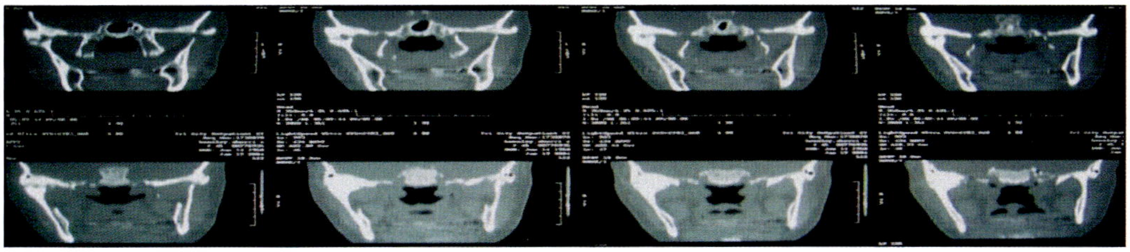

(U)

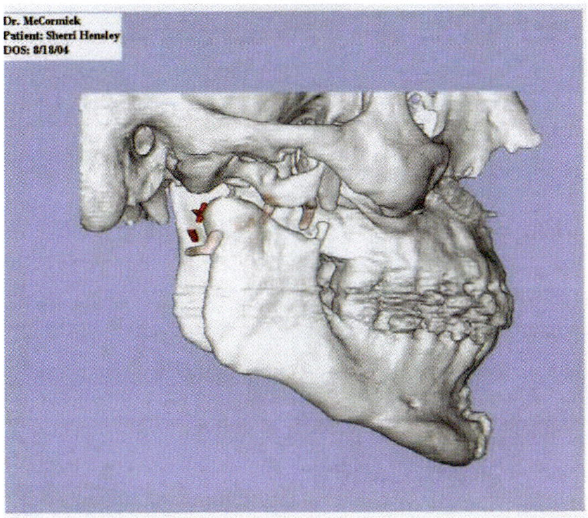

(V)

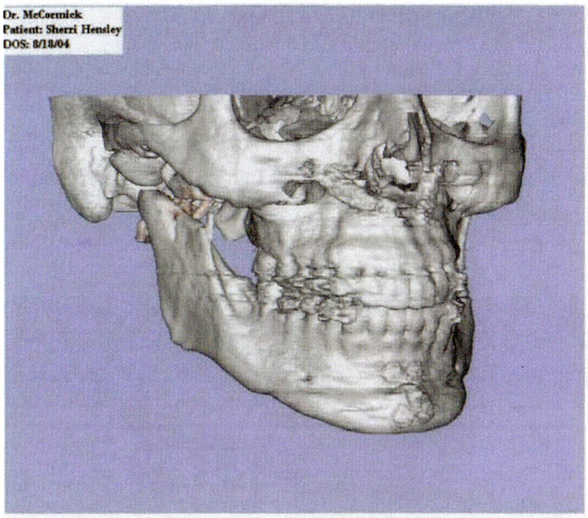

(W)

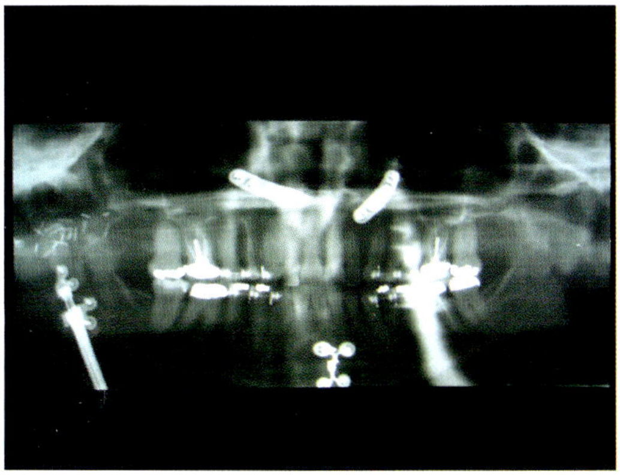

(X)

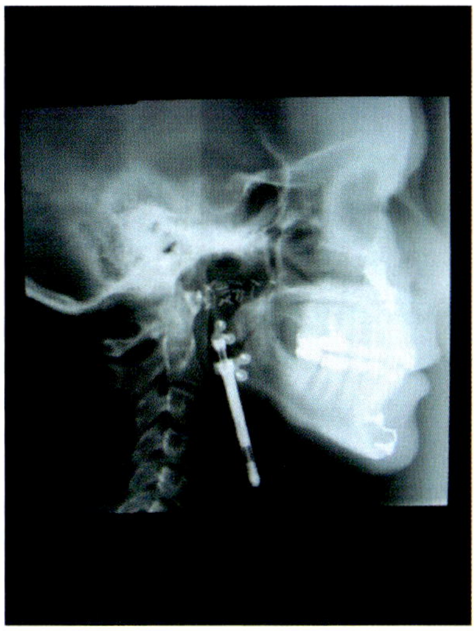

(Y)

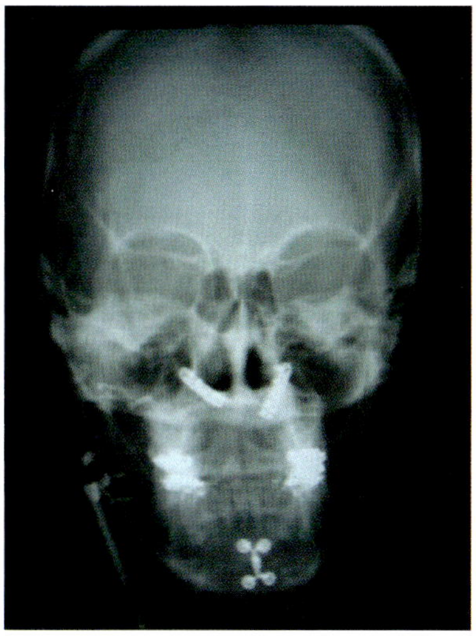

(Z)

FIGURE 13 • (*continued from previous page*) She underwent a block resection with consideration of a custom total joint replacement or ramal DO (V, W). (*continued*)

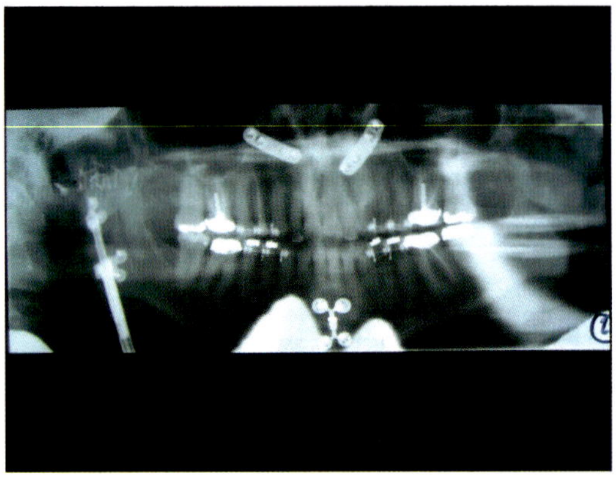

(Z-a)

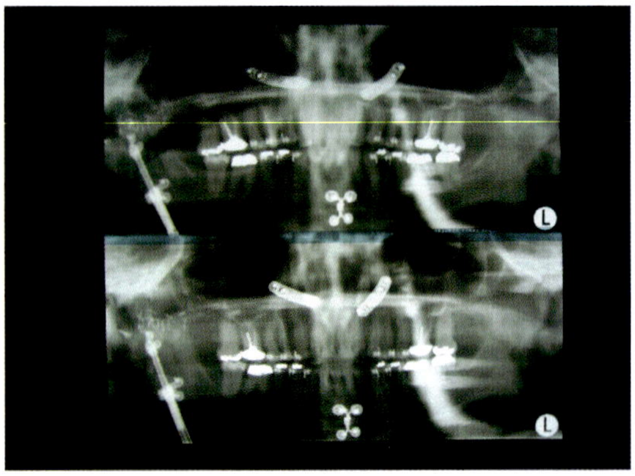

(Z-b)

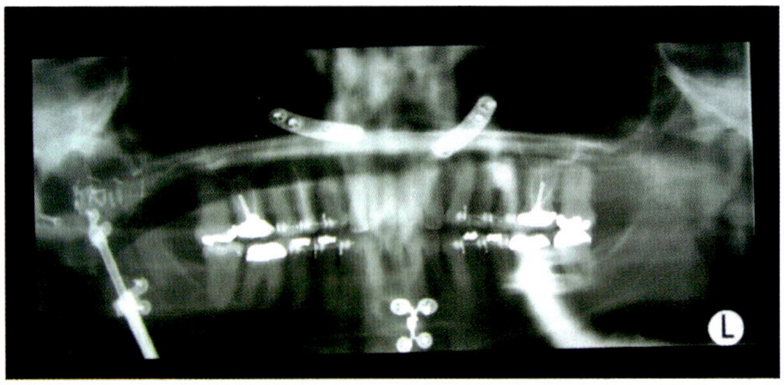

(Z-c)

FIGURE 13 • (*continued from previous page*) Unilateral Ramus distraction osteogenesis was performed (X, Y, Z). The segment was transported until the neo-condyle was seated in the fossa (Za, Zb, Zc)

Additionally, patients are instructed to wear the molding elastics in intervals allowing other time intervals for free mandibular movement (that is, 4 hours in elastics, 3 hours off). Patients are monitored very closely and the regenerate molding protocol is adjusted, should the patient complain of pre-auricular discomfort or limitation of motion. Interestingly, as a consequence of distraction, the condyle appeared radiographically to increase in size and volume, and maintained this increased dimension longer term.

CONCLUSION

Transport distraction is a powerful tool to re-create the maxilla, mandible, and temporomandibular joint. Using small incisions and tunneling techniques, significant bone grafts can be avoided. Additionally, transport DO has the advantage of re-creating all tissues: bone, nerve, muscle, skin, and mucosa.

References

1. Ilizarov GA. The principles of the Ilizarov method. Bull Hosp Jt Dis Orthop Inst 1988;48:1–5.
2. Aronson J. Biological and clinical evaluation of distraction histogenesis. Clin Orthop Relat Res 1994;301:2–3.
3. Costantino PD, Friedman CD. Distraction osteogenesis. Applications for mandibular regrowth. Otolaryngol Clin North Am 1991; 24:1433–43.
4. Stucki-McCormick SU, Fox R, Mizrahi R. Transport distraction: mandibular reconstruction. Atlas Oral Maxillofac Surg Clin North Am 1999;7:65–83.
5. McCarthy JG, Stelnicki EJ, Grayson BH. Distraction osteogenesis of the mandible: a ten-year experience. Semin Orthod 1999;5:3–8.
6. Guerrero CA, Bell WH, Contasti GI, et al. Intraoral mandibular distraction osteogenesis. Semin Orthod 1999;5:35–40.
7. Molina F, Ortiz Monasterio F. Mandibular elongation and remodeling by distraction: a farewell to major osteotomies. Plast Reconstr Surg 1995;96:825–40.
8. Sacco AG, Chepeha DB. Current status of transport-disc-distraction osteogenesis for mandibular reconstruction. Lancet Oncol 2007;8:323–30.
9. Kuriakose MA, Shnayder Y, DeLacure MD. Reconstruction of segmental mandibular defects by distraction osteogenesis for mandibular reconstruction. Head Neck 2003;25:816–24.
10. Herford AS. Use of a plate-guided distraction device for transport distraction osteogenesis of the mandible. J Oral Maxillofac Surg 2004;62:412–20.
11. Samchukov ML, Cope JB, Cherkashin AM. Basic principles of bone transport. In: Samchukov ML, Cope JB, Cherkashin AM editors. Craniofacial Distraction Osteogenesis. St Louis, Missouri: Mosby Inc; 2001. pp. 349–57.

12. Zhou HZ, Hu M, Hu KJ, Yao J, Liu YP. Transport distraction osteogenesis using nitinol spring: an exploration in canine mandible. J Craniofac Surg 2006;17:943–9.
13. Hibi H, Ueda M. New internal transport distraction device for reconstructing segmental defects of the mandible. Br J Oral Maxillofac Surg 2006;44:382–5.
14. Rachmiel A, Aizenbud D, Peled M. Enhancement of bone formation by bone morphogenetic protein-2 during alveolar distraction: an experimental study in sheep. J Periodontol 2004;75:1524–31.
15. Stucki-McCormick SU. Long term effects of TMJ distraction osteogenesis. Proceedings of the Fifth International Congress on Cranial and Facial Bone Distraction Processes; Paris, France 2006, June 21–24.
16. Guerrero C. Mandibular reconstruction using distraction osteogenesis. Proceedings of the Fifth International Congress on Cranial and Facial Bone Distraction Processes, Paris, France 2006, June 21–24.
17. Chin M. Distraction osteogenesis for dental implants. Atlas Oral Maxillofac Surg Clin North Am 1999;7:41–46.
18. Jensen OT. Distraction osteogenesis and its use with dental implants. Dent Implantol Update 1999;10:33–6.
19. Eski M, Turegun M, Deveci M, et al. Vertical distraction osteogenesis of fibular bone flap in reconstructed mandible. Ann Plast Surg 2006;57:631–6.
20. Schleier P, Hyckel P, Fried W, et al. Vertical distraction of fibula transplant in a case of mandibular defect caused by shotgun injury. Int J Oral Maxillofac Surg 2006;35:861–4.
21. Ortakoglu K, Suer BT, Ozyigit A, et al. Vertical distraction osteogenesis of fibula transplant for mandibular reconstruction: a case report. Oral Surg Oral Med Oral Pathol Oral Radiol Endod 2006;102:e8–11.
22. Rodriguez ED, Martin M, Bluebond-Langner R, et al. Multiplanar distraction osteogenesis of fibula free flaps used for secondary reconstruction of traumatic maxillary defects. J Craniofac Surg 2006;17:883–8.
23. Hikiji H, Takato T, Matsumoto S, et al. Experimental study of reconstruction of the temporomandibular joint using a bone transport technique. J Oral Maxillofac Surg 2000;58:1270–6.
24. Arun T, Kayhan F, Kiziltan M. Treatment of condylar hypoplasia with distraction osteogenesis: a case report. Angle Orthod 2002;72:371–6.
25. Cheung LK, Lo J. The long-term effect of transport distraction in the management of temporomandibular joint ankylosis. Plast Reconstr Surg 2007;119:1003–9.
26. Cavaliere CM, Buchman SR. Mandibular distraction in the absence of an ascending ramus and condyle. J Craniofac Surg 2002;13:527–32.
27. Piero C, Alessandro A, Giorgio S, et al. Combined surgical therapy of temporomandibular joint ankylosis and secondary deformity using intraoral distraction. J Craniofac Surg 2002;13:401–9.
28. Papageorge MB, Apostolidis C. Simultaneous mandibular distraction and arthroplasty in a patient with temporomandibular joint ankylosis and mandibular hypoplasia. J Oral Maxillofac Surg 1999;57:328–33.
29. Rao K, Kumar S, Kumar V, et al. The role of simultaneous gap arthroplasty and distraction osteogenesis in the management of temporo-mandibular joint ankylosis with mandibular deformity in children. J Craniomaxillofac Surg 2004;32:38–42.
30. Dean A, Alamillos F. Mandibular distraction in temporomandibular joint ankylosis. Plast Reconstr Surg 1999;104:2021–31.
31. Stucki-McCormick SU. Reconstruction of the mandibular condyle using transport distraction osteogenesis. J Craniofac Surg 1997; 8:48–52.
32. Stucki-McCormick SU, Winick R, Winick A. Distraction osteogenesis for the reconstruction of the temporomandibular joint. NY State Dent J 1998;64:36–41.
33. Kaban LB. Acquired abnormalities of the temporomandibular mandibular joint. In: Kaban LB, Troulis MJ, editors. Pediatric Oral and Maxillofacial Surgery. Philadelphia: Elsevier Science; 2004. pp. 340–76.
34. Hoffmeister B, Marcks CH, Wolff KP. The floating bone concept in intraoral distraction. J Craniomaxillofac Surg 1998;26(suppl 1): 76–81.
35. Sayan NB, Karasu HA, Uyanik LO, et al. Two-stage treatment of TMJ ankylosis by early surgical approach and distraction osteogenesis. J Craniofac Surg 2007;18:212–7.
36. De Clercq CA, Neyt LF, Mommaerts MY, et al. Condylar resorption in orthognathic surgery: a retrospective study. Int J Adult Orthodon Orthognath Surg 1994;9:233–40.
37. Hwang SJ, Haers PE, Seifert B, et al. Non-surgical risk factors for condylar resorption after orthognathic surgery. J Craniomaxillofac Surg 2004;32:103–11.

16 Alveolar Distraction Osteogenesis

Christopher J. Haggerty, DDS, MD / Michael S. Block, DMD

OUTLINE

ALVEOLAR BONE DISTRACTION: RATIONALE AND NEED

Loss of mandibular or maxillary bone, whether alveolar or basal, frequently complicates prosthetic rehabilitation including dental implant placement. Common etiologies of bone loss include acquired defects secondary to severe periodontal disease, abscess secondary to caries, trauma, tumor removal, and other such disorders.[1,2] Bone loss can include horizontal and vertical dimension, isolated, or in combination. Options to increase the quantity of bone include autogenous bone grafts in particulate or solid form, guided bone regeneration with particulate bone replacement materials, and interpositional osteotomies to acute correction of the bone defect, and the use of distraction osteogenesis to augment deficient soft and hard tissue.

Onlay autogenous bone grafts lose 25% to 33% of vertical height in the first six months after placement.[3,4] Wang and colleagues[5] reported that 89% of autogenous bone grafts from the ilium or the rib were found to have resorbed over 3 years.[5,6] Verhoeven and associates[7] reported that iliac crest onlay grafts with implant placement had a 36% decrease in mandibular bone height in 877 days. Bell and colleagues reported 33% bone loss of autogenous corticocancellous bone grafts in atrophic mandibles in 4 to 6 months.[4] Guided bone regeneration using membranes over particulate grafts works well for small bony defects,[8] but has not been shown to be predictable for larger defects.[3]

Visor osteotomies with interpositional bone grafting involves splitting of the alveolar bone either in the sagittal or axial planes, with immediate movement of the mobilized segment to achieve bone augmentation. In the past, these procedures were performed to increase the denture-bearing area. The visor osteotomy was found to result in excellent bone height improvement, but the width of the ridge was not satisfactory in the posterior quadrants. In addition, the visor osteotomy in the posterior mandible also resulted in a high incidence of damage to the inferior alveolar nerve.[9]

Because the width of the anterior mandible was not sufficient after a sagittal visor osteotomy, the technique was modified from an anterior sagittal split visor osteotomy to a horizontal interpositional osteotomy, with the inferior bone cut below the insertion of the attached lingual mucosa. This osteotomy, in combination with a bonegraft, resulted in stable augmentation of the anterior mandible.

In the maxilla, a Lefort down-fracture, with an interposition bone graft, may result in augmentation of the vertical height of the alveolus; however, graft failure and the morbidity of the procedure has decreased the incidence of the procedure over time. The etiology of tissue breakdown is attributed to the lack of resilient soft tissue. Based on past experiences, the interpositional osteotomy was found to be highly successful when the soft tissues could be moved with the mobilized segment without excessive dissection and without tension upon closure.[10]

Dental implants require sufficient bone height and width as well as an appropriate surrounding of keratinized, attached gingiva. When using traditional interpositional osteotomy techniques, often the height or width may be reconstructed, but the deficient soft tissue requires adjunctive procedures.

All of these methods (onlay grafts, interpositional osteotomies, and membrane-assisted augmentation) have their associated advantages and disadvantages. Their main disadvantage is the inability of the surgeon to achieve an excellent result when the overlying soft tissue is deficient, limiting the effective coverage of the augmentation. Distraction osteogenesis has been shown to generate soft tissue, which allows its use to augment severely deficient areas.[2] This chapter will focus on the use of distraction osteogenesis for alveolar ridge augmentation.

ALVEOLAR DISTRACTION OSTEOGENESIS (DO)

DO began in the field of orthopedics for treatment of defects secondary to trauma. In 1905, Codvilla published a paper in the American Journal of Orthopedic Surgery that advocated lengthening the femur after an osteotomy, with applied tension using nails fixed in the femur segments.[11] Dr. Ilizarov is considered the pioneer of DO in the orthopedic literature.[12,13] (See Chapter 10) Mandibular lengthening for the correction of severe mandibular retrognathism has been reported.[14,15] (See Chapter 10). Alveolar distraction was reported involving animal studies[16–21] followed by clinical reports in 1996.[22] The initial reports utilized extraosseous DO devices followed by intraosseous devices, as described below.

Advantages of alveolar DO include no risk of disease transmission as the augmented bone is native to the patient, the lack of donor site morbidity, development of bone in the appropriate dimensions for dental implant placement, stable results over time with minimum resorption, and the ability to coordinate the lengthening of both bone and its associated soft tissue into the ideal location for implant placement.[23] Alveolar DO is a minimally invasive method to generate soft tissue in preparation for bone grafting and involves relatively non-morbid surgical procedures to create the osteotomy and place the distraction devices. The DO device is activated by the patient at home without the need for an inpatient hospitalization, with monitoring of the achieved distraction by physical examination of the ridge and simple radiographs.

ANIMAL STUDIES

In order to evaluate the potential for utilizing distraction osteogenesis for alveolar bone augmentation and movement, a series of dog studies was initiated by Block and associates.[6–21] Questions to be answered included:

1. Can distraction create palatal mucosa?
2. Will bone form in the palate when the anterior maxilla is advanced?
3. How effective is vertical augmentation of the alveolar bone using distraction, since the alveolar bone is being distracted transverse to the axial plane of the mandible?

Maxillary Segmental Advancement

We hypothesized that the anterior maxilla can be advanced using distraction osteogenesis resulting in formation of new palatal mucosa and bone in the gap created by the maxillary advancement. A vestibular incision was used to approach the maxilla from the superior approach, taking care to preserve all attachments of the palatal mucosa. The horizontal osteotomy was completed and the distraction device secured (Figures 1A, 1B, 1C). After 7 days, the anterior maxilla was advanced 1.0 mm daily using 0.5-mm advancements twice a day (Figure 1D). The anterior maxilla were advanced significantly and followed histologically and radiographically (Figures 1E, 1F, 1G).

The results of these studies indicated that bone will form in the palate and alveolar bone. Soft tissue genesis did occur without the development of palatal or buccal dehiscence. The shape of the alveolar bone was slightly narrowed at the distraction gap, with mild thinning of the facial bone.

These animal studies supported the clinical use of distraction to develop soft tissue and bone within a distraction gap when the maxilla was advanced segmentally. Anterior maxillary segmental advancement was confirmed clinically in a report by Dolanmaz.[24]

Alveolar Bone Augmentation

A series of studies was performed on the dog to investigate the use of distraction to augment the mandibular alveolar bone. In the dog model, posterior teeth were extracted and a significant alveoloplasty performed to create an edentulous ridge. The hypotheses to be proved included:

1. The bone can be elevated to achieve a significant alveolar augmentation.
2. Bone will form within the distraction gap.
3. Soft tissue genesis will occur to allow for soft tissue coverage of the newly augmented ridge.
4. The augmented bone will support implant restorations and maintain alveolar bone levels over time.

The animal was consistent between the studies. After the animal healed to form an intact cortical alveolar bone at the crest, four dental implants were placed horizontally in the ridge to allow for two implants to be contained within the alveolar bone to be distracted (Figures 2A, 2B), and two implants were placed in the bone inferior to the distracted segment. After the implants integrated, they were exposed and impressions taken to fabricate an implant level distraction device. After the distraction device was completed, incisions were made, the osteotomy made, and the distraction device secured to the implants (Figure 2B). After 7 days for healing, the alveolar bone was distracted superiorly 0.5 mm every 12 hours for 1.0 mm distraction daily (Figure 2C). After the distraction was completed, time was allowed for bone formation within the distraction gap (Figure 2D). Implants were placed in the distracted areas to support a two-implant support fixed prosthesis (Figure 2E, 2F). The bone and soft tissue was followed over time.

Hypothesis 1: Bone can be elevated to achieve a significant alveolar augmentation

In all of the segmental studies, the process of distraction with 0.5 mm increments twice daily resulted in significant alveolar ridge elevation and bone formation. In these animal studies, augmentations of up to 8 mm were achieved. The shape of the superiorly repositioned segment of the alveolar crest grossly appeared similar to initial, without significant resorption. These studies did not establish the vertical limit of the augmentation. In the dog, the horizontal component of the osteotomy was easily positioned superior to the inferior alveolar canal. Neurosensory testing was not performed.

Hypothesis 2: Bone will form within the distraction gap

Radiographs were taken during the distraction process and at the conclusion of the bone formation phase (Figures 2E, 2F). Histological examination of the distraction gap was performed after

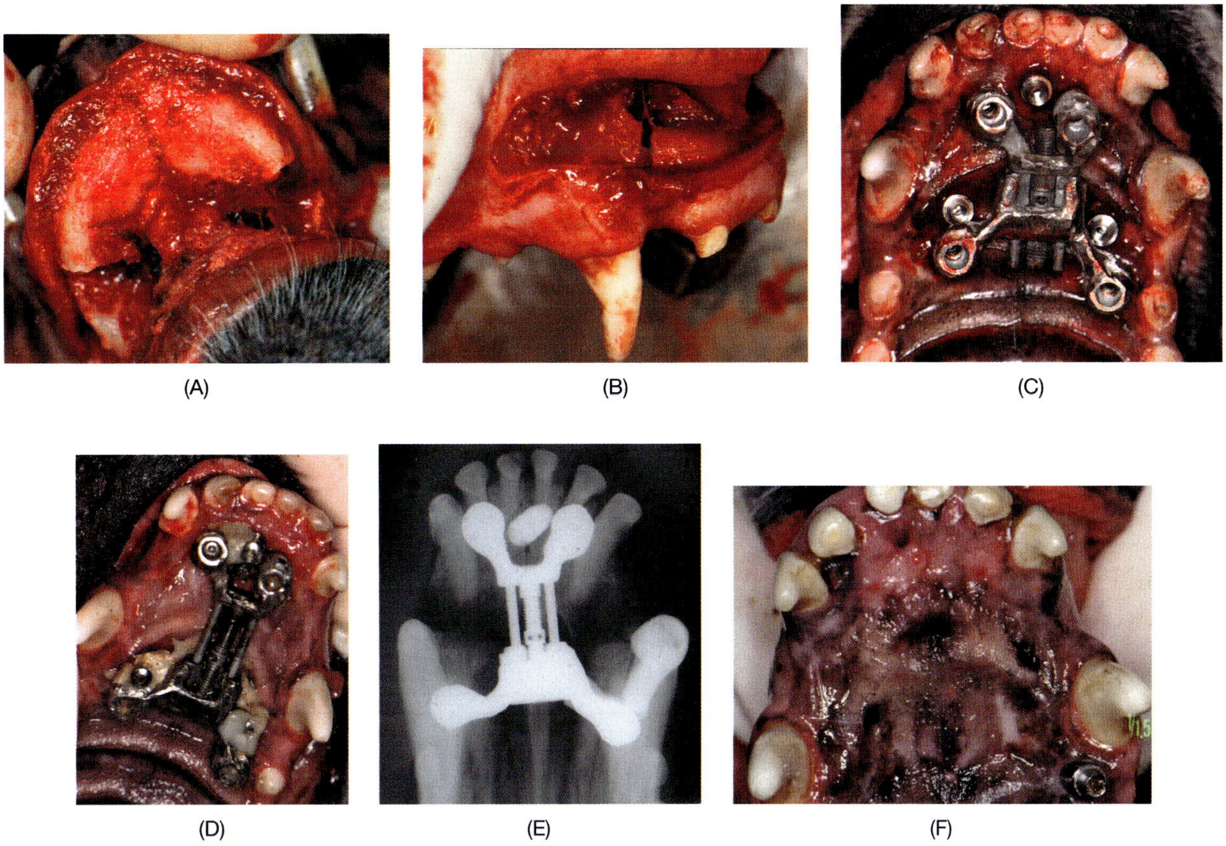

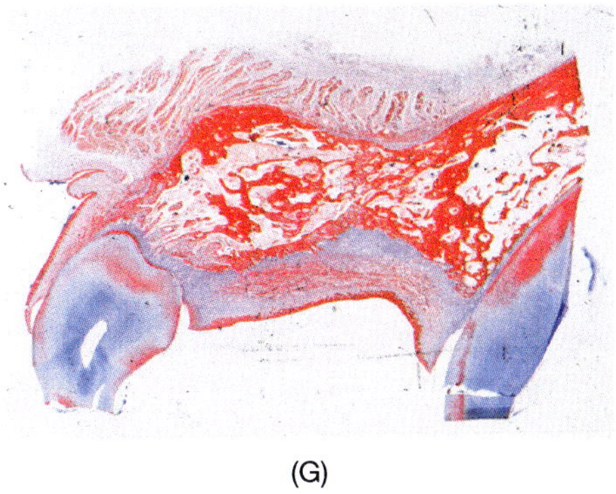

FIGURE 1 • (A) Nasal view of the canine maxilla showing the horizontal osteotomy (B) Lateral view of the maxilla showing the osteotomy cuts in the maxilla (C) The Custom distraction device is screw retained to the maxilla via the implants (D) The distraction device has been activated to advance the anterior segment of the maxilla (E) Occlusal radiograph after the distraction. (F) After healing the maxilla is advanced with excellent soft tissue genesis (G) 1:1 decalcified section stained with H and E showing bone formation across the distraction gap.

bone consolidation was confirmed at 8 weeks (Figure 2G). Based on these evaluation methods, it was concluded that bone did form within the distraction gap. Bone formed consistently along the lingual cortical plate. Bone formation was present consistently within the distraction gap. However, the facial cortex did not predictably form bone. In some areas, labial cortical bone was not present, only the marrow space and lingual plate. Cross-sections indicated small areas with definitive thinning and lack of bone width within the distraction gap.

Hypothesis 3: Soft tissue genesis will occur to allow for soft tissue coverage of the newly augmented ridge

In all specimens, the soft tissue coverage of the distraction gap was complete without dehiscence. The process of distraction of the dog alveolar crest did result in soft tissue genesis on the lingual and facial aspects to complete cover the distraction gap (Figure 2G).

Hypothesis 4: The augmented bone will support implant restorations and maintain alveolar bone levels over time

After bone had formed within the distraction gap, two dental implants were placed. One was placed within the distraction gap spanning the elevated alveolar crest, through the new bone of the distraction gap, into the native bone inferior to the distraction. The second implant was placed into non-distracted bone adjacent to the distraction site (Figure 2E, 2H).

Time was allowed for implant integration, impressions were taken, and fixed-screw retained restorations were fabricated and placed into occlusion. The animals were fed a normal textured diet. Radiographs were exposed over time and after 1 year of loading, specimens were evaluated histologically (Figures 2F, 2G, 2H, 2I).

Bone levels were maintained on the implants over time. Implants were not lost. The bridges were held in position without

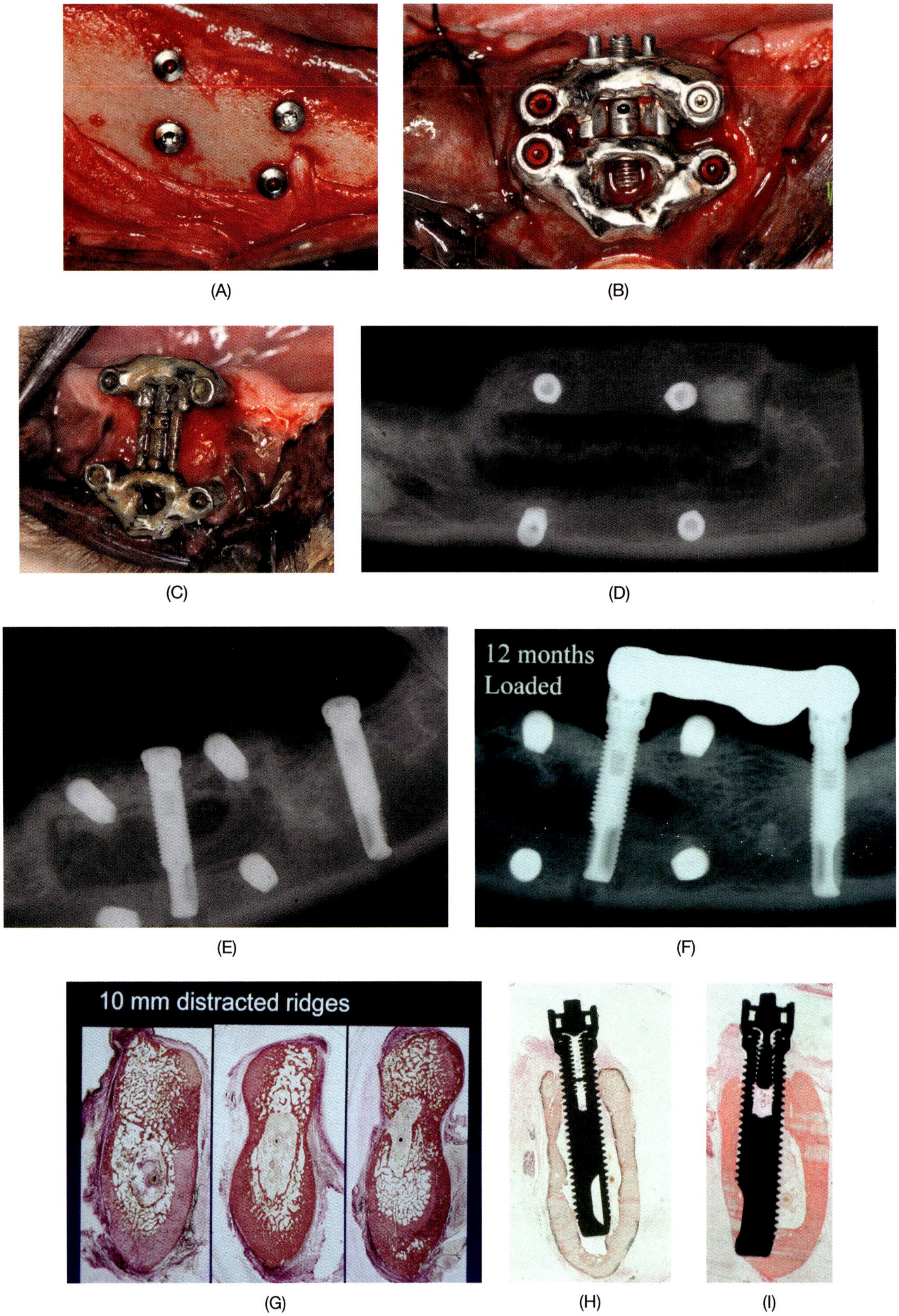

FIGURE 2 • (A) Horizontal oriented implants in dog mandibular body (B) Custom distraction device screw retained to implants after they had integrated and after the osteotomy was cut (C) The distraction device was activated ½ mm 2x each day to distract the superior portion of the ridge. (D) Radiograph 8 weeks after distraction completed showing bone formation (E) Implants were placed through the distracted ridge and the non-distracted ridge (F) Radiograph 12 months after loaded with normal diet. (G) Decalcified 1:1 H and E sections showing from left distracted ridges. Note in two sections thinning of the buccal cotex. (H) Non-decalcified plastic embedded sections of implant in non-distracted bone (I) Non-decacificed plastic embedded section of implant in distracted bone.

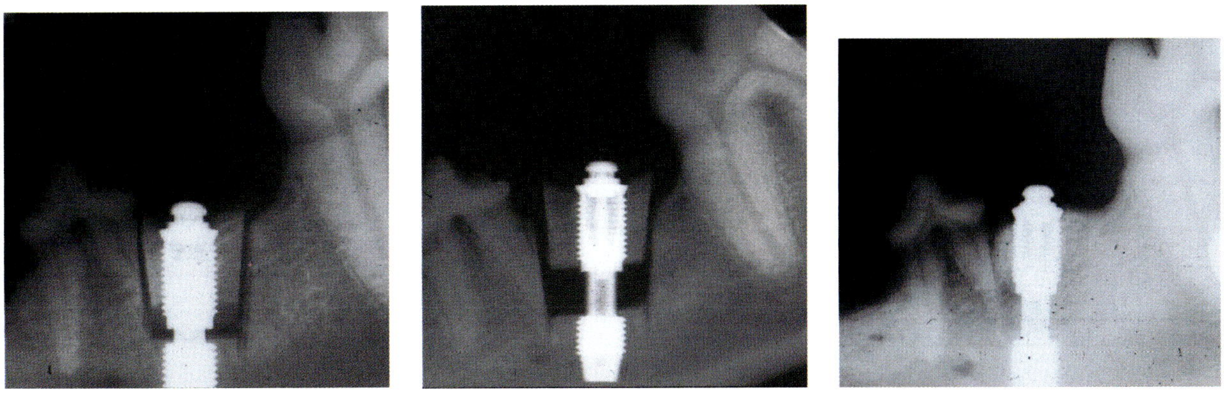

(A) (B) (C)

FIGURE 3 • (A) Intrabony distraction device in place after osteotomy was performed. (B) By turning the screw on the device, the superior portion of the alveolar bone was transported. (C) Bone formation after distraction was completed – 8 weeks

compromise. The soft tissues appeared healthy without pocket formation and without bleeding upon probing.

Histologic evaluation revealed bone remodeling with a normal trabecular bone appearance and morphology after one year of loading. Bone density within the distraction gap was similar to the native bone. Areas of thin facial bone were present but did not appear to result in adverse sequella.

Additional studies were performed using an intraosseous device. This device has an external and internal screw. It is placed by turning the external screw. When the internal screw is turned the device separates and lengthens. This device was used to investigate whether a small alveolar segment sized to match one tooth, could be distracted with bone formation within the distraction gap. As can be seen in Figures 3A, 3B, and 3C, a small segment of alveolar bone can be distracted superiorly at least 5 mm with bone formation within the distraction gap. In these animals, the labial bone was well formed without the thinning seen in the larger defect. Bone integrated to the threaded portions of this device, but not within the distraction gap. The future use of this type of distraction device doubling for the distraction device and the prosthetic part needs further clarification prior to clinical use.

These animal studies confirmed that alveolar bone could be superiorly repositioned with bone formation utilizing DO methods. The only compromise was the relatively high incidence of thin or sporadic lack of facial cortical bone width within the distraction gap. This sequence of bone formation as per Ilizarov[12,13] has been confirmed in clinical alveolar ridge distraction.[25]

Bone morphogenetic protein (BMP) and other growth factors are expressed continuously during distraction and consolidation. This concept supports the use of exogenous BMP to enhance bone formation in the distraction gap.[26–30] Investigations are ongoing to determine if bone morphogenic proteins can enhance bone development within the distraction gap. This work may result in superb bone formation without the current limitations based on clinical and animal experiences using conventional distraction methods.[31]

ALVEOLAR DISTRACTION DEVICES

Distraction devices that are designed for movement of the alveolus are either placed into the bone entering through the crest or are placed on the facial aspect of the alveolar bone with screws retaining small plates attached to the distraction apparatus. Extraosseous distraction devices are chosen when the lack of bone is severe (Figure 4B). When there is satisfactory bone width, with an isolated vertical problem, an intraosseous device can be used (Figure 3).

Extraosseous distraction devices are anchored by small bone plates connected to a rod-shaped tube with an internal screw to move the plates apart to create the distraction. The small bone plates are custom-adapted to the site, above and below the proposed osteotomy and secured in place. Turning the screw advances the distraction segment coronally. This type of distractor can provide vertical and horizontal movement depending on the vector of placement. However, horizontal augmentation of the ridge is difficult to obtain because of the initial malformed shape of the ridge. The advantages of the extraosseous device include flexibility in adaptation of the plates to the remaining bone, the lateral extensions of the plates allow for anchoring the distraction device to bone adjacent to the piriform rim of the maxilla, and the total distraction distance may be greater than with intraosseous distractors.

The plates can be contoured to allow the vector of the distraction to be labial in direction, resulting in augmentation of both the vertical and horizontal shape of the ridge. However, the resistance of the palatal mucosa to stretching can result in palatal redirection of the device. In order to minimize the palatal movement of the mobilized segment, the rod of the distraction device can be contained in a removable prosthesis worn by the patient on a daily basis.

When using an extraosseous device, the distraction rod extends from the submucosal aspect of the device through the mucosa. Over the 6 to 12 weeks following completion of the distraction, epithelial migration along the transgingival rod will occur. The resulting gingival recession exposes a portion of the distraction device. If left in position for an extended period of time (10 weeks after the distraction has been completed) secondary infection can result along with significant gingival recession and loss of soft tissue.

Pre-operative workup: The pre-operative workup should include collection of diagnostic material to accurately plan and perform the reconstruction. The final result should be understood prior to initiating the distraction stage. Imaging may include CT scans with 3D reconstruction and/or a stereolithographic model. The model of the patient's jaw can be used for pre-operative bending of the distraction device prior to surgical exposure of the bone. CT scan–generated models allow for accurate treatment planning.[32,33]

Mounted diagnostic casts are used to create a temporary appliance with teeth. Which is used to guide the surgeon during distraction to ensure that appropriate vectors are achieved. The approved, planned tooth position model can be used throughout the multiple treatment phases. The patient is educated on the need for the distraction procedure, bone grafting to achieve sufficient bone volume, implant placement surgery, and occasional soft tissue grafting to recreate keratinized and attached tissues.[(34)] If the lip line is low, which is common for these patients because of the extensive scarring from the initial trauma and subsequent surgical procedures, a fixed or fixed/removable restoration can be used.

CASE EXAMPLES

Distraction of the anterior maxilla for reconstruction of extreme atrophy secondary to avulsive injury. This woman lost a portion of her anterior maxilla including teeth and alveolar bone secondary to a motor vehicle accident. She presented with the desire to have a fixed prosthesis that would restore function and esthetics. She had less than 3 mm of bone at the anterior nasal spine region, with 5 mm of bone height at the canine regions. The soft tissue over the crest was scarred and non-mobile (Figure 4A). Because of the scarred, fixed, non-mobile crestal soft tissue, distraction osteogenesis was chosen as the method to create soft tissue for eventual bone grafting, increase the vertical dimension of her palatal and labial bone, and provide sufficient ridge form for a fixed prosthesis that would also restore missing bone in the nasal-labial region.

Under general anesthesia, a crestal incision was made with posterior vertical release, and a full thickness dissection was performed to expose the lateral aspect of the maxilla. Care was taken to avoid perforation of the nasal mucosa at the piriform rim. A distraction device was placed and the arms of the device bent to conform to the deformed maxilla, keeping the vertical vector positioned as far labially as possible. The DO device was secured with four screws, one in each arm (Figure 4B). The vertical osteotomies and a partial horizontal osteotomy were performed. The device was removed and the horizontal cut was completed with the aid of a curved osteotome (Figure 4C). Bone around the piriform rim was left intact. The DO device was replaced and secured with a total of eight screws. The incision was closed with the running rod of the device entering the mouth at the crest. After 7 days to allow for revascularization of the wound, the device was turned to achieve 1.0 mm distraction in 0.5 mm increments 2 times per day. The ridge was distracted coronally 10 mm. Eight weeks after completion of the distraction the area was re-entered via a crestal incision as before, in anticipation of device removal and bone graft to achieve further horizontal augmentation. The device was found to be stable (Figure 4D). After it was removed (Figure 4E) the gap had bone continuous with the palate but there were several isolated areas of fibrous tissue on the labial aspect. Bone was harvested from the iliac crest and placed over the labial aspect to achieve horizontal reconstruction (Figure 4F). The soft tissue was easily closed over the area. Implants were placed four months later (Figure 4G) and after an additional 4 months the implants were exposed and a final fixed, cemented porcelain-fused-to-metal prosthesis fabricated (Figures 4H, 4I, 4J).

ALVEOLAR DISTRACTION COMPLICATIONS, LIMITATIONS, AND COMPARISON TO ALTERNATIVE PROCEDURES

One of the challenges for the clinician using a distraction device is control of the vector of the distracted segment. The use of prosthetic devices to prevent palatal movement of the districted segment of bone has been useful.[34] Unique distraction devices that allow for two planes of adjustment are being developed to aid the clinician to meet this challenge.[35]

The surgeon must preserve the vascular supply to the segment of bone to be distracted. Incisions should be designed to provide a broad vascular base, and to preserve the periosteum. Intraosseous distractors require a small entrance over the crest, and a lateral reflection to create the osteotomies. Minimal reflection of the periosteum is important in order to preserve the blood supply to the distracted segment of crestal bone.[36]

For placement of an extraosseous distractor, the lateral reflection will decrease the blood supply to the bone to be distracted. Care is taken to maintain the pedicle to the bone from the lingual or palatal aspects. Often the vertical bone cuts are made without periosteal release, if possible.

Does bone form differently in the distraction gap comparing humans to animals? In animals the sequence of bone formation across the distraction gap is well described, with most authors agreeing that within 6–8 weeks bone formation has crossed the gap, creating a functionally stable continuum of bone. Histologic evaluation confirming bone formation in animals may differ from humans. Zaffe and associates reported serial biopsies in humans, in distracted alveolar bone, over a period of 88 days. They reported that, in the distraction gap, at 40 days a soft callus was present; at 60 days trabecular woven bone was present with approximately 50% of the site filled with bone; and at 88 days the bone appeared mature.[37] This bone formation in the alveolar crest may be slower in humans compared to other bones.

Alveolar distraction is an excellent method to correct gingival-level problems in the esthetic regions of the anterior maxilla.

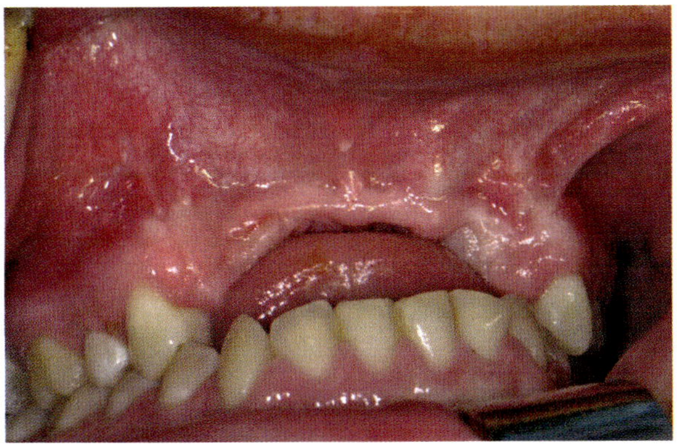

(A)

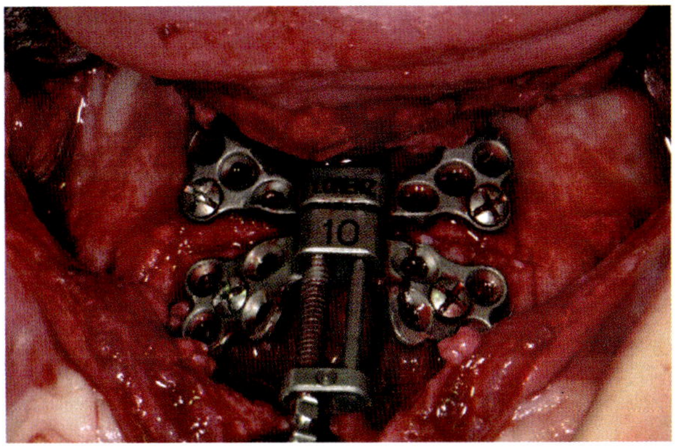

(B)

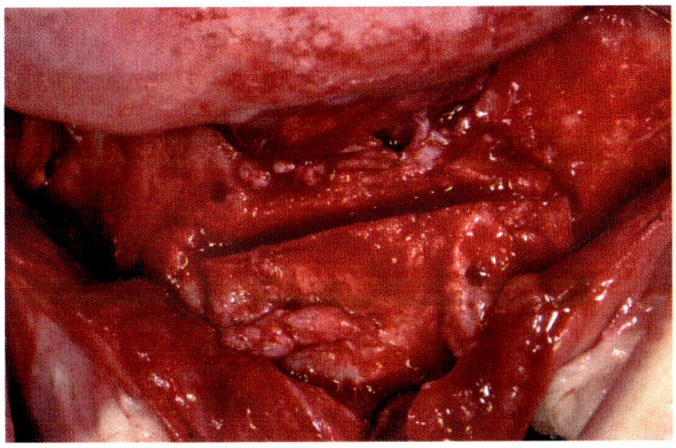

(C)

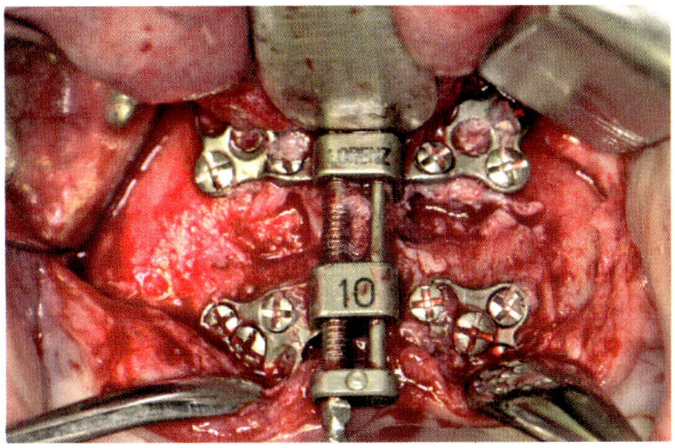

(D)

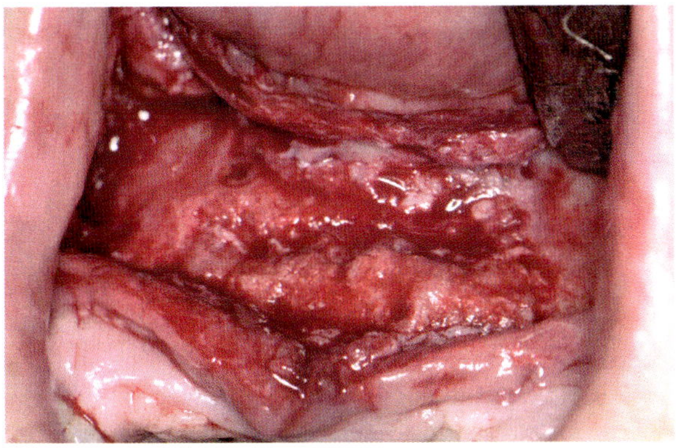

(E)

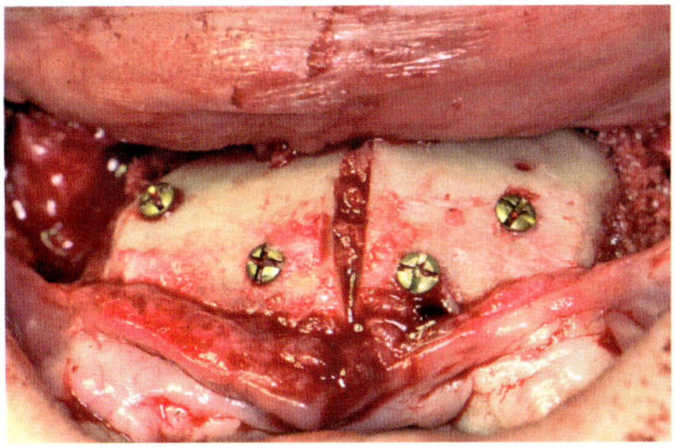

(F)

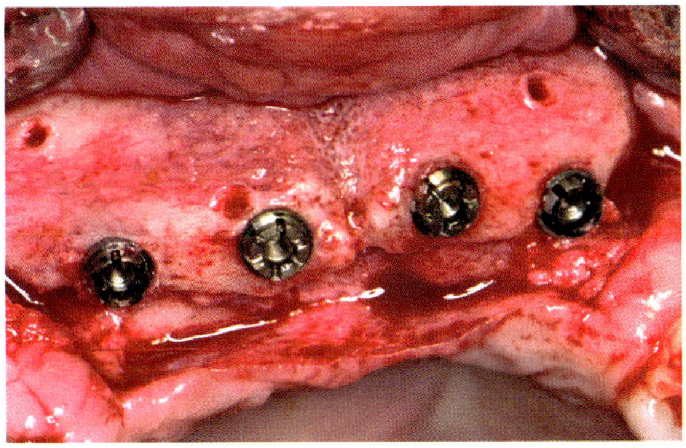

(G)

FIGURE 4 • (A) Patient missing vertical and horizontal anterior maxillary bone after car accident and avulsion. (B) A distraction device is placed and secured with 4 screws to allow for a portion of the osteotomy to be performed. (C) The device is removed and the osteotomy is completed to allow for vertical mobilization of the anterior maxilla (D) After 10 mm of distraction and 8 weeks for healing, the device is exposed for removal (E) The underlying bone formation created a new vertical positioning of the alveolus, but was thin. (F) An onlay graft harvested from the iliac crest was secured with screws. (G) 4 months later implants were placed for final restoration. (*continued*)

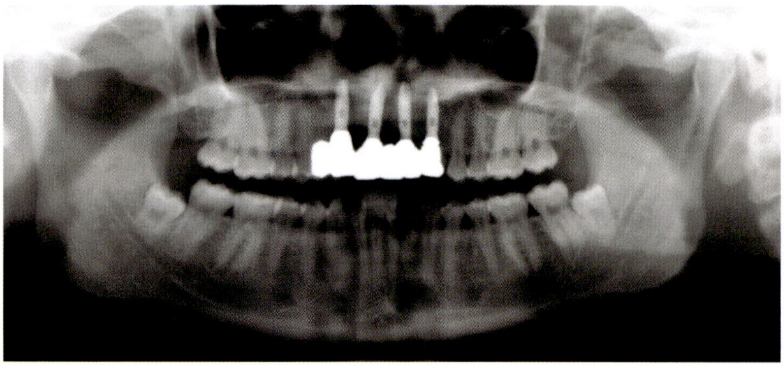

(H)

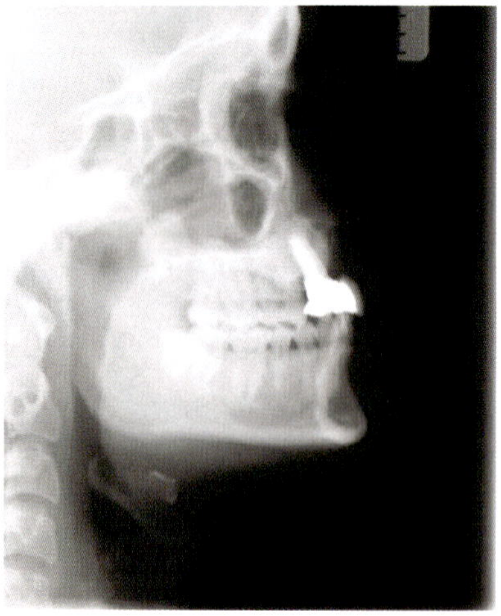

(I)

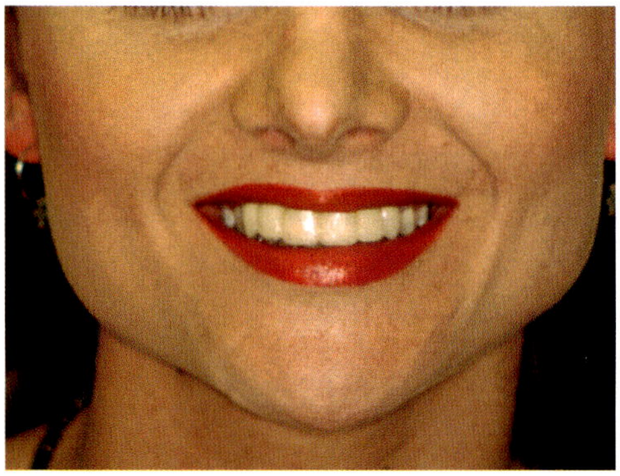

(J)

FIGURE 4 • (*continued from previous page*) (H) Implant retained fixed prosthesis – panoramic radiograph 6 months follow up. (I) The facial profile has been re-established. (J) Esthetics have been re-established with the prosthetics

When a tooth has gingival margin discrepancies compared to the adjacent tooth, then the tooth either needs to be orthodontically extruded or moved coronally with the bone using distraction osteogenesis. Distraction can be used to move the tooth and accompanying bone prior to tooth extraction. After the distraction has been completed, the tooth can be extracted and implants placed.[38]

Complications with alveolar distraction. Alveolar distraction methods do require technical expertise. Minor complications do occur in a large number of patients, with a reported incidence up to 70% of procedures. These can include lingual or palatal displacement of the distracted segment, fracture of the device, loss of screws, or incision breakdown. The incidence of major complications that totally prevent implant placement are less frequent.[39]

Jenson[40] reported on 30 consecutive cases of alveolar distraction osteogenesis in the anterior maxilla. Bone grafts were necessary in some but not all. Implant failure of 10% was believed to be related to poor bone quality. The soft tissue response was very good.

Rachmiel[41] reported on 14 patients using a distraction system that combines extraosseous and intraosseous components. He reported minimal complications and suggested that DO can result in longer implants and better esthetic bone and soft tissue position.

Fracture of the basal bone has been reported.[42] Minor complications are common, resulting in levels of morbidity that may seem minor but do play a role in the decision on the benefit-cost-morbidity analysis to choose distraction or another method.[43]

In the posterior mandible, when distraction methods were used to augment bone height, the amount of bone resorption of the distracted segment was variable, with less bone augmentation than recorded on the distraction device. Relapse was variable and not predictable.[44] Relapse was found to be 20% in a study of 11 patients, based on CT data.

Some minor complications may occur in 75% of patients. Despite minor complications, implant success was high, approximating 95%.[45,46]

In the dog, comparing onlay grafting with DO augmentation of the ridges, more evidence of bone remodeling was found with the distracted ridges. The implications are unclear. Bone contact to implants placed in both types of grafts was similar, indicating minimal difference in regard to bone response.[47,48] It is useful to monitor the amount of bone resorption that is occurring. Radiographs are useful to follow the progress of bone consolidation in a distraction gap.[49]

Malposed implants provide a problem for the restorative dentist and must be corrected either by removal or repositioning. The use of distraction methods to reposition malposed implants has been reported and is a useful tool for the clinician.[50]

Reported complication rates for alveolar DO have been high but most complications have been minor and have not prevented implant placement. Enislidis and colleagues reported a complication of 100% during the treatment of nine patients with subperiosteal devices.[45] Garcia and associates and Uckan and colleagues report overall complication rates of 70%.[51,52] Implant success in distracted alveolar bone ranges from 85% to 97%.[40,41,51] Authors conclude that the complications are minor and are due to poor surgical technique and/or incorrect device placement.

The surgical decision to utilize distraction for vertical augmentation of a small, 2 to 3 tooth-sized alveolar defect is based on several factors.

OPTIONS FOR 2 TO 3 TOOTH-SIZED DEFECTS

If the adjacent teeth are good abutments for a fixed, conventional crown and bridge restoration, a porcelain-fused-to-metal restoration with the use of pink ceramics is an excellent option. Advantages are the lack of surgical morbidity, efficient treatment in the dentist's office, and an accepted esthetic result is often obtained. Disadvantages are the use of abutment teeth and the stresses placed upon them, preparing teeth that are otherwise not in need for a restoration, and poor esthetics in the pontic area if the ceramist is not using high-quality methods.

If the treatment plan calls for implant-supported restorations, then the assessment begins with the need for bone height in the area of the defect. If there is satisfactory height for the implant but the alveolar region is deficient in height, and will be an esthetic problem, then restoration of the alveolar height is necessary. If the bone levels on the adjacent teeth are good, then either an onlay graft or a distraction procedure can be used.

Onlay grafts harvested from the symphyseal region are associated with a low incidence of sensory nerve changes in the teeth and occasionally the lips and chin. The procedure to harvest the bone and then place it to vertically augment a defect is challenging. Incisions must be closed meticulously and there is a risk of incision dehiscence, loss of graft secondary to incision dehiscence, and unpredictable graft resorption that affects the final esthetic result. The graft is done on day 1, allowed 4 months for healing, and then implants are placed and the restoration is completed approximately 9 months after the start of the procedure, if no graft complications occur.

Interpositional osteotomies with grafting at the time of osteotomy achieve the augmentation at the time of surgery. For this procedure the alveolar bone osteotomy is performed similar to that for distraction; however, the alveolar segment is mobilized and carefully moved coronally to achieve immediate vertical augmentation. A graft is placed in the gap. Plates are used as necessary. Four months after the interpositional graft is performed, implants are placed with the final restoration placed 9 months after the procedures are started. The advantage of the interpositional osteotomy with grafting is the immediate augmentation of the alveolar crest. Disadvantages include potential graft necrosis secondary to stretching the vascular pedicle, especially where the palatal tissue is attached and not easily mobilized.[(54)]

Distraction requires a surgical procedure similar to the interpositional osteotomy, with placement of either an extraosseous or intraosseous device. Most oral and maxillofacial surgeons are trained in plate and osteotomy techniques, but other specialists may require additional course training. The devices are placed and 1 week is allowed for healing. The distraction is initiated and may take 7 to 10 days. At this point, the segments are maintained in position for 8 weeks, which at that time should demonstrate bone consolidation. Then depending on the labial bone width, implants can be placed in combination of a graft to cover small labial bone defects. The implants will require 4 months for integration. The total time from the surgical procedure to final is approximately the same as the other procedures. Advantages of distraction are less chance of segment loss because of the slow movement of the segment and the soft tissue genesis that occurs. The disadvantages of this technique are the need for an additional prosthesis to guide the vector of the distraction, bi-daily attention to the device by the patient, multiple patient visits every third day during distraction to monitor the device, and the potential for the need of further grafting.

The overall success rates are similar between the three surgical augmentation procedures. The key will be the mobility of the soft tissue over the defect. If the etiology for the loss of teeth and bone was traumatic, or secondary to infectious problems, the tissue may be very tight and fibrous in nature, which would prevent immediate mobilization that would be necessary for an onlay graft or an immediate interpositional osteotomy with graft. The choice of whether to use distraction or another method is based on a full assessment of the patient's goals and needs (Figures 5 and 6).

CONCLUSION

Traditional preprosthetic reconstructive techniques, such as onlay grafts, interpositional osteotomies, and membrane-assisted augmentation have their associated advantages and disadvantages. Their main disadvantage is the inability of the surgeon to achieve an excellent result when the overlying soft tissue is deficient, limiting the effective coverage of the augmentation. Distraction osteogenesis has been shown to concurrently provide bone and generate soft tissue. The generation of mucosal alveolar coverage combined with the elimination of donor site morbidity make DO a viable, minimally invasive alternative for maxillary and mandibular ridge reconstruction.

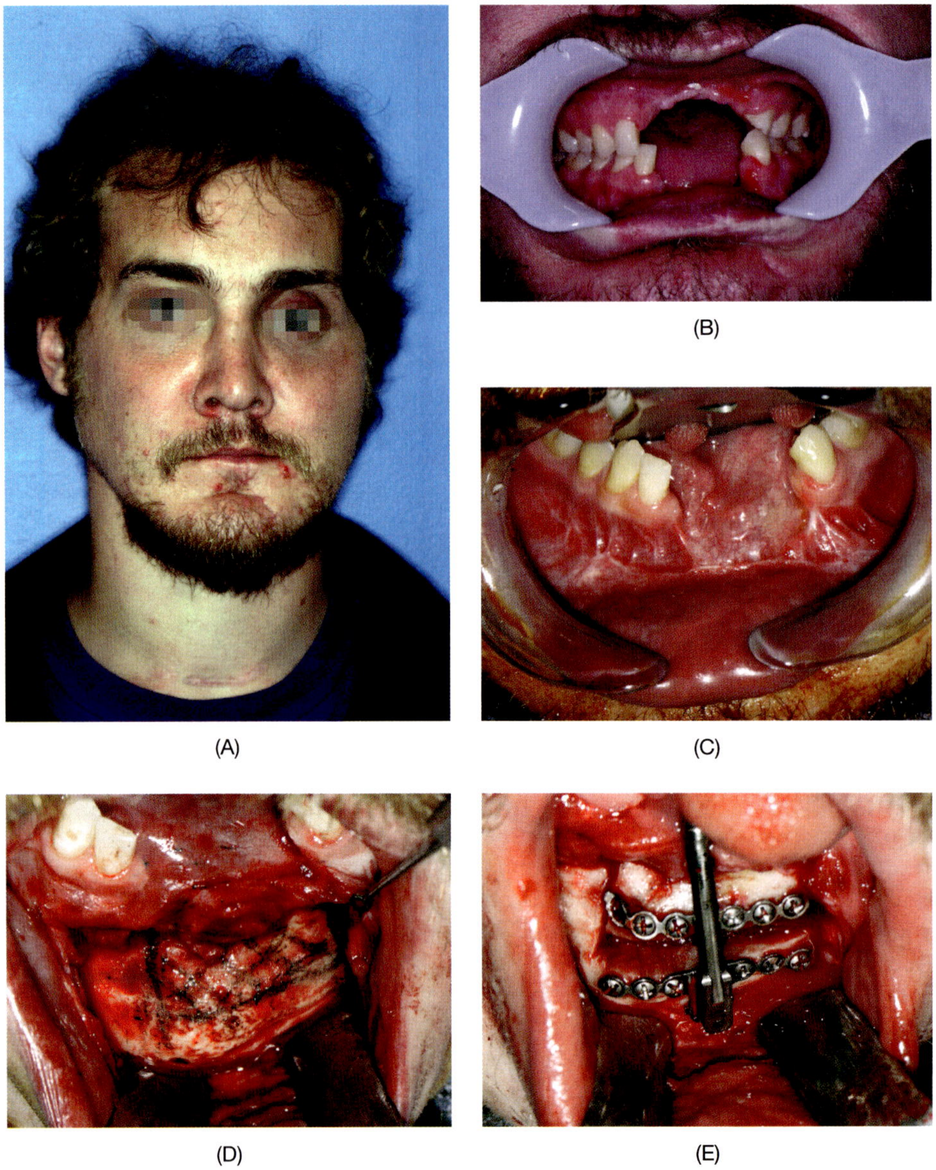

FIGURE 5 • 20 year old male involved in Motor vehicle Accident (MVA). He suffered a loss of teeth and alveolar bone. (A) Frontal and (B) intraoral and (C) close-up intraoral photograph. (D) Intraoperative photograph with proposed transport disc outlined. (E) Intraoperative photograph with distractor device in place. (Photographs courtesy of MGH Ward Service and Dr. M.J. Troulis)

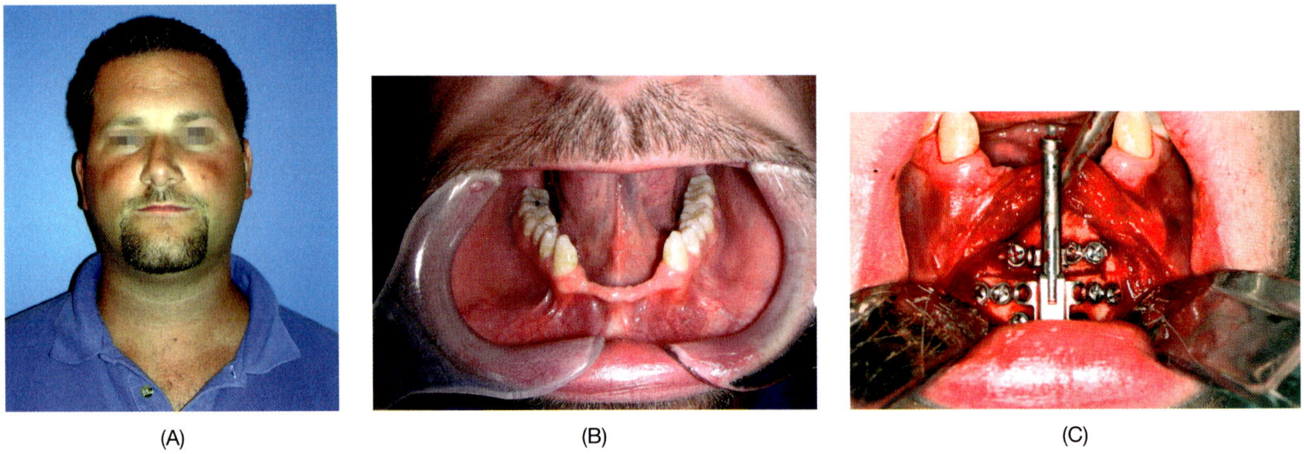

FIGURE 6 • 25 year old male with severely atrophic mandibular ridge. (A) Frontal (B) intraoral photographs. (C) Intraoperative view with Alveolar distractor in place. (*continued*)

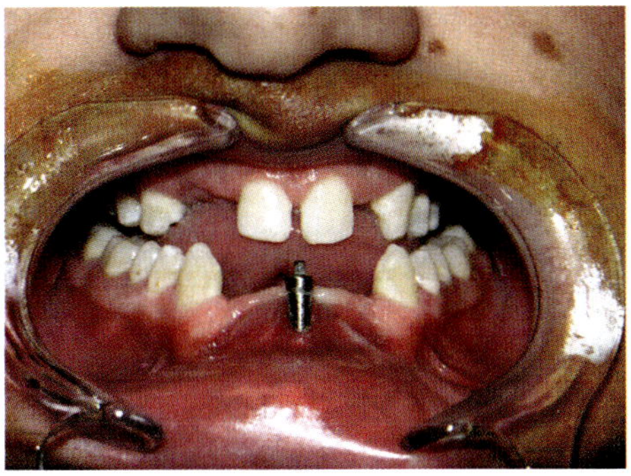

(D)

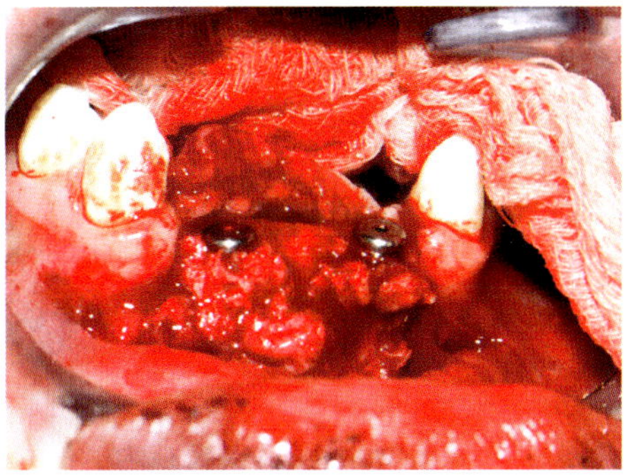

(E)

FIGURE 6 • (*continued from previous page*) (D) Intraoral view with activating arm extending transmucosally for activation. (E) Intraoperative view with 2 implants in position. (Photograph courtesy of MGH Ward Service and Dr. M.J. Troulis)

References

1. Rachmiel A, Stouji S, Peled M. Alveolar ridge augmentation by distraction osteogenesis. Int J Oral Maxillofac Surg 2001;30:510–517.
2. Block MS. Biological basis for alveolar distraction osteogenesis. In: Jensen O, editor. Alveolar Distraction Osteogenesis. Hanover Park, IL: Quintessence Co, Inc; 2002. p. 14.
3. Alkan A, Bas B, Inal S. Alveolar distraction osteogenesis of bone graft reconstructed mandible. Oral Surg Oral Med Oral Pathol Oral Radiol Endod 2005;100:e39–42.
4. Bell RB, Blakey GH, White RP, Hillebrand DG, Molina A. Staged reconstruction of the severely atrophic mandible with autogenous bone graft and endosteal implants. J Oral Maxillofac Surg 2002;60: 1135–41.
5. Wang JH, Waite DE, Steinhauser E. Ridge augmentation: an evaluation and follow-up report. J Oral Surg 1976;34:600–2.
6. Enislidis G, Fock N, Ewers R. Distraction Osteogenesis with subperiosteal devices in edentulous mandibles. British Journal of Oral and Maxillofacial Surgery 2005;43:399–403.
7. Verhoeven JI, Cune MS, Terlou M, Zoon MA, de Putter C. The combined use of endosteal implants and iliac crest onlay grafts in the severely atrophic mandible: a longitudal study. Int J Oral Maxillofac Surg 1997;26:351–7.
8. Simion M, Jovanovic SA, Tinti C, Benfenati SP. Long-term evaluation of osseointegrated implants inserted at the time or after vertical ridge augmentation. A retrospective study on 123 implants with 1–5 year follow-up. Clin Oral Implants Res 2001; 12:35–45.
9. Stoelinga PJW, de Koomen HA, Tideman H, Huybers TMJ. A reappraisal of the interposed bone graft augmentation of the atrophic mandible. J Maxillofac Surg 1983;11:107–12.
10. Fonseca RJ, Barber D, Frost DE, et al. Osseous reconstruction. In: Fonseca RJ, Davis WH, editors. Reconstructive Preprosthetic Oral and Maxillofacial Surgery. Philadelphia: WB Saunders; 1995. p. 825–52.
11. Codivilla A. On the means of lengthening in the lower limbs, the muscles, and tissues which are shortened through deformity. Am J Orthop Surg 1905;2:353–369.
12. Ilizarov G. The tension-stress effect on the genesis and growth of tissues. Part I. The influence of stability of fixation and soft tissue preservation. Clin Orthop Rel Res 1989;238:249–81.
13. Ilizarov G. The tension-stress effect on the genesis and growth of tissues. Part II. The influence of the rate and frequency of distraction. Clin Orthop Rel Res 1989;239:263–85.
14. Snyder CC, Levine GA, Swanson HM, Browne EZ. Mandibular lengthening by gradual distraction; preliminary report. Plast Reconst Surg 1973;51:506–508.
15. McCarthy JG, Schreiber J, Karp N, Thorne CH, Grayson BH. Lengthening the human mandible by gradual distraction. Plast Reconst Surg 1992;89:1–8.
16. Block MS, Brister GD. Use of distraction osteogenesis for maxillary advancement, J Oral Maxillofac Surg 1994;52:282–6.
17. Block MS, Cervini D, Chang A, Gottsegan GB. Anterior maxillary advancement using tooth supported distraction osteogenesis, J Oral Maxillofac Surg 1995;53:561–5.
18. Block MS, Akin R, Chang A, Gottsegen GB. Skeletal and dental movements after anterior maxillary advancement using implant-supported distraction osteogenesis in dogs. J Oral Maxillofac Surg 1997;55:1433–8.
19. Block MS, Chang A, Crawford CH. Mandibular alveolar ridge augmentation in the dog using distraction osteogenesis, J Oral Maxillofac Surg 1996;54:309–14.
20. Block MS, Almerico B, Crawford C, Gardiner D, Chang A. Bone response to functioning implants in dog mandibular alveolar ridges augmented with distraction osteogenesis. Int J Oral and Maxillofacial Implants 1998;13:342–51.
21. Block MS, Gardiner D, Almerico B, Neal C. Loaded HA coated and uncoated titanium threaded implants in dog alveolar ridges augmented with distraction osteogenesis. Oral Surgery, Oral Pathology, Oral Radiol 2000;89:676–85.
22. Chin M, Toth BA. Distraction Osteogenesis in maxillofacial surgery using internal devices: review of five cases. J Oral Maxillofac Surg 1996;54:45–52.
23. Erkut S, Uckan S. Alveolar Distraction osteogenesis and implant placement in a severely resorbed maxilla: A clinical report. J Prosthet Dent 2006;95:340–3.
24. Dolanmaz D, Karaman AI, Ozyesil AG. Maxillary anterior segmental advancement by using distraction osteogenesis: a case report. Angle Orthod 2003;73:201–5.
25. Raghoebar GM, Liem RS, Vissink A. Vertical distraction of the severely resorbed edentulous mandible: a clinical, histological and

electron microscopic study of 10 treated cases. Clin Oral Implants Res 2002;13:558–65.

26. Yazawa M, Kishi K, Nakajima H, Nakajima T. Expression of bone morphogenetic proteins during mandibular distraction osteogenesis in rabbits. J Oral Maxillofac Surg 2003;61: 587–92.
27. Campisi P, Hamdy RC, Lauzier D, Amako M, Schloss MD, Lessard ML. Overview of the radiology, histology, and bone morphogenetic protein expression during distraction osteogenesis of the mandible. J Otolaryngol 2002;31:281–6.
28. Campisi P, Hamdy RC, Lauzier D, Amako M, Rauch F, Lessard ML. Expression of bone morphogenetic proteins during mandibular distraction osteogenesis. Plast Reconstr Surg 2003;111:201–8.
29. Yates KE, Troulis MJ, Kaban LB, Glowacki J IGF-I, TGF-beta, and BMP-4 are expressed during distraction osteogenesis of the pig mandible. Int J Oral Maxillofac Surg 2002;31:173–8.
30. Bouletreau PJ, Warren SM, Longaker MT. The molecular biology of distraction osteogenesis J Craniomaxillofac Surg 2002; 30:1–11.
31. Terheyden H, Wang H, Warnke PH, Springer I, Erxleben A, Ludwig K, Rueger DC. Acceleration of callus maturation using rhOP-1 in mandibular distraction osteogenesis in a rat model, Int J Oral Maxillofac Surg 2003;32:528–33.
32. Poukens J, Haex J, Riediger D. The use of rapid prototyping in the preoperative planning of distraction osteogenesis of the cranio-maxillofacial skeleton. Comput Aided Surg 2003;8:146–54.
33. Gateno J, Teichgraeber JF, Xia JJ. Three-dimensional surgical planning for maxillary and midface distraction osteogenesis. J Craniofac Surg 2003;14:833–9.
34. Block MS and Baughman DG. Reconstruction of severe anterior maxillary defects using distraction osteogenesis, bone grafts, and implants. J Oral Maxillofac Surg 2005;63:291–7.
35. Garcia GA, Somoza-Martin M, Gandara Vila P, Lopez Maceiras J. Alveolar ridge osteogenesis using 2 intraosseous distractors: uniform and nonuniform distraction. J Oral Maxillofac Surg 2002; 60:1510–2.
36. Gaggl A, Schultes G, Rainer H, Karcher H. The transgingival approach for placement of distraction implants J Oral Maxillofac Surg 2002;60:793–6.
37. Zaffe D, Bertoldi C, Palumbo C, Consolo U. Morphofunctional and clinical study on mandibular alveolar distraction osteogenesis. Clin Oral Implants Res 2002;13:550–7.
38. Hurzeler MB, Zuhr O, Schenk G, Schoberer U, Wachtel H, Bolz W. Distraction osteogenesis: a treatment tool to improve baseline conditions for esthetic restorations on immediately placed dental implants—a case report. Int J Periodontics Restorative Dent 2002;22:451–61.
39. Uckan S, Haydar SG, Dolanmaz D. Alveolar distraction: analysis of 10 cases. Oral Surg Oral Med Oral Pathol Oral Radiol Endod 2002;94:561–5.
40. Jensen OT, Cockrell R, Kuhike L, Reed C. Anterior maxillary alveolar distraction osteogenesis: a prospective 5-year clinical study Int J Oral Maxillofac Implants 2002;17:52–68.
41. Rachmiel A, Srouji S, Peled M. Alveolar ridge augmentation by distraction osteogenesis Int J Oral Maxillofac Surg 2001;30:510–7.
42. Fukuda M, Iino M, Ohnuki T, Nagai H, Takahashi T. Vertical alveolar distraction osteogenesis with complications in a reconstructed mandible J Oral Implantol 2003;29:185–8.
43. Garcia AG, Martin MS, Vila PG, Maceiras JL. Minor complications arising in alveolar distraction osteogenesis J Oral Maxillofac Surg 2002;60:496–501.
44. Polo WC, Cury PR, Sendyk WR, Gromatzky A. Posterior mandibular alveolar distraction osteogenesis utilizing an extraosseous distractor: a prospective study J Periodontol 2005;76:1463–8.
45. Enislidis G, Fock N, Millesi-Schobel G, Klug C, Wittwer G, Yerit K, Ewers R. Analysis of complications following alveolar distraction osteogenesis and implant placement in the partially edentulous mandible Oral Surg Oral Med Oral Pathol Oral Radiol Endod 2005;100:25–30.
46. Saulacic N, Somoza-Martin M, Gandara-Vila P, Garcia-Garcia A. Relapse in alveolar distraction osteogenesis: an indication for overcorrection. J Oral Maxillofac Surg 2005;63:978–81.
47. Perry M, Hodges N, Hallmon DW, Rees T, Opperman LA. Distraction osteogenesis versus autogenous onlay grafting. Part I: outcome of implant integration Int J Oral Maxillofac Implants 2005;20:695–702.
48. Hodges NE, Perry M, Mohamed W, Hallmon WW, Rees T, Opperman LA. Distraction osteogenesis versus autogenous onlay grafting. Part II: biology of regenerate and onlay bone. Int J Oral Maxillofac Implants 2006;21:237–44.
49. Mazzonetto R, Allais de Maurette M. Radiographic evaluation of alveolar distraction osteogenesis: analysis of 60 cases J Oral Maxillofac Surg 2005;63:1708–11.
50. Zechner W, Bernhart T, Zauza K, Celar A, Watzek G. Multidimensional osteodistraction for correction of implant malposition in edentulous segments Clin Oral Implants Res 2001;12:531–8.
51. Uckan S, Haydar S, Dolanmaz D. Alveolar Distraction: Analysis of ten cases. Oral Surg Oral Med Oral Pathol Oral Radiol Endod 2002;94:561–5.
52. Garcia AG, Martin MS, Vila PG, Saulacic N, Rey JMG. Alveolar distraction before insertion of dental implants in the posterior mandible. British Journal of Oral And Maxillofacial Surgery 2003; 41:376–79.
53. Jensen OT, Kuhlke L, Bedard JF, White D. Alveolar segmental sandwich osteotomy for anterior maxillary vertical augmentation prior to implant placement. J Oral Maxillofac Surg. 2006; 64:290–6. Erratum in: J Oral Maxillofac Surg. 2006;64:997.

Distraction Osteogenesis for Correction of Maxillary Deformities

17

Bonnie L. Padwa, DMD, MD

OUTLINE

- Introduction
- Technique
- Conclusion

INTRODUCTION

The maxilla is a bone characterized by complex three-dimensional (3D) morphology and complex anatomic relationships with adjacent structures. In addition, the bone on either side of the standard maxillary osteotomies (Le Fort I, II, or III) is thin and without marrow. The type of corticotomy described for distraction osteogenesis by Ilizarov is not feasible. Furthermore, the LeFort I osteotomy, along the anterior and lateral walls of the maxilla, creates an open wound into the sinuses and the nose. How these differences affect osteogenesis (that is, formation of the regenerate in the maxilla during distraction) has not been documented in animal models. Nevertheless, surgeons excited by the results of distraction osteogenesis in the mandible have adapted the technique to treat maxillary deformities.[1,2]

Polley and co-workers[3,4] have documented the benefits of maxillary distraction at the LeFort I level, when compared to standard osteotomies, in cleft lip/palate patients. In addition, they have shown the results to be predictable and stable in the long term.[3,4,5] (See Chapter18). In the past, large sagittal discrepancies could not be entirely corrected by moving the maxilla forward because of the restriction of the soft tissue envelope. This was especially true in the multiply operated cleft lip and palate patient. To compensate for this, the mandible was set back to establish the desired occlusion. This treatment often compromised the esthetic result. With the use of gradual distraction, even a severely hypoplastic midface can be advanced enough to correct the entire sagittal and vertical deformity.[5] Following a LeFort I osteotomy, the maxilla can be advanced by either pushing it forward, using semi-buried devices, or pulling it forward using a rigid external device.[6,7,8,9] It is important for the surgeon to understand that there are inherent technical difficulties relating to control of vectors of movement and symmetry of advancement when using either "pushing" or "pulling" techniques. In addition, the issues of patient compliance and comfort during distraction must also be considered.[10]

The rigid external device (RED) is similar in concept to the cranial fixation devices used for treatment of maxillofacial injuries during World War I.[1,11,12] The cranial frame was connected to an intraoral appliance, fixed to the teeth with wires to put the jaw under traction forces.

The current rigid external distraction device is adjustable to allow changes in the horizontal and vertical vectors at any time during the distraction process. Because the maxilla does not have to be repositioned and rigidly fixed with plates and screws, the operative time and morbidity are decreased. Either a dental-borne appliance cemented to deciduous and/or permanent teeth (Figure 1) or bone retained retention hooks (Figure 2) can be used. The osteotomy is placed to avoid tooth buds and the level of the osteotomy is dictated by the degree and location of maxillary hypoplasia. The patient does not have to be skeletally mature because the maxilla is not acutely repositioned; hence, there is no need for bone grafts or rigid internal fixation.

As in distraction osteogenesis of the mandible and long bones, there may be soft tissue augmentation and elongation along with the skeletal expansion. Theoretically, this might reduce relapse and improve long-term stability. The limitations of rigid external maxillary distraction are: 1) the prolonged treatment time, which is usually on the order of 2–3 months; 2) the large and cumbersome devices; and 3) the inability of many patients, particularly older children and adolescents, to tolerate the devices. There have been several problems related to the cranial fixation pins, including pin loosening, frame migration, temporal scarring, CSF leakage, depressed skull fracture (after a fall), and even death resulting from migration of a fixation pin.[13,14,15]

Because of the above-mentioned drawbacks of the external device, a hybrid technique of maxillary distraction with early removal of the appliance and skeletal stabilization with rigid internal fixation has been proposed.[16] This technique permits maxillary advancement that may be unattainable by traditional methods and obviates the extended use of external hardware

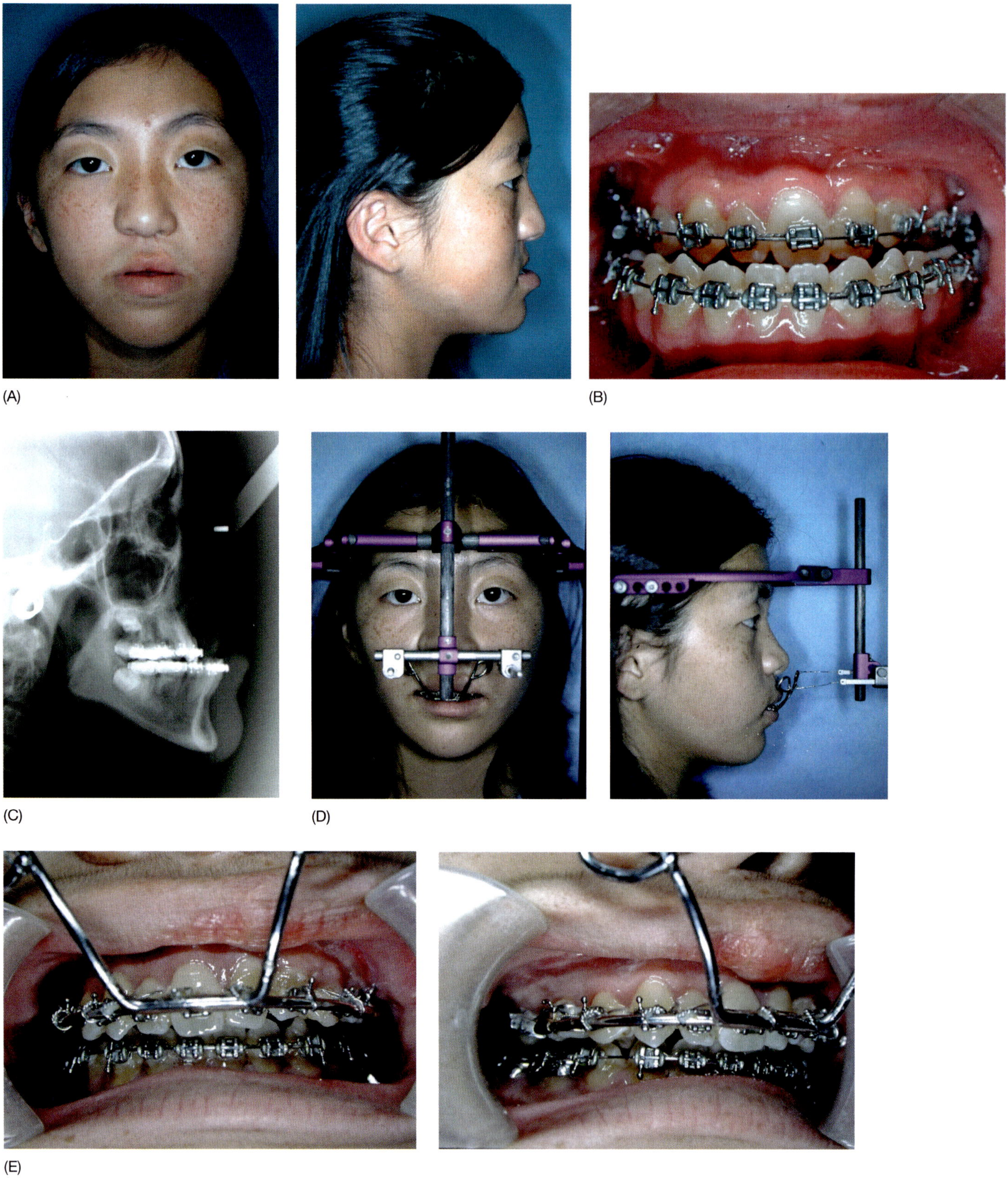

FIGURE 1 • (A) Pre-operative frontal and lateral photographs of a 15-year-old girl with left unilateral complete cleft lip/palate and midface deficiency. (B) Intraoral photograph and (C) lateral cephalogram showing 15 mm anterior crossbite and severe maxillary hypoplasia. (D) Frontal and lateral photographs of rigid external distraction device (RED, KLS Martin L.P, Kalamazoo, MI) secured to cranium. *(continued)*

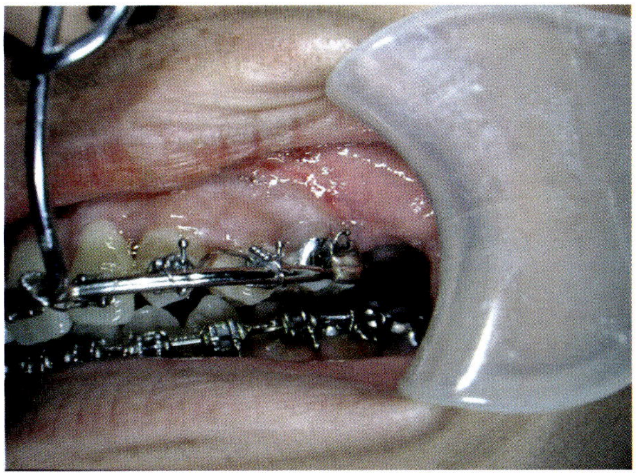

(E)

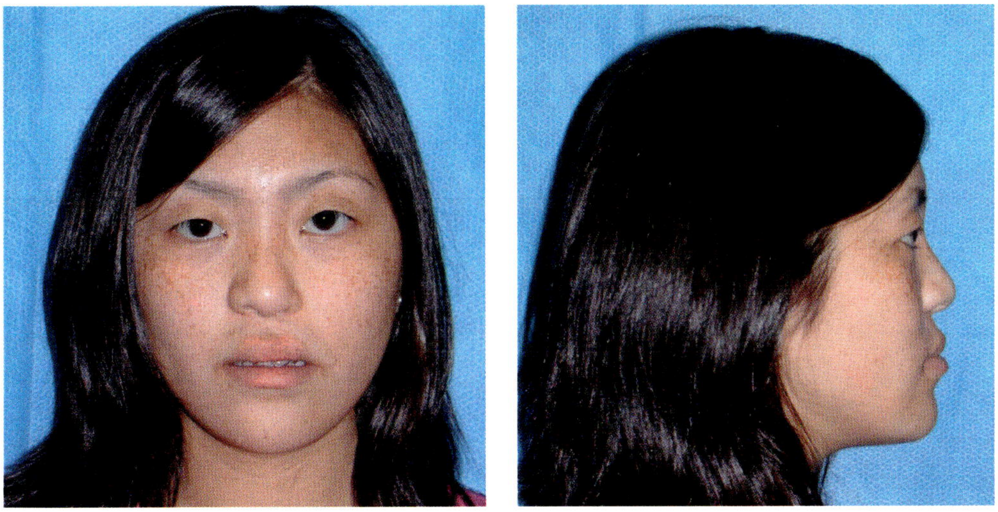

(F)

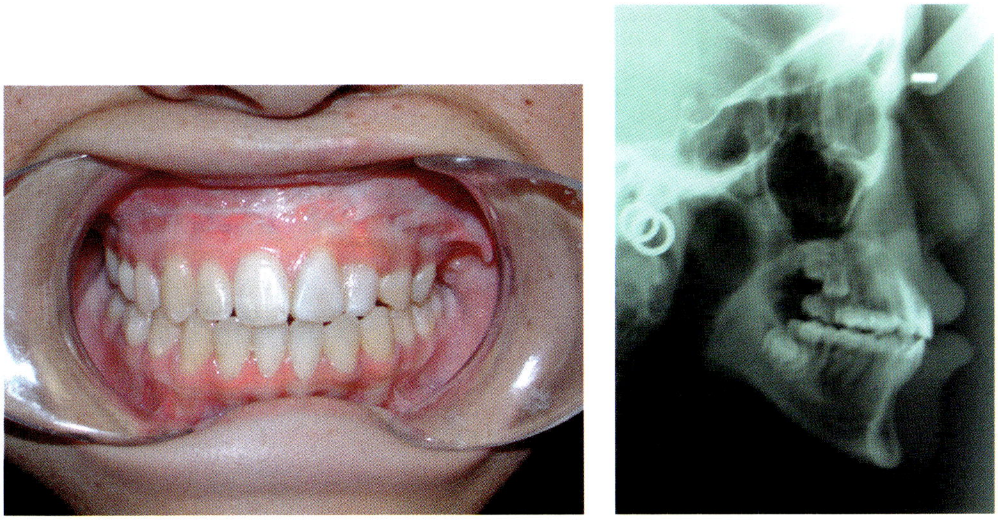

(G)

FIGURE 1 • (*continued from previous page*) (E) Intraoral photographs of tooth-borne distraction appliance soldered to molar bands and secured to teeth with 26-gauge wire. (F) Five year post-operative frontal and lateral photographs showing excellent esthetics. (G) Intraoral photograph and lateral cephalogram demonstrating stable result.

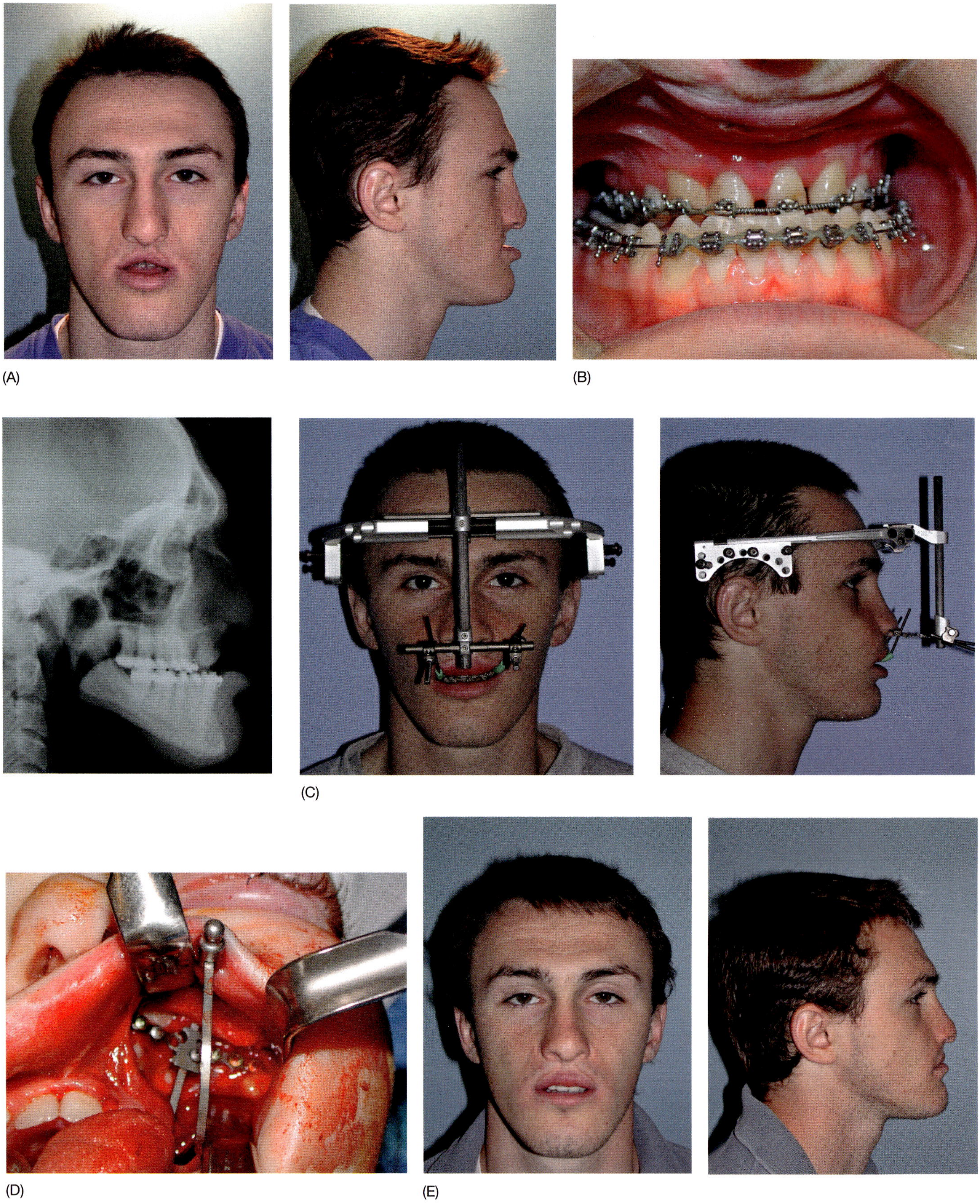

FIGURE 2 • (A) Pre-operative frontal and lateral photographs of a 17-year-old male with bilateral complete cleft lip/palate and midface deficiency. (B) Intraoral photograph and lateral cephalogram showing 14 mm anterior crossbite and severe maxillary hypoplasia. (C) Frontal and lateral photographs of rigid external distraction device (Synthes® CMF, West Chester, PA) secured to cranium. (D) Intraoral photograph of bone retained traction hooks secured to anterior wall of maxilla below osteotomy. (E) Two year post-operative frontal and lateral photographs showing excellent facial esthetics. *(continued)*

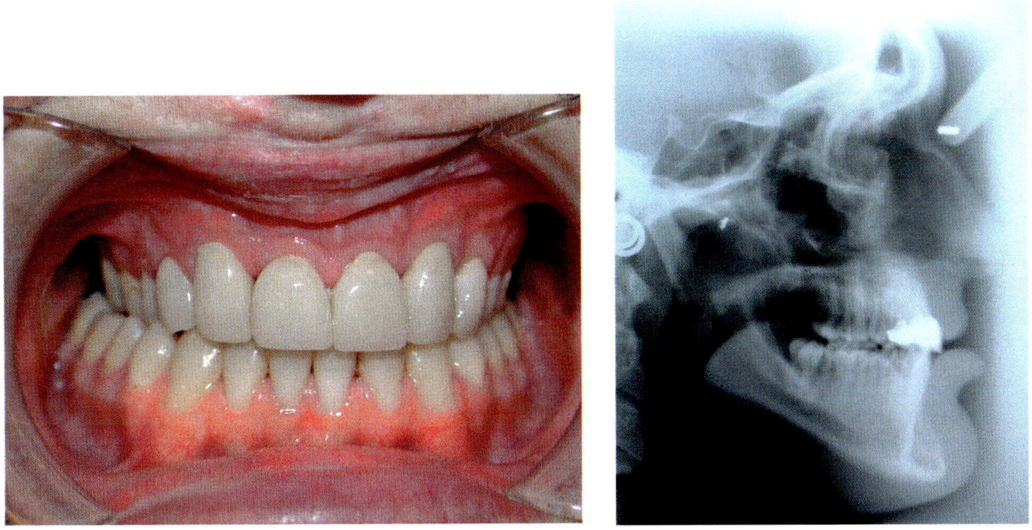

(F)

FIGURE 2 • *(continued from previous page)* (F) Intraoral photograph and lateral cephalogram demonstrating stable result.

during consolidation of the regenerate. It also allows for optimization of the occlusion, which can be difficult to do with distraction.[16] At the time the device is removed and the maxilla stabilized, ancillary procedures such as mandibular osteotomy, genioplasty, malar implants, and so on can be performed (Figure 3).

The cumbersome nature of the external device has led to considerable interest in developing internal appliances. Buried devices are designed to push the maxilla forward and they have the advantage of being low profile, hidden, and not as cumbersome as the external devices. They are rigid, so they can be used for fixation after completion of distraction. The distraction ratio with internal devices, that is, actual millimeters of bone lengthening compared to the number of activations of the device, is close to 1:1 allowing for more predictable and faster distraction.[17] However, the internal devices are not without their limitations. They have unidirectional mechanics, which reduces the ability to control vectors of movement. In fact, the surgeon cannot make any mid-course corrections of the vector with internal unidirectional devices. The anterior maxillary wall is thin, which may make placement and fixation of the device difficult. In addition, areas of thin bone, presence of teeth, and the requirements of the desired vector can make positioning the device a geometric problem. Multiple general anesthetics are required to place and remove the device. Unlike the rigid external distraction device, internal devices may be difficult to position in young patients because of unerupted teeth. Other complications include injury to the commissure, swelling of lips, and local wound infections.[18]

TECHNIQUE

A standard LeFort I osteotomy, including pterygomaxillary dysjunction, is performed and the maxilla is down-fractured and mobilized. If a rigid external device is being utilized, the patient either has a tooth-borne appliance cemented in place (Figure 1) or bone retained distraction hooks must be rigidly fixed to the inferior maxilla below the osteotomy but above the tooth roots (Figure 2). The bone plates that secure the distraction hooks are bent so as to lie passively, and are secured with several screws. The hooks are bent to go around the lip and should not put tension on the commissures. The rigid external device is secured to the cranium using four to six percutaneous screws per side. The device is positioned 3 cm to 4 cm above the superior border of the helix on each side and within the hair line. The plane of the device is set parallel to the Frankfurt horizontal. Twenty-six gauge traction wires are connected from the external hooks of the rigid intra-oral splint or bone-retained plate to the activating screws located on the horizontal bar of the device.

Semi-buried distraction devices are placed and secured proximally to the zygoma and distally to the dentoalveolar segment above the teeth and below the osteotomy (Figure 3 and 4). The devices are placed as parallel as possible to one another to control advancement in the coronal and horizontal planes. The turning element used to activate the expansion screw to advance the maxilla must be accessible and should not put undo pressure on the lip. To aid in proper positioning of the devices and decrease OR time, a stereolithographic 3D acrylic model can be created from the CT data and the footplates of the distraction devices bent pre-operatively.

Distraction commences 1 to 3 days post-operatively, at a rate of 1 full turn per day (approximately 1 mm per day). The neutral fixation period is at least 2 times the number of millimeters of distraction in days. Removal of the rigid external device can either be in the office with local anesthesia and intravenous conscious sedation or the patients are returned to the operating room for deep sedation or general anesthesia. Removal of the semi-buried intraoral devices must be done in the operating room under general anesthesia.

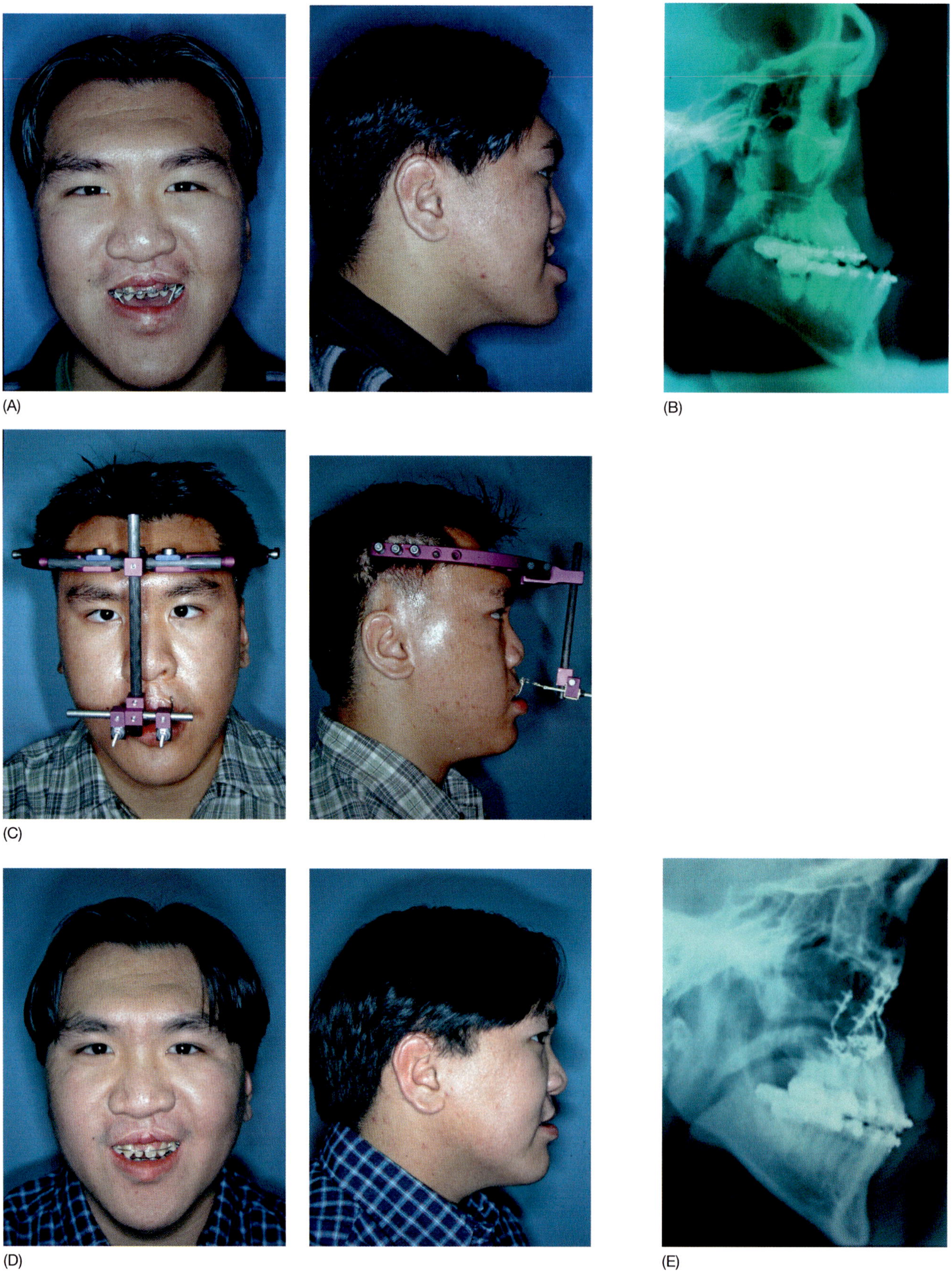

FIGURE 3 • (A) Pre-operative frontal and lateral photographs of a 17-year-old male with bilateral complete cleft lip/palate and midface deficiency. (B) Lateral cephalogram showing 17-mm anterior crossbite and severe maxillary hypoplasia. (C) Frontal and lateral photographs of rigid external distraction device (RED, KLS Martin L.P, Kalamazoo, MI) secured to cranium. (D) Two year post-operative frontal and lateral photographs following hybrid technique: removal of external distraction device and rigid fixation of maxilla and bilateral intraoral vertical ramus osteotomies for mandibular set back. (E) Lateral cephalogram documenting stable occlusion.

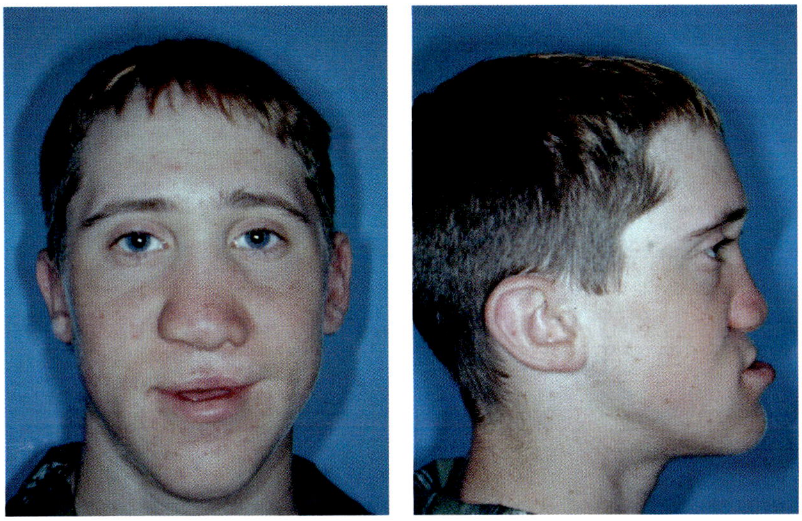

(A)

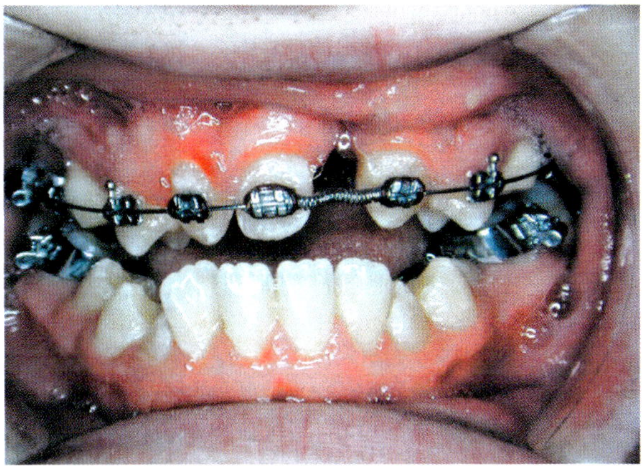

(B)

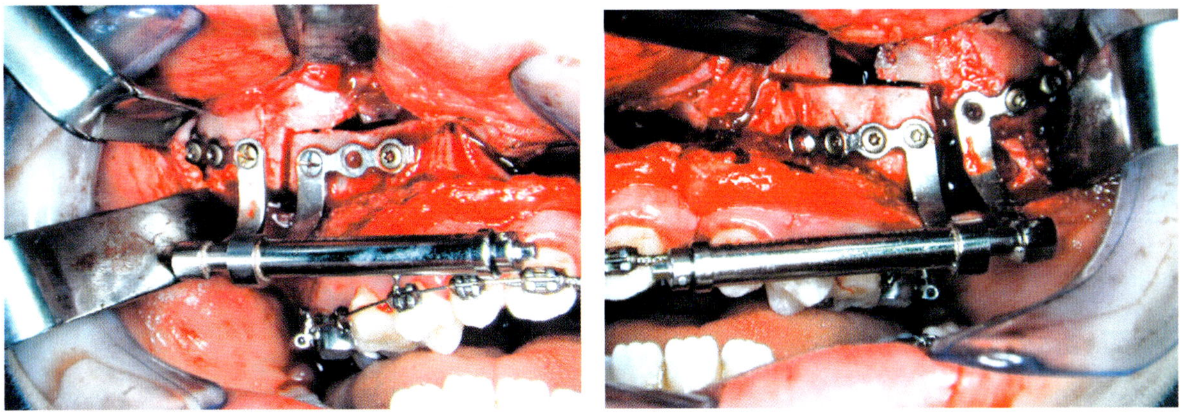

(C)

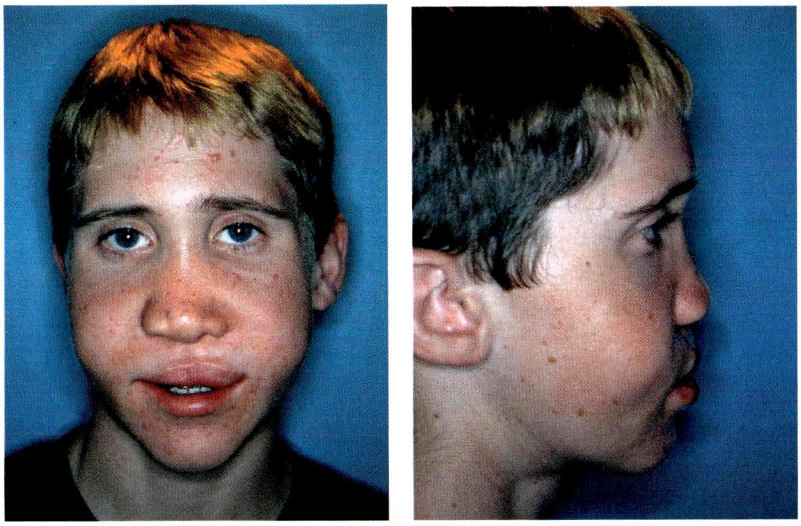

(D)

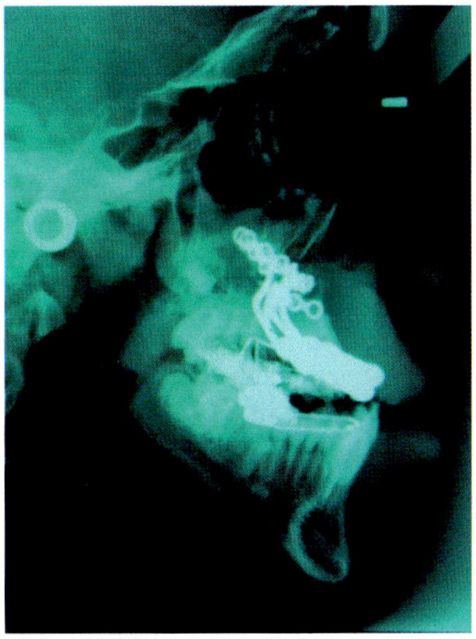

(E)

FIGURE 4 • (A) Pre-operative frontal and lateral photographs of 13-year-old boy with binderoid left unilateral complete cleft lip/palate and midface deficiency. (B) Intraoral photograph showing absent maxillary left central, left, and right lateral incisors and lateral cephalogram demonstrating 10-mm anterior crossbite and 3-mm anterior open bite. He has not had full orthodontic decompensation. (C) Intraoral photograph of internal maxillary distraction devices placed after stepped LeFort I osteotomy. (D) Frontal and lateral photographs during active distraction. (E) Lateral cephalogram shows position of distraction devices as parallel as possible to provide symmetric sagittal and vertical movement. (*continued*)

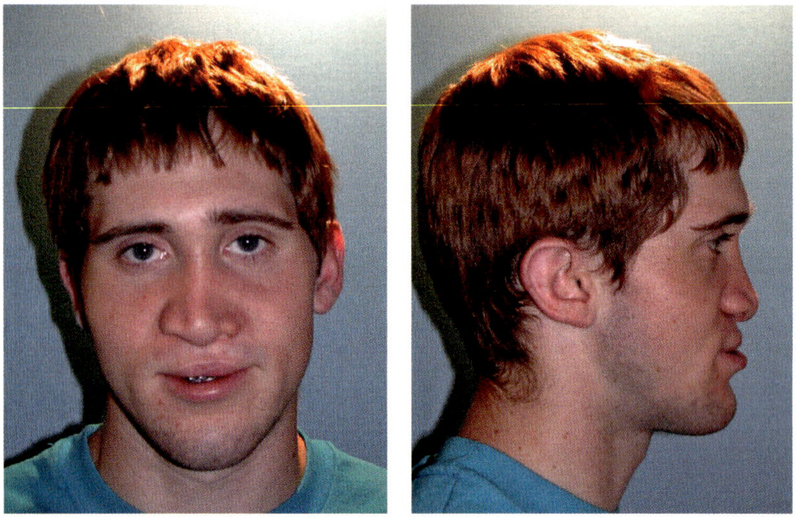

(F)

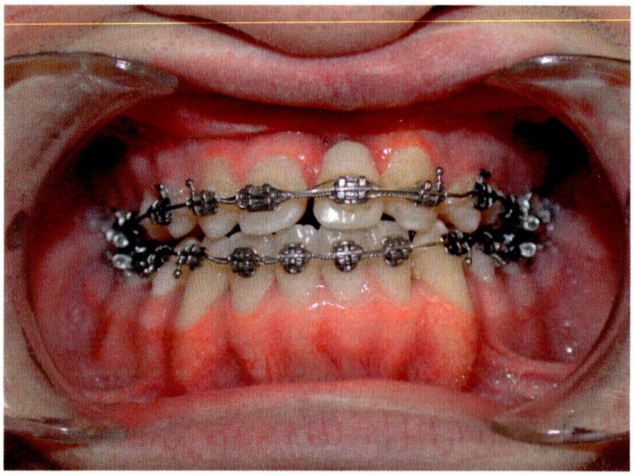

(G)

FIGURE 4 • (*continued from previous page*) (F) Frontal and lateral photographs 3 years after LeFort I distraction and cranial bone graft to nasal dorsum demonstrates stable result with improved esthetics. (G) Intraoral photograph shows occlusion near the end of orthodontic treatment.

CONCLUSION

Maxillary distraction osteogenesis is an exciting addition to the surgeon's armamentarium. The technique is currently indicated in patients with cleft lip/palate, craniofacial anomalies, and in those patients who require large magnitudes of skeletal advancement that are likely to be difficult or impossible using the standard orthognathic and craniofacial approaches.

References

1. Molina F, Ortiz Monasterio F, de la Paz Aguilar M, Barrera J. Maxillary distraction: aesthetic and functional benefits in cleft lip-palate and prognathic patients during mixed dentition. Plast Reconstr Surg 1998;101:951–63.
2. Polley JW, Figueroa AA. Management of severe maxillary deficiency in childhood and adolescence through distraction osteogenesis with an external, adjustable, rigid distraction device. J Craniofac Surg 1997;8:181–5.
3. Polley JW, Figueroa AA. Rigid external distraction: its application in cleft maxillary deformities. Plast Reconstr Surg 1998;102:1360–72.
4. Ko EW, Figueroa AA, Polley JW. Soft tissue profile changes after maxillary advancement with distraction osteogenesis by use of a rigid external distraction device: A 1-year follow-up. J Oral Maxillofac Surg 2000;58:959–69.
5. Figueroa AA, Polley JW, Friede H, Ko EW. Long-term skeletal stability after maxillary advancement with distraction osteogenesis using a rigid external distraction device in cleft maxillary deformities. Plast Reconstr Surg 2004;114:1382–92.
6. Cohen SR. Craniofacial Distraction with a modular internal distraction system: evolution of design and surgical techniques. Plast Reconstr Surg 1999;103:1592–607.
7. Cedars MG, Linck DL, Chin M, et al. Advancement of the midface using distraction techniques. Plast Reconstr Surg 1999;103:429–41.
8. Figueroa AA, Polley JW, Ko EW. Maxillary distraction for the management of cleft maxillary hypoplasia with a rigid external distraction system. Semin Orthodon 1999;5:46–51.
9. Harada K, Baba Y, Ohyama K, et al. Soft tissue profile changes of the midface in patients with cleft lip and palate following maxillary distraction osteogenesis: a preliminary study. Oral Surg Oral Med Oral Pathol 2002;94:673–7.
10. Schulten A, Lim AA, Bruun RA, Hayes C, Mulliken JB, Padwa BL. Combined push-pull distraction for correction of syndromic midfacial hypoplasia. J Oral Maxillofac Surg. 2006;64:23–30.
11. Kazanjian VH, Converse JM. Fractures of the Maxilla. In Kazanjian VH, Converse JM editors. The Surgical Treatment of Facial Injuries. Baltimore: The Williams & Wilkins Company; 1949. p. 148–168.
12. Polley JW, Figueroa AA. Rigid external distraction: its application in cleft maxillary deformities. Plast Reconstr Surg 1998; 102:1360–72.
13. Nout E, Wolvius EB, van Adrichem LN, Ongkosuwito EM, van der Wal KG. Complications in maxillary distraction using the RED II device: a retrospective analysis of 21 patients. Int J Oral Maxillofac Surg 2006;35:897–902.
14. Rieger J, Jackson IT, Topf JS, Audet B. Traumatic cranial injury sustained from a fall on the rigid external distraction device. J Craniofac Surg 2001;12:237–41.
15. Le BT, Eyre JM, Wehby MC, Wheatley MJ. Intracranial migration of Halo fixation pins. A complication of using an extraoral distraction device. Cleft Palate Craniofac J 2001;38:401–4.
16. Wong GB, Padwa BL. LeFort I soft tissue distraction: a hybrid technique. J Craniofac Surg. 2002;13:572–6.
17. Harada K, Sato M, Omura K. Maxillary distraction in patients with cleft deformity using a rigid external distraction device: a pilot study on the distraction ratio of the maxilla to the device. Scand J Plast Reconstr Surg Hand Surg 2004;38:277–80.
18. Kessler P, Wiltfang J, Schultze-Mosgau S, Hirschfelder U, Neukam FW. Distraction osteogenesis of the maxilla and midface using a subcutaneous device: report of four cases. Br J Oral Maxillofac Surg 2001;39:13–21.

18 Midface Distraction Osteogenesis

Robert J. Paresi Jr., MD, MPH / James Shoukas, MD / Alvaro Figueroa, DDS / John W. Polley, MD

OUTLINE

- Introduction
- Indications
- Methods
- Results
- Cephalometric Evaluation
  - Angular Changes
  - Linear Changes
  - Dental Changes
- Discussion
- Conclusion

INTRODUCTION

Patients with severe maxillary deficiency present multiple challenging problems for the reconstructive team. These patients often exhibit multidimensional maxillary hypoplasia and skeletal clefting with absence of maxillary and alveolar bone, as well as scarring, residual fistulas, and dental anomalies. Traditional surgical/orthodontic approaches, while sometimes successful in obtaining stable occlusal relationships, often fall short of expectations with respect to facial balance and aesthetics. The application of maxillary distraction osteogenesis in the treatment protocol of patients with severe maxillofacial anomalies offers a powerful alternative for the reconstructive team. In this chapter, we present the use of maxillary distraction osteogenesis for the treatment of patients with severe maxillary deficiency secondary to orofacial clefts also (See Chapter 17). The technique of rigid external distraction, based upon the old concept of cranial fixation, has enabled rigid control over the distraction process allowing for predictable and highly successful outcomes.

INDICATIONS

Patients with severe maxillary hypoplasia (negative overjet of 8 mm or more) are considered candidates for treatment with maxillary distraction osteogenesis. Criteria for patient selection include the following: unilateral or bilateral cleft lip and palate, severe maxillary hypoplasia (vertical, horizontal, and transverse planes) with class III malocclusion, patients requiring horizontal maxillary advancement in excess of 8 mm, normal mandibular morphology and position, patients with severe palatal and pharyngeal scarring, and patients with airway obstruction and sleep apnea.

METHODS

All patients undergo a thorough history and clinical examination as well as a complete dental and orthodontic examination. Pre-operative and post-operative photographic and cephalometric records are obtained. Time is spent with the patient, and the patient's family, explaining the distraction process in detail, utilizing photographs, video imaging, as well as discussions with other patients and their families who have undergone the procedure. The patient and parents are thoroughly familiarized with the distraction apparatus and its mechanics prior to the procedure.

A custom-made intraoral orthodontic splint, which acts as the link between the maxillary skeleton and distraction apparatus, is manufactured and inserted for each patient. This intraoral splint is constructed with rigid 0.045 or 0.050 stainless steel orthodontic wire. The splint is cemented to the first permanent molars or second primary molars and further secured at the time of surgery with circumdental stainless steel surgical wire (Figures 1A, 1B). Alternately, the splint can be secured to two maxillomandibular fixation screws placed just lateral to the canines in place of the circumdental wires. For older children and adults, the splint is inserted in the clinic prior to the surgical procedure. For young children, the splint is placed in the operating room after anesthetic induction and prior to the osteotomy. External traction hooks are bent under and in front of the upper lip with the end of the hook ending at the level of the palatal plane. The splint also includes intraoral hooks that are utilized for the required retention phase at the completion of active maxillary distraction. Certain orthodontic tooth movements as well as orthodontic expansion of the upper arch can be performed at the time of maxillary distraction by incorporation of an expansion screw, and segmentalizing the intraoral splint into two units. In order to improve rigidity, the splint can be manufactured by using an orthodontic cervical face bow. The external bow is bent to create the external traction hooks. In the anterior

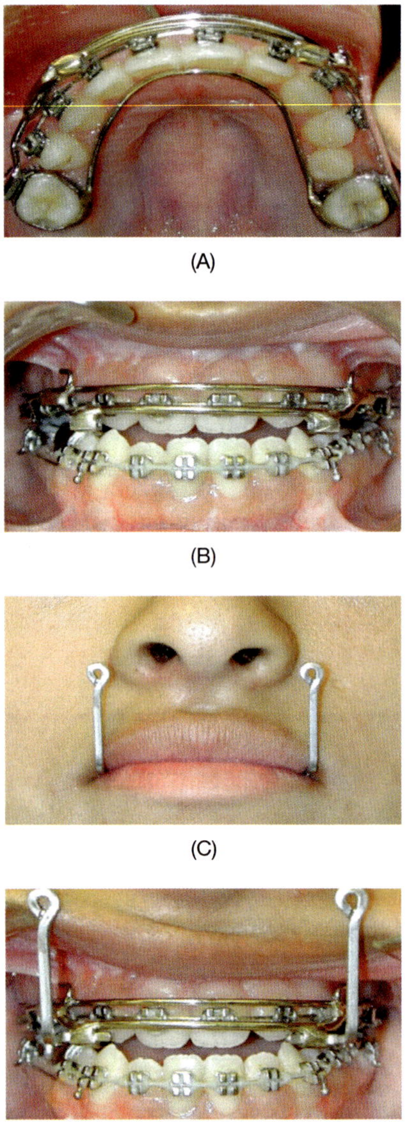

(A)

(B)

(C)

(D)

FIGURE 1 • A custom-made intraoral orthodontic splint; (A, B) intraoral splint is constructed with rigid 0.045 or 0.050 stainless steel orthodontic wire cemented to the first permanent molars or second primary molars; (C, D) removable heavy rectangular wire external traction hooks at level of palatal plane.

region two square tubes are soldered just medial to the oral commissures. These tubes are used to house removable traction hooks made of heavy rectangular wire (Figures 1C, 1D). This design offers the advantage of not having the external traction hooks present at the time of surgery facilitating anesthesia management and intraoral surgical manipulation. For ease of management, or where orthodontic support is limited, a splint can be manufactured by a neighboring dental lab (AOA Laboratories, 13931 Spring St., Sturtevant, WI 53177, 1-800-262-5221).

The Le Fort I osteotomy is the standard procedure for cleft patients undergoing distraction with the height of the transverse osteotomy varying depending on the dental age and aesthetic requirements of each patient. In patients with transitional dentition, a high transverse maxillary osteotomy is performed, just below the level of the inferior orbital rim. The osteotomy circumvents the infraorbital foramen to prevent injury to the infraorbital neurovascular bundle and avoids all permanent tooth buds. Mobilization of the maxillary segments is achieved, including pterygomaxillary and septal disjunction. Aggressive down-fracturing is not necessary. No intraoperative repositioning of the maxillary segments is achieved and no autogenous or allogenic bone grafting or internal skeletal fixation hardware is utilized. The halo portion of the distraction device is placed immediately after closure of the intraoral incision. The halo device and its use are illustrated in Figures 2 and 3C.

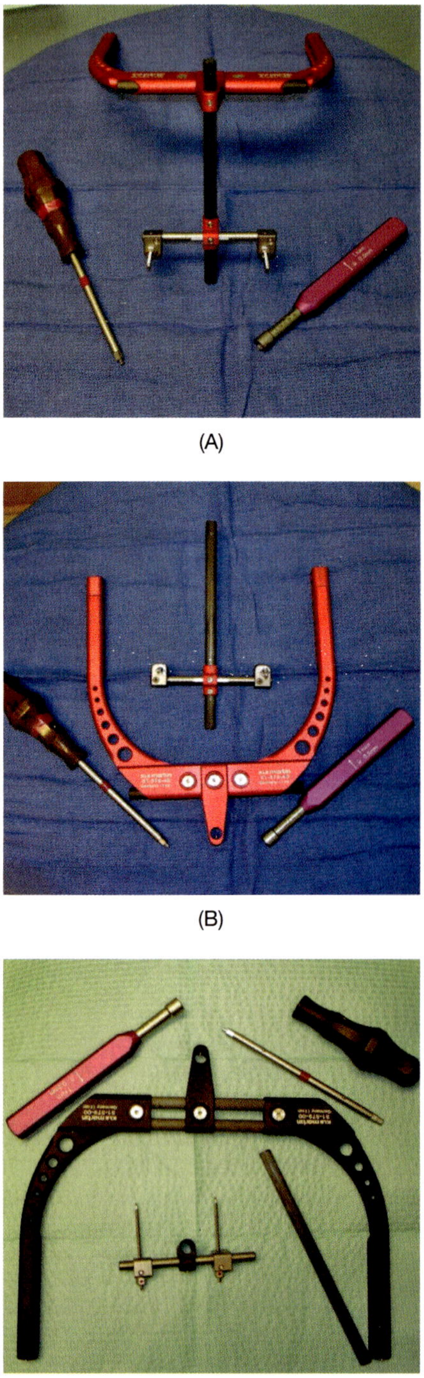

(A)

(B)

(C)

FIGURE 2 • The halo portion of the external distraction device.

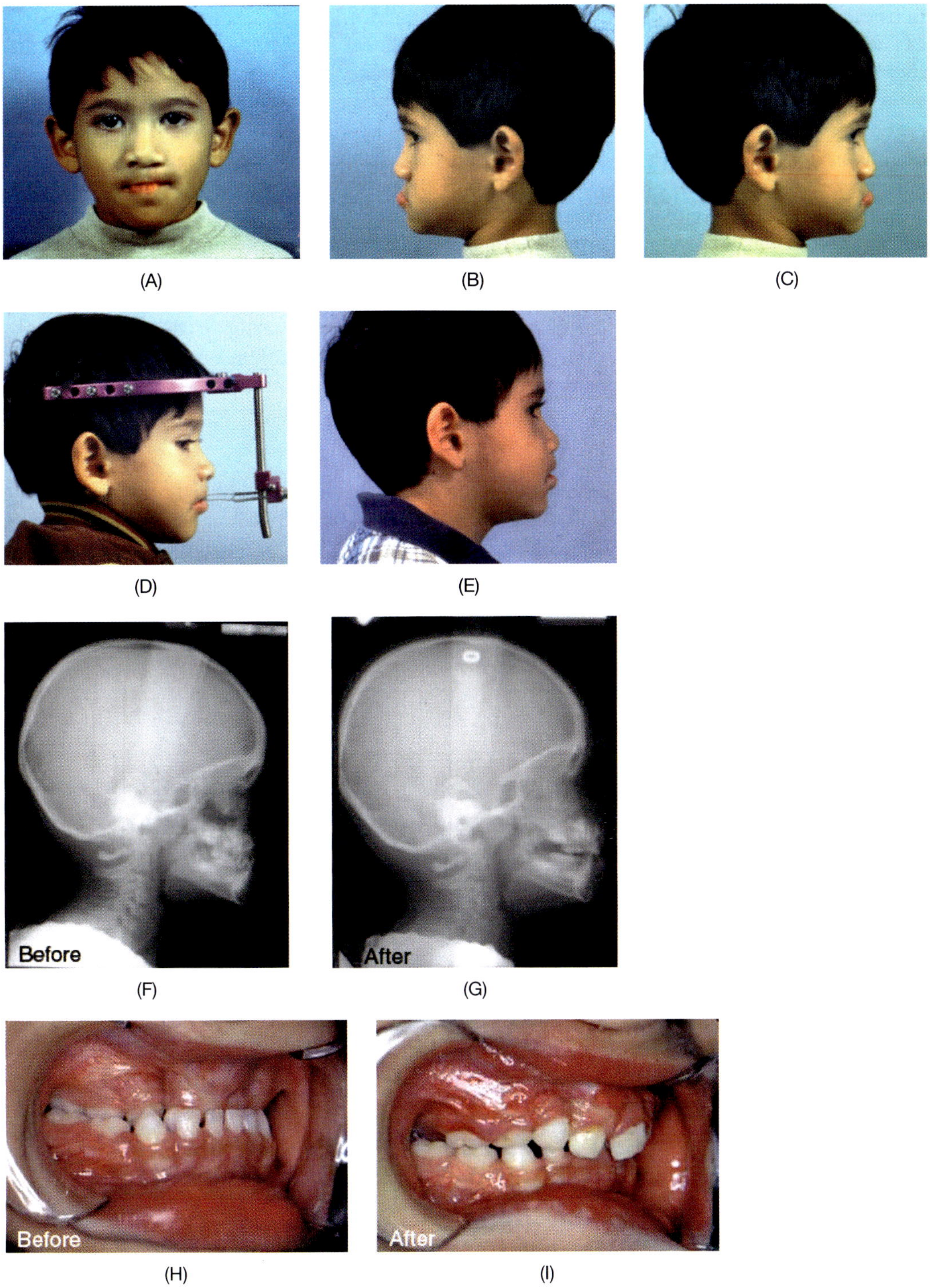

FIGURE 3 • Clinical facial, cephalometric, and occlusal photographs of 5-year-old patient before treatment with distraction (A, B, C, F, H), during distraction (D), and after distraction (E, G, I). These photographs illustrate maxillary advancement with good facial balance and occlusal relations.

RESULTS

In our first clinical series, eighteen consecutive patients were selected based on the above criteria, and were treated with maxillary distraction osteogenesis. There were 10 unilateral cleft lip and palate patients, 6 bilateral cleft lip and palate patients, and 2 patients with bilateral cleft lip and palate and severe congenital facial clefting. The patients' ages at the time of surgery ranged from 5.2 years to 25.2 years.

Fourteen patients underwent rigid external maxillary distraction.[1] These patients had the rigid device placed at the time of surgery and had their distraction performed through mechanical activation of the distraction device. Four patients in this series underwent face mask elastic distraction osteogenesis which was later abandoned because of poor results. All rigid external maxillary distraction patients had a latency period of 4 to 5 days following the osteotomy and then began distraction. The latency period is followed by the activation period. This is the period of active distraction. The activation period typically lasts 12 to 15 days, but the duration is determined clinically by the severity of the midface deficiency and anterior crossbite. The patients are followed weekly during the activation period until satisfactory and stable skeletal and occlusal changes are achieved.

After the period of active distraction, a period of rigid retention is maintained for three to four weeks. During this period, the distraction device is kept in place without any advancement. The purpose of this period is to allow the maxillary position to achieve clinical stability through bone formation and mineralization. Following the period of rigid retention, the external distraction device is removed in an office setting in most patients. For very young patients, the distraction device is removed under mild sedation.

CEPHALOMETRIC EVALUATION

The pre-operative and post-retention lateral cephalometric radiographs were utilized for analysis of our case series. The radiographs were traced on 0.003-inch acetate paper, and 12 anatomic landmarks were recorded (Figure 4). Availability of serial radiographs in all patients permitted landmark verification. All X-rays were corrected to 0% magnification. Based on the recorded anatomic landmarks, 13 measurements were calculated, 6 angular, and 7 linear (4 horizontal and 3 vertical). For the linear measurements, an x-y coordinate system utilizing the S-N plane as the horizontal was employed. Linear horizontal changes were measured relative to a line perpendicular to the S-N plane passing through the sella, and vertical changes were measured perpendicular to the S-N plane. The pre-operative and post-operative cephalometric values in the rigid external distraction group were statistically analyzed by means of a paired t-test.

Changes in the soft tissue anthopometric measurements are evident after distraction as seen in Figure 5. The nasolabial angle has increased with concurrent raising of the pronasale. The changes in the soft tissue as a result of the underlying bony framework has given this patient an impressive aesthetic improvement.

Angular Changes

In our series of patients who underwent rigid external distraction, the average pre-distraction S-N-A angle was 77.6 degrees, and the post-distraction S-N-A angle was 85.3 degrees, for an average increase in this group of 7.7 degrees (Table 1). The average pre-distraction A-N-B was –1.2 degrees and post-distraction

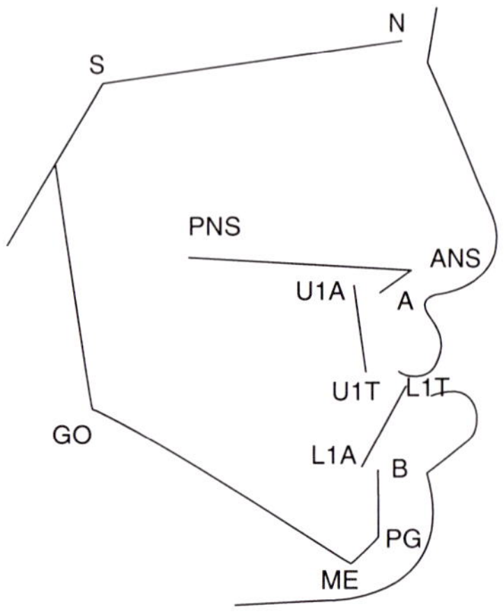

FIGURE 4 • Anatomic landmarks and reference planes. Anatomic landmarks: sella (*S*), center of sella turcica; nasion (*N*), most anterior point of the nasal frontal suture; a point (*A*), most anterior limit of the maxillary alveolar bone at the level of the incisor root apex; posterior nasal spine (*PNS*), intersection between the nasal floor and the posterior contour of the maxilla; apex of maxillary incisor root (*U1A*), uppermost point of the incisor root; tip of maxillary incisor crown (*U1T*), maxillary incisor edge; tip of mandibular incisor (*L1T*), mandibular incisor edge; apex of mandibular incisor root (*L1A*), lowermost point of the mandibular incisor root; B point (*B*), most anterior limit of the mandibular alveolar bone at the level of the incisor root apex; pogonion (*PG*), most anterior limit of the mandibular symphysis; gonion (*GO*), the point at the greater convexity of the mandibular gonial region. Reference planes: sella-nasion plane (*SN*), palatal plane (line passing through *U1A* and *U1T*): and mandibular plane (tangent to the lower border of the mandible through *ME* and *GO*).

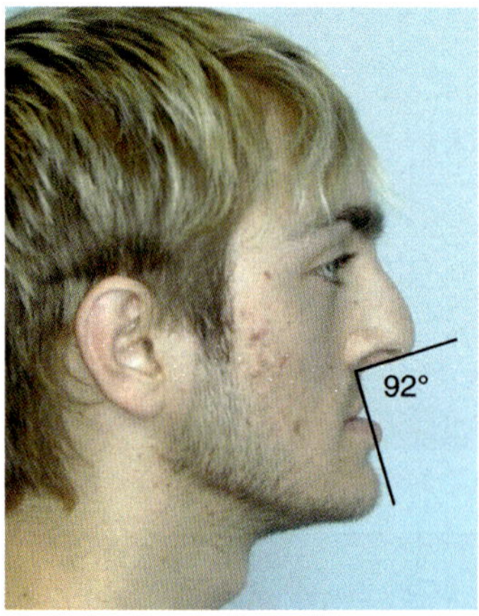

(A)

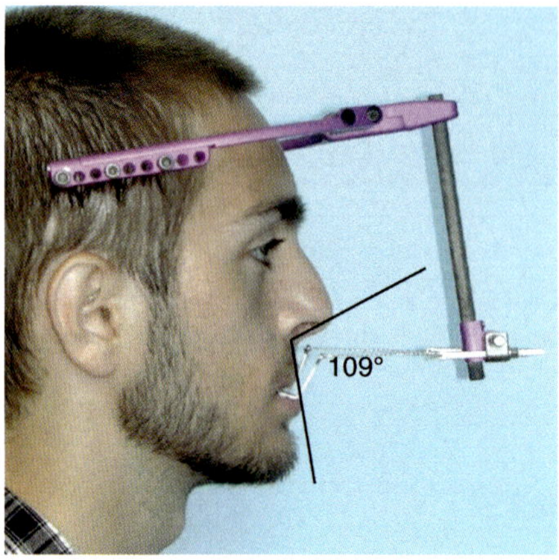

(B)

FIGURE 5 • A 24-year-old male before and after midface advancement. Notice the 17-degree increase in the nasolabial angle and the raising of the pronasale.

TABLE 1 Pre-distraction and post-distraction angular cephalometric measurements for rigid external distraction group.

DISTRACTION	MEASUREMENTS (°)	PRE-DISTRACTION	POST-DISTRACTION (4 months)	DIFFERENCE	SIGNIFICANCE
Rigid external (n = 14)	SNA	77.6 ± 5.6	85.3 ± 5.6	7.7 ± 2.9	p < 0.001
	SNB	78.8 ± 4.0	77.9 ± 4.1	−0.8 ± 1.8	NS
	ANB	−1.2 ± 3.5	7.3 ± 3.0	8.6 ± 3.6	p < 0.001
	Convexity (NAPg)	−3.5 ± 7.5	13.7 ± 6.0	17.2 ± 7.3	p < 0.001
	Mand. PL/SN angle	39.2 ± 6.7	41.4 ± 5.9	2.2 ± 2.4	p < 0.01
	UI-P.PL.angle	100.7 ± 15.7	98.8 ± 14.4	−1.2 ± 11.3	NS

7.3 degrees with an increase of 8.6 degrees. For all patients who underwent rigid external distraction, the skeletal angle of convexity increased post-distraction by 17.2 degrees. All of these three measurements were statistically significant.

Linear Changes

The ANS change between pre-distraction and post-distraction cephalometric radiographs in the rigid external distraction group was 7.1 mm. The average horizontal advancement of the A point following distraction was 8.3 mm. The horizontal advancement at the upper incisal edge averaged 11.6 mm and a positive correction of overjet by 12.7 mm post-distraction. All of the changes in the horizontal linear measurements in the rigid external group were highly significant between pre-distraction and post-distraction measurements at a *p* value of 0.001. (See Table 2.)

Dental Changes

The dental changes in both groups between pre-distraction and post-distraction cephalograms are also given in Tables 1 and 2.

TABLE 2 Pre-distraction and post-distraction linear cephalometric measurements for rigid external distraction group

LANDMARK (axis)	PRE- TO POST-DISTRACTION CHANGE RIGID EXTERNAL (mm)
ANS-x	7.1 ± 3.9*
ANS-y	−0.4 ± 3.0
A point −x	8.3 ± 3.3*
A point −y	−1.3 ± 3.4
UI-x	11.6 ± 4.6*
UI-Y	−1.8 ± 3.5
Overjet	12.7 ± 3.0*

* $P < 0.001$

In the rigid external distraction group, the change in the angle of the upper incisor edge of the palatal plane averaged 1.2 degrees for all patients. This was not statistically significant. In none of the patients in this series, were spaces created posterior to the most distal point of anchorage of the intraoral splint.

DISCUSSION

Maxillary hypoplasia is a common finding in patients with orofacial clefting. It has been estimated that 25% to 50% of all patients with complete unilateral cleft lip and palate are potential candidates for maxillary advancement to correct the functional deformities and improve aesthetic facial proportions.[2] In addition, it is recognized that the majority of patients with facial clefts have morphologically normal or slightly smaller than normal mandibles.[3,4] Patients with severe cleft maxillary deficiency are difficult patients to treat with standard surgical/orthodontic approaches. These patients present with maxillary hypoplasia (vertical, horizontal, and transverse dimensions) characterized by thin or structurally weak bone. The maxillary hypoplasia in cleft patients is also compounded by residual palatal and alveolar fistulas, absent and aberrant dentition, and scarring of the palatal and pharyngeal soft tissues.

The physical deformities exhibited by patients with severe maxillary hypoplasia contribute to multiple functional deficiencies as well. These include severe malocclusions, which result in compromised mastication, speech abnormalities, and nasal pharyngeal airway constriction.[5] The severe dish-face or concave facial profiles in these patients result in highly detrimental psychosocial ramifications as well.[6] Current protocols for the treatment of maxillary hypoplasia in cleft patients rely upon a surgical/orthodontic approach, including a Le Fort I maxillary advancement with concomitant fistula closure and maxillary and alveolar bone grafting. This surgery includes rigid internal fixation hardware for stabilization of the repositioned maxilla in the post-operative period. The long-term results of cleft patients with maxillary deficiency treated in such fashion have been

TABLE 3 Review of long-term results of sagittal maxillary relapse in cleft patients following maxillary advancement with fistula closure, bone grafting, and rigid internal fixation.

AUTHOR	ERBE ET AL. [7]	CHEUNG ET AL. [8]	POSNICK ET AL. [9]	HOCHBAN ET AL. [10]	ESKENAZI AND SCHENDEL [11]	POLLEY AND FIGUEROA [1]
N cleft Pt.	11	46	35	14	12	14
Mean follow-up	59 months	28 months	12 months	12 months	12 months	4 months
Mean max. advancement	4.6	4.5	6.9	8	7.8	11.6
Mean relapse	40%	22%	21%	25%	4%	-

reported by several authors.[7–11] The mean horizontal maxillary advancement in these reported series has averaged between 5 and 7 mm, and the mean long-term horizontal relapse averages between 20 and 25 percent (Table 3). Erbe and associates[7] presented a mean follow-up of 59 months for 11 cleft patients who underwent segmental osteotomies with maxillary advancement and simultaneous fistula closure with rigid internal fixation and bone grafting. On average, the greater maxillary segment in these patients was advanced 3.9 mm and the lesser maxillary segment 5.3 mm. At nearly 5 years post-operatively they found that the horizontal relapse of the maxilla was approximately 40%. Cheung and colleagues[8] reported a consecutive series of 46 cleft patients who underwent maxillary advancement with rigid internal fixation and simultaneous alveolar bone grafting with fistula closure. The mean horizontal maxillary advancement in this series was 4.5 mm. Relapse in the horizontal plane for the unilateral cleft patients in this series, at a mean of 28 months post-operatively, was 22%. Similar horizontal relapse rates following maxillary advancement with rigid internal fixation in cleft patients have been reported by others as well (Table 3).[9–11]

In this series, 14 consecutive patients underwent successful maxillary advancement at the Le Fort I level through maxillary distraction with rigid external distraction. Examples of our clinical experience with rigid external distraction are illustrated in Figures 3, 6, and 7. No rigid internal fixation hardware or autogenous bone grafting was utilized at the time of the osteotomy. The patients underwent distraction at the rate of 1 mm per day followed by a 3 to 4 week rigid retention period. In this group of 14 cleft patients, the mean effective maxillary advancement was 11.7 mm. The time of the post-distraction cephalometric radiographs analyzed in this group of patients averaged 4 months following completion of the distraction process. These data have been presented so that the immediate effects of the distraction process can be measured accurately (Table 3). In the immature patients, maxillofacial growth will continue, and we are currently following the development of these patients.

Long-term skeletal stability after maxillary advancement with distraction osteogenesis has been documented in our series of 17 patients when comparing them at the completion of distraction and after 2 or more years of follow-up[12] (Table 4, Figure 8). These patients demonstrated an increase in maxillary length of 1.9 mm (20%) when measured at the A' point at 2 years of follow-up. It is clear from this data and the corresponding average superimposition tracing that the regression seen with traditional Le Fort I level advancements is not a problem with distraction

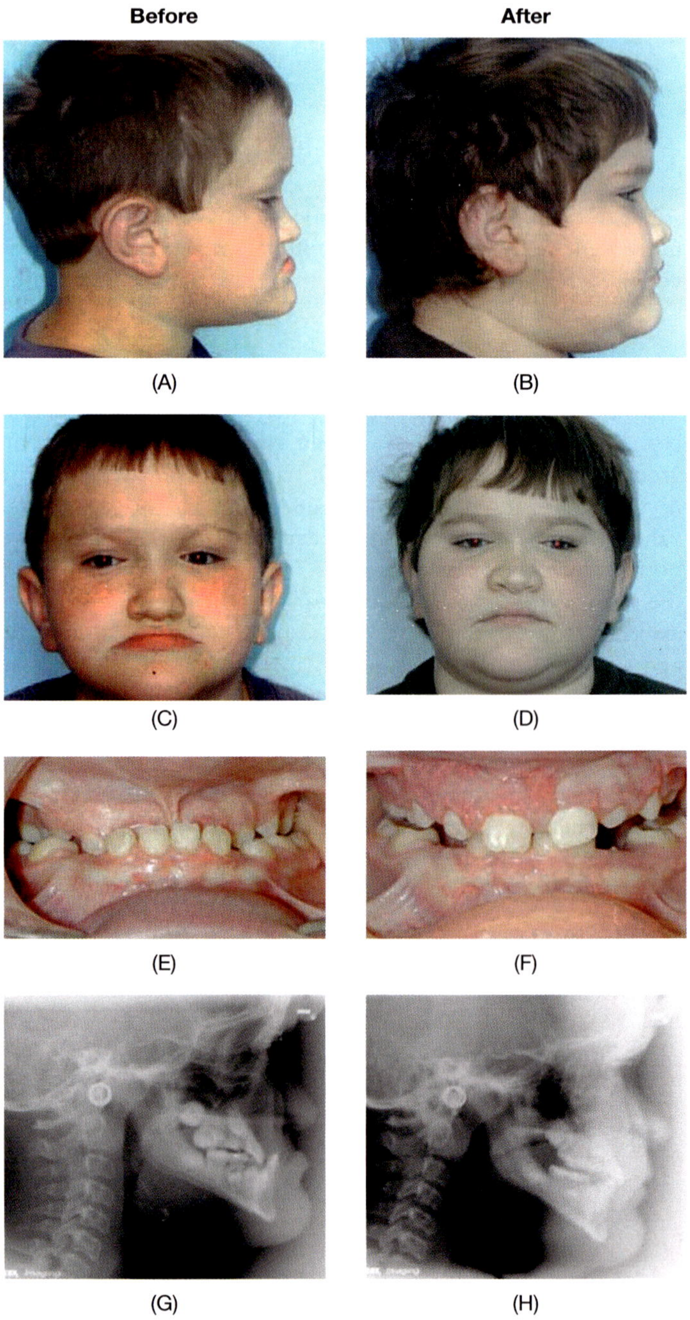

FIGURE 6 • Clinical facial, cephalometric, and occlusal photographs of a 13-year-old patient before treatment with distraction (A, C, E, G), and after distraction (B, D, F, H). These photographs illustrate maxillary advancement with good facial balance and occlusal relations.

Before:

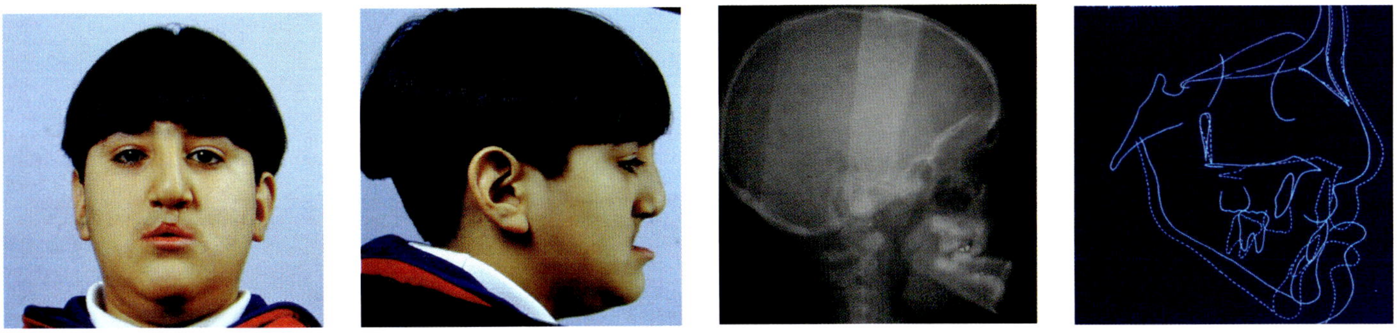

(A) (D) (G) (I)

During:

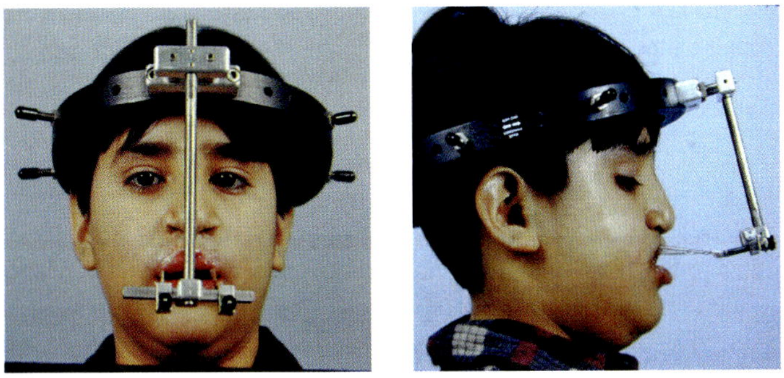

(C) (F)

FIGURE 7 • Clinical facial, cephalometric, and occlusal photographs of a 10-year-old patient before treatment with distraction (A, D, G, I), during distraction (C, F), and after distraction (B, E, H, J). The cephalogram shown in Figure 7J demonstrates the clinical stability of maxillary advancement 2 years after distraction.

After:

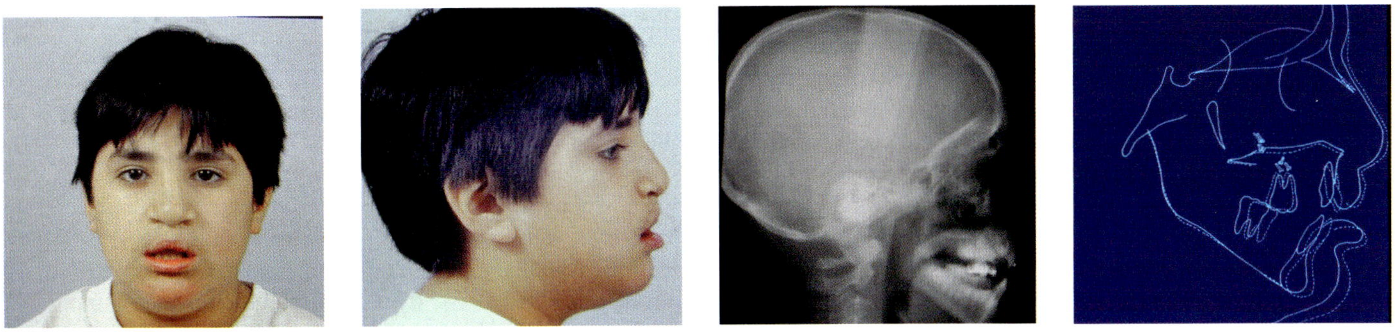

(B) (E) (H) (J)

TABLE 4 Cephalometric values immediately after and 2 or more years after distraction

	IMMEDIATELY AFTER DISTRACTION (n = 17)		TWO OR MORE YEARS AFTER DISTRACTION (n = 17)			
VARIABLE	MEAN	SD	MEAN	SD	DIFFERENCE	p
SNA (°)	83.4	4.1	81	5.4	−2.4	0.01
SN/NL (°)	9.1	7.1	10	5.3	0.9	NS
SN/ML (°)	41.4	5.5	40.9	5	−0.5	NS
A-PM (mm)	52	5.3	53.9	5.2	1.9	0
ANS-PM (mm)	54	5.3	54	5	0	NS
ANS-S (H) (mm)	65.2	5.6	65.5	6	0.3	NS
ANS-S (V) (mm)	56	9	58.6	6.5	2.6	0.05

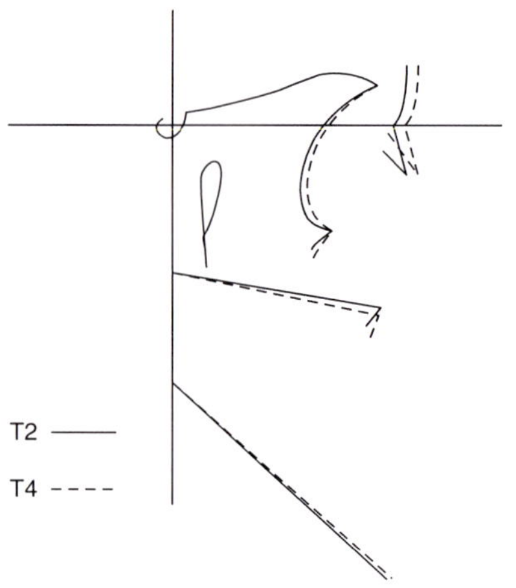

FIGURE 8 • Average superimpositions illustrating changes occurring between soon after maxillary distraction (*T2, solid line*) and 2 or more years later (*T4, dashed line*). Note anterior growth, and anterioposterior maxillary stability.

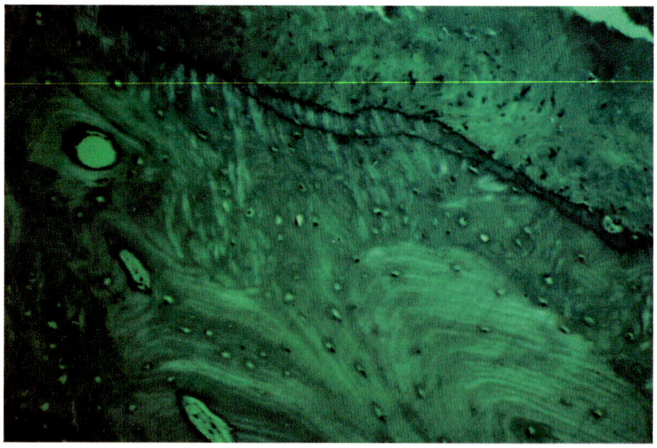

FIGURE 9 • Histologic section of a bone biopsy specimen obtained from the pterygomaxillary region at the time of finishing Le Fort I surgery in a patient who underwent previous maxillary advancement with rigid external distraction. Note the well-ossified dense lamellar bone.

osteogenesis. The regression seen in the SNA in this series is related to the forward growth at the nasion point over this period of follow-up, not a regression of point A'. This is expected given the mean age of 12.6 years of the patients in the series. Other studies have demonstrated some relapse in the horizontal plane in the first six to 12 months after distraction with no regression after 12 months.[13–15] This suggests that skeletal stability occurs at 12 months preventing relapse after that time.

In the past, it has been difficult to consistently and successfully treat patients with severe maxillary deficiency with maxillary advancement alone. Patients with large anteroposterior maxillomandibular discrepancies may require mandibular setback surgery in addition to maxillary advancement for correction of their severe horizontal deficiency. With the use of rigid external distraction, we can now gradually, and in a very stable fashion reposition a severely hypoplastic maxilla to the exact horizontal and vertical position desired. The patients create their own autogenous bone during this process, eliminating the need for a donor site as well as eliminating the need for rigid internal fixation hardware (Figure 9). The use of rigid external distraction has allowed rigid control over the distraction process and has enabled us to follow our surgical and aesthetic guidelines for the reconstruction of these patients by correcting the entire maxillary skeletal and soft-tissue discrepancy in the region of the hypoplasia only. The expansion of the soft tissue facial mask yields the most pleasing long-term aesthetic facial balance and harmony,[16,17] particularly in cleft patients.

In this initial consecutive series of patients, we utilized two different techniques for maxillary distraction. One group (face mask distraction) underwent maxillary distraction with the use of orthodontic face mask and elastic traction.[18] With the use of the elastic distraction, we were able to obtain only partial correction of the horizontal maxillary deficiencies in these patients. We considered all of our face mask patients to have failed their surgical treatment. We therefore changed the design of the distraction technique to rigid external distraction, which utilizes a skeletally (cranial) fixed distraction device that allows for rigid, predictable control over the distraction process.

Patients who underwent rigid external distraction had a mean effective horizontal maxillary movement at the upper incisal edge of 11.7 mm. A point advancement in our series of patients with rigid external distraction averaged 8.3 mm. The change in the skeletal angle of convexity in the Rachmiel and colleagues[19] series with face mask traction averaged 5.0 degrees compared with a change of 17.2 degrees in our series of rigid external distraction. Diner and associates[20] obtained only 3 mm of horizontal maxillary advancement at A point in the series of patients who underwent maxillary distraction osteogenesis with the use of a face mask and elastic traction. The experience of Hung and colleagues[21] was similar with the use of elastic forces for maxillary advancement in cleft patients. In their series, they found the mean horizontal maxillary advancement to be only 5.2 mm. Maxillary distraction at the Le Fort I level with the use of face mask and elastic traction is unpredictable and unreliable, allowing for horizontal advancement in the 4-mm to 6-mm range only. In our practice, patients who require only 4 mm to 6 mm of maxillary advancement are not routinely considered for treatment with distraction osteogenesis. In these patients, standard orthodontic surgical approaches at the appropriate age are generally performed.

Internal distraction devices have also been reported for use in midface advancement. Most of these reports have focused on advancements of the midface at the Le Fort III level or above.[22,23] Some, however, have advocated the use of an internal device for maxillary advancement at the Le Fort I level (See Chapter 17, Figure 4).[22] Internal devices may require multiple surgical approaches for their placement, a second operative procedure for their removal, and the necessity for an exit port for the activating arm of the device. The vectors of distraction with the use of internal devices will be limited by the placement of each device as well as its finite mechanics. Perhaps the greatest disadvantage of internal devices is in the design of the transverse maxillary osteotomy. With internal distraction devices, the osteotomy must be designed so that there is enough stable bone above and below the osteotomy line to allow appropriate placement and fixation of the distraction hardware. This is not required with rigid external distraction. The osteotomy design with this device is based upon the aesthetic requirements of each individual patient and not based upon the placement of internal hardware. This allows the transverse maxillary osteotomy to be carried as high as indicated along the piriform aperture as well as in the malar regions, allowing for maximal correction of the patient's pre-operative facial concavity. Maxillary advancement with rigid external distraction allows total versatility and flexibility in both the amount and direction of the distraction process. The vectors of distraction can be, and are changed at any time during the distraction process. In addition, no additional surgical procedure is required for removal of the rigid external device.

The use of the cranium as an anchorage point for the stabilization of maxillofacial surgery is not a new concept. Cranial stabilization devices for maxillofacial trauma reconstruction as well as for elective maxillofacial surgery have been reported with good success in the past.[24,25] In addition, our neurosurgical colleagues have used the cranium as a solid fixation point for the stabilization of cervical injuries and reconstructions for decades. The scalp pins (2–3 per side), which stabilize the halo component of the rigid external distraction device, are discomfort free. Not even the youngest patients have any complaints or problems with wearing of the device throughout the distraction process. No special scalp pin care is required, and the use of ointments and creams at the scalp pin interface is discouraged. The patients simply shampoo their hair in the shower with the device in place in a normal fashion. Currently, the device is readily removed without even local anesthesia in the clinic following the phase of rigid retention. In young apprehensive patients, the halo is removed with mild sedation. No secondary surgical procedure is required for removal of internal hardware.

One of the great advantages of distraction osteogenesis in the craniomaxillofacial skeleton is that, in theory, there is no limitation to the age at which patients can be treated. Contemporary surgical orthodontic approaches for the treatment of maxillary deficiency in cleft patients are dependent upon the patient having reached skeletal maturity before the reconstructive surgery can be performed. Treating patients in the transitional dentition stage with the use of autogenous bone grafting and rigid internal fixation plates and screws is technically difficult to perform without injury to the developing permanent tooth buds. The technique of rigid external distraction eliminates the negative technical factors associated with traditional orthognathic surgery in patients in transitional dental development. With maxillary rigid external distraction, only an osteotomy is performed. Repositioning skeletal segments, internal fixation hardware, and bone grafting are not required. The only limitations with rigid external distraction include adequate dentition, either primary or secondary, for fixation of the intraoral splint, as well as the ability of the patient to wear the device. Rigid external distraction now allows treatment of these patients from the age of 2 years and older. In special circumstances in patients with craniosyntosis (Crouzon and Apert syndromes) we have used the technique for monobloc advancement as early as 18 months of age. In these cases, the second or even the first primary maxillary molar have been used to support the intraoral splint. In addition, special plates have been developed to connect the edentulous maxilla to the rigid external distraction device.[26]

Our three-year follow-up of this group of patients after distraction has demonstrated outstanding stability in maxillary position.[12] We believe this stability is provided by substantial bone creation in the pterygomaxillary region with distraction as evidenced by Figure 9.[12] However, we caution all young patients undergoing rigid external distraction that they may require a final "finishing" Le Fort I procedure at skeletal maturity for final arch alignment.

CONCLUSION

The advantages of maxillary rigid external distraction are numerous. This technique allows an excellent modality for correcting the severe midface concavities, at any age, in patients with facial clefting and other hypoplastic anomalies (ectodermal dysplasia, Johansson Blizzard syndrome, and so on). Rigid external distraction follows the important principles of aesthetic maxillofacial surgery treating only the affected jaw and offers the multiple benefits of distraction osteogenesis, including not only orthotopic bone creation with expansion of the facial skeleton but soft tissue expansion as well. The technical advantages of rigid external distraction are numerous. The surgery is minor as compared with traditional orthognathic surgery. Only the osteotomy is performed, eliminating repositioning of skeletal segments, bone grafting, splints, intermaxillary fixation, and internal fixation hardware. Operative times are significantly decreased. Morbidity with rigid external distraction is very low (in our series zero). Blood transfusions are not required. Rigid external distraction can be performed on an outpatient or 23-hour admission basis. Patients begin a soft diet and normal oral hygiene the morning following surgery. The rigid external device is removed in the clinical setting or with mild sedation for the young children. All of the above has resulted in a significantly decreased cost of care for these challenging patients.

In patients with severe maxillary hypoplasia, distraction osteogenesis with rigid external distraction now offers the ability to fully restore facial convexity through a minimal procedure at almost any age. In our practice, this has dramatically changed our treatment philosophies and success for these patients who in the past have been extremely difficult to treat. The technique is currently applied with extreme success to those patients with severe maxillary non-syndromal dentofacial deformities, as well as patients with severe syndromic conditions such as Apert's, Crouzon, and Pfeiffer syndrome.[13]

References

1. Polley JW, Figueroa AA. Management of severe maxillary deficiency in childhood and adolescence through distraction osteogenesis with an external, adjustable, rigid distraction device. *J Craniofac Surg* 1997;8:181–5.
2. Ross RB. Treatment variables affecting facial growth in complete unilateral cleft lip and palate: Part 7. An overview of treatment and facial growth. *Cleft Palate J* 1987;24:5–77.

3. Da Silva Filho OG, Normando AD, Capelozza Filho L. Mandibular growth in patients with cleft lip and/or cleft palate-the influence of cleft type. *Am J Orthod Dentofacial Orthop* 1993;104: 269–75.
4. Semb G. A study of facial growth in patients with unilateral cleft lip and palate treated by the Oslo CLP team. *Cleft Palate Craniofac J* 1991;28:1–21.
5. Witzel MA, Vallino LD. Speech problems in patients with dentofacial and craniofacial deformities. In W.H. Bell (ed.), *Modern practice in orthognathic and reconstructive surgery,* Vol. 3. Philadelphia: Saunders; 1992. p. 1686.
6. Kapp-Simon K. Psychological adaptation of patients with craniofacial malformations. Psychological aspects of facial form. Monograph No. 11, Craniofacial Growth Series, Center for Human Growth and Development, University of Michigan. pp. 143–160.
7. Erbe M, Stoelinga PJ, Leenan RJ. Long-term results of segmental repositioning of the maxilla in cleft palate patients without previously grafted alveolopalatal clefts. *J Craniomaxillofac Surg* 1996;24: 109–17.
8. Cheung LK, Sammam N, Hiu E, Tideman H. The 3-dimensional stability of maxillary osteotomies in cleft palate patients with residual alveolar clefts. *Br J Oral Maxillofac Surg* 1994;32:6–12.
9. Posnick JC, Dagys AP. Skeletal stability and relapse patterns after Le Fort I maxillary osteotomy fixed with miniplates: the unilateral cleft lip and palate deformity. *Plast Reconstr Surg* 1994;94:924–32.
10. Hochban W, Ganss C, Austermann KH. Long-term results after maxillary advancement in patients with clefts. *Cleft Palate Craniofac J* 1993;30:237–43.
11. Eskenazi LB, Schendel SA. An analysis of Le Fort I maxillary advancement in cleft lip and palate patients. *Plast Reconstr Surg* 1992;90:779–86.
12. Figueroa AA, Polley JW, Friede H, et al. Long term skeletal stability after maxillary advancement with distraction osteogenesis using a rigid external distraction device in cleft maxillary deformities. *Plast Reconstr Surg* 2004;114:1382–92.
13. Figueroa, AA, Polley JW, Ko E. Distraction osteogenesis for treatment of severe cleft maxillary deficiency with the RED technique. In: Sanchukov ML, ed. *Craniofacial distraction osteogenesis, Chap. 55.* St. Louis: Mosby; 2001. pp. 485–493.
14. Suzuki EY, Motohashi N, Ohyama K. Longitudinal dento-skeletal changes in UCLP patients following maxillary distraction osteogenesis using RED system. *J Med Dent Sci* 2004;51:27–33.
15. Wen-Ching Ko E, Figueroa AA, Polley JW. Soft tissue profile changes after maxillary advancement with distraction osteogenesis by use of a rigid external distraction device: a 1-year follow-up. *J Oral Maxillofac Surg* 2000;58:959–69.
16. Rosen HM. Facial skeletal expansion: Treatment strategies and rationale. *Plast Reconstr Surg* 1992;89:798–808.
17. Rosen H. Aesthetics in facial skeletal surgery. *Perspect Plast Surgery* 1992;6:1.
18. Molina F, Ortiz-Monasterio F. Maxillary Distraction. Three years of clinical experience. In: *Proceedings of the 65th Annual Meeting of the American Society of Plastic and Reconstructive Surgeons, Plastic Surgical Forum, Vol. XIX.* 1996. p. 54.
19. Rachmiel A, Laufer D, Aizenbud D. Surgically assisted orthopedic protraction of the maxilla in cleft palate patients by distraction osteogenesis (Abstract 198). *New Orleans: American Cleft Palate-Craniofacial Association 54th Annual Meeting;* 1997.
20. Diner PA, Martinez H, Tarbadar Y, et al. Experience with distraction in maxillary deficiency at Trousseau hospital (Abstract 60). *Paris: International Congress on Cranial and Facial Bone Distraction Processes;* 1997.
21. Hung KF, Lin WY, Huang CS, Chen KT, Lo LJ. The maxillary movement distraction: Preliminary results (Abstract 55). *Paris: International Congress on Cranial and Facial Bone Distraction Processes;* 1997.
22. Cohen SR, Burstein FD, Stewart MB, et al. Maxillary-midface distraction in children with cleft lip and palate: A preliminary report. *Plast Reconstr Surg* 1997;99:1421–8.
23. Chin M, Toth BA. Le Fort III advancement with gradual distraction using internal devices (Abstract 76). *Paris: International Congress on Cranial and Facial Bone Distraction Processes*; 1997.
24. Stoelinga PJ, Van der Vijver HR, Leenen RJ, et al. The prevention of relapse after maxillary osteotomies in cleft palate patients. *J Craniomaxillofac Surg* 1987;15:326–31.
25. Houston WJ, James DR, Jones E, et al. Le Fort I Maxillary osteotomies in cleft palate cases. J. *Craniomaxillofac Surg* 1989;17:9–15.
26. Hierl T, Hemprich A. Callus Distraction of the midface in the severely atrophied maxilla: A case report. *Cleft Palate Craniofac J* 1999;36:457.

Complications with Distraction Osteogenesis of the Craniofacial Skeleton

19

Carl Bouchard, DMD / Basel Sharaf, DMD, MD / Ryan J. Smart, DMD, MD / Leonard B. Kaban, DMD, MD / Bonnie Padwa, DMD, MD / Maria J. Troulis, DDS, MSc

OUTLINE

INTRODUCTION

Since distraction osteogenesis (DO) was first described by Codivilla[1], popularized by Ilizarov in the 1950s[2,3] and described by McCarthy and associates in the craniofacial skeleton in 1992,[4] the appeal of distraction osteogenesis as an alternative to traditional osteotomies and bone grafting for correction of craniofacial dysmorphology and skeletal expansion has grown. (See Chapters 10–18.) Some of the benefits of distraction include: a less invasive procedure, less risk of nerve damage, less relapse, elimination of the need for grafts, and histogenesis of the surrounding soft tissue matrix.[5–8] (See Chapters 10–18).

Like any surgical intervention, DO bears the risk of complications. The distraction devices are cumbersome and bulky as well as complicated to use. Possible difficulties encountered by patients treated by DO include pain during activation, speech and eating problems, sleeping difficulty and disturbances of recreational activities.[9] Treatment planning is more complex and there is greater demand for patient understanding and cooperation (See Chapter 11 and 13).[10] Procedural time for placement of the distractors is usually shorter compared to alternative methods, however, more post-operative visits are generally required to monitor the progressive bone elongation.[11] The issues cited here account for many of the complications surrounding the use of distraction and can generally be avoided with careful treatment planning, patient selection, patient education and appropriate technique. In this chapter, we will review some of the common complications, the incidence of these complications, and outline methods for minimizing the incidence of these complications.

PRE-OPERATIVE PHASE (PLANNING)

Patient Selection

Distraction osteogenesis, more than with alternative surgical procedures, is highly dependant on patient and family compliance. Some factors that affect patient compliance are the bulk and complexity of the appliance, the presence of the device, prolonged treatment period, and pain associated with the device or distraction. There is a reported 4.7% complication rate related to patient compliance and a 1.0% complication rate of pain, which was severe enough to prevent or limit distraction.[12] When discussing the risks and benefits of procedures, it is particularly important with distraction to educate the patient and the family about these important issues. Because distraction requires patience and compliance on the part of the patient, careful selection of the patient is of extreme importance. During the initial interview, any hesitance by the patient or the patient's family should be thoroughly assessed before moving forward with this treatment.

Patients and families must be educated that the distractors can be bulky, unsightly, and difficult to clean. They must be instructed and evaluated in the use of the distractor, ideally prior to placement of the device (Figure 1). Social issues such as interactions with other children in school, inability to engage in physical activity, and the requirement of a blenderized diet for a prolonged time should be emphasized.

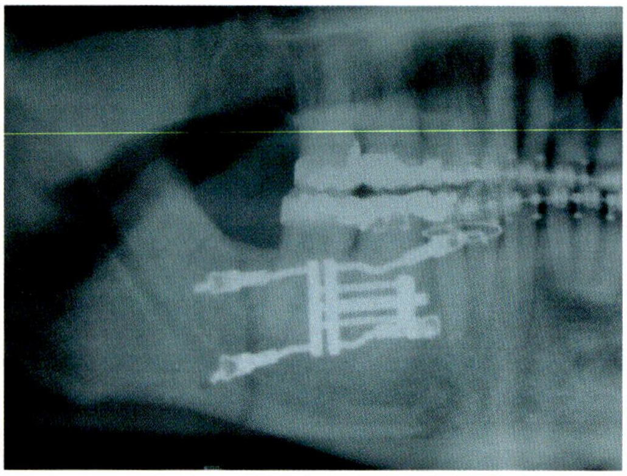

(A)

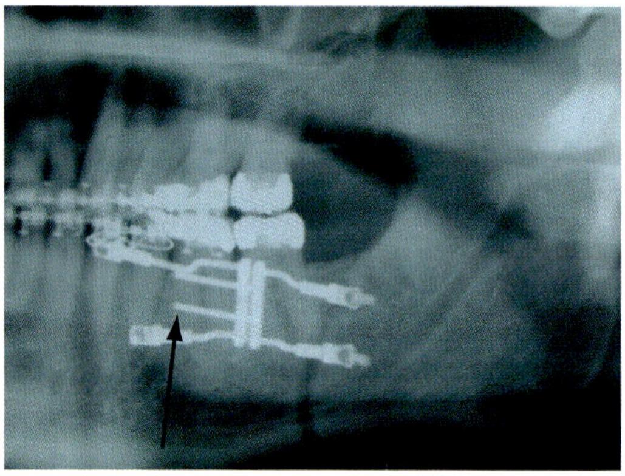

(B)

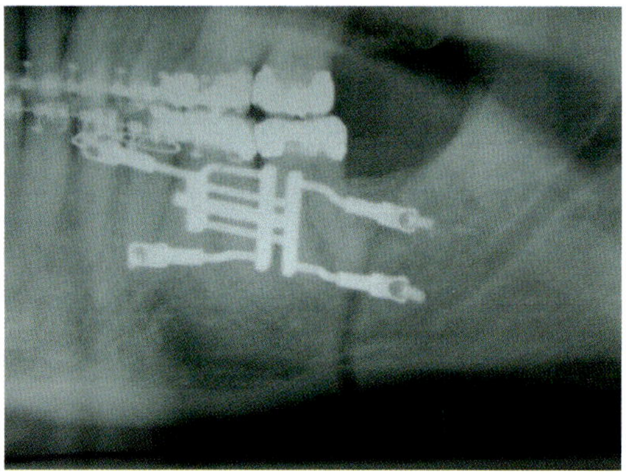

(C)

FIGURE 1 • Dismantled device. The patient activated the appliance in the wrong direction on the left side. (A) Intact device on the right side. (B) Dismantled appliance on the left side (arrow). (C) After the appliance was reassembled. The distraction continued uneventfully. Reproduced from: Troulis MJ, and Kaban LB. 2003: Complications of mandibular distraction osteogenesis. Oral Maxillofac Surg Clin North Am 15:251–264, Elsevier, with permission.

The patient and family should also be informed that they will require multiple visits to the clinic and that they must be willing to activate their devices on a daily basis.[9,10] This will prove time-consuming for the patient and the surgeon and the patient must be willing to accept this fact.

Distraction Vector

Inappropriate distraction vector resulting in malocclusion (Figures 2 and 3) and asymmetry is a common problem. Mofid and colleagues reported an inappropriate vector of distraction in

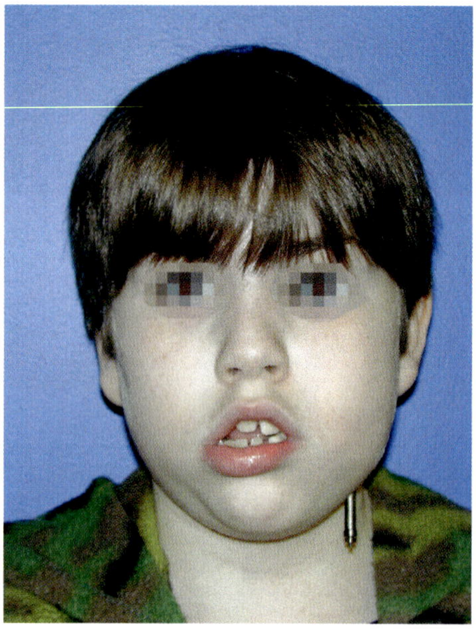

(A)

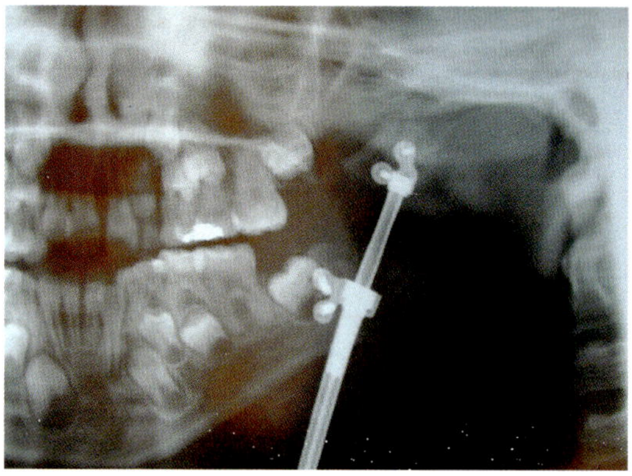

(B)

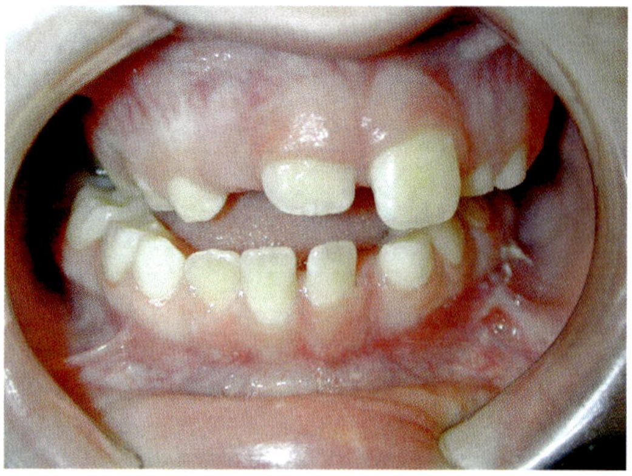

(C)

FIGURE 2 • Incorrect vector of mandibular distraction. (A) This 7 year-old-male with left hemifacial microsomia (type II-A) had distraction of the left mandibular ramus. (B), The vector of distraction was incorrect. Note the forward angulation of the device on the post-operative panorex. (B) The device should have been vertically oriented. This resulted in deviation of the mandible to the contralateral side and creation of a posterior bilateral crossbite instead of an ipsilateral open-bite.

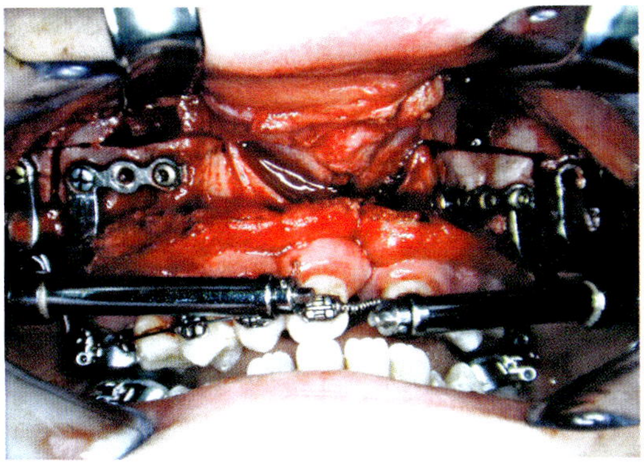

(A)

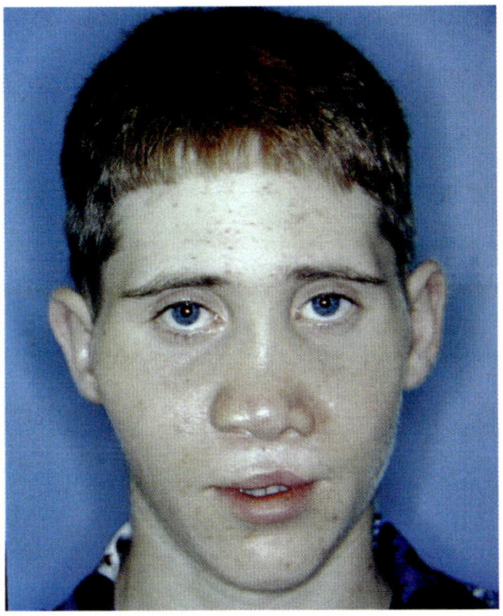

(B)

FIGURE 3 • Incorrect vector of maxillary distraction. In this case of maxillary distraction, (A) the distractors were placed with the vector converging toward each other instead of parallel to each other. (B) This resulted in excessive forces on both zygomatic bones, and the right zygoma fractured as a consequence. The lateral displacement of the right zygoma is apparent on this facial photograph.

8.8% of cases treated with a single-vector distractor in a review of 3278 cases of craniofacial distraction.[12] In a review of the literature, Swennen and associates observed an incorrect vector in 0.8% of cases of mandibular lengthening with DO.[8] As with many other techniques in orthognathic surgery a successful outcome is dependant on accuracy of pre-operative planning. There are a variety of methods for pre-operative evaluation. These include clinical examination, photographic analysis, radiographs (lateral, anteroposterior cephalograms, orthopantomogram and computed tomograms with three-dimensional (3D) reformatting. (See Chapter 11.) Computer aided analysis is also available to assimilate the radiographic, photographic and clinical information for treatment planning purposes.

Once pre-operative data gathering and analysis has been preformed, the plan must be translated into a surgical guide for positioning of the distractor as well as placement of osteotomies.

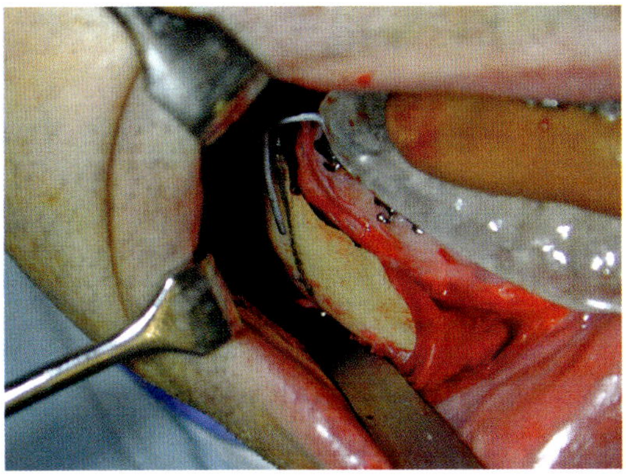

FIGURE 4 • An occlusal acrylic splint indexed to the teeth and with posterior metallic extensions allows an accurate positioning of the osteotomy.

This navigational aide is fabricated relative to the dentition using the dental casts in conjunction with pre-operative radiographs (Figure 4). Other methods, which outline the use of 3D computed tomogram scan data, have also been described.[13,14] In these cases, stable and predictable distraction results were obtained by using virtual stereolithographic models and simulated surgical movements (Figure 5; also see Chapter 11).

OPERATIVE PHASE (DISTRACTOR PLACEMENT)

Anatomic Complications

Complications such as those seen with other traditional surgical procedures include damage to adjacent tissues, bleeding, infection and unexpected reaction to medications, materials and the surgical procedure. However, there are some complications specific to placement of the distraction devices that can ultimately lead to failure of distraction, breakage of the device, increased pain associated with distraction and possible need for re-operation.[5,9,12]

Damage to teeth and the inferior alveolar nerve are possible anatomic complications of distractor placement.[12] Careful analyses of pre-treatment records, and careful operative technique can minimize risk to adjacent vital structures. However, it is impossible to completely eliminate these risks and patients must be informed of the possibility for long term or permanent damage resulting in altered post-surgical function and appearance.

Inadequate osteotomies prior to placement of the distractors can lead to inability to distract the bone segments (Figures 6, 7, 8), pain and pressure at the site of distraction and failure of the distraction devices (Figure 9). Clinically, there will be a discrepancy between the number of activation and the actual bony movements. The device appears open on radiographs but no bony gap or change in the occlusion is noticeable. This will eventually lead to premature consolidation if the problem is not identified and solved. The only solution for this complication is re-operation.

While incomplete osteotomy at the time of placement is not always the cause of distractor loosening, it remains a common cause. This can be avoided by opening the distractor once it is secured to the bone segments to insure passive motion of the

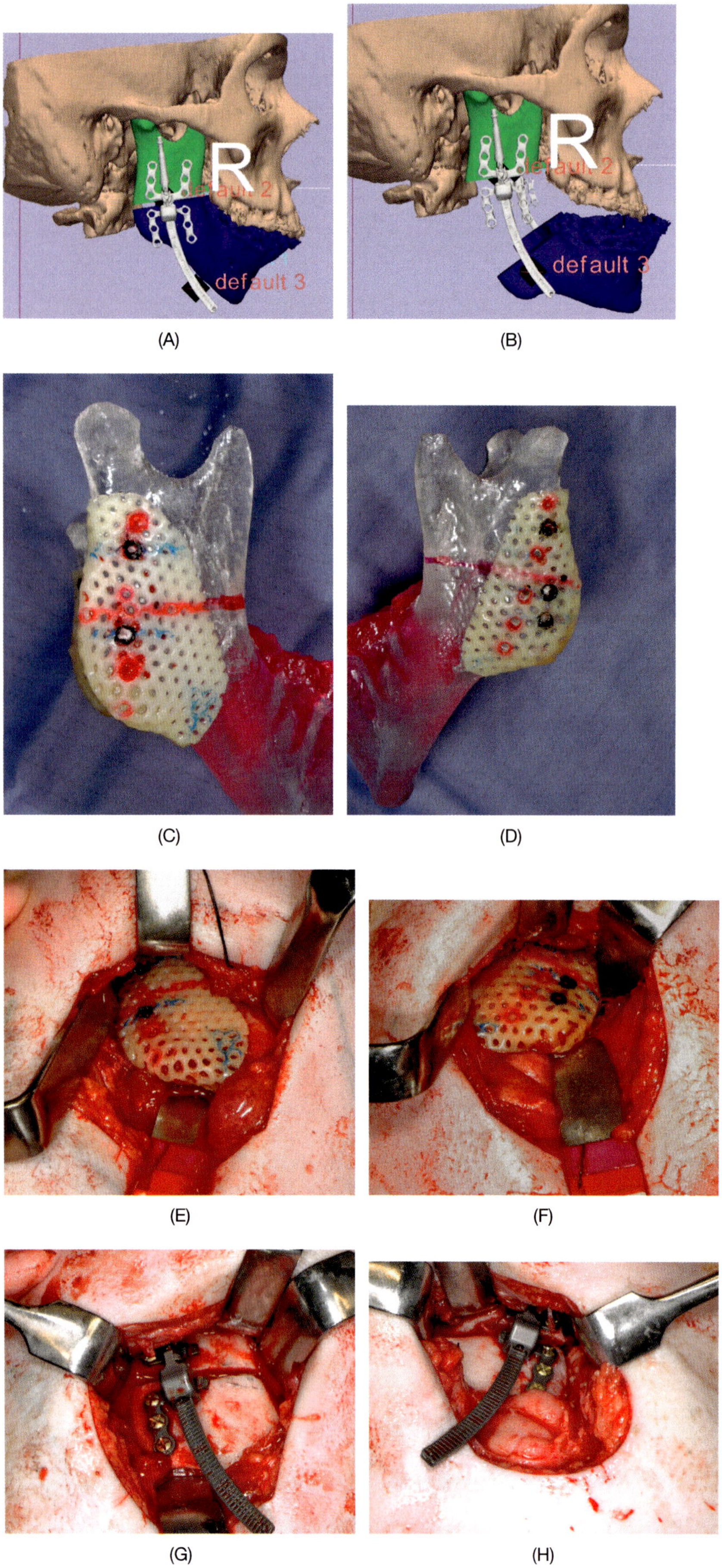

FIGURE 5 • Surgical planning. To prevent an incorrect vector of distraction and to execute precisely the planned osteotomy, a surgical guide is used intraoperatively. (A), (B) The osteotomy and the vector of distraction are planned virtually with computer software (Osteoplan, 3D Slicer). A surgical prescription is then produced. (C), (D) This information is transferred to the patient with Aquaplast surgical guides. The guides are molded on a 3D model and the osteotomy and the position of the distractor is marked on the model. (E), (F) Surgical guides in place intraoperatively. (G), (H) The osteotomies and the distractors are positioned precisely. (See Chapter 11 & 13). (Kaban et al.; Curvilinear Distraction Osteogenesis. JOMS 2009; 67:999–1000)

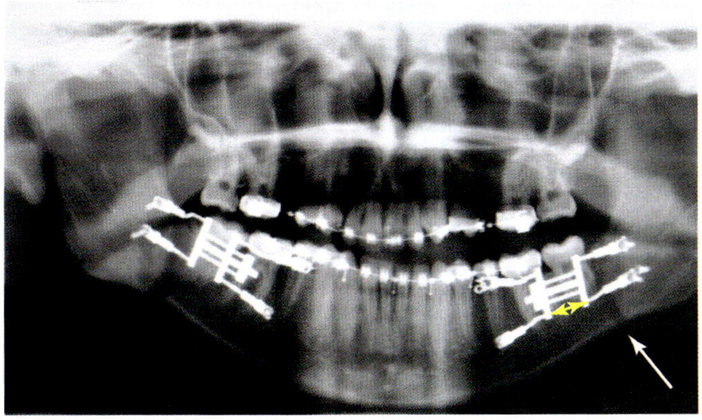

FIGURE 6 • Incomplete osteotomy. If the osteotomy is incomplete, the skeletal expansion may not occur or the vector of distraction can be distorted. This may result in pain upon device activation. In this case, the left mandibular osteotomy was incomplete. There is a discrepancy between the bony gap (white arrow) and the device opening (yellow arrow). The device opened as the foot plates bent.
Reproduced from: Troulis MJ, and Kaban LB. 2003: Complications of mandibular distraction osteogenesis. Oral Maxillofac Surg Clin North Am 15:251–264, Elsevier, with permission.

foot plates. Additionally, the foot plates should be secured and checked for stability once the device has been activated intraoperatively (Figure 10). Post-operative radiographic follow-up should be done to ensure that no screws have become loose or the distraction devices have migrated or shifted in position.

After the corticotomies are performed, the osteotomy is usually completed with osteotomes. This can lead to an aberrant fracture if the corticotomies are not completed adequately or if they were not done at the appropriate location. Ultimately, this will lead to an incorrect vector of distraction or to damage to teeth and adjacent structures. Van Strijen and colleagues reported this complication in 0.7% of the cases in a series of 70 cases of mandibular distraction.[5]

POST-OPERATIVE PHASE (DISTRACTION)

Device Failure

Distraction devices are complex, often bulky appliances that require care during placement and in the post-operative period to prevent failure. Several forms of hardware failure have been described. Mofid and associates observed 3% of hardware loosening and 4.5% of hardware fracture.[12] One literature review indicates a 33.1% rate of mechanical problems such as pin loosening or accidental trauma when evaluating cases of distraction osteogenesis of the craniofacial skeleton.[8] In a case report of a mandibular midline device, fracture of the foot plate near the screw hole (Figure 7) was noted in the consolidation phase. This complication was thought to be due to excessive manipulation of the metal during placement of the device that resulted in weakening.[15]

In order to prevent hardware failure, care must be taken during placement to minimize work hardening of the footplates by excessive plate bending during adaptation. In the post-operative period, the patients should be reminded that they must minimize physical activity so as not to disturb the distraction devices (Figure 8). This is particularly true with external devices.

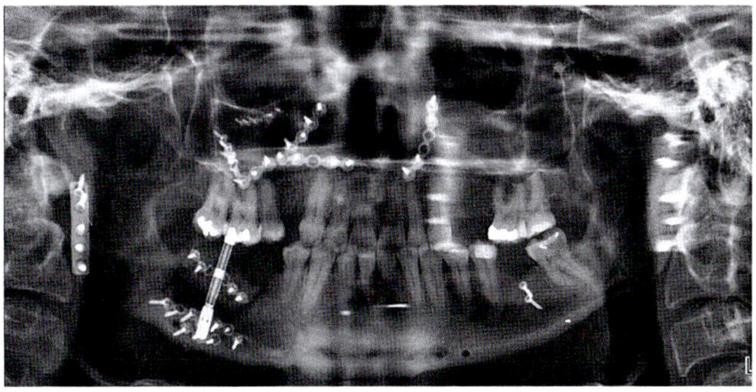

(A)

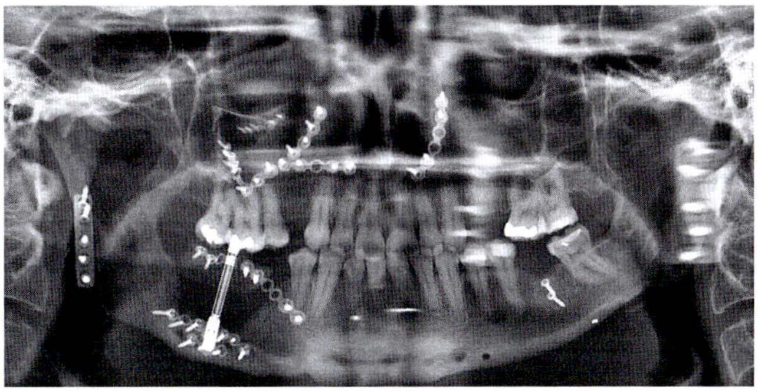

(B)

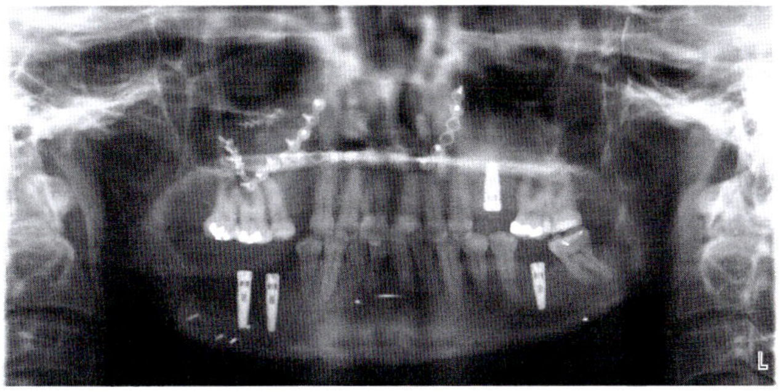

(C)

FIGURE 7 • (A) Panoramic Radiograph of an alveolar distractor in place. The transport disc is not elevating equally. Intraoperative exploration revealed fractured distal footplate (not seen on radiograph). (B) The transport device was acutely distracted into position and rigidly fixed as seen on the radiograph. (C) Panorex of same patient. Distractor has been removed and implants placed (no additional bone grathing was required).

Inadequate Distraction

In some cases, the desired amount of distraction cannot be achieved. This may be related to patient compliance, but can also be due to other problems such as interference of other anatomic structures (Figure 11), or displacement of the distraction segments.

The movement of the proximal and distal fragments during mandibular distraction osteogenesis, especially in patients with abnormal morphology of the ramus/condyle unit (eg, hemifacial microsomia), can be very complex.[14] Complications that may arise include pain as the coronoid process collides with the skull base (Figure 12). A review of cases has shown a 2.3% rate of ankylosis of the coronoid to the zygoma during mandibular distraction[8] (Figure 13). Rotation of the proximal fragment can

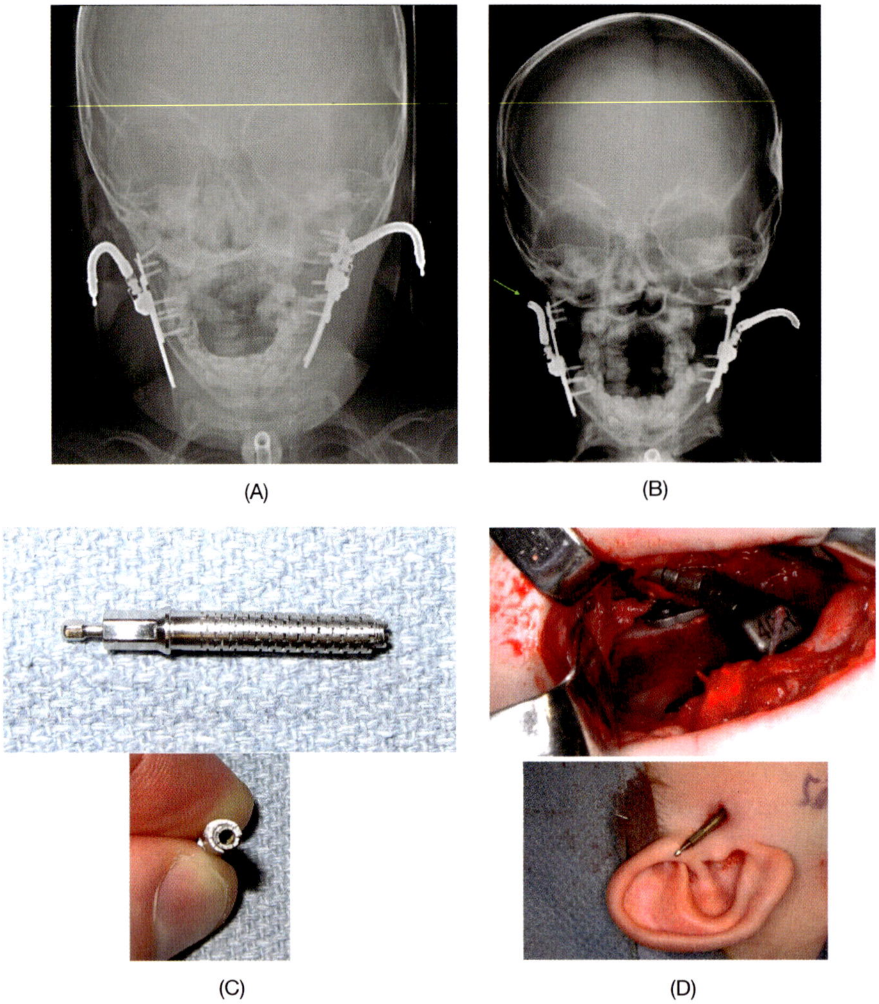

FIGURE 8 • (A) AP cephalogram of a 2 year old patient who had bilateral curvilinear distractors placed for the correction of micrognathia. (B) During active distraction the right activating arm fractured off (arrow). (C) Photograph of fractured activating arm. (D) Intraoperative photograph of replaced activating arm.

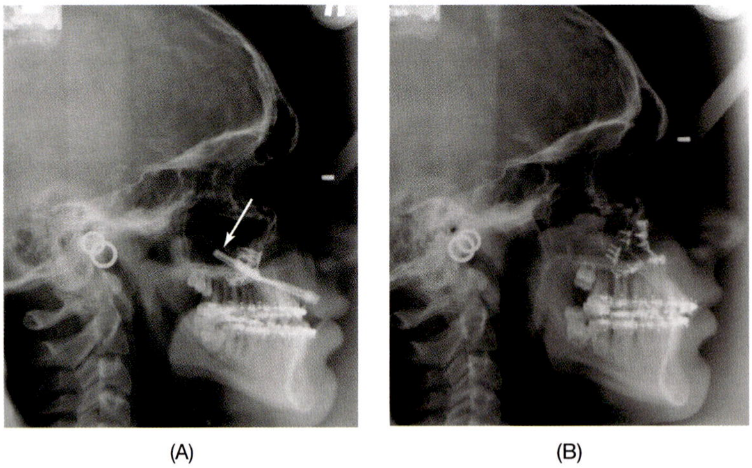

FIGURE 9 • Bending of the distraction device. When using a distractor with semi-rigid footplates, if the osteotomy is incomplete, the device can bend and this will result in failure of the treatment. (A) In this case of maxillary distraction, the appliance is activated (arrow) but the occlusion remained the same. The footplates were bent and the maxilla did not move. (B) The distractor had to be removed and the maxilla advanced manually and rigidly fixed.
Reproduced from: Troulis MJ, and Kaban LB. 2003: Complications of mandibular distraction osteogenesis. Oral Maxillofac Surg Clin North Am 15:251–264, Elsevier, with permission.

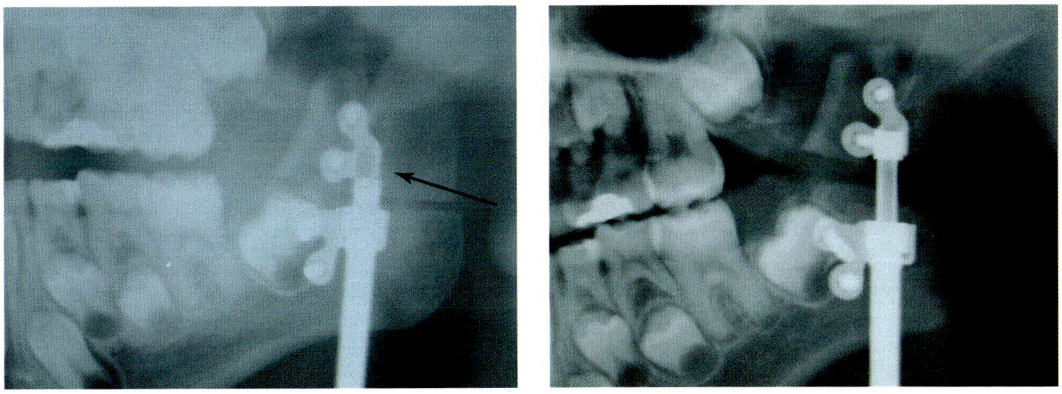

(A) (B)

FIGURE 10 • Footplate detachment. This distractor has detachable footplates and is designed to be removed under local anesthesia. (A) In this case, the proximal footplate (arrow) was detached before the completion of the distraction. (B) The distractor was reassembled and a resorbable suture was used to maintain all the components together. Reproduced from: Troulis MJ, and Kaban LB. 2003: Complications of mandibular distraction osteogenesis. Oral Maxillofac Surg Clin North Am 15:251–264, Elsevier, with permission.

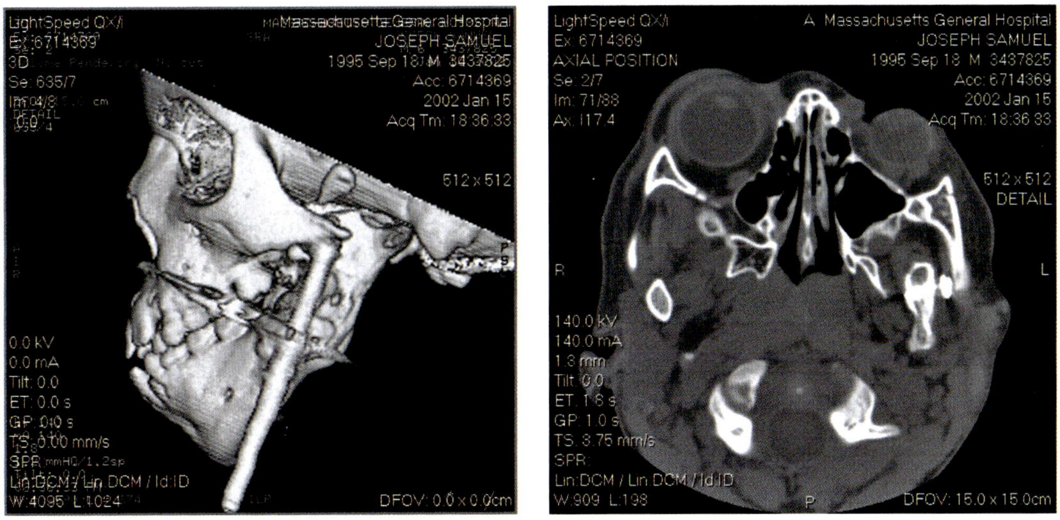

(A) (B)

FIGURE 11 • Device interference with anatomic structures. The distractor can interfere with anatomic structures and cause pain and changes in occlusion. If the device does not have detachable foot plates, its removal can be more complex than its insertion. (A), (B) CT-scan showing interference of the proximal portion of the device with the zygomatic arch (arrow). Reproduced from: Troulis MJ, and Kaban LB. 2003: Complications of mandibular distraction osteogenesis. Oral Maxillofac Surg Clin North Am 15:251–264, Elsevier, with permission.

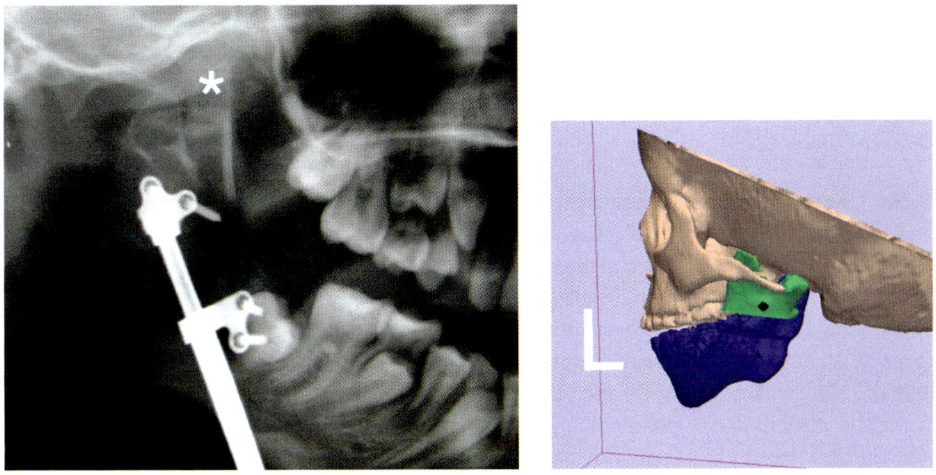

(A) (B)

FIGURE 12 • Anatomic interference. (A) The coronoid process collided with the skull base (*) during distraction in this case of hemifacial microsomia. (B) To prevent this complication, the distraction is performed virtually using computer software (Osteoplan). If the coronoid process is expected to interfere, as in this case, a coronoidectomy is performed during the placement of the distractor.

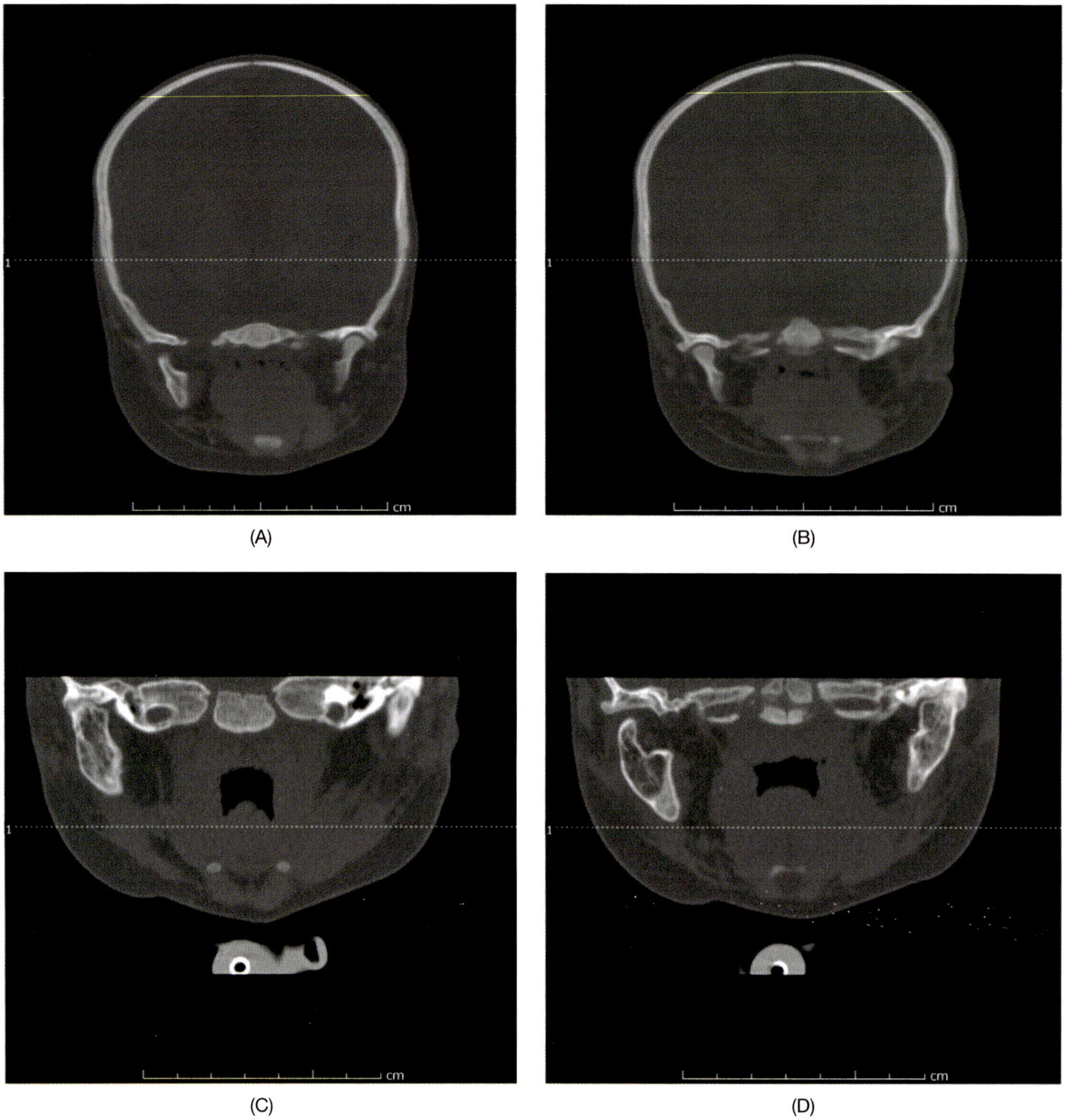

FIGURE 13 • This 3-year-old girl with severe micrognathia and airway obstruction was treated with bilateral curvilinear mandibular distractors. (A), (B) Pre-operative CT-scan demonstrating normal temporomandibular joint (TMJ) anatomy bilaterally. (C), (D) For uncertain reasons, she developed bilateral TMJ ankylosis as demonstrated in the post-distraction images.

produce a functional pseudoarthrosis at the junction of the coronoid and skull base with inadequate lengthening of the ramus and a disarticulated hypoplastic condyle remaining; although the rate of this type of complication is reported to be low (<1%).[8] Furthermore, displacement of the condyle superiorly and laterally into the temporal fossa has been reported.[13]

To avoid these complications, a 3D CT scan is obtained to help plan the distraction vector. Analysis of the CT data may predict bony interferences and collisions. Based on such predictive information, a coronoidectomy can be performed on appropriate patients, to allow for adequate ramus lengthening without bone collisions. Unfavorable rotations can also be anticipated and so that the articulation between the condyle and the skull base is not disturbed. A 3D treatment planning system also helps predict movement of the contralateral condyle during distraction. This is helpful to avoid excessive displacement and is particularly critical when widening the mandible by distraction osteogenesis.[16,17,18]

During advancement of a retrognathic mandible by DO, the patient may posture the jaw forward. This may be the result of pain, swelling in the joints, or both. The surgeon may inadvertently stop activating the device when a portion of the apparent advancement has resulted from this functional posturing and not from expansion at the osteotomy. In order to avoid this problem, panoramic radiographs should be carefully reviewed for condylar position prior to ending the active phase of distraction.

If in doubt, expansion should continue until an edge-to-edge occlusion is achieved. Then, light class III elastics are placed and the patient is observed for several days. Adjustments are then made in the device to establish the final jaw and occlusal position. If acceptable to the orthodontist, it is always safest to overcorrect and let the occlusion be finalized orthodontically.

Premature consolidation of the regenerate has also been noted to limit the amount of distraction. This is more frequent in children and in cases where the osteotomy was not completed intraoperatively. Too long a latency period or too slow a rate of distraction will produce similar results.[19] Most recent data indicate that a rate of 1 mm distraction per day is appropriate in the craniofacial skeleton.[20] An increase in the rate of distraction superior to that amount have been associated with hypoesthesia of the inferior alveolar nerve.[21] It also has been demonstrated that there is no difference in bone formation when no latency period is observed.[22]

Infection

Although reported rates of infection range between 5% to 30% in the orthopedic literature, infection rate in the craniomaxillofacial region appears to be much lower.[23] This is likely due to the angiogenesis that occurs during the distraction process.[24] Reports of infection in the setting of distraction appear to be related directly to the hardware and the fact that it is exposed and requires frequent manipulation. Most infections developed around the pins used with external distractor (Figure 14). Infections are also rare with semi-buried devices. This is probably because they are bone-borne and rarely loosen. The device and exit port allow wound drainage. Antibiotic prophylaxis during the placement and the removal of distractors along with good oral hygiene should reduce the rate of infection.[24] Van Strijen and colleagues observed 2.9% of infection after the placement of intra-oral distractor for mandibular advancement. Patient were treated with antibiotics and local irrigation.[5]

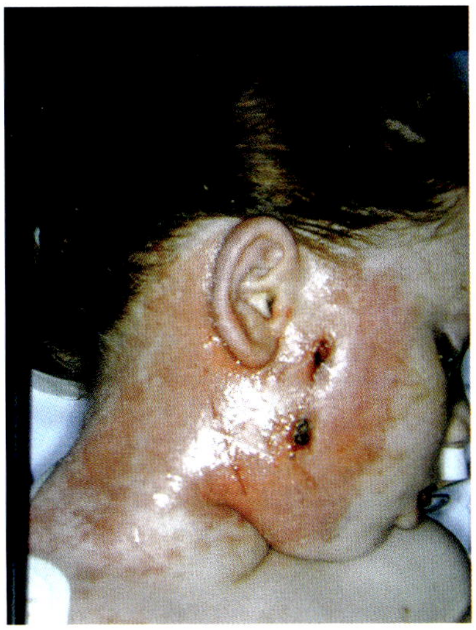

FIGURE 14 • Infection. Distractors can become infected at the time of placement or during the activation of the device. This is particularly true with extra-oral distractors with pins. This patient presented a facial cellulitis requiring appliance removal and aggressive antibiotic therapy. Reproduced from: Troulis MJ, and Kaban LB. 2003: Complications of mandibular distraction osteogenesis. Oral Maxillofac Surg Clin North Am 15:251–264, Elsevier, with permission.

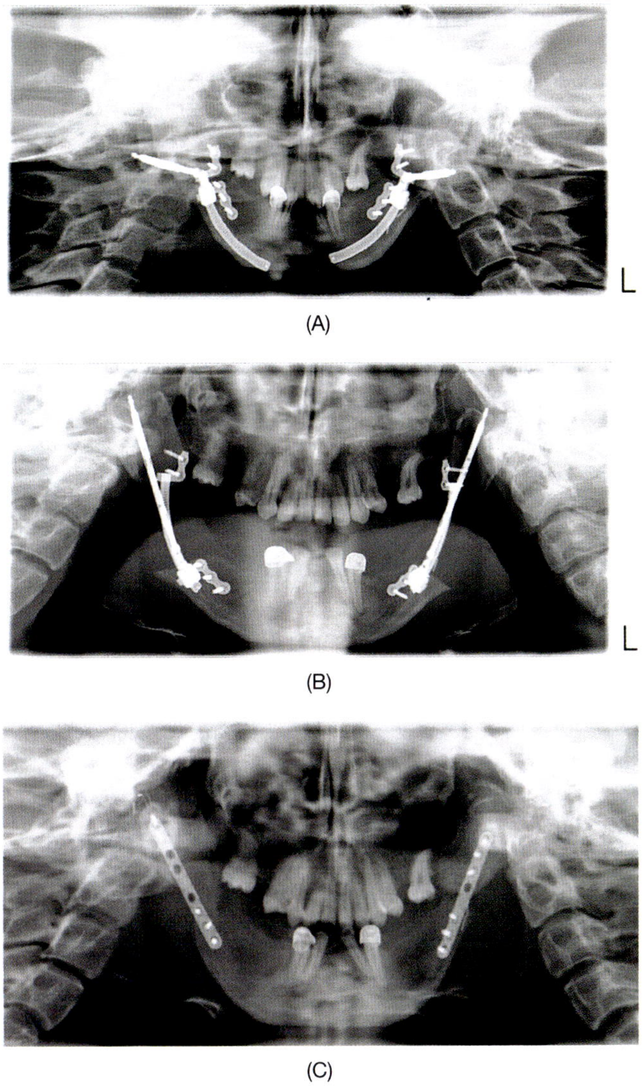

FIGURE 15 • Inadequate regenerate. This 20-year-old male patient with Nager syndrome presented severe mandibular retrognathism and was tracheostomy-dependant. The distractor was activated at a rate of 2 mm per day instead of the usual protocol. (A) Curvilinear distractors before and (B) at the end of the activation period. (C) The mandible had to be fixed with rigid internal fixation plates because of the inadequate regenerate.

Inadequate Regenerate

Lack of radiographic evidence of mineralization of the distraction gap and/or irregular density of mineralizing zones of the regenerate are signs of an inadequate or delayed bone consolidation (Figure 15). It is recommended in these situations to decrease the rate of distraction or temporary stop the distraction. Compression at the distraction site by activating the appliance in the opposite direction has also been advocated.[19] Possible causes of an inadequate regenerate includes: excessive damage to osteogenic tissue (bone marrow, periosteum), incorrect rate of distraction, displacement of bone segments, excessive initial bony gap, and instability of the distraction device.[25,26]

Inadequate bone formation is a risk in older patients, smokers, and patients having received radiation therapy and/or those with poor oral hygiene. Even under these circumstances, however, if the device is left in place for a prolonged period of time, it appears that the regenerate ultimately consolidates. If it is necessary to remove the device, a rigid fixation plate can be placed (Figure 13).[27]

CONCLUSION

Distraction ontogenesis is a very exciting and promising technique for skeletal expansion. However, it is technically demanding, requires precise planning and a significant commitment on the part of the surgeon, patient, and family. Most complications arise from poor patient preparation and compliance, inadequate pre-operative planning, and inaccurate surgical execution. In the future, technical complications will be reduced by more accurate 3D planning systems and surgical navigation.

References

1. Codivilla A. On the means of lengthening, in the lower limbs, the muscles and tissues which are shortened through deformity. Am J Orthop Surg 1904;2:353–69.
2. Ilizarov GA: The principles of the Ilizarov method. Bull Hosp Jt Dis Orthop Inst 1988;48:1–11.
3. Ilizarov GA: Clinical application of the tension-stress effect for limb lengthening. Clin Orthop Relat Res: 1990;250:8–26.
4. McCarthy JG, Schreiber J, Karp N, Thorne CH, Grayson BH: Lengthening the human mandible by gradual distraction. Plast Reconstr Surg 1992;89:1–8.
5. van Strijen PJ, Breuning KH, Becking AG, Perdijk FB, Tuinzing DB: Complications in bilateral mandibular distraction osteogenesis using internal devices. Oral Surg Oral Med Oral Pathol Oral Radiol Endod 2003;96:392–7.
6. Schreuder WH, Jansma J, Bierman MW, Vissink A: Distraction osteogenesis versus bilateral sagittal split osteotomy for advancement of the retrognathic mandible: a review of the literature. Int J Oral Maxillofac Surg 2007;36:103–10.
7. Castano FJ, Troulis MJ, Glowacki J, Kaban LB, Yates KE: Proliferation of masseter myocytes after distraction osteogenesis of the porcine mandible. J Oral Maxillofac Surg 2001;59:302–7.
8. Swennen G, Schliephake H, Dempf R, Schierle H, Malevez C: Craniofacial distraction osteogenesis: a review of the literature: Part 1: clinical studies. Int J Oral Maxillofac Surg 2001;30:89–103.
9. Primrose AC, Broadfoot E, Diner PA, Molina F, Moos KF, Ayoub AF. Patients' responses to distraction osteogenesis: a multi-centre study. Int J Oral Maxillofac Surg 2005;34:238–42.
10. Ayoub AF, Duncan CM, McLean GR, Moos KF, Chibbaro PD: Response of patients and families to lengthening of the facial bones by extraoral distraction osteogenesis: a review of 14 patients. Br J Oral Maxillofac Surg 2002;40:397–405.
11. Van Strijen PJ, Breuning KH, Becking AG, Perdijk FB, Tuinzing DB. Cost, operation and hospitalization times in distraction osteogenesis versus sagittal split osteotomy. J Craniomaxillofac Surg 2003;31:42–5.
12. Mofid MM, Manson PN, Robertson BC, Tufaro AP, Elias JJ, Vander Kolk CA. Craniofacial distraction osteogenesis: a review of 3278 cases. Plast Reconstr Surg 2001;108:1103–14.
13. Gateno J, Teichgraeber JF, Aguilar E: Computer planning for distraction osteogenesis. Plast Reconstr Surg 2000;105:873–82.
14. Troulis MJ, Everett P, Seldin EB, Kikinis R, Kaban LB. Development of a three-dimensional treatment planning system based on computed tomographic data. Int J Oral Maxillofac Surg 2002;31:349–57.
15. Uckan S, Veziroglu F, Arman A: Unexpected breakage of mandibular midline distraction device: case report. Oral Surg Oral Med Oral Pathol Oral Radiol Endod 2006;102:e21–5.
16. Perrott DH, Berger R, Vargervik K, Kaban LB: Use of a skeletal distraction device to widen the mandible: a case report. J Oral Maxillofac Surg 1993;51:435–9.
17. Samchukov ML, Cope JB, Harper RP, Ross JD: Biomechanical considerations of mandibular lengthening and widening by gradual distraction using a computer model. J Oral Maxillofac Surg 1998;56:51–9.
18. McCarthy JG, Hopper RA, Hollier LH, Jr., Peltomaki T, Katzen T, Grayson BH: Molding of the regenerate in mandibular distraction: clinical experience. Plast Reconstr Surg 2003;112:1239–46.
19. Ilizarov GA: The tension-stress effect on the genesis and growth of tissues: Part II. The influence of the rate and frequency of distraction. Clin Orthop Relat Res 1989;263–85.
20. Hollier LH, Jr., Higuera S, Stal S, Taylor TD: Distraction rate and latency: factors in the outcome of pediatric mandibular distraction. Plast Reconstr Surg 2006;117:2333–6.
21. Hu J, Tang Z, Wang D, Buckley MJ: Changes in the inferior alveolar nerve after mandibular lengthening with different rates of distraction. J Oral Maxillofac Surg 2001;59:1041–5.
22. Glowacki J, Shusterman EM, Troulis M, Holmes R, Perrott D, Kaban LB: Distraction osteogenesis of the porcine mandible: histomorphometric evaluation of bone. Plast Reconstr Surg 2004;113:566–73.
23. Cherkashin AM, Samchukov ML. Potetial mistakes and complications during distraction osteogenesis. In: *Craniofacial distraction osteogenesis* (Samchukov ML, Cope JB, Cherkashin AM, eds). St Louis, MI: Mosby, 2001: 583 p. 594.
24. Aronson J: Temporal and spatial increases in blood flow during distraction osteogenesis. Clin Orthop Relat Res 1994;301:124–31.
25. Cope JB, Samchukov ML: Mineralization dynamics of regenerate bone during mandibular osteodistraction. Int J Oral Maxillofac Surg 2001;30:234–42.
26. Aronson J, Shen X: Experimental healing of distraction osteogenesis comparing metaphyseal with diaphyseal sites. Clin Orthop Relat Res 1994;301:25–30.
27. Stuki-McCormick SU, Drew S, Mizrahi RD. Distraction osteogenesis: Overcoming the challenges of a new technique. In: *Craniofacial distraction osteogenesis* (Samchukov ML, Cope JB, Cherkashin AM, eds). St Louis, MI: Mosby, 2001:595 p. 603.

Alternative, Novel, Minimally Invasive Strategies for Treatment of Jaw Tumors: Anti-Angiogenic Adjuvant Therapy

20

Leonard B. Kaban, DMD, MD

OUTLINE

INTRODUCTION

The revolution and explosion of knowledge in biomedical sciences, particularly in molecular biology and genetics, hold tremendous promise for improving diagnosis, predicting prognosis, developing new and novel treatments, and evaluating standard treatment of problematic benign and malignant neoplasms.[1] For oral and maxillofacial surgery patients, this will result in improvements in the management of head and neck cancer as well as aggressive, destructive, and recurrent primary jaw tumors.[1] Primary aggressive jaw lesions such as giant cell tumors, myxomas, juvenile fibromatosis, desmoplastic fibromas, and aggressive fibro-osseous lesions are difficult to manage and result in considerable morbidity because of local invasion, destruction, and multiple recurrences.

Giant cell lesions are the most common benign primary bone tumors of the jaws. They are mesenchymal in origin and consist of a stroma with fibroblast-like spindle cells, multinucleated giant cells, and a dense vascular network composed of endothelial-cell–lined capillaries. There may be multiple areas of hemorrhage within the tumors and mitoses in the spindle-shaped cells. The neoplastic cell of origin has not been identified but it is probably not the giant cell per se, as these cells are present in a variety of bone tumors, particularly in children.[1–12] Giant cell lesions are classified as non-aggressive, intermediate, or aggressive based on clinical and radiographic criteria.[1–3,13–15] Non-aggressive lesions in the jaws can be successfully treated by simple curettage with an expected low recurrence rate. Aggressive lesions have reported recurrence rates as high as 70% after enucleation or curettage.[2–6,16–27] Regardless of clinical behavior, both aggressive and non-aggressive giant cell lesions have similar histologic appearances. Furthermore, there are no currently available biological markers to predict clinical behavior and standard histologic techniques do not help the clinician determine prognosis.[1–6,16]

In this chapter, as an example of a novel, biologic, minimally invasive approach, we report an update on the use of the Massachusetts General Hospital (MGH) protocol of enucleation and adjuvant interferon alpha therapy for management of aggressive giant cell lesions (giant cell tumors). This is an alternative to the standard en bloc resection required to cure these tumors. The material presented is adapted from three published papers: Biomedical Technology Revolution: Challenges and Opportunities for Oral and Maxillofacial Surgeons[1]; Adjuvant Interferon Therapy for Aggressive Giant Cell Lesions of the Jaws[2]; and Antiangiogenic Therapy of a Recurrent Giant Cell Tumor of the Mandible with Interferon-alpha-2A.[3]

Biologic behavior of benign and malignant tumors is difficult to predict. Unchanged in 40 years, the 5-year survival of oral cancer patients remains among the worst of all anatomic sites.[28] In the United States, more people die from oral cancer than melanoma, cervical, or ovarian carcinoma.[29] Current diagnostic and therapeutic modalities in head and neck cancer have many shortcomings.

Progress in understanding the molecular biology of malignant tumors may lead to improvement in current therapy and development of novel treatments. These advances may also benefit the management of patients with aggressive, locally destructive, and recurrent benign neoplasms. Finally, molecular surveillance for recurrence is a potentially powerful and important tool for clinicians.

In an effort to improve diagnosis and treatment, investigators are beginning to uncover the molecular determinants responsible for oral carcinogenesis. Virtually undefined a decade ago, there has been an explosion of genes discovered to be associated with oral cancers. Until 1997, only genes identifiable by northern blot analysis, 10 in number, were reliably associated with oral cancers. In 1998, the Cancer Genome Project identified 643 genes. Since then, the number has risen to over 2,000, and is expected to reach 3,000 to 4,000 genes ultimately.[30]

Molecular genetics has also been used to monitor patients for tumor recurrence. Brennan and associates[31] reported on the outcome in patients with histologic negative but molecular (using identification of p53) positive margins with time. The group that had negative histologic margins, but positive molecular margins (13 patients), had a 40% rate of tumor recurrence. Negative histologic and negative molecular margins strongly predicted a lack of recurrence. This type of surveillance will potentially permit early detection of recurrence or at least direct clinicians to especially close follow-up of high-risk patients.

Finally, knowledge derived from the use of molecular techniques may lead to the development of new cancer treatments. These novel therapies will be directed at the specific molecular genetic defect in cancer rather than toward a differential effect on dividing cells in general. Current chemotherapy and radiation therapy affect multiple cell types based on their rate of division and are therefore only relatively specific to the abnormal tumor cells. Consequently, the side effects of treatment are severe and may produce considerable morbidity and even mortality.[31,32]

The first pharmaceutical shown to be effective against a cancer in humans, in which the action is directed specifically at the abnormal molecular mechanism of the tumor, was reported in May 2001 in the New England Journal of Medicine.[33] Chronic myelogenous leukemia (CML) is caused by a single molecular event: a swap of sections of DNA between two chromosomes. The swap between chromosomes 9 and 22 produces an abnormal gene and gene product. The normal role of the switched portion of the gene is to turn off "abl protein," which normally modulates cell division and replication. As a result "abl protein" keeps signaling white blood cells (WBC) to divide. The new drug targets "abl protein" and prevents it from transmitting the signal to the WBC to divide. In one trial, 98% of patients in early stages of CML had blood counts revert to normal and 13% had no trace of leukemic cells. Most importantly, side effects were almost non-existent.[33,34]

Another strategy in cancer research, which may potentially result in therapies directed at the fundamental biology of the tumor, makes use of the concept of angiogenesis. In 1971, Judah Folkman, MD, a young assistant professor at Harvard Medical School, articulated his hypothesis of angiogenesis-dependent tumor growth based on his work while in the Navy. He was drafted during the Vietnam War and assigned to the National Institutes of Health. While there, he noted that melanoma cells grew only to a tumor size of 2 mm in vitro. He determined that this was not because the cells died at that stage, but because they could not recruit new blood vessels for nourishment. If he transplanted the cells into nude mice, they would grow in an unlimited fashion. The growth was paralleled by the formation of new blood vessels.[35]

Growth of new capillaries is central to the growth of cancer and all tumors. Tumors cannot grow beyond 2 mm in size if they cannot recruit new blood vessels to provide the tumor with nutrition and oxygen. Every cell lives adjacent to, or at least no further than 100 um to 200 um from, a capillary. Tumors release angiogenic growth factors such as basic fibroblast growth factor (FGF), platelet-derived growth factor (PDGF), and insulin-like growth factor (IGF) among others, which stimulate endothelial proliferation and capillary growth. New blood vessels are generated that allow nourishment and therefore growth of the tumor.[35–38] Several authors have demonstrated that FGF, an angiogenic growth factor, is produced by childhood hemangiomas, other vascular tumors and certain adult cancers.[39] In 1995, Singh and colleagues[40] reported that interferon alpha produces tumor shrinkage by inhibiting the production of the angiogenesis stimulator basic FGF.

Basic FGF has been shown to be present in proliferating hemangiomas and not in normal skin or involuted hemangiomas.[41,42] In addition, there is an inverse correlation between the presence of b-FGF and interferon beta, a natural anti-angiogenesis factor, in tissues. Based on these observations, interferon was the first anti-angiogenesis agent used to treat hemangiomas and other tumors.[36,38] Today there are a variety of clinical trials using newer and more potent anti-angiogenesis agents to treat tumors. Because these are agents directed at the fundamental tumor biology and not against dividing cells in general, the side effects are minimal.[36,43]

After 35 years of work in this field, Dr. Folkman reports on the concept of angiogenesis as an "organizing principle."[44] If a patient's disease (for example, giant cell tumor) is angiogenesis-dependent, it is conceivable that it will respond to an angiogenesis inhibitor proven to be useful for another tumor or another angiogenesis-dependent disease.

A report of the following patient will serve as an example of how the use of an anti-angiogenic agent, interferon alpha, helped change the way we manage giant cell tumors of the jaws. This patient represents the first reported case of anti-angiogenic therapy used for treatment of a solid primary bone tumor in any location.[44] Oral and maxillofacial surgeons encounter primary jaw tumors whose biologic behavior and recurrence rate are not predicted by standard histologic techniques. Hence, the surgeon is often faced with multiple recurrences of a "benign" lesion. Nobel laureate, Joseph E. Murray, MD, has called this "malignant benign disease."[45]

A 4-year-old female patient presented with a rapidly proliferating jaw lesion.[3] The tumor had grown to 6 cm in size over a

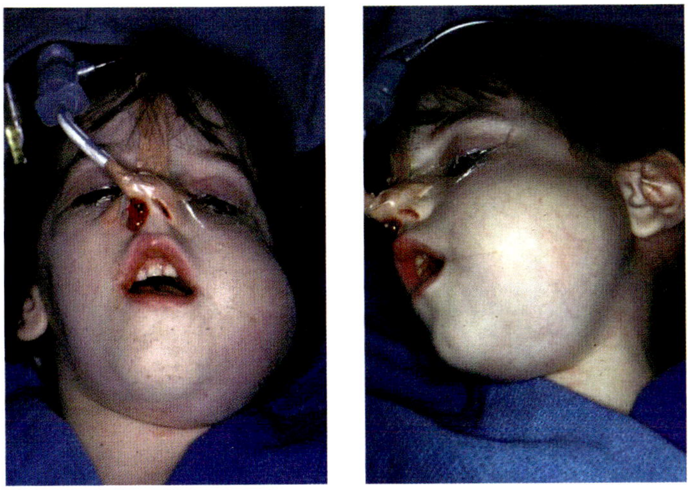

(A) (B)

FIGURE 1 • Frontal (left) and left lateral (right) photographs of a 5-year-old girl with a second massive recurrence of an aggressive giant cell lesion of the left mandible.

period of a few weeks and the child had pain, loss of appetite, and weight loss. The computed tomographic (CT) scan demonstrated a fluid-filled lesion that enhanced with contrast. The diagnosis of aneurysmal bone cyst was made. The lesion was enucleated and the child had an uneventful recovery. One month later, the lesion recurred and was decompressed and biopsied again (Figure 1). The pathology was read as: aneurysmal bone cyst with associated giant cell lesion.

The child underwent hemi-mandibulectomy, with disarticulation and placement of a titanium reconstruction plate, because of the rapid aggressive growth of the tumor (Figure 2). She again recovered uneventfully, but 6 weeks later experienced another massive recurrence. The lesion was re-biopsied and the original slides were re-examined to confirm the diagnosis of giant cell lesion. The only standard therapy available to attempt control of this extraordinary lesion would have been high dose radiation in combination with chemotherapy.

Because of the devastating effects on facial growth and secondarily, body image development, radiation if at all possible was to be avoided. For years, I had been interested in giant cell lesions and their biologic behavior.[3,5,6,16,45] In this patient, I believed I had an example of a true intraosseous vascular proliferative lesion behaving like a life-endangering proliferating hemangioma of the soft tissues in infancy.

I consulted Drs. Judah Folkman and John Mulliken, both long-term collaborators, and we decided to measure the patient's urinary FGF. We hypothesized that this tumor was an FGF-producing angiogenic tumor that would respond to interferon therapy.

Urine FGF concentration was 22,000 units/ul (normal is less than 1,000) and the patient was started on interferon-alpha therapy at a dose of 1 million units/m^2. After 3 weeks, tumor growth stabilized and she had return of energy and appetite. At 6 months, the dose was increased to 3 million units per m^2 because of her total body growth. At 8 months, the dose was increased again (6 million units per m^2) because the mother observed that the

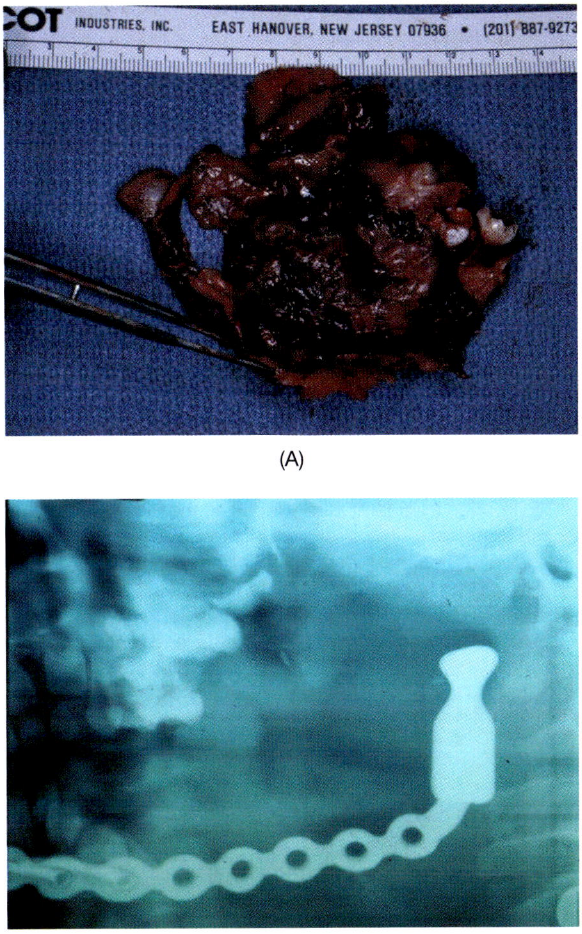

(A)

(B)

FIGURE 2 • Intraoperative (top) view of the resected specimen that included the condyle/ramus unit and mandibular body. The disk was left in place. While waiting for the final pathology report, a titanium bone plate with a condyle was used for temporary reconstruction (bottom).

tumor had stopped shrinking. She was maintained on this dose for another 4 months, during which time the tumor completely regressed and the urine FGF returned to normal levels (Figure 3). The mandible completely regenerated during this period, allowing removal of the reconstruction plate (Figure 4). A bone graft was not required to augment the regenerated jaw. She is now 9 years post-treatment and she has grown normally and has been completely rehabilitated with dental implants[2,3] (Figures 5 and 6).

DIAGNOSIS AND MANAGEMENT OF GIANT CELL LESIONS

Current Treatment Protocols

The gold standard to cure giant cell tumors is en bloc resection. This strategy produces a significant deformity and additional reconstructive operations are required to rehabilitate the patients. Because these tumors generally do not metastasize, it is agreed that less invasive therapies would be desirable.

A number of minimally invasive techniques for management of giant cell tumors have been reported. Embolization and pharmacologic agents, such as corticosteroids and calcitonin, have all

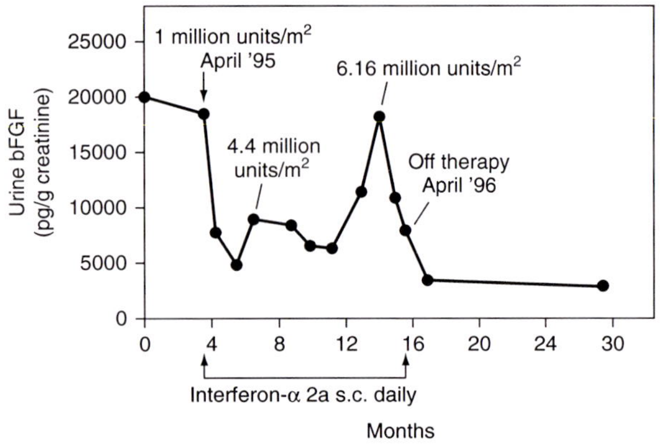

FIGURE 3 • The tumor recurred again and she was begun on interferon alpha. This figure demonstrates her dose response curve to interferon. Urine FGF dropped from 22,000 to <5,000 as the tumor became smaller in response to the drug. The dose was increased to 4.4 u/m^2 because the FGF rose and the tumor stopped shrinking. The dose was increased again to 6.16 million units per m^2 and the tumor disappeared and did not recur. The FGF went down to normal (<1,000). *(With permission from: Kaban LB, et al. Antiangiogenic Therapy of a Recurrent Giant Cell Tumor of the Mandible with Interferon Alfa-2a. Pediatrics 1999;103:1146, Figure 2.)*

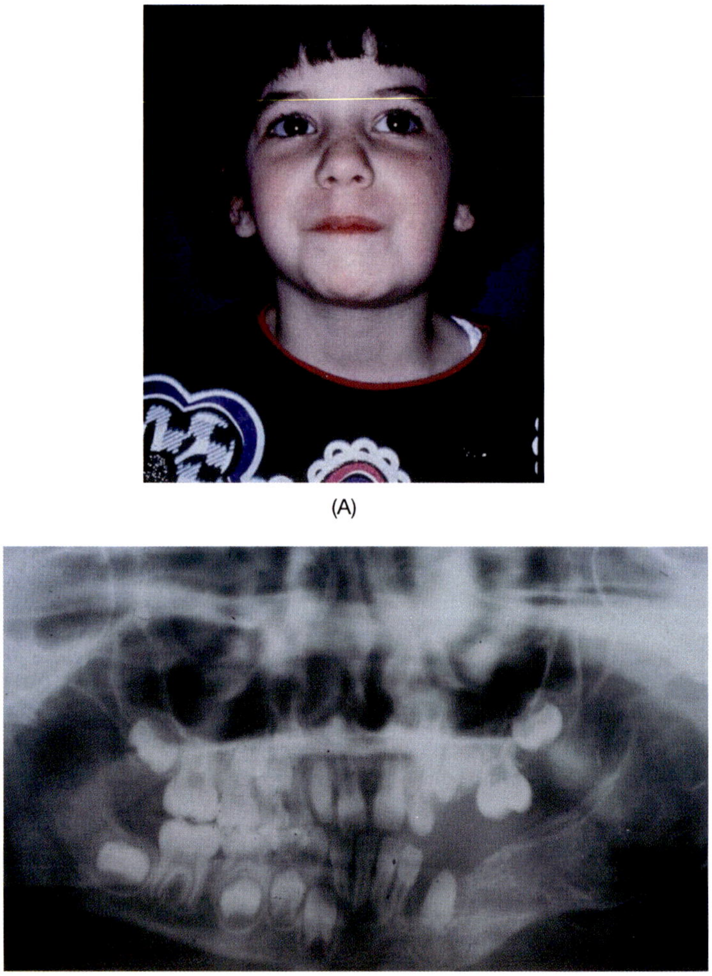

(A)

(B)

FIGURE 4 • Frontal (top) photograph and panoramic radiograph (bottom) 1 year off interferon therapy with no sign of recurrence. Note that the reconstruction plate was removed because the mandible spontaneously regenerated.

been reported for managing giant cell tumors.[17,18,22–27] Corticosteroids have been used based on the assumption that the tumors are inflammatory in nature.[22] Because corticosteroids are anti-angiogenic agents, these drugs may also have an anti-angiogenic effect on giant cell lesions, much as they do on infantile hemangiomas.

Calcitonin has been used on the theory that the osteoclast is the primary tumor cell of origin.[23–25] Results of these treatments have been variable[22–27,46] and difficult to evaluate. The authors have failed to adequately document the nature of the tumors being treated, that is, aggressive vs. non-aggressive.[46] For example, in two recently reported series on the use of calcitonin therapy,[25,28] the authors included both aggressive and non-aggressive lesions. It is critical to differentiate the two because non-aggressive lesions predictably respond to enucleation or curettage with a very low recurrence rate. The use of adjuvant or alternative therapies is unnecessary and may not be desirable. It is only the aggressive lesions that have a high recurrence rate and require en bloc resection.

MGH PROTOCOL

The hypothesis is that giant cell tumors are proliferative vascular lesions, or at least angiogenesis-dependent, and therefore would be expected to respond to anti-angiogenic therapy.[2–4]

Classification and Case Definition for Aggressive Giant Cell Lesions

Patients are classified as having aggressive giant cell lesions based on the system described by Chuong, Kaban, and Kozakowich in 1986,[5] Ficarro, Kaban, and Hansonin 1987,[6] and Kaban, and colleagues in 2002.[4] Aggressive giant cell lesions are defined as those exhibiting clinical characteristics of size greater than 5 cm, rapid growth, or recurrence after curettage. Radiographic criteria are cortical bone thinning, cortical bone perforation, tooth displacement, and root resorption. Tumors equal to or greater than 5 cm in size and/or tumors that recurred after curettage or resection are considered aggressive tumors based on these characteristics alone. Tumors with at least three of the other criteria are also classified as aggressive. Subjects who do not meet the case definition of aggressive giant cell lesions are excluded from this protocol and are treated by enucleation alone.

Patients

A sample composed of subjects with aggressive giant cell lesions, derived from the entire population of patients with giant cell lesions referred to the MGH Department of Oral and Maxillofacial Surgery between April 1995 and January 2006, was enrolled. Patients were excluded with "non-aggressive" or "intermediate" tumors, systemic disease, or any contraindication to interferon therapy. Evaluation included history, physical examination, plain

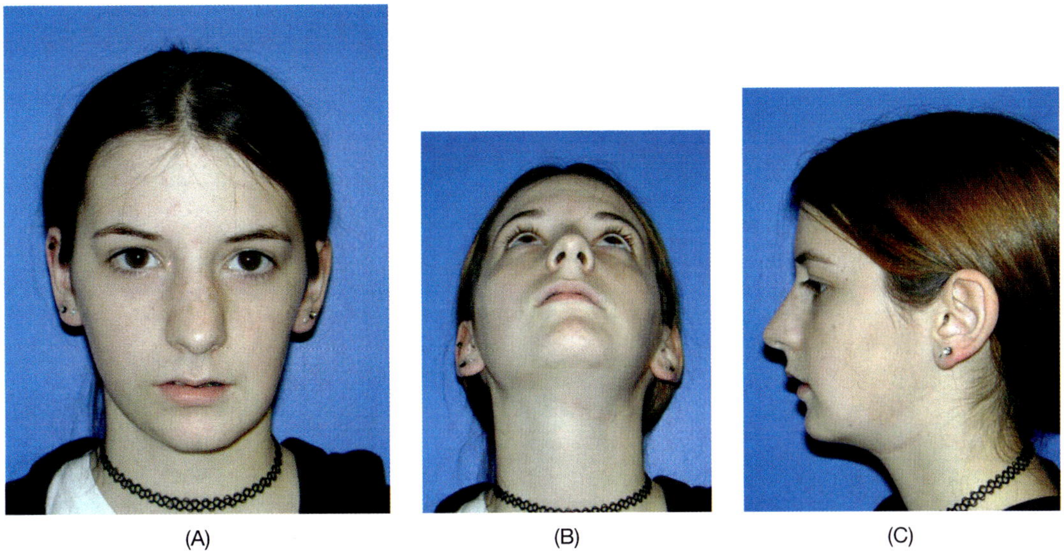

FIGURE 5 • Frontal, submental, and left lateral photographs of the patient 9 years post-treatment with no sign of recurrence.

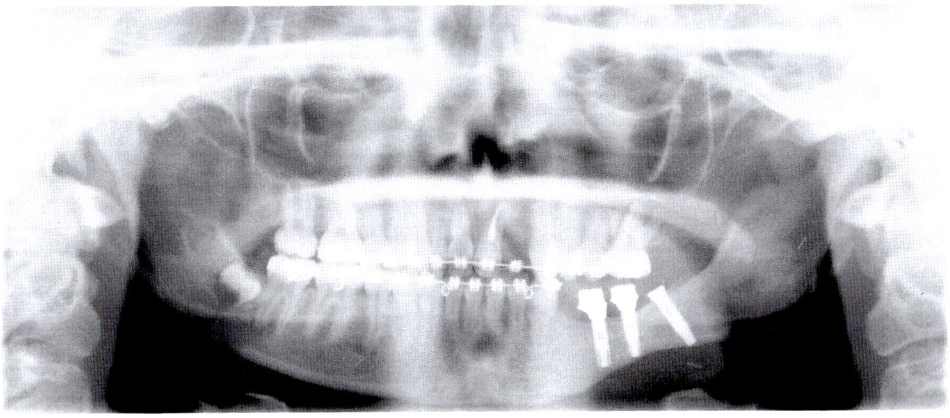

FIGURE 6 • Panoramic radiograph 9 years post-treatment. Note osseointegrated implants in the regenerated left mandible.

radiographs, and CT scans with contrast. All patients were treated by enucleation with preservation of vital structures, that is, teeth and nerves, and adjuvant interferon therapy.

Protocol (Table 1)

Baseline complete blood count and liver function tests are obtained. The lesion is enucleated, taking care to preserve teeth and nerves in the operative field (Figures 1–5). The patient is started on interferon alpha 2a (Roferon A^R, Roche Laboratories, Inc, Nutley, NJ) or interferon alpha 2b (Intron A^R, Schering, Corporation, Kenilworth, NJ) at a dose of 3 million units/meter^2 administered once per day via subcutaneous injection 48 to 72 hours post-operatively (Figure 7–11).

TABLE 1 MGH Protocol[2,4]
1. Document aggressive giant cell lesion
2. Baseline CBC, liver function tests
3. Enucleation of lesion (preservation vital structures)
4. Interferon alpha starting 48–72 hours post-operatively
5. Clinical/radiograph follow-up: immediate, 6 and 12 weeks, and every 3 months until cavity filled in with bone (panoramic or other standard plain radiographs)
6. Interferon stopped at 6–12 months (when cavity filled in with bone/confirmed by CT)

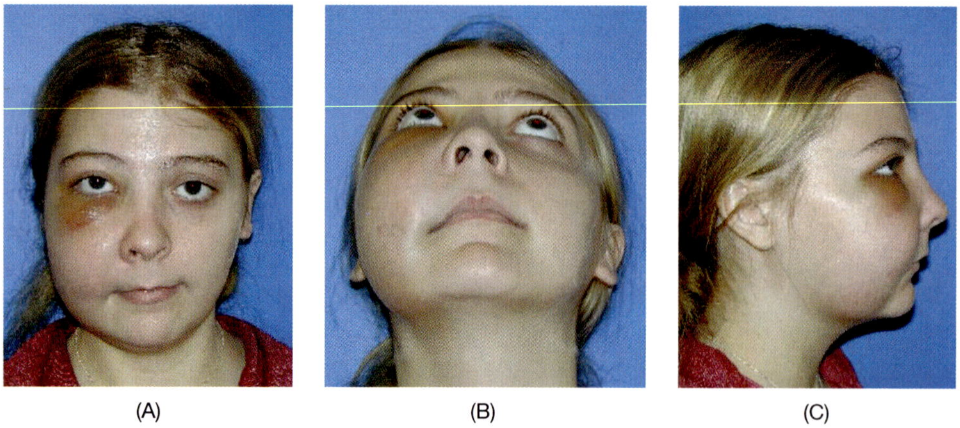

FIGURE 7 • Frontal (A), submental (B), and lateral (C) views of a 16-year-old female with a giant cell lesion of the right maxilla classified as aggressive based on size, rapid recurrence, and growth after curettage.

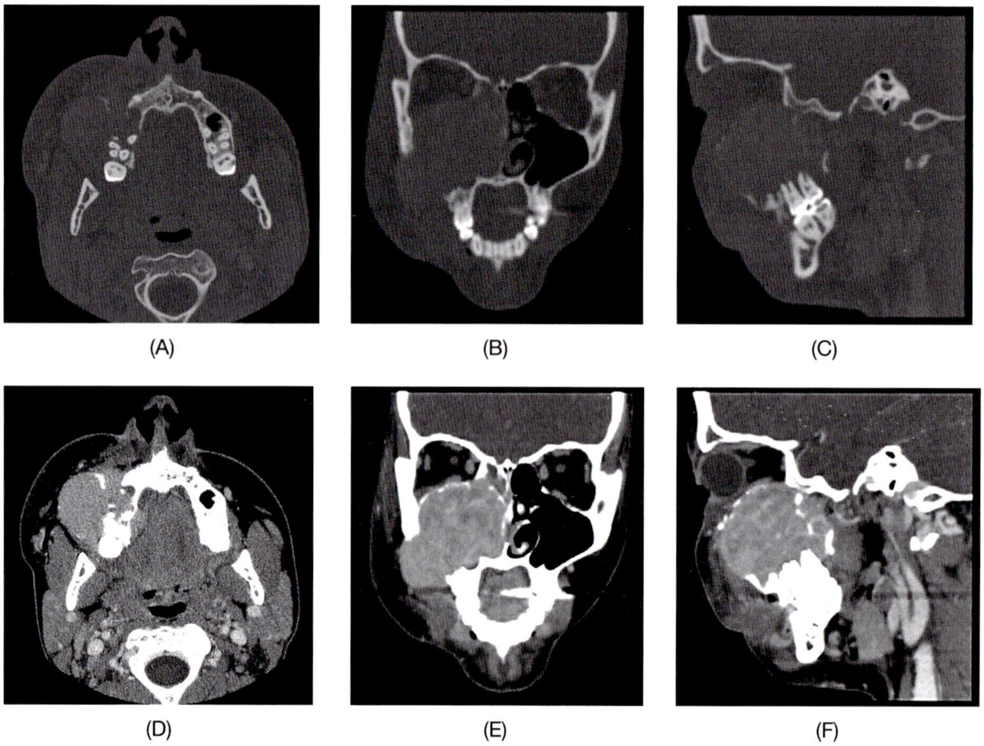

FIGURE 8 • Pre-operative CT scans. Bone (A, B, C) and soft tissue (D, E, F) windows in the axial, coronal and sagittal planes demonstrating the large tumor mass involving the entire right maxilla, nose, maxillary, and ethmoid sinuses and compressing the orbital floor.

During treatment, patients are evaluated for interferon side effects: fever, flu-like symptoms, lethargy, post-nasal drip, skin rash, hair loss, and so on. Hematocrit, hemoglobin, white blood cell and platelet counts, and liver function tests are obtained every 6 weeks (sooner if indicated). Patients are monitored for recurrence by clinical examination and radiographs.

Neutropenia (absolute neutrophil count $<1{,}000\ K/\mu m^3$), thrombocytopenia, elevated liver transaminases (>five-fold increase above normal), or development of anti-thyroid antibodies are indications to stop therapy or to reduce the interferon dose. A panoramic X-ray is obtained immediately at 6 and 12 weeks post-operatively and then at 3-month intervals until the defect appears to be filled in with bone. A CT scan is then obtained and interferon therapy is discontinued if there is CT confirmation of complete bone regeneration in the defect. If not, the protocol is continued until this occurs. Based on previously published data,[4,6,16] patients are considered cured if they have no evidence of new or recurrent tumor 2 years after completion of treatment.

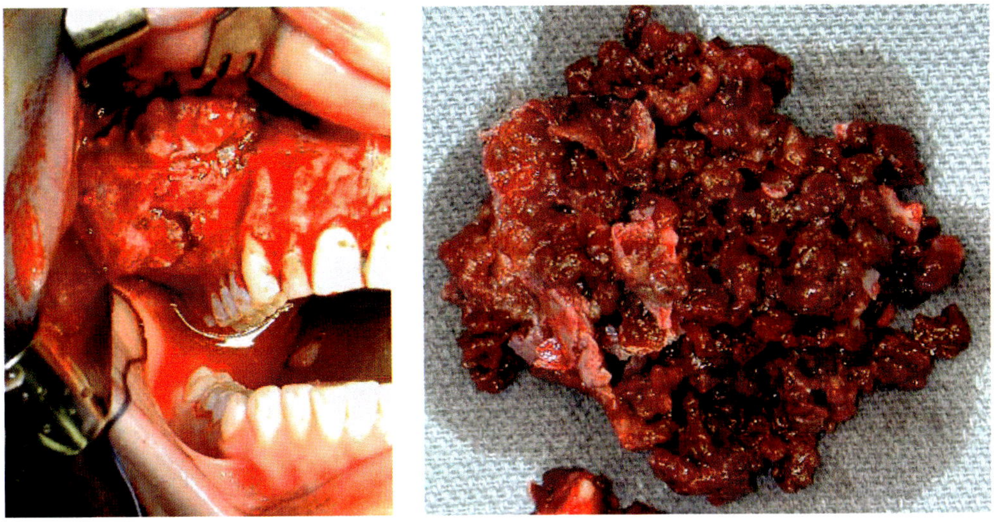

(A) (B)

FIGURE 9 • Intra-operative view (A) demonstrates the expansile lesion with bone thinning and expansion photograph of pathologic specimen (B).

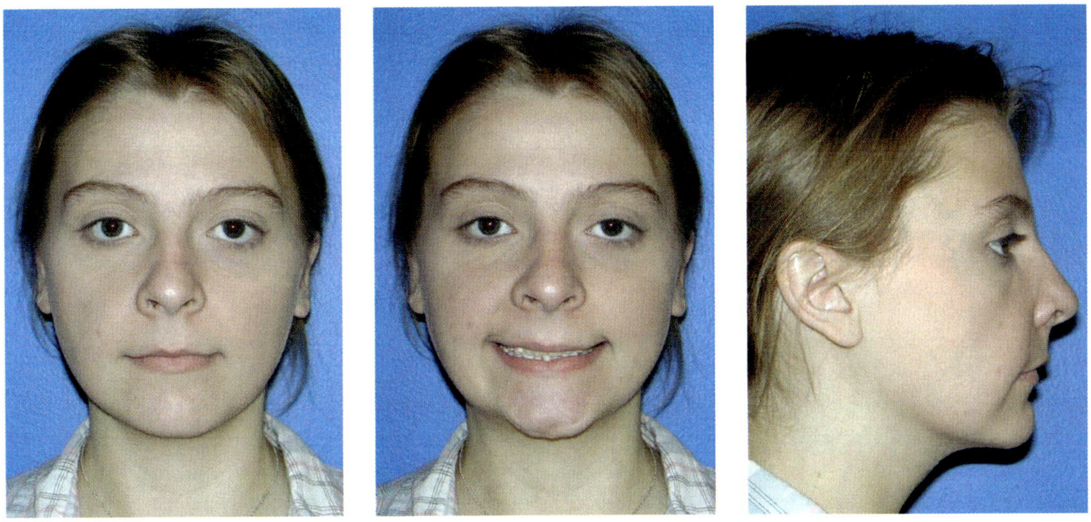

(A) (B) (C)

FIGURE 10 • Two years post-operative and 18 months off interferon, the frontal (A), smiling (B), and lateral (C) photographs demonstrate facial symmetry with no tumor present.

RESULTS

Between April 1995 and January 2006, 36 patients with giant cell lesions were evaluated and their medical records reviewed. Ten patients with non-aggressive giant cell lesions were excluded from further analyses. The study sample was composed of 26 subjects (17 female, 65 percent) with aggressive giant cell lesions. Mean age of the sample was 18.5 ± 12.1 years (range = 2.4–53.7); median age was 16 and ten subjects were ≤ 12 years of age. Tumors were classified as aggressive based on clinical behavior (recurrence, n = 16; or size and rapid growth, n = 3), radiographic findings (size, tooth displacement, root resorption, cortical thinning or perforation, n = 7), or a combination of both. Measured on CT scan, the mean tumor size in the largest dimension was 37.3 ± 15.9 mm (range = 14–70 mm). Most lesions (n = 15) were located in the posterior mandible. Five lesions were located in the anterior mandible. Of the six maxillary lesions, four were located in the anterior maxilla.

Enucleation of the giant cell lesions with preservation of teeth and nerves was performed on 23 subjects; 3 additional subjects had teeth extracted at the time the lesion was enucleated. Interferon alpha (2a or 2b) was started a median of six days after surgery at a dose of 3,000,000 units/m^2. The drug was administered daily via subcutaneous injections. The mean duration of treatment was 8.0 ± 3.1 months (range 3.2–16.6; median = 7.3 months).

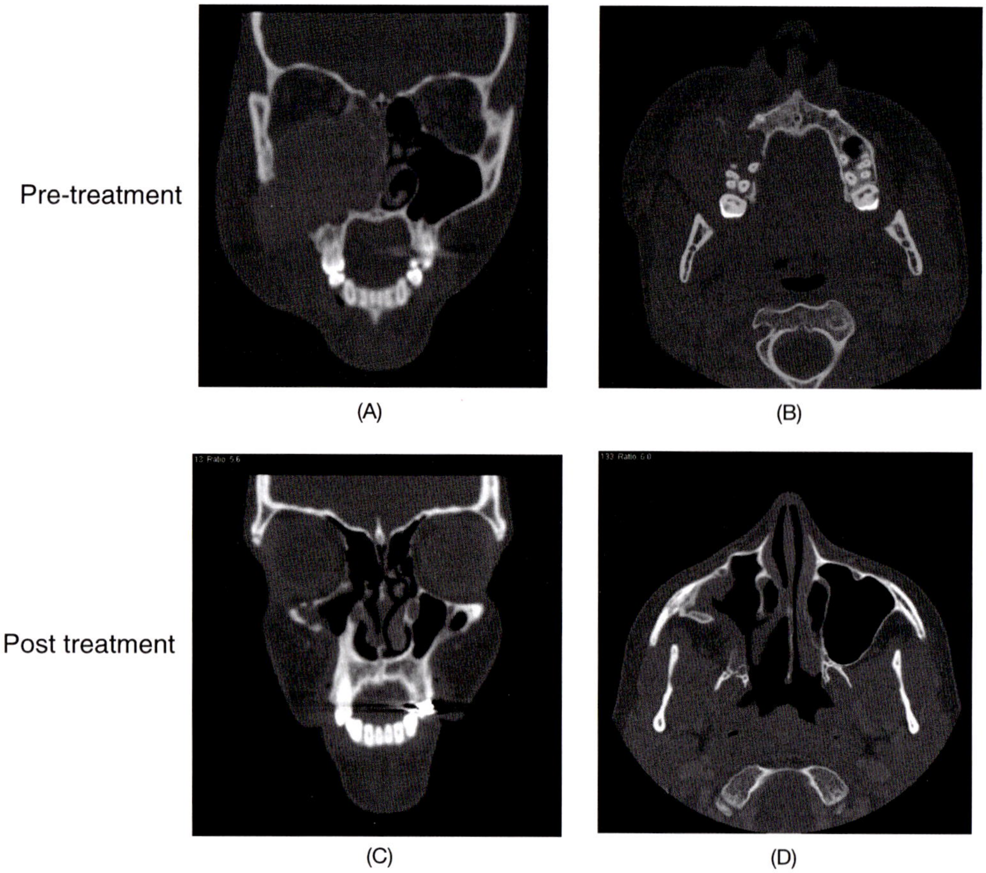

FIGURE 11 • Corresponding pre-operative (A, B) and 2 years post-operative (C, D) coronal (left) and axial (right) bone window CT scans demonstrate disappearance of tumor in maxilla, nose, and maxillary and ethmoid sinuses. *(Figures 4–11. With permission from: Kaban LB, et al. Adjuvant Antiangiogenic Therapy for Aggressive Giant Cell Lesions of the Jaws. J Oral & Maxillofac Surg. 2007)*

To date, 52 subjects enrolled in the study are considered cured or in remission or inactive treatment.

All subjects reported one or more side effects of interferon, most commonly fever or flu-like symptoms during the 24 to 48 hours after the first dose. Of these 12 required modification of the interferon dose due to rash, nausea, neutropenia, fever, or weight loss. In 11 cases, the interferon dose was reduced or discontinued temporarily. In one case, the dose was increased due to tumor growth.[2]

Side effects necessitated discontinuation of interferon therapy for 4 subjects. One patient developed anti-thyroid antibodies and a second developed unacceptable fatigue and depression. In both patients, the anti-angiogenic agent was changed to zoledronate and both responded to treatment; one patient is cured and the other in remission. In a third case, although the tumor was responding as expected, the patient developed brachial plexus palsy and refused any further adjuvant therapy. He underwent en bloc resection and is cured more than 3 years post-treatment. The fourth patient stopped interferon 5 months after initiating treatment due to an acute psychotic episode. Concurrently, she had been receiving medication for depression. The lesion had responded as evidenced by bone fill and no additional treatment was rendered. This subject is currently cured more than 5 years after cessation of interferon.[2]

All of the subjects responded to the treatment protocol as evidenced by tumor regression clinically and radiographically. The protocol was altered only because of intolerable interferon side effects.

DISCUSSION

Since our first report on the use of interferon therapy for the management of a multiply recurrent aggressive giant cell tumor of the mandible,[3] our group has treated and reported on a total of 26 patients without any recurrences using this protocol. After completion of the study, 1 patient developed a late recurrence in the anterior maxilla and has been managed successfully with intralesional steroid injections. An additional 12 patients with aggressive giant cell lesions have responded to this protocol bringing the total number of patients without recurrence to 38.

We have used a consistent definition for aggressive giant cell lesions and a consistent protocol for management. The reported recurrence rate of aggressive giant cell tumors (as defined in this paper) ranges from 30% to 70%.[4–6,18–23,25] Assuming recurrence frequencies of 30% to 70%, the expected number of recurrences after curettage alone in the sample reported here would be 7.8 and 18.2, respectively. Most patients treated with interferon experience side effects that are minor and that do not require permanent modification of the dose. However, 4 patients

(15.3%) in our study did discontinue the drug, 1 each because of fatigue, anti-thyroid antibodies, brachial plexus palsy, and psychosis. In the future, it may be possible to reduce the interferon dose while maintaining its anti-angiogenic effects.[47] Improved tolerability and quality of life was shown, with maintained efficacy, using twice daily low-dose interferon alpha 2b: a randomized phase II trial of low-dose versus intermediate-dose interferon alpha 2b in patients with metastatic renal cell carcinoma. We plan to evaluate the use of a combination of low-dose interferon (1 million units per meter^2) and a short period of intravenous bisphosphonate (3 doses) in the next phase of this study.

Abukawa,[48] in our laboratory, has demonstrated that interferon has a positive effect on osteoblast function. In cell culture, he demonstrated that porcine mesenchymal stem cells (treated with osteogenic supplements to differentiate into osteoblasts) produced denser alkaline phosphatase staining when exposed to interferon than cells not exposed to interferon. Takayanagi and associates[49] demonstrated that interferon beta inhibits osteoclast differentiation by interfering with the RANKL-induced expression of c-FOS, an essential transcription factor for osteoclast development. Mice deficient in interferon beta signaling exhibit loss of bone mass secondary to enhanced osteoclastogenesis. The combination of interferon's anti-angiogenic effects, as well as its effects on both osteoclasts and osteoblasts, probably explains the combination of tumor suppression and bone formation seen in this series of patients. We hypothesize that there may be a spectrum of mesenchymal, angiogenic, primary jaw tumors, including giant cell lesions and recurrent aneurysmal bone cysts, which would be expected to respond to anti-angiogenic therapy.[3] Giant cell tumors in long bones and the spine are also known to be richly vascular and as a result, transcatheter arterial embolization has been used, alone and in combination with chemotherapy, to treat these lesions.[17]

Since our original report,[3] interferon treatment for giant cell lesions of the mandible has been reported by others.[51,52] The use of interferon for treatment of giant cell lesions of the long bones and spine has also been reported.[17,53,54] This experience has been limited to patients with metastatic disease or locally advanced, unresectable giant cell tumors or those that have failed radiation therapy.[18,20,21] Recently, Marler and colleagues[50] reported successful anti-angiogenic therapy for giant cell angioblastomas in two patients, one with a maxillary lesion and the other a lesion of the hand. These tumors were FGF-producing angiogenic lesions with some similarity to giant cell tumors.

The observation that the enucleated bone defect regenerates and fills in with bone more rapidly than would be expected based on its size has remained consistent during this study. The possible relationship of this regenerative phenomenon, and the nature of the angiogenic tumor and interferon treatment is not understood. However, as noted previously, interferon does appear to inhibit osteoclastic bone resorption and stimulate osteoblasts and pre-osteoblasts at least in cell culture.

CONCLUSION

The results of this study indicate that the use of anti-angiogenic therapy, in combination with enucleation, is a promising strategy for the management of aggressive giant cell lesions. Our laboratory is currently studying candidate biomarkers to differentiate aggressive and non-aggressive lesions and the response to treatment.

Acknowledgments

This chapter is adapted from entries 1, 2, and 3 in the References section. I want to acknowledge my co-workers and co-authors in this ongoing project: Thomas B. Dodson, DMD, MPH; Maria J. Troulis, DDS, MSc; Meredith August, DMD, MD; Michael Wilkinson, DMD; David Ebb, MD; Francis Hornicek, MD; Judah Folkman, MD; and John B. Mulliken, MD.

References

1. Kaban LB. The biomedical technology revolution: challenges and opportunities for oral and maxillofacial surgery. Int J Oral Maxillofac Surg 2002;31:1–12.
2. Kaban LB, Wilkinson, M, Troulis MJ, Ebb D, Dodson TB. Adjuvant antiangiogenic therapy for giant cell tumors of the jaw. J Oral Maxillofac Surg 2007;65:2018–24.
3. Kaban LB, Mulliken JB, Ezekowitz RE, Ebb D, Smith PS, Folkman J. Anti- angiogenic therapy of a recurrent giant cell tumor of the mandible with interferon alfa-2A. Pediatrics 1999;103:1145–1149.
4. Kaban LB, Troulis MJ, Ebb D, August M, Hornicek FJ, Dodson TB. Antiangiogenic therapy with interferon alpha for giant cell lesions of the jaws. J Oral Maxillofac Surg 2002;60:1103–11.
5. Chuong R, Kaban LB, Kozakewich H, Perez-Atayde A. Central giant cell lesions of the jaws: a clinicopathologic study. J Oral Maxillofac Surg 1986;44:708–13.
6. Ficarra G, Kaban LB, Hansen LS. Central giant cell lesions of the maxilla and mandible: a clinicopathologic study. Oral Surg Oral Med Oral Pathol 1987;64:44–9.
7. Marx RE, Stern D. Oral and Maxillofacial Pathology. A rational for diagnosis and treatment. 1st ed. Chicago, IL: Quintessence; 2002. p. 783–9.
8. Stewart JCB. Benign nonodontogenic tumors. In Regezi JA, Sciubba JJ, editors. Oral Pathology. Clinical pathologic correlations. 1st ed. Philadelphia, WB Saunders; 1983. p. 381–3.
9. de Lange J. Central giant cell granuloma of the jaw: epidemiology, therapy, and related disorders: general introduction and aims of the study. PhD thesis. Department of Oral and Maxillofacial Surgery, Academic Medical Center. University of Amsterdam, Amsterdam: 2006. p. 14.
10. Bo L, Shi-Feng Y, Tie-Jun L. Multinucleated giant cells in various forms of giant cell containing lesions of the jaws express features of osteoclasts. J Oral Pathol Med 2003;32:367–75.
11. Tiffee JC, Aufdemorte TB. Markers for macrophage and osteoclast lineages in giant cell lesions of the oral cavity. J Oral Maxillofac Surg 1997;55:1108–12.
12. Miyamoto N, Higuchi Y, Tajima M, Ito M, Tsurudome M, Nishio M, et al. Spindle-shaped cells derived from giant-cell tumor of bone support differentiation of blood monocytes to osteoclast-like cells. J Orthop Res 2000;18:647–54.
13. Itonaga I, Schulze E, Burge PD, Gibbons CLMH, Ferguson D, Anthanasou NA. Phenotypic characterization of mononuclear and multinucleated cells of giant cell reparative granuloma of small bones. J Pathol 2002;198:30–6.
14. Nagasawa T, Kobayashi H, Kiji M, Aramaki M, Mahanonda R, Kojima T, et al. LPS-stimulated human gingival fibroblasts inhibit the differentiation of monocytes into osteoclasts through the production of osteoprotegerin. Clin Exp Immunol 2002;130: 338–44.

15. Itonaga I, Hussein I, Kudo O, Sabokbar A, Watt-Smith S, Ferguson D, et al. Cellular mechanisms of osteoclast formation and lacunar resorption in giant cell granuloma of the jaw. J Oral Pathol Med 2003;32:224–31.
16. Chuong R, Kaban LB. Diagnosis and treatment of jaw tumors in children. J Oral Maxillofac Surg 1985;43:323–32.
17. Yasko AW. Inteferon therapy for vascular tumors of bone. Current Opinion in Orthopaedics 2001;12:514–8.
18. Rock MG, Pritchard DJ, Unni KK. Metastasis from histologically benign giant cell tumors of bone. J Bone Joint Surg An 1984;66:269–74.
19. Kaban LB: Aggressive jaw tumors in children. Oral & Maxillofac Clin N Amer 1993;5:249–66.
20. Sulh MA, Greco MA, Jiang T, Goswami SB, Present D, Steiner G. Proliferation index and vascular density of giant cell tumors of bone. Cancer 1996;77:2044–51.
21. Oda Y, Miura H, Tsuneyoshi M, Iwamoto Y. Giant cell tumor of bone: oncological and functional results of long-term follow-up. Jpn J Clin Oncol 1998;28:323–8.
22. Terry BC, Jacoway JR. Management of central giant cell lesions: An alternative to surgical therapy. Oral & Maxillofac Clin N Am 1994;6:579.
23. Harris M. Central giant cell granulomas of the jaws regress with calcitonin therapy. Brit J Oral & Maxillofac Surg 1993;31:89–94.
24. Pogrel MAP, Regezi J, Harris ST, Goldring SR. Calcitonin treatment for central giant cell granulomas of the mandible: report of two cases. J Oral Maxillofac Surg 1999;57:848–53.
25. Pogrel MA. Calcitonin therapy for giant cell granuloma. J Oral Maxillofac Surg. 2003;61:649–53.
26. Kaban LB, Dodson TB. Discussion. Calcitonin therapy for central giant cell granuloma. J Oral Maxillofac Surg 2003;61:653–4.
27. de Lange J, van den Akker HP, Veldhuijzen van Zanten GO, Engelshove HA, van den Berg H, Klip H. Calcitonin therapy in central giant cell granuloma of the jaw: a randomized double-blind placebo-controlled study. Int J Oral Maxillofac Surg 2006;35:791–5.
28. Parkin DM, Pisani P, Ferlay J: Global cancer statistics. CA Cancer J Clin 1999;49:33–64.
29. Sidransky D. Nucleic acid-based methods for the detection of cancer. Science 1997;278:1054–9.
30. Todd R, Gutkind JS, Shillito EJ, Wong DT. Solid tumors: Microarray analysis of oral cancers. In: Warrington JA, Wong DT, Todd R, editors. Microarray and cancer Research. 1st. Westboro, MA: Eaton Publishing; 2002. p. 139–53.
31. Brennan JA, Mao L, Hruban RH, Boyle JO, Eby YJ, Koch WM, Goodman SN, Sidransky D. Molecular assessment of histopathological Staging in squamous-cell carcinoma of the head and neck. New Eng J Med 1995;332:429–35.
32. Folkman J: Antiangiogenesis agents, In: DeVita VT Jr., Hellman S, Rosenberg SA, eds. CANCER Principles & Practice of Oncology, 6th edition, Lippincott Williams & Wilkins, Philadelphia, 2001; pp. 509–519.
33. Goldman JM, Melo JV. Targeting the BDR-ABL tyrosine kinase in chronic myeloid leukemia [editorial]. New Eng J Med 2001; 344:1084–6.
34. Druker BJ, Talpaz M, Resta DJ, Peng B, Buchdunger E, Ford JM, Lydon NB, Kantarjian H, Capdeville R, Ohno-Jones S, Sawyers CL. Efficacy and safety of a specific inhibitor of the BCR-ABL tyrosine kinase in chronic myeloid leukemia. New Eng J Med 2001;344:1031–7.
35. Cooke R: Dr. Folkman's War: Angiogenesis and the Struggle to Defeat Cancer. New York: Random House; 2001. p. 44–57.
36. Carmeliet P, Jain RK. Angiogenesis in cancer and other diseases: From genes to function to therapy. Nature 2000;407:249–57.
37. Folkman J. Tumor angiogenesis. In Holland JF, Frei E III, Bast RC Jr, Kufe DW, Pollock RE, Weichselbaum RR, eds. Cancer Medicine, 5th ed. Ontario, Canada, B.C. Decker Inc. 2000; p. 132–52.
38. Folkman J. Angiogenesis. In Braunwald E, Fauci AS, Kasper DL, Hauser SL, Longo DL, Jameson JL, eds: Harrison's Textbook of Internal Medicine, 15th Ed. Philadelphia, McGraw-Hill; 2001. p. 517–30.
39. Takahashi K, Mulliken JB, Kozakewich HP, Rogers RA, Folkman J, Ezekowitz RA. Cellular markers that distinguish the phases of hemangioma during infancy and childhood. J Clin Invest 1994;93:2357–64.
40. Singh RK, Gutman M, Bucana CD, Sanchez R, Llansa N, Fidler IJ: Interferons alpha and beta down-regulate the expression of basic fibroblast growth factor in human carcinomas. Proc Natl Acad Sci USA 1995;92:4562–6.
41. Ezekowitz R, Mulliken J, Folkman J. Interferon alfa-2a therapy for life-threatening hemangiomas of infancy. New Eng J Med 1992;326:1456–63.
42. Ezekowitz R, Mulliken J, Folkman J. Additional Corrections: Interferon for hemangiomas of infancy. New Eng J Med 1995;333: 595–6.
43. Singhal S, Mehta J, Desikan R , et al: Antitumor activity of thalidomide in refractory multiple myeloma. N Engl J Med 1999;341:1565–71.
44. Folkman JM: Angiogenesis: an organizing principle for drug discovery? Nature Reviews April 2007;6:273–86.
45. Murray, Joseph E., Personal communication.
46. Kaban LB, Dodson TB. Management of giant cell lesions. Int J Oral Maxillofac Surg 2006;35:1074–5.
47. Tannir NM, Cohen L, Wang X, Thall P, et al. Improved tolerability and quality of life with maintained efficacy using twice daily low dose interferon alpha 2b: results of a randomized phase II trial of low dose versus intermediate dose interferon alpha 2b in patients with metastatic renal cell carcinoma. Cancer 2006;107: 2254–61.
48. Abukawa H, Kaban LB, Williams WB, Terada S, Vacanti JP, Troulis MJ. Effect of interferon alpha-2b on porcine mesenchymal stem cells. J Oral Maxillofac Surg 2006;64:1214–20.
49. Takayanagi H, Kim S, Matsuo K, Suzuki H et al. RANKL maintains bone homeostasis through c-Fos-dependent induction of interferon-beta. Nature 2002;416:774–9.
50. Marler JJ, Rubin JB, Trede NS, Connors S, Grier H, Upton J, Mulliken JB, Folkman J. Successful antiangiogenic therapy of a giant cell angioblastoma with Interferon Alfa 2b: report of 2 cases. Pediatrics 2002;109:37–41.
51. Wong M, Herzog C, Flaitz C, Marchena J, Eftekhari F. Alpha interferon in the management of central giant cell granuloma: early experienced. J Oral & Maxillofac Surg 2002;69 (suppl):86–7.
52. Collins A. Experience with antiangiogenic therapy of giant cell granuloma of the facial bones. Ann R Australas Coll Dent Surg 2000;15:170–5.
53. Folkman J. Clinical applications of research on angiogenesis. N Engl J Med 1995;333:1757–63.
54. Strander H, Boethius J, Erhardt K. Does successful interferon treatment of tumor patients require life-long treatment? J Interferon Res 1987;7:619–26.

Minimally Invasive Treatment Approaches to the Diagnosis and Treatment of Keratocystic Odontogenic Tumors

21

Meredith August, DMD, MD

OUTLINE

- Introduction
- Fine Needle Aspiration Biopsy—Utility in the Head and Neck
- Treatment Considerations for KCOTs
- Decompression/Irrigation and Cytokeratin-10 Evaluation
- Summary of the MGH Experience and Case Discussions
- Conclusion

INTRODUCTION

The keratocystic odontogenic tumor (KCOT) is a pathologic entity that generates discussion and controversy. What is appropriate treatment? What is appropriate follow-up? What is an acceptable rate of recurrence when considering treatment options? Is en bloc resection too "radical" an option for a "benign" entity? How can treatment be modified to reduce expected recurrence? Do teeth involved with the cyst need to be removed pro forma? If teeth are left in position, is endodontic therapy imperative? Is there malignant or transformative potential associated with the KCOT? What imaging studies are best for diagnosis and for subsequent follow-up?

If a room is filled with experienced oral and maxillofacial surgery (OMFS) practitioners, the answers to these and other queries regarding the KCOT will be disparate and often contradictory. To wit is the number of times that OKC discussion is part of a "controversies" discussion within this specialty.[1,2] Perhaps the conclusion must be that there is no correct treatment but rather an evolving body of literature dealing with various diagnostic and treatment options, all associated with both positive aspects and limitations.

Minimally invasive is an interesting term. Does it imply simplicity? Less work for the doctor? An easier time for the patient? Smaller incisions? Access from a distance using fiber-optic technology? It is a commonly used term and probably has different meanings to different subspecialists. When one is formulating a chapter on minimally invasive approaches to the KCOT, both diagnostic maneuvers and treatment options come to mind. Our department has expanded the use of fine needle aspiration biopsy techniques (FNAB) to include sampling tissue from intrabony pathologic entities and we have reported good success in distinguishing the KCOT from dentigerous and other odontogenic cysts.[3] Treatment with decompression and longitudinal irrigation is now an important part of our treatment protocol for large KCOTs and is based largely on the successful long-term results reported by Marker and associates[4,5] and on our own clinical and histologic follow-up data.[6] To describe this type of treatment as minimally invasive, however, may undervalue the requirement of patient compliance as well as a protracted treatment time.

This chapter will focus on pre-operative diagnosis of odontogenic cysts and tumors using modification of FNAB techniques and immunocytochemical markers that help distinguish the KCOT from the dentigerous cyst. Treatment of the KCOT with decompression and irrigation, though not a new concept, will be revisited in the context of the histologic changes noted in the epithelium using the identical cytokeratin markers used for the FNAB evaluation. This change in cyst lining may well relate to the low recurrence rate reported with this technique by our group and others.

FINE NEEDLE ASPIRATION BIOPSY—UTILITY IN THE HEAD AND NECK

The use of the fine needle to aid in the diagnosis of head and neck pathology is quite common.[7,8,9] Many soft tissue masses in this region are readily palpable and the sampling procedure is straightforward. Common anatomic sites amenable to FNAB evaluation and cytologic study are parotid masses, thyroid nodules, and enlarged lymph nodes. Deeper and less palpable entities may also be sampled using CT, MRI, or ultrasound guidance.[10,11] Early in its use, FNAB was described as a highly sensitive technique to distinguish malignant from non-malignant pathology.[12] However, due to a lack of tissue architecture when only isolated cells are extracted, diagnostic acumen was limited. Thus, a salivary gland tumor might be identified as malignant but the precise histologic diagnosis was often speculative and highly dependent upon sampling.

FNAB can be technically challenging when sampling intrabony pathology. Typically, a 23-gauge needle is used and this is often inadequate to penetrate cortical bone. If successful, however, the reliability for diagnosis of malignant bone tumors has been extensively reported in both cytopathology and orthopedic journals.[13,14] The technique has become an important and cost-effective alternative to open biopsy, especially in university-based cancer centers. Reported accuracy of FNAB for diagnosis of non-benign bone pathology is lower than that for malignant lesions.[15] Giant cell and cartilaginous tumors such as chondroblastomas are the most frequently diagnosed.[16] Cases of eosinophilic granuloma, aneurysmal bone cyst, and osteomyelitis have also been reported.[17]

Recent reviews of head and neck FNAB have included few intrabony samples.[18,19,20] Difficulty in obtaining sufficient cellular material, histologic heterogeneity of jaw tumors, and relative inexperience of cytopathologists in this area likely account for the limited use. Platt and colleagues reviewed 89 cases of FNAB of head and neck lesions over a 2-year period.[18] The majority of specimens were malignant tumors. Diagnoses made by FNAB were in agreement with the final histopathology in 91% of cases. However, only 2 specimens were of mandibular origin and no maxillary tumors were included. Schelkun and associates reviewed 213 head and neck FNABs.[19] Histopathologic specimens were available in 110 cases and an overall correlation between pathologic and FNAB diagnosis was 90%. The majority of cases were masses in the neck and thyroid gland. Only 4 specimens were from sites in the oral cavity and their exact locations were not specified. Daskalopoulou and colleagues reported the results of 1,022 FNABs in the head and neck.[20] There were 67 perioral soft tissue or facial bone specimens but no odontogenic, benign fibro-osseous or distinctly intrabony lesions.

Ramzy reported 23 FNABs of radiolucent lesions of the jaws.[21] Cytologic diagnosis was correct in 62% of cases, a valuable differential diagnosis was provided in an additional 4 cases, and 4 aspirates were non-diagnostic. The range of pathology included both malignant and benign lesions. The authors described characteristic cytologic findings that, if not independently diagnostic, when coupled with the history and clinical presentation, facilitated correct diagnosis. They concluded that FNAB is a simple, inexpensive technique that is valuable in diagnosing many radiolucent changes in the jaws.

We reported the use of FNAB on 21 cases of intrabony jaw pathology evaluated and treated at MGH.[3] Standard technique using either 23- or 25-gauge needles for sampling was employed. Papanicolou, Wright, and special stains were used as indicated. Eighteen of 21 cases had sufficient cells for analysis. Five were positive for malignant cells and 13 for non-malignant cells. The diagnosis rendered by the cytopathologist was confirmed by the final histopathology in all malignant cases (100% accuracy). This is in agreement with orthopedic studies showing high levels of accuracy with malignant disease. However, only 6/13 non-malignant FNAB diagnoses were subsequently confirmed (46% accuracy). We concluded that improvements in both technique and diagnostic ability were required for FNAB to become a useful and minimally invasive option to open biopsy in the majority of benign intrabony lesions. Technical improvements such as the use of bony introducers or larger-gauge (core biopsy) needles for better penetration of cortical bone, aspiration of multiple areas of heterogeneous lesions, and radiologic and ultrasound guidance have been recommended.[22] In addition, interpretational improvements might be expected if cytopathologists and surgeons work collaboratively.

The accuracy of FNAB for characterization of cystic lesions of the jaws is disappointing and probably relates to the lack of architectural context on FNAB material.[23] For example, to distinguish a dentigerous cyst from a KCOT on routine histology, the characteristics of the lining cells are of paramount importance. Keratocysts are classically described as having a lining that is uniform in thickness, a parakeratinized surface that is corrugated, and a prominently polarized basal layer. This is distinct from the dentigerous cyst, which most often has a thin lining composed of stratified squamous cells with a generalized absence of rete peg formation. By FNAB, these architectural features are unavailable for evaluation, thereby making distinction between the two difficult. Cytologic diagnosis of a KCOT is based upon sampling of the lining cells and demonstrating polygonal anucleated squamous cells with well-defined cell borders, an abundance of keratinizing cells, and keratin lamellae.[24]

To improve the yield of lining cells in the aspirate, we proposed using a larger needle (19-gauge) and orienting it in tangential fashion, contacting the opposite bony wall and scraping the cyst-bone interface with the bevel of the needle (Figure 1). Rapid interpretation can be used to ensure inclusion of epithelial cells in the aspirate (Figure 2).

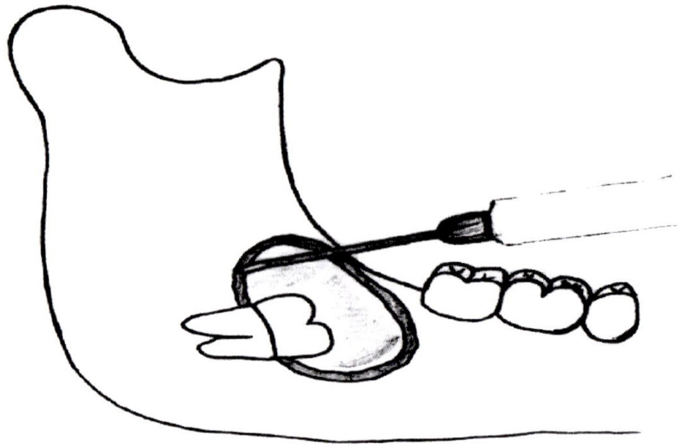

FIGURE 1 • Schematic of FNA technique, showing how the needle must "scrape" cyst lining off bony wall.

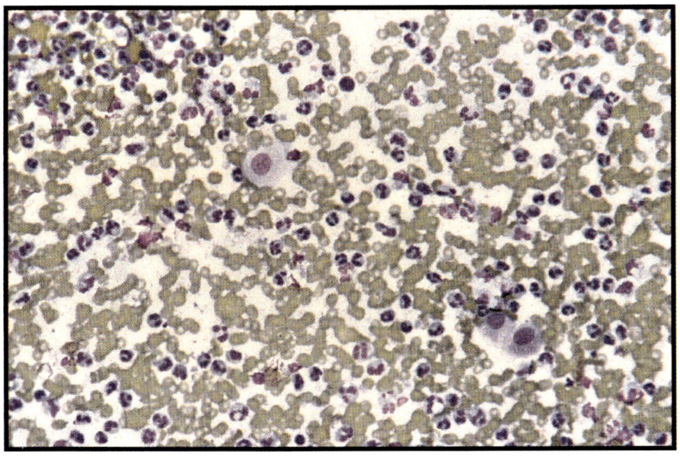

FIGURE 2 • Rapid interpretation to insure the presence of epithelial cells (noted with arrowheads).

Various epithelial markers have been identified in the lining of odontogenic cysts, including cytokeratins of various molecular weights, epithelial membrane antigen (EMA), carcinoembryonic antigen (CEA) and proliferating cell nuclear antigen (PCNA).[25,26] Several authors have described specific "keratin expression profiles" that aid in distinguishing the various types of odontogenic cysts.[27,28] Matthews and associates have demonstrated, by indirect immunoperoxidase methods, that the more ordered and cornified epithelium lining the KCOT produces low molecular-weight keratin (keratins 10/11) in abundance compared to dentigerous cysts in which it is rarely produced.[27] A characteristic band-like uptake of anti-cytokeratin 10 antibody is nicely demonstrated in the KCOT lining (Figure 3). Expression of intermediate molecular-weight cytokeratin using the monoclonal antibody LP34 in the basal layers of dentigerous cysts but not KCOTs was demonstrated by MacDonald and colleagues.[28] These distinct patterns of keratin expression likely reflect the differentiation state of the lining cells of the two cystic entities and provide an opportunity for using specific immunohistochemical markers to distinguish one from the other. Immunoperoxidase techniques can be applied to cytologic material as well as histologic samples, allowing for the establishment of a keratin profile provided that epithelial cells are present in the sample.

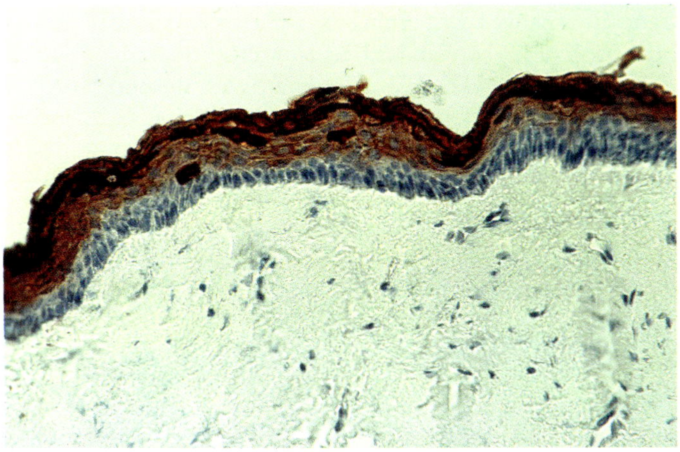

FIGURE 3 • Band-like uptake of anti-cytokeratin 10 antibody seen in the lining of a KCOT.

At our institution, we have demonstrated the efficacy of combining the modified FNAB technique described above and cytokeratin-10 immunocytochemical staining on sampled epithelial cells to allow for accurate diagnosis of KCOTs and to distinguish them from dentigerous cysts. In a prospective review of 18 odontogenic cystic lesions, the technique was found to be 100 accurate.[29] This has provided us with a straightforward and minimally invasive tool for establishing pre-operative diagnosis and has enabled timely discussion of treatment options with patients.

TREATMENT CONSIDERATIONS FOR KCOTs

Recommended treatment for the KCOT is disparate indeed. If the cyst is found at an early stage and has not caused extensive bone demineralization, cortical expansion/perforation or marked tooth displacement and root resorption, initial therapy often involves thorough enucleation and curettage.[30] This is especially true if the anatomic location of the cyst makes it amenable to this type of treatment and if routine longitudinal follow-up is feasible. Maxillary cysts are notoriously difficult to follow with the use of plain radiographs. Obtaining a post-enucleation CT scan can be invaluable when following a patient and differentiating possible recurrence from post-surgical changes.

Enucleation alone is associated with the highest recurrence rates (range of 17%–56%), especially when the cyst is removed in a fragmented fashion.[31,32] The difficulty of removing the lining in toto becomes increasingly problematic when teeth are present. In addition, concern regarding damage to the inferior alveolar nerve can result in less-than-complete enucleation. The presence of bony perforation will also predispose to recurrence if cyst lining or daughter cysts involve contiguous mucosa. Various adjunctive therapies have been advocated to decrease recurrence potential. These for the most part involve additional treatment of the underlying bone and include peripheral ostectomy or treatment of the surgical site with Carnoy's solution, electrocautery, or cryotherapy.[33,34] Marsupialization has also been reported but is associated with recurrences in the range of 25% to 100%.[35] Large or recurrent lesions in difficult anatomic locations have been treated by en bloc or marginal resection.[36]

In 1991, Brondum and colleagues reported long-term follow-up of 44 KCOTs.[4] The overall recurrence rate for this group was 18%. However, in a subset of patients (12/44) treated by decompression and irrigation, there were no reported recurrences during an observation period of 7 to 17 years. In addition, the authors reported both substantial reduction in cyst volume and a significant change in the character of the residual epithelium. At the time of residual cystectomy, the thin and friable quality associated with KCOT epithelium had changed to a thickened lining with little adherence to the surrounding bone. This same group reported long-term follow-up of 23 patients treated by decompression and irrigation.[5] Recurrence was reported in 2/23 patients (8.7%). Substantial histologic change (dedifferentiation) was observed in the epithelium of 19/23 (83%) cysts. This alteration in cystic epithelium was also noted by other authors and seemed to be associated with lesions demonstrating inflammatory change, presumably secondary to oral cavity communication.

DECOMPRESSION/IRRIGATION AND CYTOKERATIN-10 EVALUATION

Based on the promising results with decompression and irrigation, we began to use this method of treatment for large KCOTs at our institution.[6] The report of histologic changes in cyst lining over time also piqued our curiosity as to whether cystectomy specimens would show a distinct cytokeratin profile from the initial KCOT lining. If so, this so-called "dedifferentiation" might well contribute to the decreased recurrence rate reported with this treatment. The surgical procedure for stent placement to allow for irrigation is straightforward and can often be accomplished on an outpatient basis with either local anesthesia or light sedation. One caveat with the use of this treatment involves large cysts with loculations and ramifications that make thorough irrigation difficult. In such cases, every attempt is made to separate septations within the cyst proper, although this can be difficult in anatomically inaccessible sites. Various irrigation portals have been described. We have modified a pediatric nasal trumpet and after placement have sutured it to the surrounding gingiva for initial stability (Figure 4). On occasion, multiple portals are required, particularly with large loculated lesions. Patients are instructed to irrigate the cavity twice daily with 10 cc of 0.12% chlorhexidine through a curved-tip syringe.

If impacted teeth are present within the cystic lesion, they are left in situ during the treatment phase. As the cyst volume decreases, the tooth generally migrates to its premorbid alveolar position where it can be easily removed at end of treatment or at time of cystectomy with minimal concern about pathologic fracture or neurosensory alteration (Figures 5 and 6).

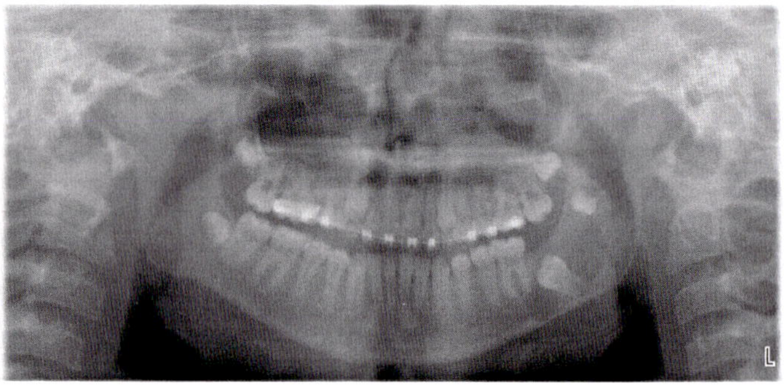

FIGURE 5 • Preoperative panoramic film showing large lucent lesion in the left mandibular ramus. The lesion is associated with teeth.

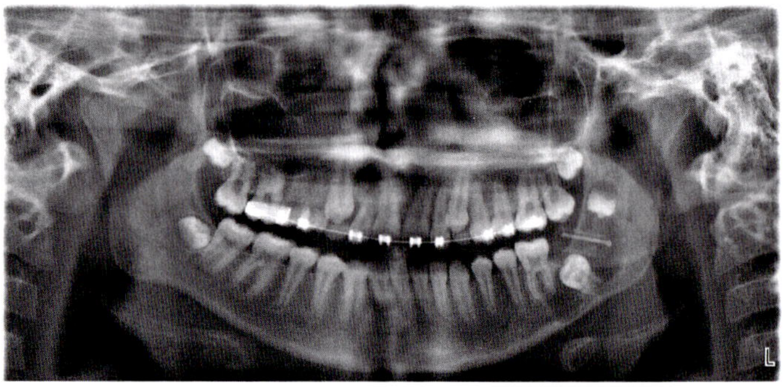

FIGURE 6 • Same patient as in Figure 5. A irrigation port in place (opaque line). The teeth have not been removed.

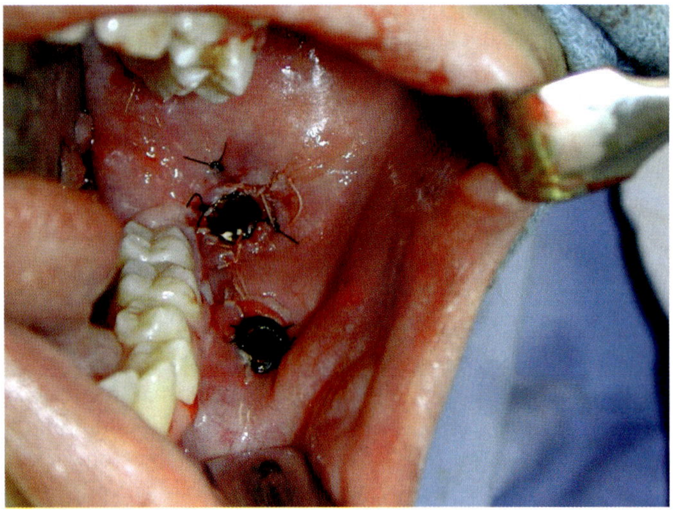

FIGURE 4 • Intraoperative view showing canulas (modified pediatric trumpet) in place.

Treatment time at our institution has averaged 9 months and during this period, serial panoramic films are obtained every 3 months looking for signs of dimensional shrinkage. When the radiolucent lesion persists beyond 9 to 12 months or if no diminution in cyst size is noted at recall, a residual cystectomy is then performed. At the time of initial biopsy or decompression, the removed cyst lining is evaluated for cytokeratin-10 expression using a standard immunohistochemical protocol. This specimen is then used to compare with the end treatment cyst lining with is likewise evaluated. During treatment, interim cytologic samples are obtained through the irrigation portal at 3-month intervals and epithelial cells, if obtained, are evaluated for cytokeratin-10 positivity.

SUMMARY OF THE MGH EXPERIENCE AND CASE DISCUSSIONS

The results of our initial series of 14 KCOTs treated in this fashion are summarized as follows: There were 8 female and 6 male patients with an average age of 32 years. Ten cysts were mandibular and 4 maxillary. The average maximal diameter of these lesions measured on plain radiograph was 5.6 cm. Cortical perforation was documented on CT in 8/14. Nine lesions were multilocular. The average irrigation time before cystectomy was 8.4 months.

Interim cytologic smears on 13/14 patients performed at 3 and 6 months were interpreted as predominantly cytokeratin-10 positive and consistent with the diagnosis of KCOT without sign of epithelial change or differentiation. Examples are shown below (Figures 7 and 8). At the time of residual cystectomy, 9/14 cases no longer showed characteristic features of KCOT and immunohistochemistry was interpreted as cytokeratin-10 negative. Of note is that the lining of these cysts was hyperplastic compared to the initial biopsy and mild-to-moderate inflammation was noted in each (Figures 9 and 10). These 9 cases were irrigated for an average of 9 months. The remaining 5 cases did not demonstrate such changes and both on routine histology and immunohistochemically, were consistent with a diagnosis of KCOT. We noted that these cases were irrigated for an average of 7 months. During a follow-up period that averaged 34 months, no cysts recurred in this group. In keeping with the appellation "minimally invasive," no cases of persistent neurosensory change were found. No endodontic therapy has been required and no secondary infection or pathologic fractures occurred.

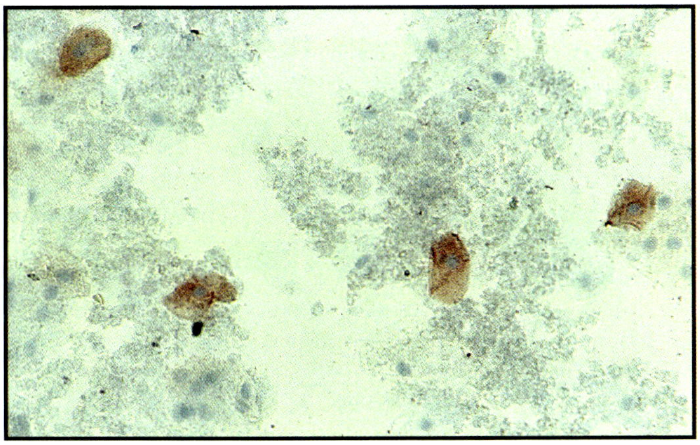

FIGURE 7 • Interim 3-month cytological sample showing cytokeratin-10 positive epithelial cells.

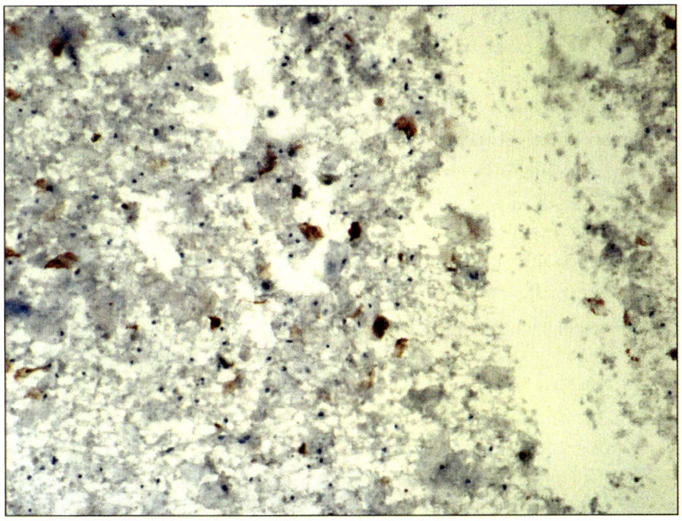

FIGURE 8 • Interim 6-month cytological sample. Cytokeratin-10 positive cells are still noted in this sample.

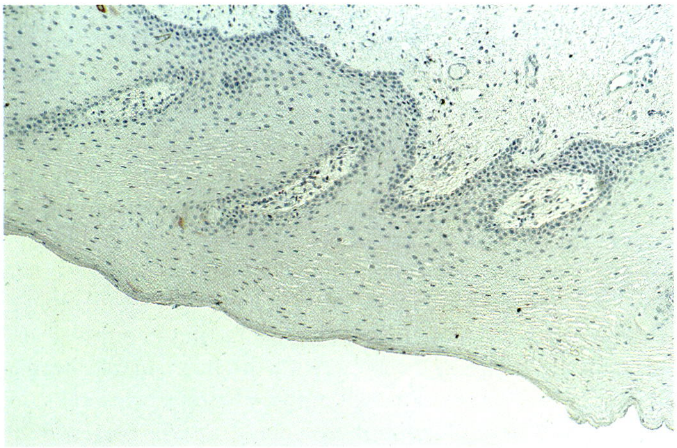

FIGURE 9 • Post-treatment immunohistochemical analysis of cystectomy lining showing no cytokeratin-10 positivity and a hyperplastic appearance.

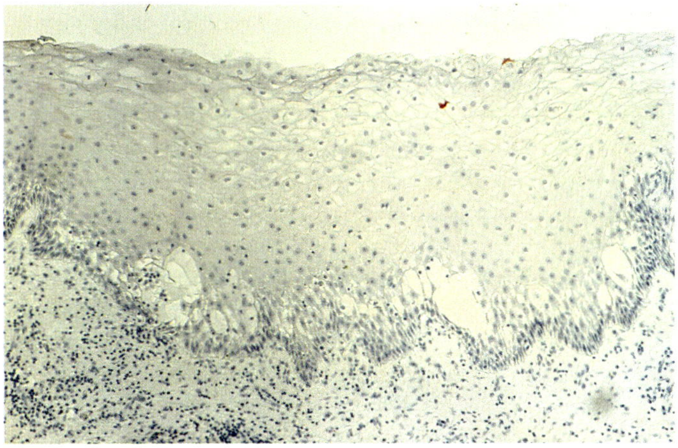

FIGURE 10 • Post-treatment cystectomy lining showing scant cytokeratin-10 positivity and moderate inflammatory infiltrate.

The change in cystic epithelium appears to be a gradual process and the biologic mechanism for this is unclear. Careful histologic evaluation of the residual cyst lining occasionally shows transformation in some areas and not in others. In addition, daughter or satellite cysts found in the surrounding wall at the time of cystectomy may continue to express cytokeratin-10 and KCOT features despite the fact that the main portion of the lesion does not. This is also a concern in areas of loculations separate from the main body of the cyst. Our results also suggest that if this type of treatment is contemplated, at least 9 months of irrigation may be required to induce dedifferentiation; therefore, patient compliance issues are of great importance and pretreatment patient discussion needs to focus on the considerable time commitment that will be required.

CONCLUSION

The techniques described above are not new. Fine needle aspiration biopsy has been part of our diagnostic armamentarium for many years. Orthopedic literature nicely discusses its use in intrabony pathology. We present a modification in technique using a more substantial needle to allow for jaw penetration without the need for an introducer, especially in cases where the bone has been expanded and thinned due to cystic enlargement. The key to successful diagnosis is to sample cells representative of the lining of the cyst. This requires contacting a bony wall and mechanically scraping the bone with the bevel of the needle. Rapid chairside evaluation will ensure that epithelial cells are present for interpretation. If not, additional passes will be required. The use of cytokeratin-10 immunocytochemistry has greatly facilitated diagnosis and has provided a minimally invasive technique for pre-operative diagnosis in cases of large KCOTs. We have found this of great benefit in terms of patient counseling and treatment planning. The rapidity with which results are obtained is also beneficial for patients.

Treatment of KCOTs using the above protocol of decompression/irrigation is similarly not a new or novel notion, but rather one that has enjoyed renewed enthusiasm based on long-term follow-up studies showing a low associated recurrence rate. As mentioned, appropriate treatment of this lesion is the subject of debate and controversy. The ability to offer treatment that is surgically well-tolerated and that preserves bone, teeth, and neurosensory function is desirable. The long-term time commitment required may be problematic for some patients, and the likelihood of compliance needs to be assessed on a case-by-case basis. We have found this technique invaluable in cases of large and destructive cysts involving difficult anatomic sites and those in which the likelihood of pathologic fracture and neurosensory alteration is high.

References

1. Swanson AE. The recalcitrant keratocyst. Int J Oral Maxillofac Surg 1986;15:451–6.
2. Regezi JA, Courtney RM, Batsakis JG. The pathology of head and neck tumors: Cysts of the jaws. Part 12. Head Neck Surg 1981;4:48–57.
3. August M, Faquin WC, Ferraro NF, Kaban LB. Fine-needle aspiration biopsy of intraosseous jaw lesions. J Oral Maxillofac Surg 1999;57:1282–6.
4. Marker P, Brondum N, Clausen PP et al. Treatment of large odontogenic keratocysts by decompression and later cystectomy: a long-term follow-up and histologic study of 23 cases. Oral Surg Oral Med Oral Pathol Oral Radiol Endod 1996;82:122–31.
5. Brondum N, Jensen VJ. Recurrence of keratocysts and decompression treatment. A long-term follow-up of forty-four cases. Oral Surg Oral Med Oral Pathol Oral Radiol Endod 1991;72:265–9.
6. August M, Faquin WC, Troulis MJ, Kaban LB. Dedifferentiation of odontogenic keratocyst epithelium after cyst decompression. J Oral Maxillofac Surg 2003;61:678–83.
7. Brennan PA, Mackenzie N, Oeppen RS, Kulamarva G et al. Prospective randomized clinical trial of the effect of needle size on pain, sample adequacy and accuracy in head and neck fine-needle aspiration cytology. Head Neck 2007;29:919–22.
8. Weinberger MS, Rosenberg WW, Meurer WT, Robbins KT. Fine-needle aspiration of parotid gland lesions. Head Neck 1992;14:483–7.
9. Cusumano RJ, Kaufman D, Weiss M, Gallo L et al. Needle aspiration biopsy of the pre-epiglotic space. Head Neck 1989; 11:41–5.
10. Lo CP, Chen CY, Chin SC, Lee KW et al. Detection of suspicious malignant cervical lymph nodes of unknown origin: diagnostic accuracy of ultrasound-guided fine-needle aspiration biopsy with nodal size and central necrosis correlate. Can Assoc Radiol J 2007;58:286–91.
11. Weber AL, Randolph G, Aksoy FG. The thyroid and parathyroid glands. CT and MR imaging and correlation with pathology and clinical findings. Radiol Clin North Am 2000;38:1105–29.
12. Sismanis A, Strong MS, Merriam J. Fine needle aspiration biopsy diagnosis of neck masses. Otolaryngol Clin North Am 1980;13:421–9.
13. Wahane RN, Lele VR, Bobhate SK. Fine needle aspiration cytology of bone tumors. Acta Cytol 2007;51:711–20.
14. Britt T, Clifford C, Barger A, Moroff S et al. Diagnosing appendicular osteosarcoma with ultrasound-guided fine-needle aspiration: 36 cases. J Small Anim Pract 2007;48:145–50.
15. Soderlund V, Skoog L, Kreicbergs A. Combined radiology and cytology in the diagnosis of bone lesions: a retrospective study of 370 cases. Acta Orthop Scand 2004;75:492–9.
16. Yamamoto T, Nagira K, Marui T, Akisue T et al. Fine-needle aspiration biopsy in the initial diagnosis of bone lesions. Anticancer Res 2003;23:793–7.
17. Handa U, Bal A, Mohan H, Bhardwaj S. Fine needle aspiration cytology in the diagnosis of bone lesions. Cytopathology 2005;26:360–1.
18. Platt JC, Davidson D, Nelson CL et al. Fine-needle aspiration biopsy: an analysis of 89 head and neck cases. J Oral Maxillofac Surg 1990;48:702–6.
19. Schelkun PM, Grundy WG. Fine-needle aspiration biopsy of head and neck lesions. J Oral Maxillofac Surg 1991;49:262–7.
20. Daskalopoulou D, Rapidis AD, Maounis N et al. Fine-needle aspiration cytology in tumors and tumor-like conditions of the oral and maxillofacial region: diagnostic reliability and limitations. Cancer 1997;81:238–52.
21. Ramzy I, Aufdemorte TB, Duncan DL. Diagnosis of radiolucent lesions of the jaw by fine needle aspiration biopsy. Acta Cytol 1985;29:419–24.
22. Yang YJ, Damron TA. Comparison of needle core biopsy and fine-needle aspiration for diagnostic accuracy in musculoskeletal lesions. Arch Pathol Lab Med 2004;128:759–64.
23. Minami M, Kaneda T, Ozawa K, Yamamoto H et al. Cystic lesions of the maxillomandibular regions: MR imaging distinction of odontogenic keratocysts and ameloblastomas from other cysts. AJR Am J Roentgenol 1996;166:943–9.
24. Gunhan O, Dogan N, Celasun B, Sengun O et al. Fine needle aspiration cytology of oral cavity and jaw bone lesions. A report of 102 cases. Acta Cytol 1993;37:135–41.
25. Li TJ, Browne RM, Matthews JB. Quantification of PCNA+ cells within odontogenic jaw cyst epithelium. J Oral Pathol Med 1994;23:184–9.
26. Cordell J, Richardson TC, Pulford K et al. Production of monoclonal antibodies against human epithelial membrane antigen for use in diagnostic immunocytochemistry. Br J Cancer 1985;52: 347–54.

27. Matthews JB, Mason GI, Browne RM. Epithelial cell markers and proliferating cells in odontogenic jaw cysts. J Pathol 1988; 156:283–90.
28. MacDonald AW, Fletcher A. Expression of cytokeratin in the epithelium of dentigerous cysts and odontogenic keratocysts: an aid to diagnosis. J Clin Pathol 1989;42:736–9.
29. August M, Faquin WC, Troulis M, Kaban LB. Differentiation of odontogenic keratocyst from nonkeratinizing cysts by use of fine-needle aspiration biopsy and cytokeratin-10 staining. J Oral Macillofac Surg 2000;58:935–40.
30. Blanus N, Freund B, Schwartz M, Furst IM. Systematic review of the treatment and prognosis of the odontogenic keratocyst. Oral Surg Oral Med Oral Pathol Oral Radiol Endod 2000;90:553–8.
31. Meara JG, Shah S, Li KK et al. The odontogenic keratocyst: a 20 year clinicopathologic review. Laryngoscope 1998;108:280–3.
32. Chow HT. Odontogenic keratocyst: a clinical experience in Singapore. Oral Surg Oral Med Oral Pathol Oral Radiol Endod 1998;86:573–7.
33. Salmassy DA, Pogrel MA. Liquid nitrogen cryosurgery and immediate bone grafting in management of aggressive primary jaw lesions. J Oral Maxillofac Surg 1995;53:784–90.
34. Tagesen J, Jensen J, Sindet-Pedersen S. Comparative study of treatment of keratocysts by enucleation, enucleation combined with cryotherapy or fixation of the cyst membrane with Carnoy's solution followed by enucleation: a preliminary report. Tandaegebladet 1990;94:674–9.
35. Nakamura N, Misuyasu T, Mitsuyasu Y, Taketomi T et al. Marsupialization for odontogenic keratocysts: long-term follow-up analysis of the effects and changes in growth characteristics. Oral Surg Oral Med Oral Pathol Oral Radiol Endod 2002;94: 543–53.
36. Zhao YR, Wei JX, Wang SP. Treatment of odontogenic keratocysts: a follow-up of 255 Chinese patients. Oral Surg Oral Med Oral Pathol Oral Radiol Endod 2002;94:151–6.

Index

Note: Page numbers in *italics* indicate figures or tables.